Water Footprints and Sustainable Development

Current Directions in Water Scarcity Research

VOLUME 6

Series Editors

Robert Mcleman

Michael Brüntrup

Volume 1: From Catchment Management to Managing River Basins
ISBN: 9780128148518

Volume 2: Drought Challenges
ISBN: 9780128148204

Volume 3: Water Productivity and Food Security
ISBN: 9780323914512

Volume 4: Indigenous Water and Drought Management in a Changing World
ISBN: 9780128245385

Volume 5: Water Resources
ISBN: 9780323853781

Current Directions in Water Scarcity Research

Water Footprints and Sustainable Development

Volume 8

Edited by

Suhaib A. Bandh
Department of Higher Education, Sri Pratap College, Government of Jammu and Kashmir, Srinagar, India

Fayaz A. Malla
Department of Environmental Science, Government Degree College Tral, Jammu and Kashmir, India

Series Editors

Robert McLeman

Michael Brüntrup

Elsevier
Radarweg 29, PO Box 211, 1000 AE Amsterdam, Netherlands
125 London Wall, London EC2Y 5AS, United Kingdom
50 Hampshire Street, 5th Floor, Cambridge, MA 02139, United States

ISBN: 978-0-443-23631-0

ISSN: 2542-7946

For Information on all Elsevier publications
visit our website at https://www.elsevier.com/books-and-journals

Publisher: Candice Janco
Acquisitions Editor: Maria Elekidou
Editorial Project Manager: Namrata Lama
Production Project Manager: Paul Prasad Chandramohan
Cover Designer: Christian Bilbow

Typeset by MPS Limited, Chennai, India

Contents

List of contributors

Nurul Huda Abd Kadir BIOSES RIG Group, Faculty of Science and Marine Environment, Universiti Malaysia Terengganu, Kuala Nerus, Terengganu, Malaysia

Yasar Abdullah Sustainable Development Study Centre, Government College University, Lahore, Pakistan

Mahmood Adeel Department of Environmental Sciences, GC Women University Sialkot, Pakistan

Anubhav Agrawal Electronics Engineering Department, School of Engineering & Technology, BML Munjal University, Gurugram, Haryana, India

Mukhtar Ahmed Meteorological Centre, India Meteorological Department Rambagh Srinagar, India

Mohamed F. Alajmi Department of Pharmacognosy, College of Pharmacy, King Saud University, Riyadh, Saudi Arabia

Tabinda Amtul Bari Sustainable Development Study Centre, Government College University, Lahore, Pakistan

Ankur Mechanical Engineering Department, School of Engineering & Technology, BML Munjal University, Gurugram, Haryana, India

Tawakkal Baharuddin Department of Government Studies, Universitas Muhammadiyah Makassar, Indonesia

Suhaib A. Bandh Department of Higher Education, Sri Pratap College, Government of Jammu and Kashmir, Srinagar, India

Suhail Bashir ENS Environmental Consultancy Sharjah, United Arab Emirates

Ananya Roy Chowdhury Department of Botany, Chakdaha College, Nadia, West Bengal, India

Tiziana Crovella Department of Economics, Management and Business Law, University of Bari Aldo Moro, Largo Abbazia Santa Scolastica, Bari, Italy

Raúl Alejandro Cuevas-Jacquez Campus Region of the Llanos, National Technological of Mexico, Guadalupe Victoria, Durango, Mexico

Achintya Das Department of Physics, Mahadevananda Mahavidyalaya, Barrackpore, West Bengal, India

Abhishek Gautam Department of Mechanical Engineering, Indian Institute of Technology Bombay, Mumbai, India

Sonia Grover Department of Pharmacognosy, College of Pharmacy, King Saud University, Riyadh, Saudi Arabia

Afzal Hussain Department of Pharmacognosy, College of Pharmacy, King Saud University, Riyadh, Saudi Arabia

Esmail Karamidehkordi Department of Agricultural Extension and Education, Faculty of Agriculture, Tarbiat Modares University (TMU), Tehran, Iran

Vahid Karimi Department of Agricultural Extension and Education, Faculty of Agriculture, Tarbiat Modares University (TMU), Tehran, Iran

Ashish Kumar Department of Science, Technology and Technical Education, Nalanda College of Engineering, Bihar Engineering University, Patna, Bihar, India; Division of Research and Development, Lovely Professional University, Punjab, India

Ravinder Kumar Mechanical Engineering Department, School of Engineering & Technology, BML Munjal University, Gurugram, Haryana, India

Giovanni Lagioia Department of Economics, Management and Business Law, University of Bari Aldo Moro, Largo Abbazia Santa Scolastica, Bari, Italy

Foozia Majeed Conservation Biology Lab Sos in Zoology Jiwaji University Gwalior, Gwalior, Madhya Pradesh, India

Afaan A. Malla Department of Environmental Science, Government Degree College Baramulla, Indira Gandhi National Open University, Baramulla, Jammu and Kashmir, India

Fayaz A. Malla Department of Environmental Science, Govt. Degree College Tral, Jammu and Kashmir, India

Ruben Ivan Marin-Tinoco Department of Chemical area Environmental Technology, Technological University of Rodeo, Rodeo, Durango, Mexico; Rural Hospital No. 162, Mexican Social Security Institute, Rodeo, Durango, Mexico

Cesar Alberto Meza-Herrera Regional Universitary Unit on Arid Lands, Chapingo Autonomous University, Bermejillo, Durango, Mexico

Zainab Muhammad Faculty of Science and Marine Environment, Universiti Malaysia Terengganu, Kuala Nerus, Terengganu, Malaysia

Ladan Naderi School of Environmental Studies, University of Victoria, Victoria, British Columbia, Canada

Yaman Ahmed Naji Faculty of Science and Marine Environment, Universiti Malaysia Terengganu, Kuala Nerus, Terengganu, Malaysia

Cayetano Navarrete-Molina Department of Chemical area Environmental Technology, Technological University of Rodeo, Rodeo, Durango, Mexico

Annarita Paiano Department of Economics, Management and Business Law, University of Bari Aldo Moro, Largo Abbazia Santa Scolastica, Bari, Italy

Andi Luhur Prianto Department of Government Studies, Universitas Muhammadiyah, Makassar, Indonesia

Abdullah Kaviani Rad Department of Environmental Engineering and Natural Resources, College of Agriculture, Shiraz University, Shiraz, Iran

Farhana Rahman Department of Environmental Science, Higher Education Department Govt. of Jammu and Kashmir, Government Degree College Tral, India

Anand B. Rao Interdisciplinary Program in Climate Studies, Indian Institute of Technology Bombay, Mumbai, India; Centre for Technology Alternatives for Rural Areas, Indian Institute of Technology Bombay, Mumbai, India

Showkat Rashid Department of Economics, Higher Education Department Govt. of Jammu and Kashmir, Government Degree College Tral, India

Javed Rimsha Sustainable Development Study Centre, Government College University, Lahore, Pakistan

José Luis Rodríguez-Álvarez Campus Region of the Llanos, National Technological of Mexico, Guadalupe Victoria, Durango, Mexico

María de los Ángeles Sariñana-Navarrete Department of Chemical area Environmental Technology, Technological University of Rodeo, Rodeo, Durango, Mexico

Reshma Shinde Interdisciplinary Program in Climate Studies, Indian Institute of Technology Bombay, Mumbai, India

Gerald Singh Agricultural Extension, Communication and Rural Development Department, Faculty of Agriculture, University of Zanjan, Zanjan, Iran

Ranbir Singh School of Mechanical Engineering, Lovely Professional University, Phagwara, Punjab, India

Nazir A. Sofi Department of Agriculture Research Information System, Sher-e-Kashmir University of Agricultural Sciences and Technology, Shalimar Campus, Srinagar, Jammu and Kashmir, India

Manoj Sood Department of Hydro & Renewable Energy, Indian Institute of Technology Roorkee, Roorkee, India

Chinmaya Kumar Swain ICAR-National Rice Research Institute, Cuttack, Odisha, India

Mir Tamana Department of Environmental Science, Higher Education Department Govt. of Jammu and Kashmir, Government Degree College Tral, India

Yan Tan Department of Geography, Environment and Population, School of Social Sciences, The University of Adelaide, Adelaide, South Australia, Australia

Abhinay Thakur Department of Science, Technology and Technical Education, Nalanda College of Engineering, Bihar Engineering University, Patna, Bihar, India; Division of Research and Development, Lovely Professional University, Punjab, India

Lipsa Tripathy ICAR-National Rice Research Institute, Cuttack, Odisha, India

Luis Manuel Valenzuela-Nuñez Faculty of Biological Sciences, Juarez University of the State of Durango, Gomez Palacio, Durango, Mexico

Waseem A. Wani Department of Chemistry, Govt. Degree College Tral, Kashmir, Jammu and Kashmir, India

Shastri Yogendra Interdisciplinary Program in Climate Studies, Indian Institute of Technology Bombay, Mumbai, India; Department of Chemical Engineering, Indian Institute of Technology Bombay, Mumbai, India

Nina Yuslaini Department of Government Science, Universitas Islam Riau, Indonesia

Chapter 1

Increasing water resources: a combined approach of water footprint and geographic information system for the sustainability challenge in the agricultural context

Tiziana Crovella* **and Annarita Paiano**
Department of Economics, Management and Business Law, University of Bari Aldo Moro, Largo Abbazia Santa Scolastica, Bari, Italy
**Corresponding author. e-mail address: tiziana.crovella@uniba.it.*

1.1 Introduction

The agricultural sector is responsible for global freshwater consumption equal to 70% and generates the major impact on the shortage of water resources. Particularly, water pressure and scarcity affect the socioeconomic development of the countries (Li et al., 2022). Intending to address the need for food and water of a growing global population - that in 2050 it could reach 9.7 billion people, for the agricultural context is fundamental the estimation of the water footprint (WF), crop yield, and evapotranspiration for evaluating the pressure on water resources (Chanu & Oinam, 2023).

For this reason, it is crucial to recommend some effective methods for decreasing irrigation water consumption considering several elements such as crop varieties with higher yields, regions, clusters of cultivation and growth potential. All these methods are necessary to potentially reduce WF (Garofalo et al., 2019; Mokarram et al., 2021; Zhuo et al., 2016). Among several environmental indicators studies applied to agricultural context in a combined approach with other tools, WF applications are very limited in the literature (Bilge Ozturk et al., 2022). Above all, the combined applications with other tools, such as life cycle assessment (LCA) or geographic information system (GIS), are very few. Most studies have been conducted to calculate WF focusing on the relationship between WF and other agricultural parameters, such as yield of products and type of cultivated plants (Mokarram et al., 2021).

The past studies underlined that the integration of the WF methodology with high-resolution local data deriving from georeferencing through GIS supports agricultural processes with low environmental impact, reduction of water resource consumption, and increase in yield (Shtull-Trauring et al., 2016).

Therefore, GIS tool is contemplated amongst the efficient IT-based technological tools for displaying complex information that include several geographical representation of the agricultural system. Besides, the integration of other programs (such as CROPWAT by FAO for the WF estimation) within GIS can improve the implementation and analysis of spatial data deriving from more solutions (Haji et al., 2020; Monika & Srinivasan, 2015).

In the last 10 years, some authors have used GIS to extract and combine relevant information from climate, soil, and land use to develop a water balance model with high-resolution grids (Daccache et al., 2014) to build a model of input data with WF calculation.

Accordingly, when multiple approaches are combined (e.g., WF and GIS), the output data are digitized and integrated, also according to computerized approaches, into a unified geoprocessing platform for enabling the visualization and further processing of risk factors (Haji et al., 2020).

Water Footprints and Sustainable Development. DOI: https://doi.org/10.1016/B978-0-443-23631-0.00001-7

GIS, which is already very popular for the specific location of current and future uses of land for energy crops, has been applied to analyze the suitability of territory in the geographical context (Viccaro et al., 2022). Hence, a current and updated application can satisfy the data collection for building a dataset model of water consumption in the agricultural sector. Similarly, Feng et al. (2017) proposed a GIS-based model to evaluate the suitability of marginal lands for some plants as switchgrass, miscanthus, and hybrid poplar in the Upper Mississippi River Basin (United States) to produce biomass resources.

Consequently, to date, after the first evaluation of the feasibility of the combinate application of GIS and WF tools, the authors proposed a systematic literature review (SLR) of the scientific interest for testing this combined application in the agricultural context, particularly in the cultivation of crops for food production and processing.

The general research question of this chapter is the understanding of whether the WF indicator combined with the GIS tool has been applied in the past, particularly in the sector of agricultural production and crops cultivation for nutrition and regarding food-related human benefits, and can be used in the management of water resources too.

In particular, the authors tested whether, within a GIS environment, the output of a vegetation model can be combined with other data sources to conduct a more comprehensive and spatially explicit analysis of water consumption for growing crops for human nutrition at certain geographic locations.

1.2 Materials and methods

In order to analyze the scientific interest, the authors analyzed different methods for managing a review and chosen the systematic literature review (SLR) whereas is suitable for producing reliable outputs consistent with the general research question. Particularly, this is a process that allows to collect relevant evidences on a specific topic and meets all eligibility criteria stated earlier (Mengist et al., 2020).

The typical process of a SLR is based on four stages: 1) searching (express the searching string and choose the types of databases for the products quantification such as Scopus, Web of Science or others), 2) assessment (predefinition of the products of scientific literature included and excluded, and definition of the quality of assessment criteria), 3) synthesis of the results (extracting and categorizing the metadata and building the dataset with the identified sample) and, lastly, 4) analysis process (description of the results and output and finalizing the conclusion (Mengist et al., 2020).

Particularly, the authors investigated the time range 2000–2023 on Scopus and Web of Science (WoS) platforms, conducting a structured TITLE–ABS–KEY search for selecting the scientific production. The keywords combination, used on the two main databases managed respectively by Elsevier and Clarivate, is represented by the following scheme: *Agriculture AND water AND footprint AND GIS*.

Firstly, the results identified on Scopus consisted of 23 products, which included 17 articles, 4 conference papers, 1 book chapter, 1 conference review, and 0 reviews. Secondly, on WoS, it has been quantified 44 products that included 39 articles, 3 proceeding papers, 1 data paper, and 1 review article.

In the first stage, the authors excluded 52 duplicates and 10 products (conference papers, book chapters, and conference reviews).

In this stage, the authors screened 42 records for developing the sample to be analyzed as shown in Table 1.1.

Later, the scholars excluded 7 records for title, abstract, and editors inconsistent with the purpose of the current study (Table 1.2).

Particularly, the sample of articles excluded comprises 7 articles that are not focused on crop cultivation and have not applied GIS in combination with the WF approach. Particularly, some scholars quantified the effects of irrigation-induced surface heterogeneity, evaluating soil moisture, and vegetation growth variability through a measurement model with a large aperture scintillometer. Although the application was on over drip-irrigated vineyards cluster, other studies (Geli et al., 2020) did not directly use GIS in combination with the WF indicator. Furthermore, in their study, Wangda et al. (2019) did not use GIS, but GeoEye-1 image and lidar data were collected using a regional growth approach with the aim to delineate individual tree canopies in forest and carbon biomass context. Another application carried out by Mondal et al. (2017) consisted of a scan indicating spherical and granular structures of bricks, using aerial shots instead of applying GIS in combination with WF.

Other scholars proposed another combined approach based on the use of the metabolic theory, and in particular the application of organic biological metabolism including socioeconomic factors to analyze the regional water resources Notwithstanding, GIS tool was used by the scholars merely to present the investigated area (Ren et al., 2016).

In only one case GIS was used to estimate the proportions and cover type of the land and to process water maps (Oni et al., 2015). However, this article has been excluded because it focused on the water resource and did not analyze

TABLE 1.1 List of articles assessed by the abstracts.

Platform	No.	Author (Year)	Journal	Method	Cluster	Operation
Scopus	1	Chanu and Oinam (2023)	*Songklanakarin Journal of Science and Technology*	AquaCrop GIS	Paddy production	Quantification of the paddy yield and WF under different rainfall conditions
	2	Bilge Ozturk et al. (2022)	*Environmental Science and Pollution Research*	DEM. ArcGIS-ArcMap 10.7.1	Kiwi fruit production	Quantification of WF
	3	Hanson et al. (2022)	*Science of the Total Environment*	ArcGIS ver 10.3	Atmospheric application	Onsite detection of atmospheric and vegetation variables
	4	Montealegre et al. (2022)	*Science of the Total Environment*	GIS	Rooftops of buildings in an urban district	Extraction suitable rooftop areas
	5	Bhat et al. (2021)	*Environmental Pollution*	GIS—Landsat-8 OLI/TIRS satellite image, 30 m resolution	Riverine water quality	Quantification of percentage of landscape pattern composition and configuration metrics of river basin
	6	Bontinck et al. (2021)	*International Journal of Life Cycle Assessment*	QGIS version 3.4.11	River basin	For mapping the information at basin level
	7	Longo et al. (2021)	*Journal of Environmental Management*	DayCent (daily version) model-GIS platform	Maize wheat barley, soybean, sunflower, rapeseed, potato, sugar beet crops, pastures, and meadows	For assessing the identified environmental benefits resulting from implementing Rural Development Programme
	8	Karim et al. (2020)	*Water*	GIS-based MCDM FFP	Freshwater aquifers	Visualization of the suitability of an aquifer for a given use across a region
	9	Fu et al. (2019)	*Sustainability*	GIS	Wheat, maize, cotton, and groundnut	Quantification WF combining CROPWAT and GIS
	10	Casella et al. (2019)	*Sustainable Production and Consumption*	ESRI's ArcGIS software	Temporary river catchment	Processing climate, land use, and soil type data combining CROPWAT and GIS
	11	Kourgialas et al. (2018)	*Science of the Total Environment*	GIS decision making tool	Aquifer systems	Including information of contamination of the main aquifers
	12	Shtull-Trauring et al. (2016)	*Frontiers in Plant Science*	GIS technology in high-resolution data	Banana, avocado, and palm-dates	Data regarding crop distribution and field size retrieved integrating GIS and WF

(Continued)

TABLE 1.1 (Continued)

Platform	No.	Author (Year)	Journal	Method	Cluster	Operation
	13	Cazcarro et al. (2016)	*Journal of Cleaner Production*	ArcGIS software and GIS layers	Vulnerable areas	Combination of mesoeconomic input–output model with GIS localized information
	14	Daccache et al. (2014)	*Environmental Research Letters*	GIS	Olives, citrus, vineyards, and cereals,	For extracting and combine relevant information from climate, soil, and land use database to run the water balance model
	15	Atkinson et al. (2014)	*Ecological Applications*	GIS	Unionid mussels in watersheds	For determining land cover within the study basins, and estimating net anthropogenic nitrogen inputs
	16	Multsch et al. (2013)	*Geoscientific Model Development*	MapWinGIS library and ESRI's ArcGIS	Cereals, vegetables, fodder crops, and fruits	For interpolating climatic stations data
	17	Schaldach et al. (2009)	*Advances in Geosciences*	ArcGIS 9.2	Silage maize for biogas production	For preparing spatial input data
WoS	2	Elkamel et al. (2023)	*Sustainable Cities and Society*	GIS maps	Urban agriculture	For leading to more realistic results from a spatial aspect for our case study
	10	Pulighe and Pirelli (2023)	*Renewable and Sustainable Energy Reviews*	WebGIS	Marginal and underutilized lands	For identifying the scenario that minimizes the use of input resources and optimizing the nexus combining GIS and life cycle assessment
	22	Carlson et al. (2023)	*Landscape and Urban Planning*	ArcGIS Pro	Natural areas and wildland vegetation	For developing databases of protected areas and for defining national forest boundaries
	12	Viccaro et al. (2022)	*Renewable Energy*	GIS	Two energy crops (rapeseed and cardoon)	Combining GIS with water–energy–food nexus approach
	1	Mokarram et al. (2021)	*Theoretical and applied Technology*	GIS	Cereal crops	Combining multiple regression and Artificial Neural network models in GIS
	3	Soula et al. (2021)	*Groundwater for Sustainable Development*	QGIS	Aquifer management	For spatially analyzing water availability
	5	Norizan et al. (2021)	*Water*	ArcGIS 10.3	Irrigated area	For making an interpolation map
	7	Rizvi et al. (2021)	*Water Resource Management*	ArcGIS	Irrigated crop area	For analyzing spatial data

(*Continued*)

TABLE 1.1 (Continued)

Platform	No.	Author (Year)	Journal	Method	Cluster	Operation
	21	Krishnan et al. (2021)	*Ocean & Coastal Management*	ArcGIS software	Coral reef	For determining the total area of coral reef zones
	6	Haji et al. (2020)	*Computer & Chemical Engineering*	ArcGIS software	Open fields agriculture	For digitizing and integrate data into a unified system for the processing of risk factors
	14	Terrado et al. (2016)	*Environmental Research Letters*	GIS	River basins	For spatial data acquisition
	17	Lamastra et al. (2016)	*Science of the Total Environment*	Web GIS software	Viticulture	For measuring the environmental impact of viticulture in a holistic way
	18	Root-Bernstein et al. (2017)	*Regional Environmental Change*	GIS analysis	Silvopastoral savanna ("espinal")	For identifying least-cost paths between areas of high and low Espinal condition
	20	Arodudu et al. (2013)	*Biomass and Bioenergy*	ArcGIS version 10.0.	Rural area	For estimating the energy consumption combining Life Cycle Inventory and GIS
	23	Field et al. (2017)	*Global Change Biology Bioenergy*	GIS	Dedicated bioenergy feedstock crops on marginal agricultural lands	For determining spatial data
	24	McKean et al. (2009)	*Remote Sensing*	GIS	Channels and riparian zones	For automatically measuring channel geometry
	4	Centofanti et al. (2008)	*Science of the Total Environment*	ESRI ArcGIS 9.1	Crops and area	For identifying the full range of unique combinations of climate, soil, and crop types
	8	Cai et al. (2007)	*Publication Cover. New Zealand Journal of Agricultural Research*	ArcGIS 9.0	Nature reserve	For developing spatial characteristics

*Legend: : a) crops production articles: Scopus (1, 2, 7, 9, 12, 14, 16, 17), WoS (12, 1, 5, 7, 6, 17, 4); b) urban agriculture articles: Scopus (4), WoS (2); c) river basins articles: Scopus (5, 6, 8, 10, 11, 15), WoS (3, 14); d) atmospheric application article: Scopus (3), WoS (n.a); e) vulnerable areas article: Scopus (13), WoS (10, 22, 21, 18, 20, 23, 24, 8). *GIS*, geographic information system; *WF*, water footprint; *WoS*, Web of Science.
Source: Authors' elaboration on data Scopus and WoS.

the results following the application of a combined approach for measuring the sustainability of a food crop. In addition, another study focused on the reflection variability associated with changing soil moisture, used for examining its impact on soil moisture recovery and focussed on tundra lakes, excluding the combined use of WF and GIS software, was not considered (Högström et al., 2014).

Moreover, a study built to project the future global distribution of water resource and to plan its use of tree- and grass-based bioenergy systems, using only literature data and databases, and not models elaborated by GIS software, was excluded (King et al., 2013). The aforementioned studies were excluded because they did not adhere to the topic as they neither used GIS in combination with WF nor focused on cultivation for food production.

TABLE 1.2 List of records excluded by titles, abstracts, and editors.

No.	Author (Year)	Journal	Method	Cluster	Operation
1	Geli et al. (2020)	*Water*	Large aperture scintillometer (LAS)	Over drip-irrigated vineyards	For evaluating the variability of soil moisture and vegetation growth
2	Wangda et al. (2019)	*International Journal of Remote Sensing*	GeoEye-1 imaging and small-scale lidar	Forest and carbon biomass	For performing species stratification
3	Mondal et al. (2017)	*Science of the Total Environment*	Geostatistical tools	Brick kiln sites	Use statistical tools to analyze the spatial distribution of toxic metals
4	Ren et al. (2016)	*Journal of Environmental Management*	Water resources metabolism theory	Regional water resources	For presenting the concept of biological metabolism in the carrying capacity of regional water resources
5	Oni et al. (2015)	*Hydrological Processes*	ArcGIS	Lake watershed	For estimating and elaborating images of soil and water resources
6	Högström et al. (2014)	*Remote Sensing*	ENVISAT and advanced synthetic aperture radar	Tundra lakes	For investigating soil moisture
7	King et al. (2013)	*BioScience*	Data from literature	Lignocellulosic sources	For analyzing the water use efficiency of tree- and grass-based bioenergy systems

Source: Authors' elaboration on data Scopus and WoS.

Also, after this analysis, the authors carried out a metaanalysis overview. Particularly, after matching the scientific interest of institutions on the Scopus platform and using the keywords combination *Agriculture AND water AND footprint AND GIS*, it emerged that most studies associated with this topic were carried out in the United States and Asia, and only 3 in Italy (Table 1.3).

This result represents another reason that pushed the authors to conduct this literature review with the subsequent objective of proposing a combined application of WF and GIS on some typical crops of the Italian territory.

Later, the authors followed the steps of the PRISMA model and carried out the analysis (Fig. 1.1).

Specifically, after the quantification of 23 papers on Scopus and 44 on WoS, the authors excluded 14 duplicates. Subsequently, the sample of 53 papers was cleaned up by chapters, reviews, and conference papers. At this stage, the authors screened 42 papers and at the end of this process they excluded 7 papers (Table 1.2) because they were not totally in line with the topic and objective of the study.

Therefore, 35 papers were analyzed in depth from which a further 20 articles were removed as they were not consistent with applications in the agricultural cultivation sector for food production but fell within other macroclusters.

Particularly, the 35 articles assessed by abstracts, before being excluded, have been divided by the macro areas of GIS application (Table 1.4): crop production, river basins, vulnerable areas, urban agriculture, and atmospheric application. After this further assessment, the number of papers that were full-text eligible and relevant to the topic was 15, those that were included in the crops production cluster.

Generally, the authors used the SLR method to generate a synthetic and replicable snapshot of existing knowledge, quantifying trends, and interests, and highlight the gaps observed on the topic covered. Particularly, in this SLR the authors carried out a structured analysis related to the scientific interest the combined use of WF-GIS for agricultural purposes. As Mengist et al. (2020) focussed in their methodological article published a frew years ago, this systematic literature review is based on the analytical method suitable for studies on environmental sciences, agricultural, and biological sciences, and also for other social science fields. For this reason, the authors carried out this SLR, similar to the approach used recently (Crovella et al., 2024) with the aim to carried out a complete analysis of the past studies related to the scientific interest the combined use of WF-GIS for agricultural purposes, as mentioned above, and for providing to practitioners and stakeholders - involved in che agricultural chain - new perspectives for the application of combined tools towards environmental sustainability.

TABLE 1.3 Scientific interest for institution.

Institution	Country	Number of matching documents
Eskisehir Technical University	Turkey	3
Shandong University of Science and Technology	China	2
National Institute of Technology Manipur	India	2
Texas Tech University	United States	2
University of Padua	Italy	3
University of Worcester	United Kingdom	2
Ordu Üniversitesi	Turkey	1
Universidad de Zaragoza	Spain	5
ENEA	Italy	1
Nakhon Si Thammarat Rajabhat University	Thailand	1
University of Kashmir	India	4
Qingdao Agricultural University	China	1
Nakhon Si Thammarat Rajabhat University	Thailand	1
Prince of Songkla University	Thailand	1
Hebei Institute of Water Science	China	1
Danish Technological Institute	Denmark	1
Hellenic Open University	Greece	1
Lifecycles	Australia	2
Institute of Genetics and Developmental Biology Chinese Academy of Sciences	China	1
Henan University	China	1
Eskisehir Technical University	Turkey	1
Banaras Hindu University	India	1
Technical University of Crete	Greece	1
Politecnico di Bari	Italy	1
Ondokuz Mayis Üniversitesi	Turkey	1

Source: Authors' elaboration on Scopus (2023).

1.3 Analytical approach

According to Mengist et al. (2020), the authors developed this SLR for analyzing the results, building a framework of GIS and WF combined application in crop production, and exploring general data regarding the research topic. Particularly, the interest in the field increased in the last three years as shown in Fig. 1.2, with the peak reached in 2021 (4 articles).

Of the 14 journals in which the 15 articles selected for the crop production cluster were published (Table 1.4), only *Science of the Total Environment* presents more than one contribution (Fig. 1.3). These 2 studies mainly concern environmental and sustainability issues in the wine sector, considering an extensive approach and the best possible environmental combination for climate, soil, and cultivation.

Moreover, regarding the methodologies applied by several authors in their study, the GIS model (33% equal of 6 papers) was the most applied software, followed by ArcGIS in 5 analyses (Fig. 1.4). Specifically, Viccaro et al. (2022)

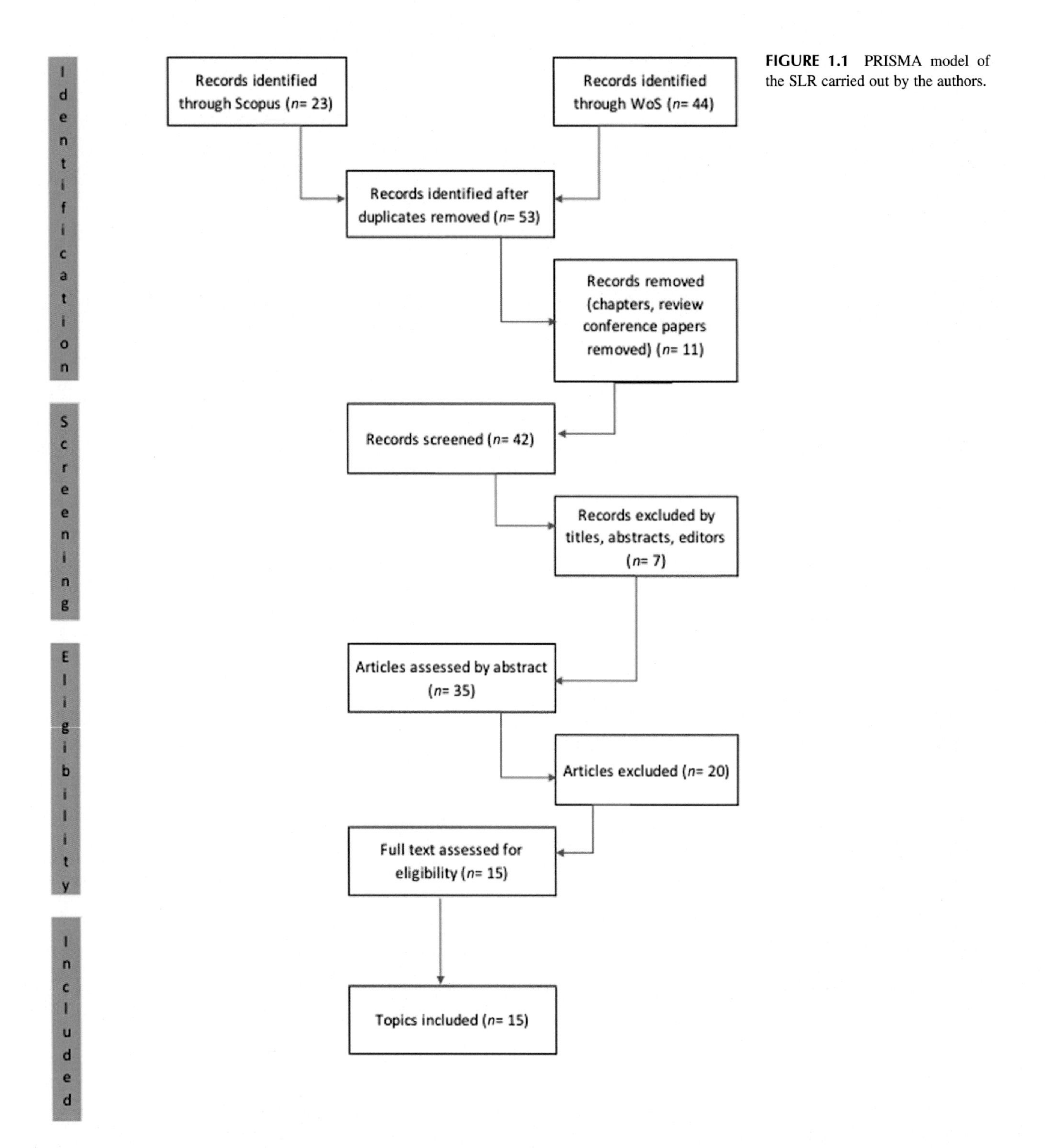

FIGURE 1.1 PRISMA model of the SLR carried out by the authors.

operated in a GIS environment using QGIS software, creating georeferenced raster layers with a resolution of 100 m and using Monte Mario (located in Rome in Italy) as a geographical reference point. Instead, Rizvi et al. (2021) used the geometric correction of the images which was carried out through an orthorectification operation with the georeferencing tool called ArcGIS. Among the strengths of this tool, it emerged that ArcGIS allows the visualization of steps including coordinate transformations, better cartographic projection, resampling, and interpolation of data up to mosaicking.

TABLE 1.4 Clusters identification after eligibility analysis.

Clusters	Authors	Numbers
Crops production	Chanu and Oinam (2023); Bilge Ozturk et al. (2022); Longo et al. (2021); Fu et al. (2019); Shtull-Trauring et al. (2016); Daccache et al. (2014); Multsch et al. (2013); Schaldach et al. (2009); Mokarram et al. (2021); Norizan et al. (2021); Haji et al. (2020); Rizvi et al. (2021); Viccaro et al. (2022); Lamastra et al. (2016)	15
River basins	Bhat et al. (2021); Bontinck et al. (2021); Karim et al. (2020); Casella et al. (2019); Kourgialas et al. (2018); Atkinson et al. (2014); Soula et al. (2021); Terrado et al. (2016)	8
Vulnerable areas	Cazcarro et al. (2016); Cai et al. (2007); Pulighe and Pirelli (2023); Root-Bernstein et al. (2017); Arodudu et al. (2013); Krishnan et al. (2021); Carlson et al. (2023); Field et al. (2017); McKean et al. (2009)	9
Urban agriculture	Montealegre et al. (2022); Elkamel et al. (2023)	2
Atmospheric application	Hanson et al. (2022)	1

Source: Authors' elaboration.

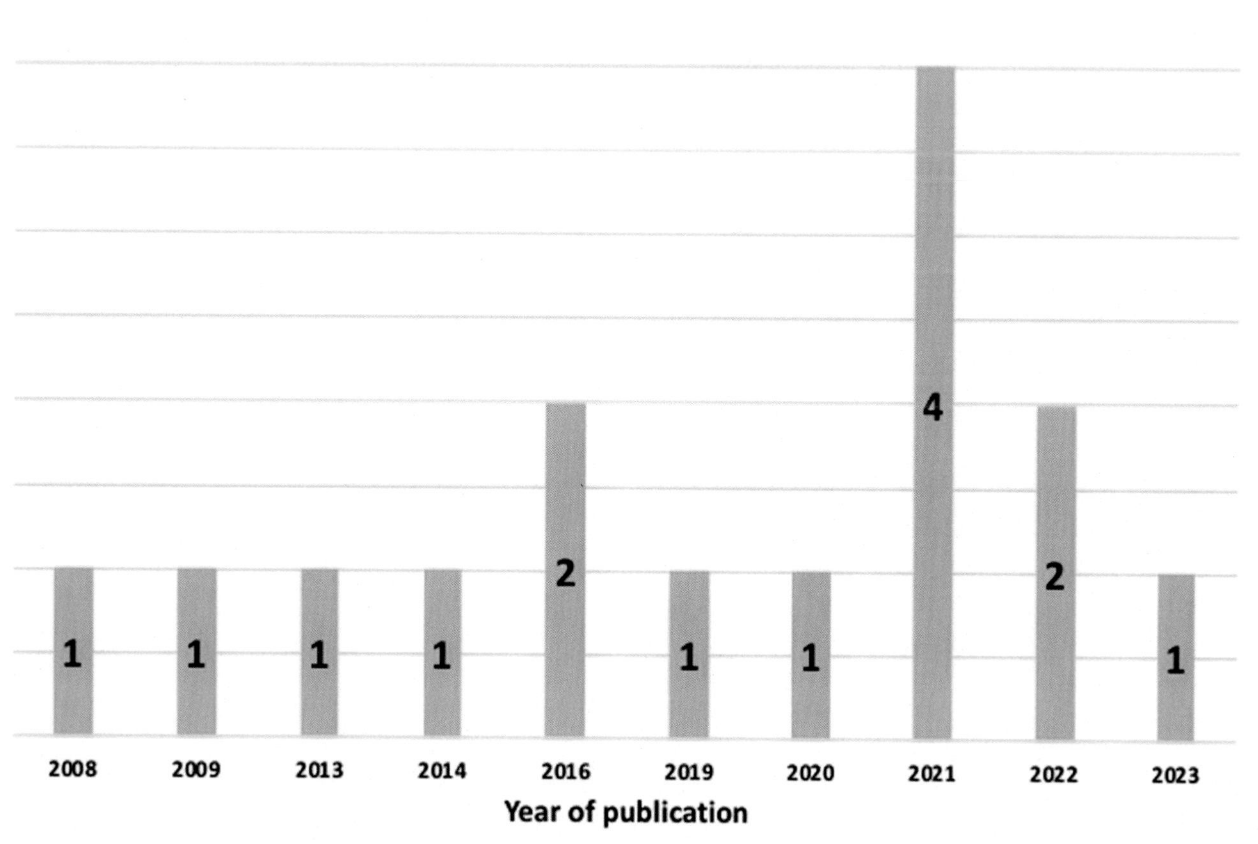

FIGURE 1.2 Research timeline. *Source: Authors' elaboration.*

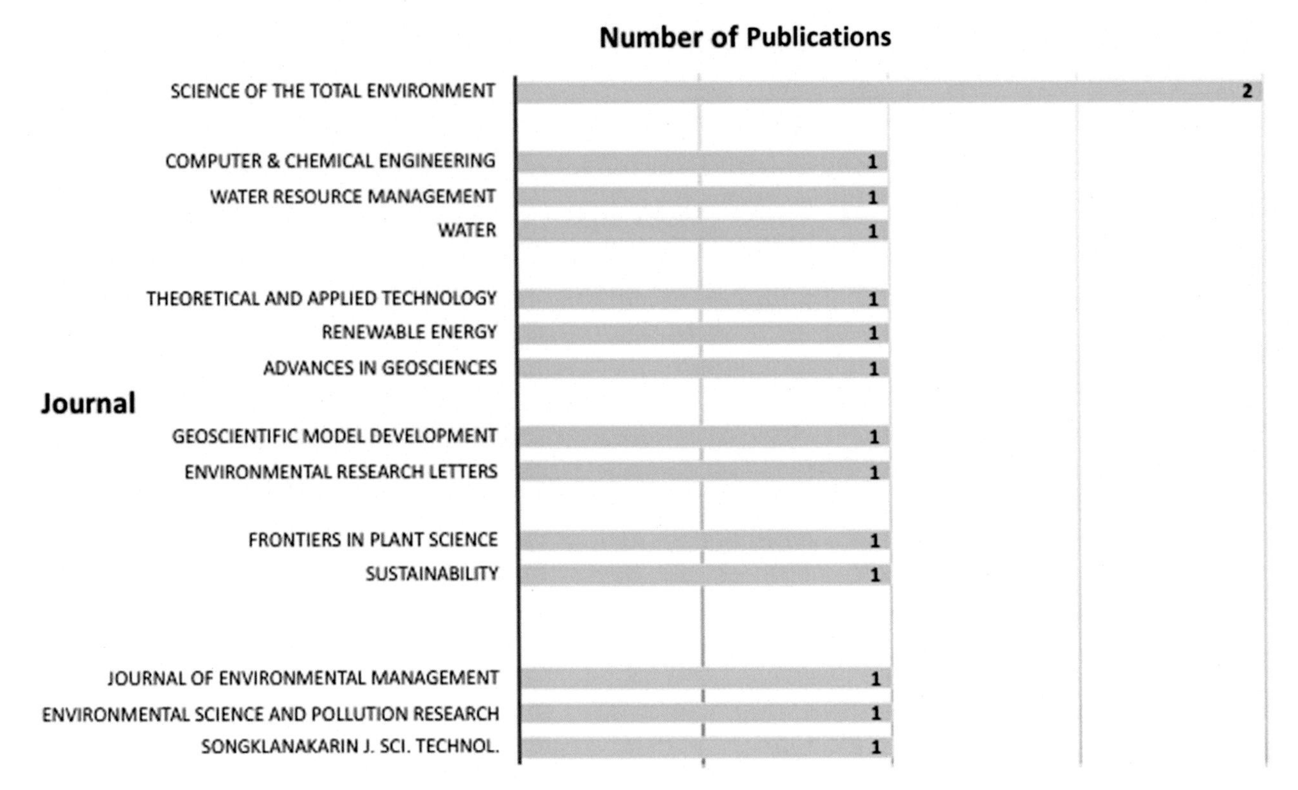

FIGURE 1.3 Publishing journal. *Source: Authors' elaboration.*

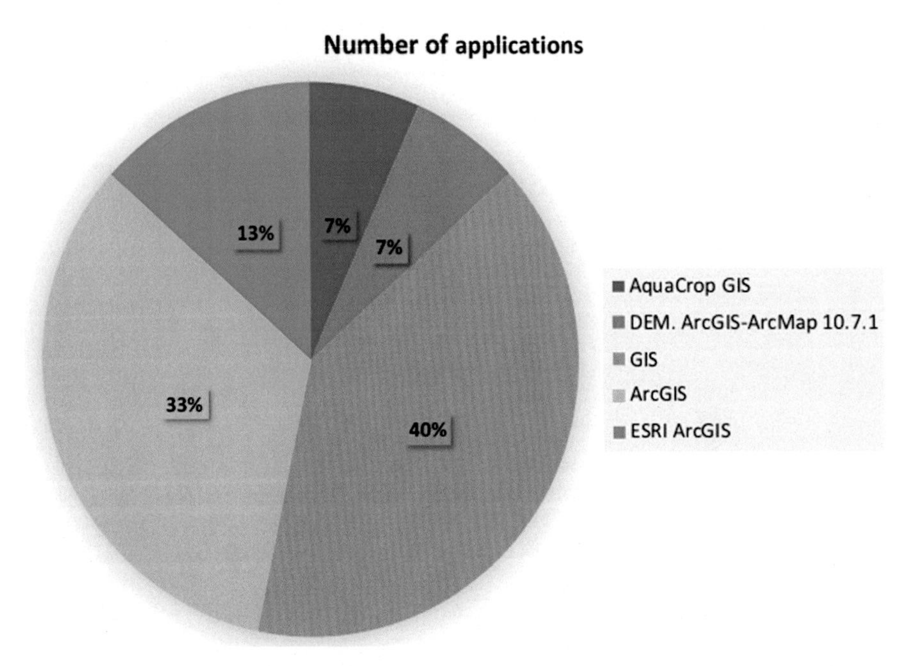

FIGURE 1.4 Methodologies applied by authors. *Source: Authors' elaboration.*

1.4 Role of geographic information system in crops production

After the build sample of 15 papers included in the analysis, the authors of this chapter carried out an in-depth analysis of the contents and applications. Among these studies, emerge some applications which focussed on some crops, in particular rice, fruit, maize, wheat, barley, soybean, sunflower, rapeseed, potato, sugar beet crops, pastures and meadows, cotton and groundnut, oil palm, and vineyard. Moreover, other applications dealt with fertilizers, pesticide use and irrigation systems.

As mentioned earlier, the evaluation of the WF of a rice field is important above all for planning the correct use of water resources. For this reason, Chanu and Oinam (2023) first quantified the paddy yield and subsequently quantified the WF, under different rainfall conditions, over a period of 10 years. In this application, AquaCrop GIS software simulated paddy yield and evapotranspiration. Through this analysis, the value of the green WF of the rice field was found to be higher than the value of the blue WF. This result revealed that rainwater increases the yield of the rice field through greater use of green water and reduced use of blue water (Chanu & Oinam, 2023). Lastly, GIS combined with WF has also been used in fruit cultivation. Particularly, kiwi crop flow data was digitized and ArcGIS-ArcMap 10.7.1 was used for analysis (Bilge Ozturk et al., 2022).

Similarly, Fu et al. (2019) combined CROPWAT 8.0 by FAO and GIS technology to calculate green, blue, and grey WFs of wheat, maize, cotton, and groundnut, focussing on the spatial variations of WFs for crops in different rainfall years.

Moreover, few years ago a GIS model platform was developed for integrating geographic and alphanumeric data to estimate the environmental benefits resulting from the implementation of the agri-environmental measures (AEM) for Rural Development Program in Veneto Region, in Northeast Italy. In particular, GIS was used for comparing cropping systems (maize, wheat, barley, soybean, sunflower, rapeseed, potato, sugar beet crops, and pastures and meadows) in the preceding and anticipating phases the implementation of the program and for investigating the effect on soil, water, atmosphere, and biota. At the end of this evaluation, it was possible to generate short- and long-term scenarios (Longo et al., 2021).

However, more complex applications have also been produced. Mainly, Shtull-Trauring et al. (2016) combined high resolution with GIS data to analyze the impact of agricultural practices, considering crop type, spatial factors (drainage basins, climate, and soil type), and WF of crops. An integrated evaluation, by associating water data in high-resolution grids with a geodatabase created by GIS, also allows for the evaluation of the necessary demand of water and energy footprint of irrigation production (Daccache et al., 2014). In most applications like Multsch et al. (2013), WF spatial data management and analysis are achieved by integrating all calculations into a GIS tool.

In addition, GIS can be applied in a multiple-application model by combining GIS, decision theory methods, process-based vegetation model output, and spatial analysis (Schaldach et al., 2009). In the study by Mokarram et al. (2021), multiple regression and artificial neural network models in a GIS environment have been used.

Considering the importance of this study and the application in the context of the Funding Programme for Recovery and Resilience of Italy, as a European Union (EU) Member State, the authors focussed on the territorial application concentrating on the Mediterranean and European context. In their study, Centofanti et al. (2008) described the elaboration of a pan-European dataset for the characterization of soil, climate, crops, and land cover considering the diversity of European agricultural and environmental conditions related to pesticide use. Each dataset was intersected using GIS to identify the full range of unique combinations of climate, soil, and crop types.

After this assessment, the scholars retrieved multiple combinations of GIS and WF tools: the quantification of water-holding capacity values was obtained using GIS and the Kriging method. Subsequently, after collecting the estimated irrigation depth data for the oil palm plantation, the total net irrigation data was simulated in FAO's CROPWAT model and compared (Norizan et al., 2021). Moreover, Norizan et al. (2021) analyzed data for each irrigation management zone using one-way ANOVA and the Tukey–Kramer test using SAS Software Version 9.4. Even with some new methodologies called "Node" based on the energy, water, and food nexus, it was possible to apply GIS (Norizan et al., 2021). Indeed, this methodology included a decentralization process using GIS-based approaches, developed some composite geospatial risk indicators using the analytical hierarchy process, and conducted a resource utilization assessment (Haji et al., 2020). Other applications dealt with operations correlated to crop cultivations: for example, Rizvi et al. (2021) presented a methodology for assessing the irrigation system performance based on satellite remote sensing and ground data using GIS tools. Viccaro et al. (2022) developed a model of analysis using multicriteria decision analysis techniques in GIS, combining the water–energy–food nexus approach. Lamastra et al. (2016) applied fuzzy logic and implemented this nexus in a web GIS software with the aim to obtain a structured opinion on the agronomic sustainability of vineyard management, using an observation scale from "A" as excellent to "E" as completely unsustainable. Particularly, in another study (Lamastra et al., 2016) a fuzzy logic (deriving from mathematical application) was linked to GIS (a mapping system) as a useful instrument for measuring the environmental impact of viticulture, and more in general for observing several agricultural crops in a holistic way.

These assessments underlined that the combination of several tools, in particular WF and GIS, can be applied in the agricultural context to manage water resources and reduce the pressure linked to its use and the impacts of water scarcity.

In conclusion, this analysis carried out through an SLR pushes toward studies that evaluate the existing knowledge in a specific topic, such as crops cultivation in agri-food industry stimulating future application in other contexts for the implementation by policymakers, practitioners, and the scientific community.

1.5 Geographic information system models applied

The analyses of the 15 papers included in the crop production cluster highlighted the several kinds of GIS tools used. Particularly, Chanu and Oinam (2023) have chosen AquaGIS as a crop growth model for estimating evapotranspiration and crop yield. AquaCrop-GIS is designed using Excel files and the input data required are crop, soil, groundwater, and resource management files. This model provides the evapotranspiration and crop yield values necessary to estimate the WF, also in manual form (Ignacio et al., 2015). Bilge Ozturk et al. (2022) used ArcGIS-ArcMap 10.7.1: this GIS model aggregates large quantities of point features into dynamic polygons called bins. It is a visualization tool that lets to explore large datasets (such as crop fields) while optimizing the drawing performance of layers that contain thousands or millions of point features (ESRI, 2023).

Moreover, the model—GIS platform applied by Longo et al. (2021) proved to be a suitable tool for evaluating the effects of the implementation of a rural development program on some crops in the Veneto Region, in Italy.

Nevertheless, in a context of an agricultural irrigated system located in the Mediterranean area, through the GIS model, the rainwater fallen over a month was quantified through an integrated model of spatial and statistical data with the aim to estimate water capacity and carbon footprint (CF) (Daccache et al., 2014).

To apply GIS for carrying out a water consumption analysis, some software, based on an open-source spatial programming library, called MapWinGIS Active, can be used for the management of grid and shape files (Multsch et al., 2013).

Anyhow, input spatial data can be derived from the GIS software ArcGIS 9.2. and as a basis for the analysis the CORINE 2000 database with land cover data for EU Member Countries, can be used. Furthermore, the one developed by Heistermann (2006) for non-Member Countries can be used as a global crop map. Moreover, the analysis was carried out with the ArcGIS software using Map Algebra operations.

Among the several models available, the GIS model ESRI ArcGIS 9.1 was also used in 2008 (Centofanti et al., 2008). Moreover, the spatial distribution of the 16-footprint climatic zones was digitized to provide a polygon dataset for GIS operations. Although to date a first attempt has been identified to quantify variations on a pan-European scale through the combination of WF and GIS, in the future this application could be replicated for European Member Countries such as Italy. In addition, Norizan et al. (2021) used GIS in ArcGIS 10.3 to make an interpolation map. Recently, Rizvi et al. (2021) used geometric correction of all images by orthorectification using the "geo-reference" tool of ArcGIS. Furthermore, all operational steps included coordinate transformation, map projection, resampling, interpolation, and mosaicking.

This snapshot offered the different applicable GIS models and provides ideas for analysis and georeferenced data collection as required by GRINS - PNRR Program.

1.6 Discussion

Generally, the benefits of the combined application of WF and GIS have been demonstrated by Shtull-Trauring et al. (2016). Above all, the integration of the WF calculation methodology with the different local data on climate, yields, crops, and soils at high resolution processed with GIS can favor the modeling of crops and agricultural practices more suitable for one site rather than for another. Practically, the application of this combined approach can support choices to practice crops that are more productive and have less impact on the water resources of one place rather than another.

Furthermore, this integration of WF and GIS approaches allows us to identify the optimal combinations between high production yield, low total WF, and reduction in management costs of water withdrawal from sources managed by public administrations. Therefore, this approach could also reduce the management costs of production, especially for crops such as wheat, barley, and corn, which are fundamental to address the needs of growing populations, as highlighted by (Mokarram et al., 2021).

Therefore, this study also supports analysis of the relationship between environmental performance, crop traceability, forecasting of water resource consumption, and reduction of impacts in the agricultural sector, stimulating the necessary resilience by practioners and stakeholders.

However, in order to apply this combined approach, as highlighted by Elkamel et al. (2023) attention must be paid to the model design and construction. Hence, it is necessary to operate according a holistic approach that consider model and simultaneously data as input values for the model: particularly, georeferenced data, level of precision of the cartographic surveys, distribution, dimensions of the data, prices for water withdrawal, consumption of electricity, weather conditions, and crop yield.

However, this SLR assessment has demonstrated that there are numerous opportunities to apply a combined model of analytical approaches in agriculture, intending to conserve land, increase production yields, reduce impacts, and conserve natural resources such as water. For example, the study by Carlson et al. (2023) provided a broad assessment of the potential conservation of forests starting from some detailed maps of wild areas close to buildings.

Finally, this study can provide ideas for applications of the model of joint approaches, not only between WF and GIS but also more generally among other tools to indicate every possible alternative for the conservation of natural resources and with the achievement of sustainability objectives.

From this SLR, it also emerged that good digitalization of data in agriculture, using Computer Science Principles and Internet of Things (IoT), is fundamental for setting up a combined model of tools, comparing the documented information and designing a transition towards a sustainable, circular and resilient approach (Crovella et al., 2021). Furthermore, the planning of reports and documents can also support the various public and private stakeholders and practitioners in planning production, trade, and transformation activities, as underlined by Schwab et al. (2023).

Finally, all the information generated following the application of a combined WF and GIS approach are essential for planning irrigation activities with fresh water, scheduling the quantity of water needed, and the times and methods of operation of the systems. All maps generated with evapotranspiration data using remotely sensed multispectral vegetation indices represent a useful tool for quantifying crop water consumption, crop water requirements on a regional and local scale and for providing sustainable alternatives as underlined by Parmar et al. (2023).

1.7 Conclusions

The results related to this SLR contain several applications of the combination of WF and GIS tools in the agricultural context. Although there are only 15 articles that are relevant to the topic of this study, the authors presented some replicable applications for the crops cultivation that stakeholders and practitioners can take into consideration.

Specifically, considering the application of GIS for paddy crops, the most cultivated crops in India, this tool can be used similarly in Northern Italy for rice production or in Southern Italy for grape cultivation. Moreover, the different models, developed to balance the water impact in the Mediterranean regions, are replicable in the regions of Southern Italy under the same pedoclimatic conditions.

Technically, ArcGIS is useful to map large cultivated fields such as fruit fields as apple orchards, peach orchards, vineyard for Italy. However, it is possible to achieve changes in agricultural production models starting by reducing water consumption, implementing circular economic models of resource reuse, and combining multiple approaches for impact mapping.

Furthermore, the combination of CROPWAT by the FAO and GIS allows us to know the variations of the different components of the WF at different precipitation conditions. Therefore, the integration of the WF approach and high-resolution local data retrieved with GIS leads to a reduction in water consumption, increase in yield, reduction of WF values, and minimization of negative environmental impacts.

The main difficulties and limitations of applying the combined WF and GIS approach are linked to the difficulties in finding detailed information and to a lack of digitized information on agriculture.

Finally, this chapter represents a guide for the agricultural economy of the country and for all involved, and represents one of the first outputs of the GRINS - Growing, Resilient, INclusive and Sustainable Project.

Funding

"This study was funded by the European Union - NextGenerationEU, in the framework of the GRINS - Growing Resilient, INclusive and Sustainable project (GRINS PE00000018 – CUP H93C22000650001). The views and opinions expressed are solely those of the authors and do not necessarily reflect those of the European Union, nor can the European Union be held responsible for them."

Acknowledgments

"This study was funded by the European Union - NextGenerationEU, in the framework of the GRINS -Growing Resilient, INclusive and Sustainable project and thanks to a PNRR M4C2 - Investiment 1.3 Enlarged Partnership GRINS - Growing Resilient, INclusive and Sustainable by University of Bari Aldo Moro, Department of Economics, Management and Business Law and to address the objectives of

Spoke 1. CUP: H93C22000650001, Code PNRR_PE_71, Identified Code PE0000018, Thematic: Building of datasets for the circular economy of the main Italian production systems. The views and opinions expressed are solely those of the authors and do not necessarily reflect those of the European Union, nor can the European Union be held responsible for them".

References

Arodudu, O., Voinov, A., & van Duren, I. (2013). Assessing bioenergy potential in rural areas—A NEG-EROEI approach. *Biomass and Bioenergy*, *58*, 350–364. Available from https://doi.org/10.1016/j.biombioe.2013.07.020.

Atkinson, C. L., Christian, A. D., Spooner, D. E., & Vaughn, C. C. (2014). Long-lived organisms provide an integrative footprint of agricultural land use. *Ecological Applications*, *24*(2), 375–384. Available from https://doi.org/10.1890/13-0607.1, http://www.esajournals.org/doi/pdf/10.1890/13-0607.1.

Bhat, S. U., Khanday, S. A., Islam, S. T., & Sabha, I. (2021). Understanding the spatiotemporal pollution dynamics of highly fragile montane watersheds of Kashmir Himalaya, India. *Environmental Pollution*, *286*, 18736424. Available from https://doi.org/10.1016/j.envpol.2021.117335, https://www.journals.elsevier.com/environmental-pollution.

Bilge Ozturk, G., Ozenen Kavlak, M., Cabuk, S. N., Cabuk, A., & Cetin, M. (2022). Estimation of the water footprint of kiwifruit: In the areas transferred from hazelnut to kiwi. *Environmental Science and Pollution Research*, *29*(48), 73171–73180. Available from https://doi.org/10.1007/s11356-022-21050-y, https://link.springer.com/journal/11356.

Bontinck, P. A., Grant, T., Kaewmai, R., & Musikavong, C. (2021). Recalculating Australian water scarcity characterisation factors using the AWARE method. *International Journal of Life Cycle Assessment*, *26*(8), 1687–1701. Available from https://doi.org/10.1007/s11367-021-01952-8, http://www.springerlink.com/content/0948-3349.

Cai, H.-S., Zang, X. L., Zhu, D.-H. (2007). Study of ecological capacity change and quantitative analysis of ecological compensation in a nature reserve based on RS and GIS: A case study on Po-yang Lake Nature Reserve, China. *New Zealand Journal of Agricultural Research, 50*(5), 757–766. Available from https://doi.org/10.1080/00288230709510348.

Carlson, A. R., Radeloff, V. C., Helmers, D. P., Mockrin, M. H., Hawbaker, T. J., & Pidgeon, A. (2023). The extent of buildings in wildland vegetation of the conterminous U.S. and the potential for conservation in and near National Forest private inholdings. *Landscape and Urban Planning*, *237*. Available from https://doi.org/10.1016/j.landurbplan.2023.104810, http://www.elsevier.com/inca/publications/store/5/0/3/3/4/7.

Casella, P., De Rosa, L., Salluzzo, A., & De Gisi, S. (2019). Combining GIS and FAO's crop water productivity model for the estimation of water footprinting in a temporary river catchment. *Sustainable Production and Consumption*, *17*, 254–268. Available from https://doi.org/10.1016/j.spc.2018.11.002, http://www.journals.elsevier.com/sustainable-production-and-consumption/.

Cazcarro, I., Duarte, R., & Sánchez-Chóliz, J. (2016). Downscaling the grey water footprints of production and consumption. *Journal of Cleaner Production*, *132*, 171–183. Available from https://doi.org/10.1016/j.jclepro.2015.07.113.

Centofanti, T., Hollis, J. M., Blenkinsop, S., Fowler, H. J., Truckell, I., Dubus, I. G., & Reichenberger, S. (2008). Development of agro-environmental scenarios to support pesticide risk assessment in Europe. *Science of the Total Environment*, *407*(1), 574–588. Available from https://doi.org/10.1016/j.scitotenv.2008.08.017.

Chanu, B. N., & Oinam, B. (2023). Assessment of crop yield and water footprint of kharif paddy production under different rainfall years. *Songklanakarin Journal of Science and Technology*, *45*(3), 451–455. Available from https://sjst.psu.ac.th/journal/45-3/20.pdf, Prince of Songkla University, India.

Crovella, T., Paiano, A., Falciglia, P. P., Lagioia, G., & Ingrao, C. (2024). Wastewater recovery for sustainable agricultural systems in the circular economy – A systematic literature review of Life Cycle Assessments. *Science of The Total Environment*, *912*, 169310. Available from https://doi.org/10.1016/j.scitotenv.2023.169310.

Crovella, T., Paiano, A., Lagioia, G., Cilardi, A. M., & Trotta, L. (2021). Modelling Digital Circular Economy framework in the Agricultural Sector. An Application in Southern Italy. *Engineering Proceedings*, *9*(1), 15. Available from https://doi.org/10.3390/engproc2021009015.

Daccache, A., Ciurana, J. S., Rodriguez Diaz, J. A., & Knox, J. W. (2014). Water and energy footprint of irrigated agriculture in the Mediterranean region. *Environmental Research Letters*, *9*(12), 124014. Available from https://doi.org/10.1088/1748-9326/9/12/124014.

Elkamel, M., Valencia, A., Zhang, W., Zheng, Q. P., & Chang, N. B. (2023). Multi-agent modeling for linking a green transportation system with an urban agriculture network in a food–energy–water nexus. *Sustainable Cities and Society*, *89*. Available from https://doi.org/10.1016/j.scs.2022.104354, http://www.elsevier.com/wps/find/journaldescription.cws_home/724360/description#description.

ESRI. (2023). EsriItalia. ArcGIS Enterprise requisiti d'installazione. 2023. https://doc.arcgis.com/it/insights/2023.1/get-started/arcgis-enterprise-requirements.htm.

Feng, Q., Chaubey, I., Engel, B., Cibin, R., Sudheer, K. P., & Volenec, J. (2017). Marginal land suitability for switchgrass, Miscanthus and hybrid poplar in the Upper Mississippi River Basin (UMRB). *Environmental Modelling and Software*, *93*, 356–365. Available from https://doi.org/10.1016/j.envsoft.2017.03.027, http://www.elsevier.com/inca/publications/store/4/2/2/9/2/1.

Field, D. J., Morgan, C. L., McBratney, A. B. (2017). Global soil security. Springer. ISBN: 978-3-319-43393-6. Available from https://doi.org/10.1007/978-3-319-43394-3.

Fu, M., Guo, B., Wang, W., Wang, J., Zhao, L., & Wang, J. (2019). Comprehensive assessment of water footprints and water scarcity pressure for main crops in Shandong Province, China. *Sustainability*, *11*(7), 1856. Available from https://doi.org/10.3390/su11071856.

Garofalo, P., Ventrella, D., Kersebaum, K. C., Gobin, A., Trnka, M., Giglio, L., Dubrovský, M., & Castellini, M. (2019). Water footprint of winter wheat under climate change: Trends and uncertainties associated to the ensemble of crop models. *Science of the Total Environment*, *658*, 1186–1208. Available from https://doi.org/10.1016/j.scitotenv.2018.12.279, http://www.elsevier.com/locate/scitotenv.

Geli, H. M. E., González-Piqueras, J., Neale, C. M. U., Balbontín, C., Campos, I., & Calera, A. (2020). Effects of surface heterogeneity due to drip irrigation on scintillometer estimates of sensible, latent heat fluxes and evapotranspiration over vineyards. *Water*, *12*(1), 81. Available from https://doi.org/10.3390/w12010081.

Haji, M., Govindan, R., & Al-Ansari, T. (2020). Novel approaches for geospatial risk analytics in the energy−water−food nexus using an EWF nexus node. *Computers & Chemical Engineering*, *140*, 106936. Available from https://doi.org/10.1016/j.compchemeng.2020.106936.

Hanson, M. C., Petch, G. M., Ottosen, T.-B., & Skjøth, C. A. (2022). Climate change impact on fungi in the atmospheric microbiome. *Science of The Total Environment*, *830*, 154491. Available from https://doi.org/10.1016/j.scitotenv.2022.154491.

Heistermann, M. (2006). Modelling the global dynamics of rain-fed and irrigated croplands. Reports on Earth System Science.

Högström, E., Trofaier, A. M., Gouttevin, I., & Bartsch, A. (2014). Assessing seasonal backscatter variations with respect to uncertainties in soil moisture retrieval in Siberian Tundra regions. *Remote Sensing*, *6*(9), 8718−8738. Available from https://doi.org/10.3390/rs6098718, http://www.mdpi.com/2072-4292/6/9/8718/pdf.

Ignacio, J. L., Margarita, G. V., & Elias, F. (2015). *AquaCrop-GIS Reference Manual*. Food and Agriculture Organization.

Karim, A., Gonzalez Cruz, M., Hernandez, E. A., & Uddameri, V. (2020). A GIS-based fit for the purpose assessment of brackish groundwater formations as an alternative to freshwater aquifers. *Water*, *12*(8), 2299. Available from https://doi.org/10.3390/w12082299.

King, J. S., Ceulemans, R., Albaugh, J. M., Dillen, S. Y., Domec, J. C., Fichot, R., Fischer, M., Leggett, Z., Sucre, E., Trnka, M., & Zenone, T. (2013). The challenge of lignocellulosic bioenergy in a water-limited world. *BioScience*, *63*(2), 102−117. Available from https://doi.org/10.1525/bio.2013.63.2.6.

Kourgialas, N. N., Karatzas, G. P., Dokou, Z., & Kokorogiannis, A. (2018). Groundwater footprint methodology as policy tool for balancing water needs (agriculture & tourism) in water scarce islands—The case of Crete, Greece. *Science of the Total Environment*, *615*, 381−389. Available from https://doi.org/10.1016/j.scitotenv.2017.09.308, http://www.elsevier.com/locate/scitotenv.

Krishnan, P., Abhilash, K. R., Sreeraj, C. R., Samuel, V. D., Purvaja, R., Anand, A., Mahapatra, M., Sankar, R., Raghuraman, R., & Ramesh, R. (2021). Balancing livelihood enhancement and ecosystem conservation in seaweed farmed areas: A case study from Gulf of Mannar Biosphere Reserve, India. *Ocean and Coastal Management*, *207*. Available from https://doi.org/10.1016/j.ocecoaman.2021.105590, http://www.elsevier.com/inca/publications/store/4/0/5/8/8/9.

Lamastra, L., Balderacchi, M., Di Guardo, A., Monchiero, M., & Trevisan, M. (2016). A novel fuzzy expert system to assess the sustainability of the viticulture at the wine-estate scale. *Science of the Total Environment*, *572*, 724−733. Available from https://doi.org/10.1016/j.scitotenv.2016.07.043, http://www.elsevier.com/locate/scitotenv.

Li, M., Cao, X., Liu, D., Fu, Q., Li, T., & Shang, R. (2022). Sustainable management of agricultural water and land resources under changing climate and socio-economic conditions: A multi-dimensional optimization approach. *Agricultural Water Management*, *259*, 107235. Available from https://doi.org/10.1016/j.agwat.2021.107235.

Longo, M., Dal Ferro, N., Lazzaro, B., & Morari, F. (2021). Trade-offs among ecosystem services advance the case for improved spatial targeting of agri-environmental measures. *Journal of Environmental Management*, *285*, 112131. Available from https://doi.org/10.1016/j.jenvman.2021.112131.

McKean, J., Nagel, D., Tonina, D., Bailey, P., Wright, C. W., Bohn, C., & Nayegandhi, A. (2009). Remote sensing of channels and riparian zones with a narrow-beam aquatic-terrestrial LIDAR. *Remote Sensing*, *1*(4), 1065−1096. Available from https://doi.org/10.3390/rs1041065, http://www.mdpi.com/2072-4292/1/4/1065/pdf, United States.

Mengist, W., Soromessa, T., & Legese, G. (2020). Method for conducting systematic literature review and meta-analysis for environmental science research. *MethodsX*, *7*, 100777. Available from https://doi.org/10.1016/j.mex.2019.100777.

Mokarram, M., Zarei, A. R., & Etedali, H. R. (2021). Optimal location of yield with the cheapest water footprint of the crop using multiple regression and artificial neural network models in GIS. *Theoretical and Applied Climatology*, *143*(1−2), 701−712. Available from https://doi.org/10.1007/s00704-020-03413-y, https://rd.springer.com/journal/volumesAndIssues/704.

Mondal, A., Das, S., Sah, R. K., Bhattacharyya, P., & Bhattacharya, S. S. (2017). Environmental footprints of brick kiln bottom ashes: Geostatistical approach for assessment of metal toxicity. *Science of the Total Environment*, *609*, 215−224. Available from https://doi.org/10.1016/j.scitotenv.2017.07.172, http://www.elsevier.com/locate/scitotenv.

Monika., Srinivasan, D., & Reindl, T. (2015). GIS as a tool for enhancing the optimization of demand side management in residential microgrid. *Proceedings of 2015 IEEE Innovative Smart Grid Technologies − Asia (ISGT ASIA)*, 1−6. Available from https://doi.org/10.1109/ISGT-Asia.2015.7387041.

Montealegre, A. L., García-Pérez, S., Guillén-Lambea, S., Monzón-Chavarrías, M., & Sierra-Pérez, J. (2022). GIS-based assessment for the potential of implementation of food−energy−water systems on building rooftops at the urban level. *Science of the Total Environment*, *803*, 149963. Available from https://doi.org/10.1016/j.scitotenv.2021.149963.

Multsch, S., Al-Rumaikhani, Y. A., Frede, H. G., & Breuer, L. (2013). A Site-sPecific agricultural water requirement and footprint estimator (SPARE: WATER 1.0). *Geoscientific Model Development*, *6*(4), 1043−1059. Available from https://doi.org/10.5194/gmd-6-1043-2013, http://www.geosci-model-dev.net/volumes_and_issues.html.

Norizan, M. S., Wayayok, A., Abdullah, A. F., Mahadi, M. R., & Karim, Y. A. (2021). Spatial variations in water-holding capacity as evidence of the need for precision irrigation. *Water (Switzerland)*, *13*(16), 20734441. Available from https://doi.org/10.3390/w13162208, https://www.mdpi.com/2073-4441/13/16/2208/pdf.

Oni, S. K., Futter, M. N., Buttle, J., & Dillon, P. J. (2015). Hydrological footprints of urban developments in the Lake Simcoe watershed, Canada: A combined paired-catchment and change detection modelling approach. *Hydrological Processes*, *29*(7), 1829−1843. Available from https://doi.org/10.1002/hyp.10290, http://onlinelibrary.wiley.com/journal/10.1002/(ISSN)1099-1085.

Parmar, S. H., Patel, G. R., & Tiwari, M. K. (2023). Assessment of crop water requirement of maize using remote sensing and GIS. *Smart Agricultural Technology*, *4*, 100186. Available from https://doi.org/10.1016/j.atech.2023.100186.

Pulighe, G., & Pirelli, T. (2023). Assessing the sustainability of bioenergy pathways through a land-water-energy nexus approach. *Renewable and Sustainable Energy Reviews*, *184*, 113539. Available from https://doi.org/10.1016/j.rser.2023.113539.

Ren, C., Guo, P., Li, M., & Li, R. (2016). An innovative method for water resources carrying capacity research—Metabolic theory of regional water resources. *Journal of Environmental Management*, *167*, 139–146. Available from https://doi.org/10.1016/j.jenvman.2015.11.033, https://www.sciencedirect.com/journal/journal-of-environmental-management.

Rizvi, S. A., Ahmad, A., Latif, M., Shakir, A. S., Khan, A. A., Naseem, W., & Gondal, M. R. (2021). Implication of remote sensing data under GIS environment for appraisal of irrigation system performance. *Water Resources Management*, *35*(14), 4909–4926. Available from https://doi.org/10.1007/s11269-021-02979-0, http://www.wkap.nl/journalhome.htm/0920-4741.

Root-Bernstein, M., Guerrero-Gatica, M., Piña, L., Bonacic, C., Svenning, J. C., & Jaksic, F. M. (2017). Rewilding-inspired transhumance for the restoration of semiarid silvopastoral systems in Chile. *Regional Environmental Change*, *17*(5), 1381–1396. Available from https://doi.org/10.1007/s10113-016-0981-8, http://springerlink.metapress.com/app/home/journal.asp?wasp = 64cr5a4mwldrxj984xaw&referrer = parent&backto = browsepublicationsresults,451,542;.

Schaldach, R., Fïrke, M., & Lapola, D. (2009). A model-based assessment of the potential role of irrigated cropland for biogas production in Europe,. *Advances in Geosciences*, *21*, 85–90. Available from https://doi.org/10.5194/adgeo-21-85-2009, http://www.adv-geosci.net/volumes.html, Copernicus GmbH, Germany.

Schwab, A. K., Halberstadt, J., Besieda, A., Kraft, M., & Ortland, A. (2023). Organizational aspects of digitalization in the context of agriculture: Exemplary results from analyzing data flows in German dairy farming. *Procedia Computer Science*, *219*(1), 1006–1011. Available from https://doi.org/10.1016/j.procs.2023.01.378, 18770509 Elsevier B.V. Germany January 2023, http://www.sciencedirect.com/science/journal/18770509.

Shtull-Trauring, E., Aviani, I., Avisar, D., & Bernstein, N. (2016). Integrating high resolution water footprint and GIS for promoting water efficiency in the agricultural sector: A case study of plantation crops in the Jordan Valley. *Frontiers in Plant Science*, *7*, 2016. Available from https://doi.org/10.3389/fpls.2016.01877, http://journal.frontiersin.org/article/10.3389/fpls.2016.01877/full.

Soula, R., Chebil, A., McCann, L., & Majdoub, R. (2021). Water scarcity in the Mahdia region of Tunisia: Are improved water policies needed? *Groundwater for Sustainable Development*, *12*, 100510. Available from https://doi.org/10.1016/j.gsd.2020.100510.

Terrado, M., Sabater, S., & Acuña, V. (2016). Identifying regions vulnerable to habitat degradation under future irrigation scenarios. *Environmental Research Letters*, *11*(11), 114025. Available from https://doi.org/10.1088/1748-9326/11/11/114025.

Viccaro, M., Caniani, D., Masi, S., Romano, S., & Cozzi, M. (2022). Biofuels or not biofuels? The "Nexus Thinking" in land suitability analysis for energy crops. *Renewable Energy*, *187*, 1050–1064. Available from https://doi.org/10.1016/j.renene.2022.02.008, http://www.journals.elsevier.com/renewable-and-sustainable-energy-reviews/.

Wangda, P., Hussin, Y. A., Bronsveld, M. C., & Karna, Y. K. (2019). Species stratification and upscaling of forest carbon estimates to landscape scale using GeoEye-1 image and lidar data in sub-tropical forests of Nepal. *International Journal of Remote Sensing*, *40*(20), 7941–7965. Available from https://doi.org/10.1080/01431161.2019.1607981, https://www.tandfonline.com/loi/tres20.

Zhuo, L., Mekonnen, M. M., Hoekstra, A. Y., & Wada, Y. (2016). Inter-and intra-annual variation of water footprint of crops and blue water scarcity in the Yellow River basin. *Advances in Water Resources*, *87*, 29–41.

Chapter 2

Assessment of water footprints in different sectors: utilization, safety and challenges

Nurul Huda Abd Kadir[1,*], Yaman Ahmed Naji[2], Zainab Muhammad[2] and Suhail Bashir[3]

[1]BIOSES RIG Group, Faculty of Science and Marine Environment, Universiti Malaysia Terengganu, Kuala Nerus, Terengganu, Malaysia, [2]Faculty of Science and Marine Environment, Universiti Malaysia Terengganu, Kuala Nerus, Terengganu, Malaysia, [3]ENS Environmental Consultancy Sharjah, United Arab Emirates

**Corresponding author. e-mail address: nurulhuda@umt.edu.my.*

2.1 Introduction

Water is an essential natural resource that plays a crucial role in the living world. The escalating demands on water resources, driven by factors such as industrial production, agricultural irrigation, and everyday domestic activities like drinking and washing, pose significant challenges to its sustainability (Çankaya, 2023; Dey et al., 2023; Gosling & Arnell, 2016).

To mitigate the risk of water scarcity, it is imperative to gain a deep understanding of efficient water utilization and the minimization of losses (Azzam et al., 2022). Water is universally recognized as one of the most vital and valuable resources, essential for the existence and development of all living organisms and ecosystems (Emam, 2022; Ma et al., 2022). It is widely acknowledged that the increasing water demand, coupled with the ever-changing climatic conditions, significantly impacts the sustainability of this precious resource (Jeyrani et al., 2021).

Recent research on water security has depicted the increasing scarcity of water, attributed to factors such as population growth, economic expansion, climate change, and inadequate water management (Vörösmarty et al., 2021). Falkenmark (1995) mentioned a practical categorization of freshwater resources into blue water resources (BWR; surface water and groundwater) and green water resources (precipitation) to assess freshwater availability (Dey et al., 2023). Freshwater availability has been classified by Falkenmark (1995) into BWR and green water resources (Mekonnen & Gerbens-Leenes, 2020). Blue water (BW) refers to ground and surface water which includes lakes, reservoirs, rivers, and wetlands among others.

The concept of "blue water footprint" (BWF) refers to the utilization of BWR during the production of goods and services for individuals or communities, a concept discussed by Çankaya and others. (Çankaya, 2023; Liu et al., 2023).

However, the research conducted by Mekonnen & Gerbens-Leenes, has identified regions with higher BWF levels, including India, China, the United States, Pakistan, and Iran. Notably, these regions are associated with the larger production of specific crops like wheat, rice, cotton, sugar cane, fodder, and maize, each of which has a significant BWF (Mekonnen & Gerbens-Leenes, 2020).

It is important to note that there is a typical trade-off between ecological and socioeconomic benefits (the ability to sustainably provide freshwater for the production and life of human activities and ecosystems) This trade-off is inherent in the concept of sustainability. As the demand for blue water usage (BWU) has surged alongside socioeconomic growth, it has outpaced local blue water availability (BWA), resulting in local blue water scarcity (BWS), as noted by Huang et al. (2023).

The global concern of BWS is an alarming concern, given its impact on aquatic ecosystems and the potential future implications for human health, food security, economic development, and social stability (Gao et al., 2021).

Researchers like Mekonen and Hoekstra (2020) reported and indicated that around four billion people experience severe water scarcity for at least 1 month out of every year in different river basins across the growing world.

Water Footprints and Sustainable Development. DOI: https://doi.org/10.1016/B978-0-443-23631-0.00002-9

Moreover, numerous studies have concentrated on concerns related to increasing water pollution issues (Guan et al., 2021; Ke et al., 2024; Wang et al., 2018), in addition to the water usage disparities between downstream and upstream regions (Fu et al., 2018) and between water source areas and intake locations (Fang et al., 2021).

It is unfortunate to mention that until now, there is a lack of available information regarding strategies to mitigate BWU associated with various human activities, products, and other resources. Such strategies are crucial to ensure the continued security and availability of these resources for future generations, employing more sustainable approaches. It is worth noting that one of the United Nations' Sustainable Development Goals (SDGs 6), outlined by the United Nations emphasizes ensuring freshwater supply for basic human needs and well-being of aquatic ecosystems. This goal calls for improvements in sustainable freshwater withdrawal and supply to substantially reduce the number of people impacted by water scarcity (ORGANIZAÇÃO DAS NAÇÕES UNIDAS, 2015; UN/DESA, World Population Prospects, 2017).

On the contrary, green water relates to rainwater that falls on land, gets absorbed into the soil and is utilized by plants and microorganisms. It is subsequently released back into the atmosphere through processes like transpiration or evaporation (Ma et al., 2022; Zhao et al., 2021).

The availability of green water is influenced by a multitude of factors, including precipitation, vegetation type, soil type, water infiltration, air and soil temperatures, soil nutrient content, and atmospheric CO_2 concentration (Khalili et al., 2023). Considering that the majority of water usage worldwide is attributed to agricultural production, understanding the extent of water consumption and losses in agriculture plays a critical role in mitigating the risk of water scarcity (Ma et al., 2022). Cultivated plants draw water from both green water and BW sources, and this water is subsequently returned to the atmosphere through a process known as evapotranspiration (ET) (Azzam et al., 2022).

2.2 Blue water

BW sources, which include ground and surface water, are usually considered safe and clean for human consumption. BW comprises water that has evaporated, water contained within products, water not returned to the same catchment area, or water not released during the same time frame (Hoekstra, 2019).

BW can be generated through three primary methods: the first method involves runoff from the soil surface, which enters surface storage structures. The second method entails water moving vertically through soil layers, recharging aquifers, and potentially being withdrawn at a later point. The final method involves water flowing laterally within soil layers and entering surface storage structures (Zisopoulou & Panagoulia, 2021).

BW is also a product of rain that does not permeate the ground, while the part that does infiltrate becomes soil moisture in the unsaturated zone or goes deeper, merging with the groundwater in the saturated zone. In contrast, the rainwater that is retained in the unsaturated zone of the soil is called green water. Furthermore, when BW accumulates in small depressions or potholes, it does not slowly seep into the soil or evaporate into the atmosphere but returns as precipitation. Moreover, to cater to the water requirements of ecosystems, green water can be generated from BW through both natural and human-induced processes (Mao et al., 2020).

The demand for freshwater is on the rise, and the rate of increase is closely linked to socioeconomic factors like population growth and consumption patterns. In another aspect, both higher income levels and an increasing global population play a role in the amplified consumption of freshwater resources. In addition, climate change, as forecasted to diminish and render regional freshwater supplies more unpredictable (Chang, 2019; Liu et al., 2023), could further exacerbate the issue of local BWS. Also the use of surface water for various purposes involves energy consumption and the release of greenhouse gases (GHGs), potentially contributing to climate change. A prime example is irrigated agriculture, which not only accounts for 18% of the world's GHG emissions but also utilizes a substantial 90% of the world's freshwater resources.

Moreover, quantifying BW is of utmost importance for assessing the level of water security in a given region because it directly impacts human requirements such as domestic water supply, irrigation demands, industrial production, and power generation. In contrast, groundwater is essential for plant growth, rain-fed agriculture, and terrestrial ecosystem services (Veldkamp et al., 2017).

2.2.1 The uses of blue water

BW represents ground and surface water that meets the standards for human and environmental well-being, serving a variety of purposes including agriculture, industrial processes, and domestic water supply. It is replenished by precipitation, such as rainfall or snowfall. Some of the common and basic uses for BW include the following.

2.2.1.1 Agriculture

Freshwater is largely used for agriculture, accounting for 99% of global water utilization. This allocation is essential to meet the needs of a growing population and the surging demand for food, fiber, and biofuels. Agricultural output could increase by nearly half by 2050 in comparison to 2012 (Mekonnen & Gerbens-Leenes, 2020; Mekonnen & Hoekstra, 2020).

The general BWF for agriculture was 8362 km^3/year in 2011 (80% green, 11% blue, and 9% gray water). Burek et al. (2016) forecasted a substantial 20%–30% surge in water demand between 2010 and 2050 (Burek et al., 2016). Given the overwhelming demand, both land and water resources are already scarce and are expected to become even scarcer, both locally and globally. In light of these challenges, the sustainable management of agricultural water (BW) plays a pivotal role in providing for our food needs and fostering social development (Beltran-Pea et al., 2020).

To achieve food security and effectively reduce poverty and hunger, it is crucial to manage the agricultural water demand. The development of efficient management strategies, centered on the performance of agricultural water (BW), has been widely recognized as a fundamental approach for managing these available limited water resources. (Foster et al., 2020; H. Gao et al., 2022).

Furthermore, there is no doubt that global warming has led to an increased demand for agricultural water resources on a global scale (Luan et al., 2018). As a result of climatic and socioeconomic shifts, regions in mid- and high-latitude areas will confront increased agricultural water strains in the 21st century, including bigger regional discrepancies in the distribution of agricultural water supply and increasing agricultural water demand (Fabre et al., 2015).

Both the Food and Agriculture Organization of the United Nations (FAO) and the Intergovernmental Panel on Climate Change have affirmed that "agriculture is one of the sectors most susceptible to the impacts of climate change," (Veettil et al., 2022). Moreover, certain factors, including rising temperatures, increased CO_2 levels, and GHG emissions add to more extreme climate events and can potentially impact crop growth on a larger scale. These factors contribute to a reduction in available agricultural water resources, intensifying the conflict between crop water supply and demand (Azzam et al., 2022). BW serves a variety of purposes in agriculture, including crop irrigation, providing drinking water for livestock, and supporting aquaculture.

2.2.1.2 Irrigating crops

Irrigation is the technique to provide optimum water to plants and crops to stimulate growth and enhance productivity Presently, irrigation constitutes a substantial 70% of all global BWRs consumed by humankind (FAO Licence, 2018). The productivity has been significantly increased by the use of irrigation methods to make up for soil moisture shortages, notably in arid and semiarid locations where rainfall is infrequent or unpredictable; However, it can also result in challenges such as groundwater depletion, salinization, and waterlogging (Dikeogu et al., 2021).

There are several irrigation methods available and outlined, including drip irrigation (microsprinkler or emitter), overhead or center pivot (sprinkler irrigation), and surface irrigation (flooding or furrowing; Sable et al., 2019). Notable factors that have contributed to enhanced crop yields comprise the adoption of improved crop varieties, increased irrigation, pesticide applications, and more effective soil and water management. However, it is crucial to recognize that the continuous increase in crop production is not expected to persist indefinitely. In many regions around the world, the production of essential agricultural goods has begun to stagnate.

Furthermore, the research conducted by Mekonnen & Hoekstra (2020) reveals that a substantial portion, 57% of the world's BWF is unsustainable. This unsustainability is particularly evident in certain crops that have a significant water footprint, while over 70% of the global BWF that is unsustainable is attributed to the cultivation of just five crops: wheat (27%), rice (17%), cotton (10%), sugarcane (8%), and fodder (7%).

These crops, which range from 43% for rice to 68% for wheat, have a critical BWF proportion that cannot be maintained. Notably, the Middle East and Central Asia have the largest percentage of unsustainable BWF used in crop cultivation, accounting for over 71% of it. In terms of specific countries, Qatar is at the top of the list, followed by Uzbekistan and Pakistan (each at 68%), and Turkmenistan (67%) (Mekonnen & Hoekstra, 2020).

2.2.1.3 Livestock

Livestock have a substantial impact on various aspects including food security, agricultural sustainability, and climate change (Romero-Ruiz et al., 2023). Increasingly, livestock is recognized as an essential component of the regenerative agricultural system and plays a significant role in the circular bioeconomy by recycling low-quality feed sources (Teague & Kreuter, 2020). Livestock encompasses domesticated animals that are raised for food, fiber, and other products, and they include cows, sheep, goats, pigs, chickens, and fish. They need the BW for various purposes, such as drinking, cooling, and cleaning, among others, and they also indirectly consume BW by consuming plants that require significant water for growth (Wilkinson & Lee, 2018).

2.2.1.4 Aquaculture

Aquaculture describes the farming of species that live in water, including fish, shellfish, plants, and algae. Aquaculture is a potential solution to the challenge of providing environmentally sustainable protein sources to the growing human population. It can also help to protect and restore the environment by improving effectively biodiversity and ecosystem services (Vasquez-Mejia et al., 2023). Aquaculture primarily relies on BW (both ground and surface water) that flows into lakes, ponds, rivers, aquifers, and oceans for the growth of fish. In 2018, global aquaculture consumed 201 km^3 and 122.6 km^3 of water. Due to the projected increase in the cultivation of fish species and crop production, this demand is expected to surge by 2.3 times by 2050 (Jiang et al., 2022).

2.2.1.5 Industrial water

Globally, industries represent the second largest consumers of BW, accounting for 20% of all BWU. This water is employed in various industrial processes, including those in the pharmaceutical, textile, paper, and other sectors. Given the limited availability of water resources for all industrial activities, the establishment of effective water use strategies is essential for achieving the environmental sustainability of these resources. The demand for water supply continues to rise along with industrial development (Bailey et al., 2022). Generally, as highlighted by Aziz et al., heavy industrial activities tend to have a more significant water usage footprint compared to other sectors. Compared to 3 years ago (2015–2017), when only the manufacturing sector scored 30% for all years, the manufacturing sector had the largest percentage of existing industries in 2018 (32%). This depicts that there was significant growth in the manufacturing sector in 2018 (Aziz et al., 2021).

BWF can serve as a valuable tool for investigating total water consumption and establishing connections between anthropogenic water use, consumption patterns, and industrial operations. The allocation of water resources across different industrial sectors should be planned logically and scientifically to ensure the sustainable use of these vital resources (Aziz et al., 2021).

2.2.1.6 Pharmaceutical industry

The pharmaceutical industry is one of the most significant sectors that extensively utilizes BW for the manufacture of active medicine substances and finished products such as tablets, capsules, injections, syrups, and others. This industry is a highly water-dependent economic sector and requires different degrees of water purity. As a result, there are stringent guidelines and standards for medical products, distribution, and storage of water for pharmaceutical applications. Furthermore, the level of filtration and treatment required depends on the intended use of the water. For example, water intended for oral solutions should be free of contaminants and potable, while water for injections must meet standards for sterility and freedom from pyrogens. Moreover, various processes, such as distillation, reverse osmosis, ion exchange, ultrafiltration, and ultraviolet radiation, are employed to provide water suitable for medicinal use. In addition, BW can also be utilized for cleaning and sterilizing tools and containers. It is essential to note that the active chemicals and solvents used in pharmaceutical operations have the potential to contaminate BW (Strade et al., 2020).

2.2.1.7 Textile industry

The textile industry is chosen as an illustrative example because it not only consumes a substantial volume of water but also stands as a significant contributor to water pollution. Given the textile industry's direct relevance to people's daily lives, extensive efforts have been focused on water conservation and waste reduction within this sector. Moreover, the continuous economic growth of the textile industry takes a toll on the environment and water resources, leading to an annual increase in textile production when compared to previous years (Bailey et al., 2022; Li et al., 2017).

2.2.1.8 Paper industry

The third industrial sector under consideration is the pulp and paper industry, which happens to be one of the oldest and most significant industrial sectors. Paper manufacturing, in particular, is a costly and resource-intensive endeavor that relies heavily on water. The substantial demand for water in the pulp and paper industries has led to an escalating strain on water resources. This strain has resulted in declining groundwater levels, rising pumping costs, and, most significantly, widespread water shortages in many regions (Nandan et al., 2017).

2.2.1.9 *Domestic water*

Municipal water use, sometimes referred to as domestic water use, public water supply, or household water consumption, pertains to the utilization of water provided by water utilities and distributed through the public water distribution network. The domestic sector represents the ultimate end-consumer in the socioeconomic water cycle and consumes a substantial volume of water, both in physical and virtual ways (Chai et al., 2020).

Consequently, the world's rapidly growing population has led to an increased demand for water resources. According to projections by the United Nations, the global population is expected to exceed 8.6 billion by 2030, compared to 7.5 billion in 2017. A significant portion of this population growth is associated with developing countries, where ensuring access to an adequate water supply is becoming progressively more challenging (UN/DESA, World Population Prospects, 2017).

Furthermore, in recent decades, the surge in water demand has been exacerbated by urbanization and increasing incomes, which have propelled consumer demand to higher levels (Bao & Chen, 2015). For instance, a population of 1.4 billion, witnessed an 18-fold growth in household final consumption expenditure between 1978 and 2016.

The increasing demand and overconsumption of water resources have already posed serious threats to the ecosystem and public health, and have hindered economic development in water-scarce regions (Chai et al., 2020; Lin et al., 2015).

BW serves various purposes, including facilitating the transportation of people and goods across different regions and continents via ships, boats, ferries, and other watercraft. In addition, it plays a role in energy production, as BW can be harnessed to generate electricity through hydropower plants that utilize dams or turbines to capture the kinetic energy of flowing water.

2.2.2 Safety of the blue water

The safety of BW depends on its source, quality, quantity, and purpose of use. Although BW is generally clean for human use, it could also contain some contaminants and toxins that pose threats to water security and safety. In various regions, the scarcity and safety of BW are undergoing significant changes due to factors like urbanization, nonpoint sources of pollution originating from agricultural fields, and the discharge of untreated water from industries and households (Mishra et al., 2021; Mishra et al., 2021; Van Vliet et al., 2017). For instance, Wuhan and Chongqing are suffering from pollution and wastewater treatment challenges. While on the other hand, Karachi is facing issues related to water supply, climate change, revenue recovery, industrial pollution, and contamination.

Shanghai is similarly contending with problems related to pollution, flooding, the impacts of major hydraulic projects, and saltwater intrusion. All these countries were found to be extremely vulnerable to the land change components. Various factors can influence the safety of BW, including the level of local BWS and the presence of chemical hazards (Keys et al., 2017; Zisopoulou & Panagoulia, 2021).

It is important to note that, Hoekstra (2019) reported that BWS values are divided into four categories: (1) miner (the BWF is less than 20% of natural runoff and falls within BWA; river runoff is unaltered or only slightly modified and presumptive environmental flow limits are not broken), (2) medium (the BWF is between 20% and 30% of natural runoff; the runoff has undergone moderate modification and the environmental flow standards are not being satisfied), (3) major (the BWF is between 30% and 40% of the natural runoff; the runoff is extensively modified and the environmental flow standards are not met), and (4) severe (the monthly BWF is greater than 40% of natural runoff; the runoff has been significantly altered and the environmental flow criteria have not been met).

Furthermore, a city that faces BWS overuses its BWR at the expense of regional ecological benefits and may even threaten those of its neighbours. Undoubtedly, it will worsen regional inequality, stifle development (Yang et al., 2019), and ultimately lead to an unsustainable future. Moreover, the quality of the water at both its source and its final destination might be impacted by BW; irrigation, for instance, can degrade soil and drain water supplies, which can lower agriculture production and harm natural habitats. Similar to domestic and industrial water consumption, inadequate wastewater treatment can result in the contamination of receiving waters by a number of hazardous chemicals which may contribute to toxicity incidence to consumers. The wastewater which contain hazardous chemicals should be treated, bioremediated and highly biodegradable in nature before distribution to ensure it is safe to be consumed (Mekonnen & Hoekstra, 2020). Thus, ensuring the safety of BW remains a great and challenging concern.

2.2.3 Challenges faced by blue water

All the countries around the world are grappling with significant water shortages due to the effects of rapid industrialization, population growth, agricultural expansion, climate change, and pollution. Although human interventions in river systems over the past few decades have facilitated to some extent (in comparison to scenarios without interventions) to reduction of water shortage, they have also made downstream areas more affected by water scarcity than upstream areas (Veldkamp et al., 2017). A central focus of current research is finding solutions to address the challenges of water resource scarcity and contamination while promoting sustainable and responsible water resource management.

2.3 Green water

Studies have highlighted a critical fact that 80% of the world's cropland relies on rain-fed agriculture, generating 60%–80% of the global food supply while utilizing only green water. Green water, which can be naturally replenished through processes like rainfall, is thus recognized as a pivotal component in ensuring food security (A. Zhao et al., 2014). In addition, some researchers say that green water plays an indispensable role in sustainable development. This is because traditional methods of increasing food production, such as land expansion and irrigation areas, are unfavorable to the ecological environment and sustainability (Ma et al., 2022; Sposito, 2013).

In response to escalating global food demand, in future interdisciplinary research is required to enhance the utilization of green water and significantly increase the biomass of rain-fed crops multiple times (Ma et al., 2022). Moreover, green water resources serve not only as a key supply of water for agricultural crop production but also as a critical component of the health and survival of grasslands and forests, the primary source of water for timber, feeds, fiber, and bioenergy production, as well as a potential source of freshwater consumption for human society (Li et al., 2022). However, inadequate access to green water resources over an extended period of time may result in detrimental ecological deterioration, increased competition for water resources between human activities and ecosystems, and ultimately affecting the availability of water resources and human well-being (Nie et al., 2023). Despite the growing significance of green water in crop production and the ecological environment, investigations and awareness of the uses and effective management of green water are still limited. Therefore, there is a need for comparative studies on the utilization, challenges, and protection of green water to ensure its optimal usage and sustainability.

2.3.1 Utilization of green water for agriculture

Globally, the utilization of green water accounts for a substantial 80% of the total water usage, making it an indispensable resource for promoting sustainable water development. The primary consumers of green water resources are crops and vegetation, primarily through the process of ET (Zhao et al., 2021). The ET process (Fig. 2.1) consists of soil evaporation and transpiration. Transpiration refers to the water lost to the atmosphere from small openings on the leaf surfaces, while evaporation is the water lost from the wet soil and plant. ET is a key component of the water and agricultural water cycle (Massmann et al., 2019).

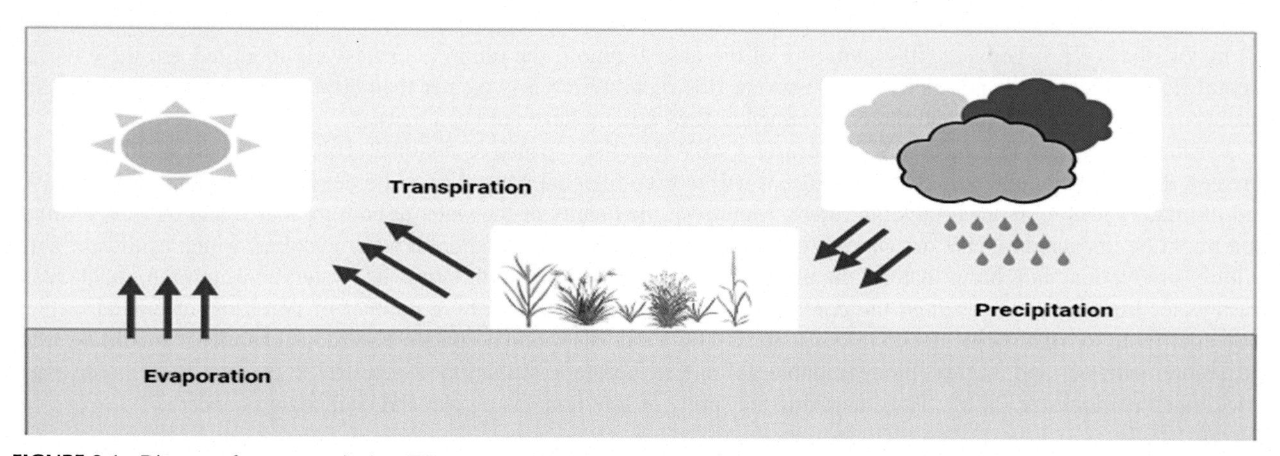

FIGURE 2.1 Diagram of evapotranspiration (ET) process which composed of evaporation (water lost fro the wet soil) and transpiration (the water lost to the atmosphere through water evaporation process from plants).

It is another name for crop water usage, which encompasses soil evaporation and the water utilized by plants for cooling and growth. Water is essential for plants to uptake and absorb nutrients to support life and subsequent growth. In the context of agriculture and food production, especially in arid and semiarid regions with infrequent rainfall, green water holds great significance. Farmers can boost their resistance to drought and other climate-related hazards by employing green water, which can reduce dependence on surface and groundwater (Ma et al., 2022). Green water is extracted by the root system of plants through transpiration, thus it is no longer available as stored water in the soil. Furthermore, the majority of weed plants that grow along with the cultivated plants serve as animal feeds, emphasizing the importance of green water in the production of food, livestock, and industrial crops. Due to the substantial amount of water used in agriculture as well as the impact that water scarcity can have on agricultural productivity, the relationship between water, food, and other environmental resources (such as climate, soils, and energy) has gained prominence as a critical area of focus for both environmental conservation and development policy research (Ma et al., 2022).

2.3.2 Green water—a source of water for soil microorganisms

Soil microorganisms such as bacteria, fungi, protozoa, nematodes, and arthropods rely on the supply of green water, which is crucial for maintaining soil health and promoting sustainable agriculture. These microorganisms play a crucial role in breaking down organic matter, recycling nutrients, and enhancing soil quality. Moreover, soil microbes contribute to the formation of soil aggregates, which in turn enhance soil aeration, water infiltration, and porosity for optimum production (Kaviya et al., 2019).

2.3.3 Green water: use in phytotherapy

Green water plays a pivotal role in crop production by providing water for both cultivated and wild plants. Many of the plants used in medicine or phytotherapy exist as wild or weed plants, where their primary source of water is precipitation. Phytotherapy, a complementary medicine practice, utilizes substances derived from plants or herbs to address health conditions and prevent disease (Chen et al., 2016).

2.3.4 Green water: use in horticulture

Green water is of significant importance not only in food and industrial production but also in horticulture, which encompasses garden crops such as fruits, vegetables, and ornamental plants. According to FAO Land and Water report, groundwater, surface water, and municipal water are the primary sources of water for horticultural practices and it is worth noting that collecting rainwater from rooftops has been identified as a cost-effective and less-polluted alternative for horticultural purposes.

2.3.5 Green water harvest

Rainwater harvest is an effective practice of collecting rainwater from various surfaces, including rooftops, paved areas, and even directly from the atmosphere. The collected water is then directed into a reservoir or storage tank made of different materials, including metal, fiberglass, concrete, or plastic. However, if the harvested rainwater is properly treated and filtered, it can be utilized for various purposes like irrigation for farms and gardens, domestic use, and even as a source of drinking water. Collecting rainwater from rooftops or other surfaces is an easy and cost-effective method to harness green water for a wide range of uses including domestic use. Rainwater harvesting systems enable the collection and storage of rainwater for future use to reduce dependence on the available limited freshwater resources and mitigate the need for freshwater from alternative sources (Bottom, 2000).

2.3.6 Factors affecting green water availability

2.3.6.1 Climate

The amount of freshwater can be significantly influenced by detrimental climate factors such as air temperature, relative humidity, wind speed, and atmospheric carbon dioxide concentration. These factors can alter and impact both the potential and actual ET, which in turn affects the amount of green water available for plant growth. An increase in ET can lead to a significant decrease in the amount of green water available for plant growth. Generally, an increase in air temperature enhances the moisture-holding capacity in the atmosphere, thus intensifying evaporation and accelerating the hydrological cycle (Khalili et al., 2023; Massmann et al., 2019). On the other hand, higher relative humidity can lower the ET rates, while an increased wind speed can result in more plant transpiration (Khalili et al., 2023). Furthermore, it is important to note that the change in the atmospheric concentration of CO_2 also influences the water cycle and plant physiology in addition to the stomatal opening effect, photosynthesis, and respiration (Massmann et al., 2019). Furthermore, rising temperature contributed to increased humidity which refers to an increase in the amount of water vapor in the atmosphere (humidity). Humidity has a significant impact on cloud formation and precipitation development, while it is worth noting that in case the humidity is high, there is more water vapor in the atmosphere, increasing the likelihood of cloud and potential rain formation (Dai et al., 1999).

2.3.6.2 Vegetation type

Green water, which refers to the portion of precipitation retained in the soil and used by plants for growth and transpiration, is significantly influenced by the type of vegetation. Deep-rooted vegetation can uptake water out of deeper soil layers, increasing the amount of green water available for plant growth and reducing the quantity of water that percolates into groundwater (Kong et al., 2023). The vegetation type also affects the rate of ET, which in turn affects green water availability. Plants with large leaf area index (LAI) can enhance the rate of ET by receiving more solar radiation while providing more surface area for transpiration. In addition, as per the available data and research the vegetation can influence the soil structure and nutrient cycling, both of which can have an impact on the quality of green water. For instance, plants with significant root biomass can improve the structure of the soil and increase the organic matter content, which in turn improves the capacity of the soil to retain water and nutrients (Wang et al., 2020).

2.3.6.3 Soil nutrients

Soil nutrients, particularly nitrogen, phosphorus and potassium can impact green water in various ways. Adequate nutrient availability can stimulate the growth of biomass and plants, which in turn increases the need for moisture from the vegetation. In addition, nutrient like nitrogen plays a fundamental role in the formation of chlorophyll, which is essential for photosynthesis and plant growth. Plants with higher nitrogen concentrations tend to produce more leaves and have a larger LAI, which can enhance evapotranspiration rates and the amount of green water they utilize (Khalili et al., 2023; Wang et al., 2020).

2.3.6.4 Soil temperature

The strong relationship between soil temperature and plant growth is explained by the fact that temperature encourages crop development through significantly increased water and nutrient uptake, whereas cold hinders water uptake due to reduced water viscosity and slows down the process of photosynthesis. In conclusion, low soil temperatures can reduce plant root uptake of nutrients and water, resulting in lower rates of green water consumption (Khalili et al., 2023; Onwuka, 2018).

2.3.6.5 Soil type

Soil type is regarded as an important factor in soil functioning, as it significantly influences the ability of soil to sustain plants and animal life as well as to regulate environmental quality. Soil texture, structure, and organic matter content have an impact on the rate of infiltration, the capacity of soil to retain water, and the availability of nutrients (Pulido Moncada et al., 2014). For instance, sandy soil features larger pore sizes, enabling water to move through it quickly but reducing its water-holding capacity. In contrast, clay soils have smaller pores that can retain water but may limit water circulation.

2.3.6.6 Green water pollution

The release of active pharmaceutical substances and their metabolites into the environment commonly occurs through various ways like agricultural applications of wastewater and sewage, and biosolids containing pharmaceuticals originating from wastewater discharged by households, hospitals, and other medical facilities. In addition, the release of natural fertilizers (manure and slurry) (Gworek et al., 2021) pesticides, and industrial waste into the soil transferred into various parts of plants that could be detrimental to the plants and also lead to contamination of water that unsafe for plants and microorganisms to use.

2.3.6.7 Socioeconomic activities

The availability and quality of water are significantly impacted by socioeconomic activities. For example, population growth, urbanization, and industrialization can lead to a significant increase in water consumption and pollution, ultimately resulting in a decline in the availability of water resources. (Saketa, 2022). Agriculture could be a major contributor to water scarcity, as it consumes the largest share of global water resources. The challenge of water scarcity can be aggravated by the use of irrigation systems, fertilizers, and pesticides which can result in soil degradation and water pollution (Goswami & Bisht, 2017). Efficient water management is essential to minimize the detrimental effects of these socioeconomic activities. This can be achieved by effective development and implementation of policies and practices that encourage the efficient use of water resources while simultaneously limiting environmental deterioration which is a global concern (A. Mishra et al., 2021; B. Mishra et al., 2021). Crop rotation, integrated pest control, precision agriculture, and conservation tillage are few examples of effective agricultural management strategies (Nasir et al., 2021).

2.4 Conclusions

Green water is indeed a critical element in maintaining food security as it is a renewable resource that can be restored through natural means such as precipitation. However, it sustains life, supports agriculture, industry, and energy production, and provides various other ecological benefits. Therefore, it can be impacted by various factors like climate change, mismanagement, soil type, temperature, and nutrients. It is worth noting that effective water management is a critical issue that requires careful planning, coordination, and implementation of policies and practices to ensure sustainability.

References

Aziz, E. A., Moni, S. N., & Yussof, N. (2021). Blue water footprint−industrial water consumption nexus: A case of water supply for industrial activities in Semambu Water Treatment Plant. *IOP Conference Series: Materials Science and Engineering*, *1092*(1), 012045. Available from https://doi.org/10.1088/1757-899x/1092/1/012045, 1757−8981.

Azzam, A., Zhang, W., Akhtar, F., Shaheen, Z., & Elbeltagi, A. (2022). Estimation of green and blue water evapotranspiration using machine learning algorithms with limited meteorological data: A case study in Amu Darya River Basin, Central Asia. *Computers and Electronics in Agriculture*, *202*, 107403. Available from https://doi.org/10.1016/j.compag.2022.107403, 01681699.

Bailey, K., Basu, A., & Sharma, S. (2022). The environmental impacts of fast fashion on water quality: A systematic review. *Water (Switzerland)*, *14* (7). Available from https://doi.org/10.3390/w14071073. 20734441. MDPI, Canada. https://www.mdpi.com/2073-4441/14/7/1073/pdf.

Bao, C., & Chen, X. (2015). The driving effects of urbanization on economic growth and water use change in China: A provincial-level analysis in 1997−2011. *Journal of Geographical Sciences*, *25*(5), 530−544. Available from https://doi.org/10.1007/s11442-015-1185-8, 1009637X. Science in China Press, China. http://www.springer.com/sgw/cda/frontpage/0,11855,1-40391-70-66229542-0,00.html?changeHeader = true\.

Beltran-Pea, A., Rosa, L., & D'Odorico, P. (2020). Global food self-sufficiency in the 21st century under sustainable intensification of agriculture. *Environmental Research Letters*, *15*(9). Available from https://doi.org/10.1088/1748-9326/ab9388, 17489326. IOP Publishing Ltd, United States, https://iopscience.iop.org/article/10.1088/1748-9326/ab9388.

Bottom, D. W. (2000). Rainwater harvesting. *Landscape Architecture*, *90*(4), 40−137, 00238031.

Burek, P., Langan, S., Cosgrove, W., Fischer, G., Kahil, T., Magnuszewski, P., Satoh, Y., Tramberend, S., Wada, Y., & Wiberg D. (2016). The water futures and solutions initiative of IIASA. 23−26.

Çankaya, S. (2023). Evaluation of the impact of water reclamation on blue and grey water footprint in a municipal wastewater treatment plant. *Science of The Total Environment*, *903*, 166196. Available from https://doi.org/10.1016/j.scitotenv.2023.166196, 00489697.

Chai, L., Han, Z., Liang, Y., Su, Y., & Huang, G. (2020). Understanding the blue water footprint of households in China from a perspective of consumption expenditure. *Journal of Cleaner Production*, *262*, 121321. Available from https://doi.org/10.1016/j.jclepro.2020.121321, 09596526.

Chang, H. (2019). Water and climate change. *International Encyclopedia of Geography*. https://doi.org/10.1002/9781118786352.wbieg0793.pub2.

Chen, S. L., Yu, H., Luo, H. M., Wu, Q., Li, C. F., & Steinmetz, A. (2016). Conservation and sustainable use of medicinal plants: Problems, progress, and prospects. *Chinese Medicine (United Kingdom)*, *11*(1). Available from https://doi.org/10.1186/s13020-016-0108-7. 17498546. BioMed Central Ltd., China. http://www.cmjournal.org/.

Dai, A., Trenberth, K. E., & Karl, T. R. (1999). Effects of clouds, soil moisture, precipitation, and water vapor on diurnal temperature range. *Journal of Climate*, *12*(8), 2451−2473. Available from https://doi.org/10.1175/1520-0442. 08948755. American Meteorological Soc, United States. http://journals.ametsoc.org/loi/clim.

Dey, A., Remesan, R., & Kumar, R. (2023). Blue and green water re-distribution dependency on precipitation datasets for a tropical Indian River basin. *Journal of Hydrology: Regional Studies*, *46*, 101361. Available from https://doi.org/10.1016/j.ejrh.2023.101361, 22145818.

Dikeogu, T. C., Okeke, O. C., Nwachukwu, H. G. O., & Agbo, C. C. (2021). Salinization and waterlogging in agricultural lands: Causes, effects and mitigation. *International Journal of Innovative Environmental Studies Research*, *9*(1), 45−57.

Emam, H. E. (2022). Accessibility of green synthesized nanopalladium in water treatment. *Results in Engineering, 15*, 100500. Available from https://doi.org/10.1016/j.rineng.2022.100500, 25901230.

Fabre, J., Ruelland, D., Dezetter, A., & Grouillet, B. (2015). Simulating past changes in the balance between water demand and availability and assessing their main drivers at the river basin scale. *Hydrology and Earth System Sciences, 19*(3), 1263–1285. Available from https://doi.org/10.5194/hess-19-1263-2015, 16077938. Copernicus GmbH, France., http://www.hydrol-earth-syst-sci.net/volumes_and_issues.html.

Falkenmark, M. (1995). *Coping with Water Scarcity under Rapid Population Growth. Conference of SADC Ministers* (pp. 23-24). Pretoria.

Fang, Z., Chen, J., Liu, G., Wang, H., Alatalo, J. M., Yang, Z., Mu, E., & Bai, Y. (2021). Framework of basin eco-compensation standard valuation for cross-regional water supply—A case study in northern China. *Journal of Cleaner Production, 279*, 123630. Available from https://doi.org/10.1016/j.jclepro.2020.123630, 09596526.

FAO Licence. (2018). The State of Food and Agriculture. CC BY-NC-SA 3.0 IGO. Rome.

Foster, T., Mieno, T., & Brozović, N. (2020). Satellite-based monitoring of irrigation water use: Assessing measurement errors and their implications for agricultural water management policy. *Water Resources Research, 56*(11). Available from https://doi.org/10.1029/2020WR028378. 19447973. Blackwell Publishing Ltd, United Kingdom. http://agupubs.onlinelibrary.wiley.com/hub/journal/10.1002/(ISSN)1944-7973/.

Fu, Y., Zhang, J., Zhang, C., Zang, W., Guo, W., Qian, Z., Liu, L., Zhao, J., & Feng, J. (2018). Payments for Ecosystem Services for watershed water resource allocations. *Journal of Hydrology, 556*, 689–700. Available from https://doi.org/10.1016/j.jhydrol.2017.11.051. 00221694. Elsevier B. V., China. http://www.elsevier.com/inca/publications/store/5/0/3/3/4/3.

Gao, H., Xue, Y., Wu, L., Huo, J., Pang, Y., Chen, J., & Gao, Q. (2022). Protective effect of *Lycium ruthenicum* polyphenols on oxidative stress against acrylamide induced liver injury in rats. *Molecules, 27*(13). Available from https://doi.org/10.3390/molecules27134100. 14203049. MDPI, China. https://www.mdpi.com/1420-3049/27/13/4100/pdf?version = 1656151935.

Gao, T., Wang, X., Wei, D., Wang, T., Liu, S., & Zhang, Y. (2021). Transboundary water scarcity under climate change. *Journal of Hydrology, 598*, 126453. Available from https://doi.org/10.1016/j.jhydrol.2021.126453, 00221694.

Gosling, S. N., & Arnell, N. W. (2016). A global assessment of the impact of climate change on water scarcity. *Climatic Change, 134*(3), 371–385. Available from https://doi.org/10.1007/s10584-013-0853-x. 15731480. Springer Netherlands, United Kingdom. http://www.wkap.nl/journalhome.htm/0165-0009.

Goswami, K. B., & Bisht, P. S. (2017). The role of water resources in socio-economic development. *International Journal for Research in Applied Science & Engineering Technology (IJRASET), 887*, 2321–9653.

Guan, X., Liu, M., & Meng, Y. (2021). A comprehensive ecological compensation indicator based on pollution damage—Protection bidirectional model for river basin. *Ecological Indicators, 126*, 107708. Available from https://doi.org/10.1016/j.ecolind.2021.107708, 1470160X.

Gworek, B., Kijeńska, M., Wrzosek, J., & Graniewska, M. (2021). Pharmaceuticals in the soil and plant environment: A review. *Water, Air, and Soil Pollution, 232*(4). Available from https://doi.org/10.1007/s11270-020-04954-8. 15732932. Springer Science and Business Media Deutschland GmbH, Poland. http://www.kluweronline.com/issn/0049-6979/.

Hoekstra, A. Y. (2019). Green–blue water accounting in a soil water balance. *Advances in Water Resources, 129*, 112–117. Available from https://doi.org/10.1016/j.advwatres.2019.05.012. 03091708. Elsevier Ltd, Netherlands. http://www.elsevier.com/inca/publications/store/4/2/2/9/1/3/index.htt.

Huang, H., Zhuo, L., Wang, W., & Wu, P. (2023). Resilience assessment of blue and green water resources for staple crop production in China. *Agricultural Water Management, 288*, 108485. Available from https://doi.org/10.1016/j.agwat.2023.108485, 03783774.

Jeyrani, F., Morid, S., & Srinivasan, R. (2021). Assessing basin blue–green available water components under different management and climate scenarios using SWAT. *Agricultural Water Management, 256*, 107074. Available from https://doi.org/10.1016/j.agwat.2021.107074, 03783774.

Jiang, Q., Bhattarai, N., Pahlow, M., & Xu, Z. (2022). Environmental sustainability and footprints of global aquaculture. *Resources, Conservation and Recycling, 180*, 106183. Available from https://doi.org/10.1016/j.resconrec.2022.106183, 09213449.

Kaviya, N., Upadhayay, V. K., Singh, J., Khan, A., Panwar, M., & Singh, A. V. (2019). *Role of microorganisms in soil genesis and functions. Mycorrhizosphere and Pedogenesis* (pp. 25–52). India: Springer Singapore, India Springer Singapore. Available from https://link.springer.com/book/10.1007/978-981-13-6480-8, https://doi.org/10.1007/978-981-13-6480-8_2.

Ke, Z., Xiaoqi, L., Chuanfu, Z., Yiwen, L., Xintong, Q., & Miaolin, D. (2024). Transformation characteristics and mechanism of blue and green water flows at watershed and typical ecosystem scale in China. *Ecohydrology & Hydrobiology, 24*(1), 201–216. Available from https://doi.org/10.1016/j.ecohyd.2023.09.002, 16423593.

Keys, P. W., Wang-Erlandson, L., Gordon, L. J., Galaz, V., & Ebbesson, J. (2017). *Approaching moisture recycling governance, Global Environmental Change, 45*, 15–23. Available from https://doi.org/10.1016/j.gloenvcha.2017.04.007.

Khalili, P., Razavi, S., Davies, E. G. R., Alessi, D. S., & Faramarzi, M. (2023). Assessment of blue water–green water interchange under extreme warm and dry events across different ecohydrological regions of western Canada. *Journal of Hydrology, 625*. Available from https://doi.org/10.1016/j.jhydrol.2023.130105. 00221694. Elsevier B.V., Canada. http://www.elsevier.com/inca/publications/store/5/0/3/3/4/3.

Kong, M., Li, Y., Zang, C., & Deng, J. (2023). The impact mechanism of climate and vegetation changes on the blue and green water flow in the main ecosystems of the Hanjiang River Basin, China. *Remote Sensing, 15*(17), 20724292. Available from https://doi.org/10.3390/rs15174313. Multidisciplinary Digital Publishing Institute (MDPI), China. http://www.mdpi.com/journal/remotesensing.

Li, Y., Cai, Y., Wang, X., Li, C., Liu, Q., Sun, L., & Fu, Q. (2022). Classification analysis of blue and green water quantities for a large-scale watershed of southwest China. *Journal of Environmental Management, 321*. Available from https://doi.org/10.1016/j.jenvman.2022.115894. 10958630. Academic Press, China. https://www.sciencedirect.com/journal/journal-of-environmental-management.

Li, Y., Lu, L., Tan, Y., Wang, L., & Shen, M. (2017). Decoupling water consumption and environmental impact on textile industry by using water footprint method: A case study in China. *Water (Switzerland), 9*(2). Available from https://doi.org/10.3390/w9020124. 20734441. MDPI AG, China. http://www.mdpi.com/2073-4441/9/2/124/pdf.

Lin, X., Sha, J., & Yan, J. (2015). Exploring the impacts of water resources on economic development in Beijing-Tianjin-Hebei Region.

Liu, L., Hu, X., Zhan, Y., Sun, Z., & Zhang, Q. (2023). China's dietary changes would increase agricultural blue and green water footprint. *Science of the Total Environment*, *903*, 165763. Available from https://doi.org/10.1016/j.scitotenv.2023.165763, 00489697.

Luan, X. B., Yin, Y. L., Wu, P. T., Sun, S. K., Wang, Y. B., Gao, X. R., & Liu, J. (2018). An improved method for calculating the regional crop water footprint based on a hydrological process analysis. *Hydrology and Earth System Sciences*, *22*(10), 5111–5123. Available from https://doi.org/10.5194/hess-22-5111-2018. 16077938. Copernicus GmbH, China. http://www.hydrol-earth-syst-sci.net/volumes_and_issues.html.

Ma, W., Wei, F., Zhang, J., Karthe, D., & Opp, C. (2022). Green water appropriation of the cropland ecosystem in China. *Science of the Total Environment*, *806*, 150597. Available from https://doi.org/10.1016/j.scitotenv.2021.150597, 00489697.

Mao, G., Liu, J., Han, F., Meng, Y., Tian, Y., Zheng, Y., & Zheng, C. (2020). Assessing the interlinkage of green and blue water in an arid catchment in Northwest China. *Environmental Geochemistry and Health*, *42*(3), 933–953. Available from https://doi.org/10.1007/s10653-019-00406-3. 15732983. Springer, China. http://www.wkap.nl/journalhome.htm/0269-4042.

Massmann, A., Gentine, P., & Lin, C. (2019). When does vapor pressure deficit drive or reduce evapotranspiration? *Journal of Advances in Modeling Earth Systems*, *11*(10), 3305–3320. Available from https://doi.org/10.1029/2019MS001790. 19422466. Blackwell Publishing Ltd, United States. http://agupubs.onlinelibrary.wiley.com/agu/journal/10.1002/(ISSN)1942-2466/.

Mekonnen, M. M., & Gerbens-Leenes, W. (2020). The water footprint of global food production. *Water (Switzerland)*, *12*(10). Available from https://doi.org/10.3390/w12102696. 20734441. MDPI AG, United States. http://www.mdpi.com/journal/water.

Mekonnen, M. M., & Hoekstra, A. Y. (2020). Sustainability of the blue water footprint of crops. *Advances in Water Resources*, *143*. Available from https://doi.org/10.1016/j.advwatres.2020.103679. 03091708. Elsevier Ltd, United States. http://www.elsevier.com/inca/publications/store/4/2/2/9/1/3/index.htt.

Mishra, A., Alnahit, A., & Campbell, B. (2021). Impact of land uses, drought, flood, wildfire, and cascading events on water quality and microbial communities: A review and analysis. *Journal of Hydrology*, *596*. Available from https://doi.org/10.1016/j.jhydrol.2020.125707. 00221694. Elsevier B.V., United States. http://www.elsevier.com/inca/publications/store/5/0/3/3/4/3.

Mishra, B., Kumar, P., Saraswat, C., Chakraborty, S., & Gautam, A. (2021). Water security in a changing environment: Concept, challenges and solutions. *Water*, *13*(4), 490. Available from https://doi.org/10.3390/w13040490, 2073–4441.

Nandan, A., Yadav, B., Baksi, S., & News, D.B. (2017). Assessment of water footprint in paper & pulp industry & its impact on sustainability. psjd.icm.edu.pl.

Nasir, J., Ashfaq, M., Baig, I. A., Punthakey, J. F., Culas, R., Ali, A., & Hassan, F. U. (2021). Socioeconomic impact assessment of water resources conservation and management to protect groundwater in Punjab, Pakistan. *Water (Switzerland)*, *13*(19). Available from https://doi.org/10.3390/w13192672. 20734441. MDPI, Pakistan. https://www.mdpi.com/2073-4441/13/19/2672/pdf.

Nie, N., Li, T., Miao, Y., Zhang, W., Gao, H., He, H., Zhao, D., & Liu, M. (2023). Asymmetry of blue and green water changes in the Yangtze river basin, China, examined by multi-water-variable calibrated SWAT model. *Journal of Hydrology*, *625*. Available from https://doi.org/10.1016/j.jhydrol.2023.130099, 00221694. Elsevier B.V., China., http://www.elsevier.com/inca/publications/store/5/0/3/3/4/3.

Onwuka, B. (2018). Effects of soil temperature on some soil properties and plant growth. *Advances in Plants & Agriculture Research*, *8*(1). Available from https://doi.org/10.15406/apar.2018.08.00288, 23736402.

ORGANIZAÇÃO DAS NAÇÕES UNIDAS. (2015). Resolution adopted by the General Assembly on 25 September 2015. Transforming our world: the 2030 Agenda for Sustainable Development.

Pulido Moncada, M., Gabriels, D., Lobo, D., Rey, J. C., & Cornelis, W. M. (2014). Visual field assessment of soil structural quality in tropical soils. *Soil and Tillage Research*, *139*, 8–18. Available from https://doi.org/10.1016/j.still.2014.01.002, 01671987.

Romero-Ruiz, A., Rivero, M. J., Milne, A., Morgan, S., Meo-Filho, P. D., Pulley, S., Segura, C., Harris, P., Lee, M. R., Coleman, K., Cardenas, L., & Whitmore, A. P. (2023). Grazing livestock move by Lévy walks: Implications for soil health and environment. *Journal of Environmental Management*, *345*. Available from https://doi.org/10.1016/j.jenvman.2023.118835, 10958630. Academic Press, United Kingdom., https://www.sciencedirect.com/journal/journal-of-environmental-management.

Sable, R., Kolekar, S., Gawde, A., Takle, S., & Pednekar, A. (2019). A review on different irrigation methods. *International Journal of Applied Agricultural Research*, *14*(1), 49–60.

Saketa, Y. (2022). Assessment of future urban water demand and supply under socioeconomic scenarios: a case of Assosa town. *Water Supply*, *22*(10), 7405–7415. Available from https://doi.org/10.2166/ws.2022.329. 16070798. IWA Publishing, Ethiopia. https://watermark.silverchair.com/ws022107405.pdf?.

Sposito, G. (2013). Green water and global food security. *Vadose Zone Journal*, *12*(4), 1–6. Available from https://doi.org/10.2136/vzj2013.02.0041, 1539–1663.

Strade, E., Kalnina, D., & Kulczycka, J. (2020). Water efficiency and safe re-use of different grades of water—Topical issues for the pharmaceutical industry. *Water Resources and Industry*, *24*, 100132. Available from https://doi.org/10.1016/j.wri.2020.100132, 22123717.

Teague, R., & Kreuter, U. (2020). Managing grazing to restore soil health, ecosystem function, and ecosystem services. *Frontiers in Sustainable Food Systems*, *4*. Available from https://doi.org/10.3389/fsufs.2020.534187. 2571581X. Frontiers Media S.A., United States. http://www.frontiersin.org/journals/sustainable-food-systems#.

UN/DESA, World Population Prospects. (2017). Data booklet (ST/ESA/SER.A/401). Population Division, 1–24.

Van Vliet, M. T. H., Florke, M., & Wada, Y. (2017). Quality matters for water scarcity. *Nature Geoscience*, *10*(11), 800–802. Available from https://doi.org/10.1038/NGEO3047. 17520908. Nature Publishing Group, Netherlands. http://www.nature.com/ngeo/index.html.

Vasquez-Mejia, C. M., Shrivastava, S., Gudjónsdóttir, M., Manzardo, A., & Ögmundarson, Ó. (2023). Current status and future research needs on the quantitative water use of finfish aquaculture using Life Cycle Assessment: A systematic literature review. *Journal of Cleaner Production*, *425*. Available from https://doi.org/10.1016/j.jclepro.2023.139009. 09596526. Elsevier Ltd, Iceland. https://www.journals.elsevier.com/journal-of-cleaner-production.

Veettil, A. V., Mishra, A. K., & Green, T. R. (2022). Explaining water security indicators using hydrologic and agricultural systems models. *Journal of Hydrology, 607*. Available from https://doi.org/10.1016/j.jhydrol.2022.127463. 00221694. Elsevier B.V., United States. http://www.elsevier.com/inca/publications/store/5/0/3/3/4/3.

Veldkamp, T. I. E., Wada, Y., Aerts, J. C. J. H., Döll, P., Gosling, S. N., Liu, J., Masaki, Y., Oki, T., Ostberg, S., Pokhrel, Y., Satoh, Y., Kim, H., & Ward, P. J. (2017). Water scarcity hotspots travel downstream due to human interventions in the 20th and 21st century. *Nature Communications, 8*. Available from https://doi.org/10.1038/ncomms15697. 20411723. Nature Publishing Group, Netherlands. http://www.nature.com/ncomms/index.html.

Vörösmarty, C. J., Stewart-Koster, B., Green, P. A., Boone, E. L., Flörke, M., Fischer, G., Wiberg, D. A., Bunn, S. E., Bhaduri, A., McIntyre, P. B., Sadoff, C., Liu, H., & Stifel, D. (2021). A green−gray path to global water security and sustainable infrastructure. *Global Environmental Change, 70*. Available from https://doi.org/10.1016/j.gloenvcha.2021.102344. 09593780. Elsevier Ltd, United States. http://www.elsevier.com/inca/publications/store/3/0/4/2/5.

Wang, X., Gard, W., Borska, H., Ursem, B., & van de Kuilen, J. W. G. (2020). Vertical greenery systems: From plants to trees with self-growing interconnections. *European Journal of Wood and Wood Products, 78*(5), 1031−1043. Available from https://doi.org/10.1007/s00107-020-01583-0. 1436736X. Springer Science and Business Media Deutschland GmbH, Netherlands. http://www.springer.com/life + sci/forestry/journal/107.

Wang, X., Shen, C., Wei, J., & Niu, Y. (2018). Study of ecological compensation in complex river networks based on a mathematical model. *Environmental Science and Pollution Research, 25*(23), 22861−22871. Available from https://doi.org/10.1007/s11356-018-2316-4. 16147499. Springer Verlag, China. http://www.springerlink.com/content/0944-1344.

Wilkinson, J. M., & Lee, M. R. F. (2018). Review: Use of human-edible animal feeds by ruminant livestock. *Animal, 12*(8), 1735−1743. Available from https://doi.org/10.1017/S175173111700218X. 1751732X. Cambridge University Press, United Kingdom. http://www.cambridge.org/journals/journal_catalogue.asp.

Yang, Z., Song, J., Cheng, D., Xia, J., Li, Q., & Ahamad, M. I. (2019). Comprehensive evaluation and scenario simulation for the water resources carrying capacity in Xi'an city, China. *Journal of Environmental Management, 230*, 221−233. Available from https://doi.org/10.1016/j.jenvman.2018.09.085. 10958630. Academic Press, China. https://www.sciencedirect.com/journal/journal-of-environmental-management.

Zhao, A., Zhu, X., Chen, S., Li, M., Liu, X., & Pan, Y. (2014). The 3rd International Conference on Agro-Geoinformatics, Agro-Geoinformatics 2014 10.1109/Agro-Geoinformatics.2014.6910620 9781479941575 Institute of Electrical and Electronics Engineers Inc. China Trend analysis for blue and green water resources in the Weihe River basin of Northwest China during the past thirty years and in the near future. September 25.

Zhao, D., Liu, J., Yang, H., Sun, L., & Varis, O. (2021). Socioeconomic drivers of provincial-level changes in the blue and green water footprints in China. *Resources, Conservation and Recycling, 175*, 105834. Available from https://doi.org/10.1016/j.resconrec.2021.105834, 09213449.

Zisopoulou, K., & Panagoulia, D. (2021). An in-depth analysis of physical blue and green water scarcity in agriculture in terms of causes and events and perceived amenability to economic interpretation. *Water (Switzerland), 13*(12). Available from https://doi.org/10.3390/w13121693, 20734441. MDPI AG, United Kingdom., https://www.mdpi.com/2073-4441/13/12/1693/pdf.

Chapter 3

Comparative water footprint analysis of rural and urban areas

Cayetano Navarrete-Molina[1,*], María de los Ángeles Sariñana-Navarrete[1], Cesar Alberto Meza-Herrera[2], Luis Manuel Valenzuela-Nuñez[3] and Ruben Ivan Marin-Tinoco[1,4]

[1]*Department of Chemical area Environmental Technology, Technological University of Rodeo, Rodeo, Durango, Mexico,* [2]*Regional Universitary Unit on Arid Lands, Chapingo Autonomous University, Bermejillo, Durango, Mexico,* [3]*Faculty of Biological Sciences, Juarez University of the State of Durango, Gomez Palacio, Durango, Mexico,* [4]*Rural Hospital No. 162, Mexican Social Security Institute, Rodeo, Durango, Mexico*

**Corresponding author. e-mail address: tiziana.crovella@uniba.it.*

3.1 Introduction

Of the renewable and nonrenewable resources that humans extract from the environment, water is the most critical component (Uddin et al., 2021), being an indispensable part of life and the development of economic and social activities (Shi et al., 2022). However, the current pressure on water resources (WRs), aggravated by population growth and climate change (CC), has led to global water scarcity (WS) (Wen et al., 2019; Yan et al., 2021). Against this backdrop, the need arose to assess the environmental impact (EI) caused by indiscriminate abstractions of WRs from surface and groundwater basins, additionally, to understand the fate of the resource and management strategies (Q. Huang et al., 2021; H. Huang et al., 2021; Jackson & Head, 2020). Consequently, Arjen Hoekstra, PhD (June 28, 1967—November 18, 2019), a Professor at the University of Twente, defined the concept of water footprint (WF) (Hoekstra, 2003), which laid the foundation for a better understanding of the relationship between human activities involving water consumption (WC) and available WRs. With advancing research, marked CC, and the need to assess the human footprint, WF was defined as an "*indicator of the direct and indirect water use, behind all goods and services consumed by an individual or individuals in a country, as well as, the appropriation of freshwater resources, alongside the traditional and restricted measure of water abstraction*" (Hoekstra et al., 2011).

The early efforts to understand WF focused only on consultative water usage. However, to improve understanding of the concept, the study was complemented by dividing WF into three main components: green water footprint (GnWF), blue water footprint (BlWF), and grey water footprint (GyWF) (Hoekstra, 2020; Holmatov et al., 2019). Previously, the models applied included the study of BlWF and GnWF as a single component, and GyWF was interpreted as "dilution water volume" (Hoekstra, 2020). In this sense, each of the components was defined as follows: GnWF represents the consumption of rainwater, BlWF stands for the consumption of groundwater and surface water, and GyWF is the volume of water needed to dilute pollutants in water bodies to meet water quality standards (Harris et al., 2020). However, the rapid increase in the world's population has led to a disproportionate extraction of resources, which is more evident in urban areas (UAs), considering that their inhabitants exert a much greater pressure on the environment than those living in developing areas, also known as rural areas (RAs) (Ruíz-Pérez et al., 2020). In this scenario, the evaluation and comparative analysis of WFs in both sectors arise from the need to understand the relationship between consumption and availability of WRs in the same way, so that it will show the best way forward in taking actions that favor sustainable water management (SWM) (Durán-Sánchez et al., 2018).

To achieve the above, the scientific community and researchers need to have a clear understanding of the importance and evolution of the WF concept. They should consider that WF is an indicator that estimates the total volume of water used to produce goods and services by an individual, a community, or a company (Hoekstra, 2015). This indicator includes water that is directly consumed, either as drinking water or water used for irrigation, and water that is indirectly consumed, such as water used in the food processing industry, products, or services (Hogeboom et al., 2018).

Water Footprints and Sustainable Development. DOI: https://doi.org/10.1016/B978-0-443-23631-0.00003-0

The WF study provides information without differentiating between geographical areas and level of development, which allows the identification of consumption patterns and those sectors that generate a much higher impact on WC (Fialkiewicz et al., 2018). This information allows the community at large to manage the available WRs more efficiently and to promote SWM practices. Similarly, knowing how much water is used in the production of goods or services and the daily consumption can motivate the community at large to take the ownership of sustainable WC practices, contributing to raising awareness on WC practices, current changes in the environment, and how WFs in RAs and UAs can be decreased (Gómez-Llanos et al., 2020).

The appropriation of sustainable policies and practices (SPP) related to WC has globally been a cause for concern for governments. For this reason, within the objectives of the United Nations 2030 Agenda, Goal 6 stands out: Ensure access to water and sanitation for all. This establishes that access to safe drinking water, sanitation, and hygiene is the most basic human need for health and well-being (Berger et al., 2021; UN, 2023), considering that the water demand has outpaced population growth to the extent of causing the increasing appearance of regions where there are records of WS (Liu et al., 2017). Given such a scenario, the assessment of WF makes it possible to identify those environments where water usage is unsustainable, and evaluate the impact of anthropogenic activities, apart from understanding the management that decision-makers carry out to ensure access to WRs in any environment, guaranteeing water security for the inhabitants of the environment under study (Mishra et al., 2021). The 2030 Agenda itself sets out 16 other goals, in which WF becomes important, as water is a vital and an important resource for the development of society and ecosystems, and can contribute to sustainable development. WF can be related to other EI indicators (Fig. 3.1) (Paterson et al., 2015).

However, these relationships do not have the same magnitude in RA's and UA's; therefore, their calculation and comparative analysis must take into account important differences. One must start from the fact that a UA, at some point in its development history, was an RA, whose needs were limited to noncumulative consumption only, which during the transition can lead to the destruction of ecosystems and overexploitation of groundwater basins (Anh et al., 2023; Hirahara, 2020; Kurnia et al., 2022; Shi et al., 2022). Therefore, analyzing WFs in both environments (RAs and UAs) aims to examine and understand the main differences in the management of WRs and around water quality. Recent research has demonstrated the obvious differences in WC between RAs and UAs, which include domestic, industrial, and agricultural consumption, in addition to the GyWF generated in both areas (Garrick et al., 2019; Hutchings et al., 2022; Lee & Lin, 2020). The comparative assessment of WFs in both areas allows the establishment of consumption patterns and potential sources of pollutant release, together with external phenomena that may affect the water load of ground and surface basins, and the dynamics of resource mobilization in general (Anh et al., 2023; Leão et al., 2022). Another advantage of conducting comparative analyses between the two areas is the opportunity to measure the efficiency of the usage of WRs, both in daily activities and for water quality recovery, as well as to estimate the main differences in sustainable resource use (Yan et al., 2023). In this sense, in the UAs, the main use of water is for domestic purposes, unlike in the RAs, where the main use of water is in the agricultural sector. So establishing water–human consumption–production relationships based on available WRs should be strengthened with WFs assessments (Cao et al., 2023).

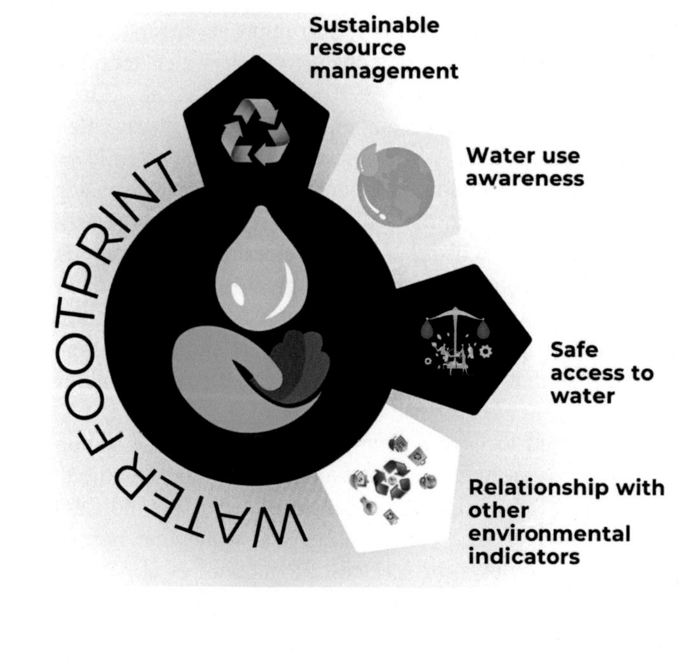

FIGURE 3.1 The water footprint and its relationship with other environmental impact indicators.

The procurement, supply, and use of WRs in rural and urban environments is influenced by the exchange of supplies for the transformation of resources as well as land conversion and ecosystem degradation in the transition from rural to urban environments (Kurnia et al., 2022). Estimating the WFs allows for identifying the sources of WRs used in agricultural activities, industry, and domestic activities, among others, accounting not only for the water abstracted from surface and groundwater basins, but also for estimating the fluctuation of virtual water involved in the process (Daloğlu Çetinkaya et al., 2022; Garrick et al., 2019). Water is one of the resources of watersheds, which has been overabstracted to meet the demands of daily activities, causing unsustainable use of WRs, which is made more evident by the very marked CC (Hirahara, 2020; Kirby & Mainuddin, 2022). In this respect, Mekonnen and Hoekstra (2020) state that the environmental flow of global BlWF in agriculture violates the sustainable environmental flow, putting ecosystems at risk. Therefore, the assessment of WFs in RAs (WFRAs) and WFs in UAs (WFUAs) with an environmental protection approach provides tools for sound decision making regarding the use and care of ecosystems (Pecl et al., 2017). In its assessment, there are important challenges, such as limited access to the availability of the required information during the methodological process (Guan et al., 2022; Tompa et al., 2022) and important differences in the legal requirements of both areas (Wiśniowska, 2023). Therefore, the objectives of the WF study (Fig. 3.2) require considering the current policies of each area, identifying which environmental management practices can be applied to rural and urban settings, and providing a deeper understanding of the difference and efficiency of water usage, along with the EI generated by each zone. The implementation of SWM policies and practices promotes water security and achievement of the 2030 Sustainable Development Goals. However, to achieve the desired effect (Fig. 3.2), it requires a better understanding of water issues, which promotes a more sustainable, efficient, fair, and equitable use of water (Meza-Herrera et al., 2022; Ornelas-Villarreal et al., 2022a, 2022b).

3.2 Assessment of the water footprint analysis of rural and urban areas

By definition, WF promotes the study of water usage in the production of goods and services, considering the water used directly or indirectly in the process using one of the following study models: Water Quality Index, Water Footprint Network, Logarithmic Mean Divisia Index Model, and Water Footprint Assessment (Ma et al., 2020). In this context, the BlWF and GnWF provide knowledge on how and in which WRs extracted from ecosystems are used. However, it is necessary to emphasize its GyWF because it shows the load of pollutants that are discharged into watersheds for the production of goods and services (Gosling & Arnell, 2016; Li et al., 2023). This is where the study of the WFRAs and WFUAs becomes relevant, as it provides insight into the current use of WRs in both areas. A detailed analysis of the WFs in these areas provides certainty about the points of vulnerability, understanding the impact at the regional/local level, and making appropriate resource management decisions . Thereupon, it is important to mention that to face the great water challenges, the calculation of the WFs of any area must be complemented with other

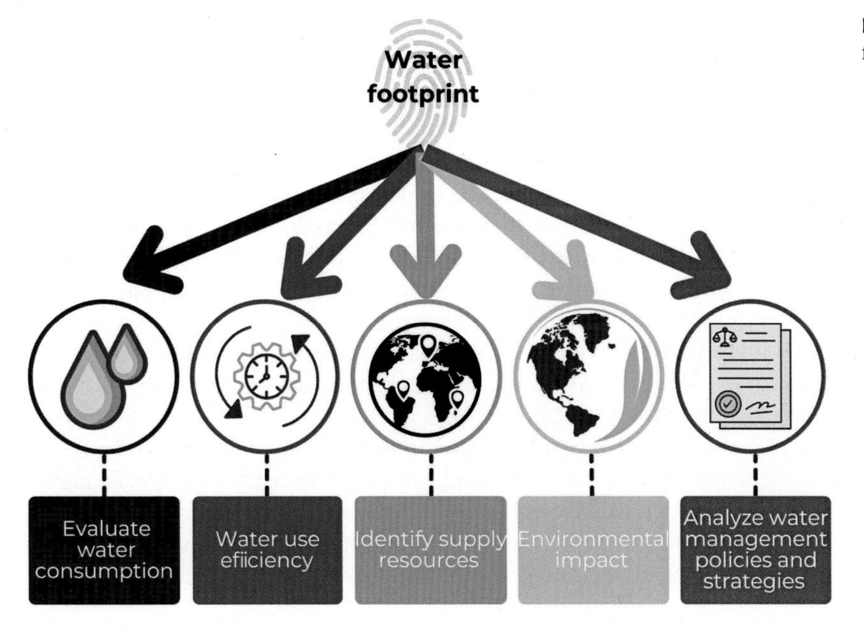

FIGURE 3.2 Main goals of calculating the water footprint.

indicators that allow to visualize the problem from a holistic perspective, which in turn will have the potential to generate a response aimed at sustainability (Navarrete-Molina et al., 2019a; 2019b; 2020; Rios-Flores et al., 2018).

Another point, which is clarified by the calculation of WFRAs and WFUAs, is the identification of WS points due to population demand and CC. However, there may be a certain degree of uncertainty in the procedures, considering that WFs are mainly taken as an indicator of efficiency and equity in the distribution of resources (Wichelns, 2017). Therefore, it is necessary to assess the potential risks of WS from a global perspective, focusing the estimation of WF on water quality degradation, virtual water flows, and possible actions to ensure access to clean, safe, and affordable water in both rural and urban settings (Wichelns, 2017; Zhang et al., 2019). Within the study of WRs management and potential scarcity points, assessing the EI for resource extraction is a component closely related to the assessment of WF. The greater the magnitude of the EI, the greater the environmental changes, which will be reflected in biological responses at the global level, as the distribution of species is altered to remain within preferred environmental conditions (Pecl et al., 2017). Therefore, assessing WFRAs and WFUAs helps to estimate the impact of human activities within ecosystems, and shows what the main differences are between environments, all aimed at the acceptance of SPP that support global sustainability (Anh et al., 2023; Garrick et al., 2019; Hirahara, 2020; Hutchings et al., 2022; Kurnia et al., 2022; Lee & Lin, 2020; Shi et al., 2022).

To develop an adequate assessment of WFRAs and WFUAs, it is necessary to consider the main differences in the characteristics of both areas. In that way, one of the main differences is population density. This characteristic refers to the fact that in UAs, the population density is high compared to RAs. In this connection, it is estimated that by 2050, the proportion of people in UAs will increase to 68% of the world's population, promoting industrialization and connection between cities and increasing pressure on WRs (Parra-Orobio et al., 2023). Globally, intensive city activity consumes approximately 75% of the available freshwater; therefore, proper and timely management of WRs at the urban level is crucial to minimize the risks of WS (Zhu et al., 2022). In contrast, activities carried out in a rural environment are less destructive because of less agglomeration among the population, preserving to a greater extent the ecosystems they are surrounded by (Kocalar, 2022).

Another characteristic that differentiates the two areas is the available infrastructure. Considering that the world is becoming increasingly urbanized, and the demographic change of urban sprawl is becoming notorious as land is being incorporated around it, the challenges toward sustainability that humans face necessarily involve contrasting rural and urban goods and services (Hutchings et al., 2022). The infrastructure of UAs comprises a higher level of complexity than RAs, but the constant growth of the areas implies a greater demand for water and natural resources, which threatens the overall livelihood and sustainability of WRs, requiring the execution of WR management strategies (Chapagain et al., 2022). Strategies such as, "*green infrastructure*" have as their main focus to increase sustainability, resilience, and well-being, and to minimize vulnerability to natural hazards (Depietri, 2022), which can be implemented in RAs and UAs, contributing to achieving an ecological, economic, technological, and institutional balance. Assessing WFs in these times of growth and technological development promotes the identification of those areas where surface water quality is decreasing, mainly due to the discharge of pollutants of urban origin, rather than agricultural practices. This is mainly evident in developing countries (Chen et al., 2022).

In the same vein, the economic activities developed in both areas are different. In the RAs, the economic flow is mostly conditioned by activities related to the interaction with the environment, including agriculture, livestock, and fishing. In contrast, in the UAs, the economic flow is mainly generated by industrial activities, trade, service provision, development of new technologies, and research in areas like biotechnology and neurobiology (Balland et al., 2020). Despite these differences, both spaces depend on each other for development, considering the interrelationship generated between them because of the goods they obtain from ecosystems (Gebre & Gebremedhin, 2019). The economic flow between sectors and zones also generates a WF, the most relevant being that generated by the agrifood industry, which has recently been forced to adopt appropriate measures for a more sustainable management of WRs and reduce water demand (Karwacka et al., 2020). In addition, the differences in economic flows between the two areas lead to significant differences in the lifestyles and environmental settings of the two areas. Therefore, it is important to keep in mind that these differences may vary from country to country and region to region, as urbanization and economic development may have different impacts from one area to another. Lifestyles in RAs and UAs are often different. Cultural diversity and fast-paced lifestyles are distinctive features in both areas. However, sometimes the limited economic opportunities in a rural environment led to increased levels of migration to more populated and economically active areas. This movement creates a higher demand for goods and services, linking to indicators of WS, and changes in the environment and ecosystems (Lyu et al., 2019). In both settings, the assessment of WFs helps to identify hotspots of environmental change and virtual water flows, relating them to everyday activities, to the lifestyle of the population (Liu et al., 2022).

3.3 Characteristics of the rural water footprint

3.3.1 Factors that contribute to the water footprint in rural areas

As mentioned earlier, WF refers to the total amount of freshwater used directly or indirectly by an individual, community, or company (Hoekstra et al., 2011). In the case of RAs, several factors contribute to the WF generated. Within these factors, the main consumer of water in RAs is agricultural activities. Agricultural production stands out, as it is considered the most water-intensive economic activity in the world; yet it is fundamental in RAs and for the supply of food to the UAs (Balland et al., 2020; Cao et al., 2023; Pecl et al., 2017; Yan et al., 2023). In this regard, the irrigation methods used in crop production, for example, flood irrigation or drip irrigation, significantly influence the amount of water used. Furthermore, the type of crop produced plays an important role in the WFRAs. While some crops require large amounts of water, particularly rice, others may have lower water requirements (He et al., 2021; Q. Huang et al., 2021; H. Huang et al., 2021). In addition to agricultural activities, livestock activities in RAs also have a significant impact on the WFRAs. This is due, to the fact that livestock rearing requires water for drinking, irrigation of pastures, and production of crops used as feed for livestock. In addition, the number and type of livestock, as well as SWM practices on livestock farms, must be considered because these factors significantly affect the WFRA (Bhagat et al., 2020; Ibidhi & Ben Salem, 2020; Navarrete-Molina et al., 2020; 2019a; 2019b; Ngxumeshe et al., 2020).

Similarly, another factor contributing to WFRAs is household water usage in RAs, which, although generally lower than in UAs, also contributes to WFs. Household water usage in RAs includes daily activities, for example, WC for flushing, cleaning, cooking food, and watering gardens (Garrick et al., 2019; He et al., 2021). Although to a lesser extent than in UAs, water infrastructure in RAs, such as wells, water distribution systems, and treatment systems, also influences availability and WC. If infrastructure is inadequate or not well managed, water wastage and a greater impact on WFRA can occur (Anh et al., 2023; Hutchings et al., 2022).

Likewise, the CC may affect the availability and quality of water in RAs. In recent years, droughts have become more frequent or intense worldwide, reducing water supply and causing greater pressure on the available WRs (Hirahara, 2020; Kirby & Mainuddin, 2022; Wen et al., 2019; Yan et al., 2021). This influences agricultural practices and also access to drinking water in RAs. However, the level of environmental education (EE), public awareness (PA), and citizen participation (CP) concerning water use efficiency in RAs is also an important and crucial factor in the efforts to reduce WFRA. It is necessary to raise awareness among the rural population about the importance of adopting SWM practices, including efficient irrigation, rainwater harvesting, and reuse, and the use of appropriate technologies, which will contribute to the reduction of WFRAs (Gómez-Llanos et al., 2020; He et al., 2021; Hogeboom et al., 2018; Q. Huang et al., 2021; H. Huang, et al., 2021; Rouse, 2016). Finally, it should be noted that the factors influencing the WFRAs may vary according to geographical location, climatic conditions, and region-specific practices. Therefore, it is necessary to consider the particular characteristics of each RA when calculating and analyzing its WFs.

3.3.2 Impact of agricultural, livestock, and forestry activities on water consumption and biodiversity

Agricultural, livestock, and forestry activities have a significant impact on WC and biodiversity in RAs. As discussed, agriculture is the main economic activity using water worldwide, mainly because of its use for crop irrigation and drinking water supply for livestock (Balland et al., 2020; Bhagat et al., 2020; Ibidhi & Ben Salem, 2020; Ngxumeshe et al., 2020). However, some inefficient irrigation methods can lead to water wastage. In addition, the expansion of agriculture can lead to overexploitation of WRs, significantly affecting groundwater aquifers and reducing river flows to an alarming point of depletion (He et al., 2021; Q. Huang et al., 2021; H. Huang, et al., 2021). Such overexploitation can trigger a range of negative consequences for aquatic ecosystems and the availability of water for other uses, such as human consumption (Hirahara, 2020; Kirby & Mainuddin, 2022; Wen et al., 2019; Yan et al., 2021). In addition, livestock farming has also shown a high WC worldwide. Livestock need water for drinking and to maintain adequate hygienic conditions. Moreover, intensive livestock production systems require large amounts of water for cleaning facilities and producing fodder and concentrates for livestock feed (Bhagat et al., 2020; Ibidhi & Ben Salem, 2020; Navarrete-Molina et al., 2020; 2019a; 2019b; Ngxumeshe et al., 2020). Typically, this can lead to pressure on local WRs, especially in RAs where the clean water supply is limited, contributing to decreased availability of water for human consumption and increased WFRAs (Gómez-Llanos et al., 2020).

In this context, the overexploitation of forest ecosystem resources also has a significant impact on WC and biodiversity. In this regard, deforestation and indiscriminate logging have a significant impact on biodiversity and water availability (Ma et al., 2019; Ulayi et al., 2021; Wang et al., 2020; 2021; 2023). For example, forest clearance reduces habitat for many species, leading to biodiversity loss and extinction of plants and animals. In addition, forests are key ecosystems for the water cycle, considering that they regulate water infiltration into the soil and promote ideal conditions for higher precipitation (Baleta et al., 2021; Wang et al., 2020; 2021; 2023). Consequently, increased deforestation can alter rainfall patterns and increase the risk of droughts or floods. In light of the above, it is important to note that there are sustainable agricultural, livestock, and forestry practices that have the potential to reduce the impact on WC and biodiversity. These include efficient use of water in agriculture, adoption of appropriate irrigation techniques, adoption of sustainable livestock systems, and responsible forest management that promotes the conservation of forest ecosystems, among others (Martínez-Martínez et al., 2023; Okumu et al., 2021; Wang et al., 2020; 2021; 2023).

3.3.3 Policies and practices sustainable related to the water footprint in rural areas

Globally, different research has reported success stories on sustainable agricultural practices and water conservation (WConser) in RAs. Some examples are mentioned below:

1. **Conservation agriculture in Latin America:** Conservation agriculture, which promotes minimizing soil disturbance and maximizing water retention, has been implemented in many Latin American RAs. This practice includes techniques like no-till, crop rotation, and permanent vegetation cover. In Brazil, Argentina, and Paraguay, farmers have been able to reduce soil erosion, improve water infiltration, and increase agricultural productivity by adopting conservation agriculture (Barrera et al., 2021; Kassam et al., 2015, 2019, 2022; Martínez-Cruz et al., 2019; Montes de Oca Munguia et al., 2021; Speratti et al., 2015; Ruzzante & Bilton, 2018; Woltering et al., 2022).
2. **Drip irrigation in Israel:** Israel is known for its expertise in the efficient use of water in agriculture. By adopting drip irrigation, farmers apply water directly to the roots of plants, minimizing evaporation and waste. This approach has enabled Israel to become a world leader in the production of crops, for instance, fruits, vegetables, and flowers, despite being located in an arid region (Eshel et al., 2022; Kassam et al., 2015, 2019, 2022; Mirlas et al., 2020; Mor-Mussery et al., 2021; Woltering et al., 2022; Yue, 2020).
3. **Agroforestry in Kenya:** Agroforestry has been promoted as a sustainable practice mainly in the RAs of Kenya. This practice combines trees and crops in the same area, benefiting both farmers and the environment. Trees help conserve water by reducing evaporation, improving infiltration, and stabilizing the soil, as well as providing shade and shelter for livestock, enhancing biodiversity, and generating additional forest products, for example, firewood and fruits (Andersson & D'Souza, 2014; Brown et al., 2017; Cordingley et al., 2015; Kassam et al., 2015; 2019; 2022; Marenya et al., 2021; Ruzzante & Bilton, 2018; Woltering et al., 2022).
4. **Integrated water management in some European countries:** In Spain, Netherlands, France, Switzerland, and Italy, an integrated water management approach to agriculture has been implemented in recent years. This approach includes the collection and reuse of rainwater, the application of efficient irrigation techniques, and the use of crops more adapted to local conditions. This has contributed to reducing water demand in agriculture and preserving WRs in a region where water is scarce (Iglesias & Garrote, 2015; Kahil et al., 2015; Kassam et al., 2015; 2019; 2022; Long et al., 2016; Pulido-Velazquez & Ward, 2017; Rossi, 2022).

The above examples are just a few. Throughout the world, numerous initiatives and projects are underway to promote more sustainable and resilient agriculture in the face of the challenges of CC and WS, contributing to reducing the WFRAs.

3.4 Characteristics of the water footprint in urban areas

3.4.1 Factors that contribute to the water footprint in urban areas

The WFUA refers to the total amount of water used directly and indirectly by a UA to meet its daily needs. This generates competition between agricultural, industrial, and domestic uses, thus becoming one of the main drivers of water conflicts between RAs and UAs (Q. Huang et al., 2021; H. Huang et al., 2021; Wen et al., 2019; Yang et al., 2016). These conflicts continue to increase, considering the global urban expansion trends, causing an increase in water demand for domestic and industrial use in UAs, resulting in overexploitation and depletion of water sources (WSo) near large population centers, forcing the use of more distant and expensive WSo, which often affects this resource in RAs close to cities, and producing serious economic, social, cultural, and environmental effects, mainly on the areas from which water is transferred (Do et al., 2022; Franci et al., 2015; Yan et al., 2021). This is considered by some researchers to be the main cause of the global increase in water insecurity (WI), which is expected to intensify in the UAs (Gesualdo et al., 2019; Van Ginkel et al., 2018).

Some causes related to WI in UAs are CC, decreasing precipitation, high population density, and poor management of available water. According to Veldkamp et al. (2017), many countries are experiencing WI conditions and by 2050, many more will face a significant reduction in their surface WRs. By the mid-2010s, 27% of the world's population inhabited areas potentially affected by severe WS (UN, 2018). However, it is estimated that by 2050, this will increase to almost double that reported in 2010 (Boretti & Rosa, 2019; UN, 2018). This has the potential to generate a rapid deterioration in the water security of urban households, considering that overurbanization rarely considers the availability of WRs, particularly in UAs where water and sanitation infrastructure is underdeveloped (Jackson & Head, 2020; Muller, 2017; Ullrich et al., 2018). To put the situation in perspective, in 2018, the United Nations estimated that two-thirds of the world's population would live in cities (UN, 2018).

Considering the above, it is necessary to understand the main factors contributing to WFUAs, and CC stands out as one of them. CC has the potential to aggravate WFUAs by causing changes in weather patterns, which can trigger more frequent droughts or intense rainfall events, affecting water availability, increasing water demand or the need to manage more efficient water storage systems (Degefu et al., 2018; Gesualdo et al., 2019; Khatibi & Arjjumend, 2019; Li et al., 2018; Velpuri & Senay, 2017). To achieve the above, it is necessary to consider another factor that also affects WFUAs, such as urban infrastructure (UI; roads, buildings, drainage systems, and paved areas). Considering that UI has the potential to alter the natural water cycle, decrease infiltration into the soil, and increase rainwater runoff, among other negative effects (Mahjabin et al., 2018; Manzardo et al., 2016; Paterson et al., 2015). Within the UI, water infrastructure management could be considered as one of the main factors contributing to WFUAs, especially related to losses and leaks in water distribution networks, considering that water supply systems usually have old or damaged pipes that cause leaks, which results in a significant loss of drinking water (Godinez-Madrigal et al., 2020; Ruíz-Pérez et al., 2020; Tortajada, 2016). Proper management and improvement of water infrastructure, aside from, early detection of leaks, are key to reducing the negative effects of water losses and leaks in cities, which would significantly contribute to reducing WFUAs.

In this context, reducing water losses and leakage in cities will increase the availability of water for household consumption (Tortajada, 2016). This consumption includes the use of water for drinking, cooking, washing, bathing, and watering gardens. However, consumption habits, water usage efficiency, and population density are other factors contributing to the amount of water used in households and therefore to increasing WFUAs (Guma et al., 2023; Rouse, 2016; Wiig et al., 2023). In addition, water usage in industrial and commercial activities in UAs also has a significant impact on WFUAs. Considering that industry, water treatment plants, commercial establishments, and public utilities demand high volumes of water for production, and cleaning, among others, these sectors need to be encouraged to implement more efficient SWM practices (Mohan et al., 2021; Sun & Kato, 2022). However, it is not only domestic, industrial, and commercial water use that affects water availability in cities. Another factor contributing to the increase in WFUAs is urban agriculture, which includes urban gardens, rooftop cultivation, and community gardening systems. This requires appropriate and thoughtful selection of crops for urban agriculture and must be complemented by an environmentally sustainable water supply to maintain or even reduce the WFUAs caused by urban agriculture (Cooper et al., 2022; Haitsma Mulier et al., 2022; Hogeboom, 2020; Langemeyer et al., 2021; Nouri et al., 2019). Therefore, reducing WFUAs is crucial to ensure SWM.

3.4.2 Impact of urban infrastructure on water consumption and wastewater generation

As mentioned in the previous point, UI generates a significant impact on WFUAs, mainly related to the WC and wastewater (WW) generation of UAs (Godinez-Madrigal et al., 2020; Ruíz-Pérez et al., 2020; Tortajada, 2016). In this context, drinking water supply and sewerage systems directly affect the WFUAs. Among the factors that affect WC in cities, the infrastructure available for drinking water supply stands out, considering that the supply of drinking water to the inhabitants of UAs depends on WSo such as reservoirs, rivers, or aquifers (Sinha et al., 2023). The design and efficiency of the infrastructure of water supply systems in cities directly affects the amount of water available and consumed. For example, if pipes leak, water is wasted. Therefore, the UI must adapt to meet the water supply needs of the UAs (Adedeji & Hamam, 2020; Balaei et al., 2020; Morris-Iveson & Day, 2021; Regmi et al., 2022; Sinha et al., 2023), considering factors like population density, appliance efficiency, and WConser practices, which affect WC in homes and commercial buildings (Aghayev & Pashayeva, 2021; Long et al., 2022; Zarei et al., 2023). Another factor to be considered at this point is the irrigation systems in cities, considering that UI uses significant amounts of water to irrigate parks, gardens, and green areas (Barrios & Teixeira De Mello, 2022; Boochani & Tabaei, 2022; Li et al., 2017; Valencia et al., 2022; Elkamel et al., 2023). Therefore, the choice of efficient irrigation systems, for example, drip irrigation, can reduce WFUAs, when compared to more traditional irrigation methods, including gravity or sprinkler irrigation.

However, UAs not only have high WFs; they also generate large amounts of WW, in this sense, the proper management of the WW generated depends directly on the available UI. The UI related to WW management includes the water network, not only for drinking WC distribution but also the drainage network and WW collection systems, which are essential to keep cities clean (Guma et al., 2023; Lu, 2016; Silver & Wiig, 2023). The efficiency of clean and polluted water distribution and collection systems affects the accumulation of waste and its potential EI (Pandit, 2016; Sinha et al., 2023). In this respect, to reduce the amount of waste sent to disposal areas or landfills, it is important to have adequate recycling infrastructures in place. This includes awareness programs to encourage CP in the proper use of water (Gómez-Llanos et al., 2020; Rouse, 2016). In addition, the UI must consider the adequate treatment of WW, especially those coming from areas that discharge hazardous pollutants or require special management, for example, medical or chemical waste (Depietri, 2022; Godinez-Madrigal et al., 2020; Rouse, 2016). This implies the implementation of waste treatment systems, for instance, WW treatment plants and incineration or composting plants. In this context, it is important to consider that urban planning and investment in sustainable infrastructure can help reduce both WFUAs and WW generation. This should include SPP, such as the development of more efficient technologies, the promotion of conservation and recycling practices, and EE, PA, and CP in the proper management of these resources (Balaei et al., 2020; Godinez-Madrigal et al., 2020; Sinha et al., 2023).

3.4.3 Sustainable policies and practices related to the water footprint in urban areas

The SWM in UAs is crucial to minimize WFs, as well as to ensure an adequate supply of clean water for present and future generations (Berger et al., 2021; Mishra et al., 2021; UN, 2023). Table 3.1 shows some SPPs related to WFUA recently promoted in some cities around the world.

As shown in Table 3.1, it is essential to promote efficient water use in the UAs. This implies the implementation of technologies and practices that reduce WC, together with the installation of water-saving devices in homes and buildings, complemented EE, PA, and CP on responsible water use, and the detection and repair of leaks in urban water infrastructures (Aghayev & Pashayeva, 2021; Gómez-Llanos et al., 2020; Long et al., 2022; Zarei et al., 2023). In this context, another key policy is to promote water reuse and recycling. This includes the implementation of systems for the collection and treatment of greywater (unpolluted WW from washing machines, sinks, showers, etc.) for use in nonpotable activities such as garden irrigation, street cleaning, or toilet flushing (Garrick et al., 2019; Hutchings et al., 2022; Lee & Lin, 2020). This should be complemented by the installation of rainwater harvesting systems, which consist of collecting and storing rainwater for later use. Rainwater harvesting and storage can be done at the individual household level or the community level in buildings or neighborhoods. Which will contribute to the reduction of the WFUAs, by promoting less reliance on conventional WSo and reducing the pressure applied by humanity on the finite WRs available (Mahjabin et al., 2018; Manzardo et al., 2016; Paterson et al., 2015).

The SPP related to WFUAs could be complemented with the creation of green and permeable spaces that allow natural infiltration of water into the soil, along with the construction of green roofs that retain rainwater and the protection and restoration of local water bodies (rivers or wetlands) (Depietri, 2022; Li et al., 2018; Nouri et al., 2019; Sinha et al., 2023). To achieve the above, it is essential to implement integrated SWM, which must involve all relevant stakeholders, for example, government, businesses, community organizations, and citizens. This implies coordinating water-related SPP, considering aspects like conservation, supply, sanitation, and water quality (Balaei et al., 2020; Godinez-Madrigal et al., 2020; Sinha et al., 2023). However, this needs to be complemented by a strong EE, PA, and CP campaign on the importance of water and the need to reduce WFs. In this regard, governments can develop awareness campaigns, which should target the community to encourage SPP and also inform about the water cycle and offer advice on responsible water utilization (Gómez-Llanos et al., 2020; Rouse, 2016). Finally, the combination of the above measures can help to ensure the availability and sustainability of WRs in cities and contribute significantly to the reduction of WFUAs.

3.5 Impacts and consequences of the rural and urban water footprint

Typically, the WFs vary significantly between RAs and UAs, mainly due to differences in consumption and production patterns. For example, in RAs, the WFs tend to be lower, due to lower population density and lower WC in domestic and industrial activities (Garrick et al., 2019; Hutchings et al., 2022; Lee & Lin, 2020; Parra-Orobio et al., 2023; Veldkamp et al., 2017). Moreover, UAs often import a significant amount of food and goods from other regions, implying a virtual WF associated with the production of those goods (Daloğlu Çetinkaya et al., 2022; Wichelns, 2017; Zhang et al., 2019; Garrick et al., 2019). However, in RAs, the WFs are mainly influenced by the development of primary activities aimed at food production. These activities require large amounts of water for crop irrigation and

TABLE 3.1 Sustainable policies and practices related to the water footprint in urban areas.

City, country	Policies or practices implemented	References
Seville, Spain	Renewal of urban infrastructure. The project involves street renewal with water-sensitive criteria, with five green areas, and road and pavement construction.	Ruiz-Pérez et al. (2022).
Jaipur, India	The study evaluates established plans and strategies to identify policy gaps on the path to water security.	Chapagain et al. (2022).
Nebraska, USA	Consumptive use of water. The study correlates total water usage and consumptive water usage in residential and landscape areas, with regression models to estimate water use, and sustainable management.	Li et al. (2017).
Eight cities in the United States	Urban agriculture. The research evaluates the environment-population–water interaction toward the sustainable management of resources for the production of spinach, carrots, and sweet corn in urban areas. Study carried out in eight cities in the United States of America, which are: Austin, Texas; Buffalo, New York; Jacksonville, Florida; Kansas City, Missouri; Portland, Oregon; Raleigh, North Carolina; Salt Lake City, Utah; and San Jose, California.	Cooper et al. (2022).
Great Miami, USA	Food–energy–water interaction. The comprehensive model proposes the integration of policy, technology, and environmental management, as well as rainwater collection systems and reuse of wastewater (recovered and rainwater).	Elkamel et al. (2023).
El Zapotillo, Mexico	Water distribution policies. The study analyzes the management of hydrological basin resources by government policies in the region. The unsustainable exploitation of underground basins is highlighted, and science–political collaboration is proposed.	Godinez-Madrigal et al. (2020).
Ten cities around the world	Water security. The study considers the development of indicators that relate pressure–state–impact–response to identify water security–infrastructure–urban development. Research carried out in 10 cities around the world, which are: Amsterdam, Netherlands; Toronto, Canada; Singapore, Singapur; Dubai, the United Arab Emirates; Beijing, China; Hong Kong, China; São Paulo, Brazil; Nairobi, Kenya; Lima, Peru; and Jakarta, Indonesia.	Van Ginkel et al. (2018).
Paris, France and Berlin, Germany	Social conscience. The research proposes digital development to link society-government-technology in specific urban water resources management practices.	Stein et al. (2023).
Cities of Africa and Asia	Vertical agriculture is proposed as a sustainable practice that guarantees resilience to climate change, the food system, and economic needs.	Oh and Lu (2023).

livestock maintenance. Furthermore, it has been reported that these activities can be directly related to the depletion of local WRs, generating negative impacts on the environment and increasing the ecological impact on ecosystems (Balland et al., 2020; Bhagat et al., 2020; Ibidhi & Ben Salem, 2020; Ngxumeshe et al., 2020). In this sense, it is necessary to highlight that the SWM is fundamental in both areas. Such management can contribute to reducing the WFs and ensuring water availability for future generations, through the appropriation of WConser actions, the efficient use of WRs and the promotion of innovative technologies (Balaei et al., 2020; Godinez-Madrigal et al., 2020; Long et al., 2022; Sinha et al., 2023; Zarei et al., 2023).

If a proper assessment of the SWM strategies of any area or region of the world is to be made, it is necessary to consider two important aspects related to the EI and sustainability of water: WC and WW generation, which are closely related, but have some differences and similarities. The main differences between these two aspects lie in (1) The nature of the resource: Water is a renewable natural resource, while WW is a waste, i.e., it is a waste product that is generated as a result of human activities and is not itself renewable (Q. Huang et al., 2021; H. Huang et al., 2021; Jackson & Head, 2020; Uddin et al., 2021; Wen et al., 2019; Yan et al., 2021); (2) The life cycle: Water is used and then can be recovered, which allows for a continuous cycle. On the other hand, WW once generated, if reuse is desired, may require different management processes, including collection and treatment; if reuse is no longer desired, final disposal or disposal systems need to be added to the management processes, which can be more complex and expensive (Garrick et al., 2019; Hutchings et al., 2022; Lee & Lin, 2020; Mohan et al., 2021; Sun & Kato, 2022); (3) Impact on ecosystems: Excessive WC can negatively affect aquatic ecosystems by decreasing river and lake flows, as well as contributing to WS. On the other hand, WW can contaminate soils, groundwater, and oceans, affecting biodiversity and generating negative impacts on ecosystems (Anh et al., 2023; Hirahara, 2020; Kurnia et al., 2022; Shi et al., 2022); (4) Consumption sources: Water is used in a wide variety of activities, such as domestic consumption, agriculture, industry, and energy generation. On the other hand, WW comes mainly from households, industrial production, and commercial activities (Garrick et al., 2019; Hutchings et al., 2022; Lee & Lin, 2020; Parra-Orobio et al., 2023; Veldkamp et al., 2017); among other differences.

However, the WC and WW generations have some similarities, just to mention a few: (1) Environmental impact: Both WC and WW generation have a significant impact on the environment. Accordingly, excessive water usage can deplete WRs and cause shortages, while excessive WW generation can pollute air, water, and soil, negatively affecting the quality of the environment (Godinez-Madrigal et al., 2020; Ruíz-Pérez et al., 2020; Tortajada, 2016); (2) The need for proper management: These two aspects that contribute to WFs require proper management to minimize their negative impact. That is, proper SWM through conservation, efficient use, and reuse are important to ensure its long-term availability. Similarly, proper WW management involves quantity reduction, recycling, recovery, and safe final disposal (Durán-Sánchez et al., 2018; Q. Huang et al., 2021; H. Huang et al., 2021; Jackson & Head, 2020; Meza-Herrera et al., 2022; Ornelas-Villarreal et al., 2022a; 2022b); (3) Lack of individual and collective awareness: WC and WW generation require a real change of awareness in people and industries. The EE, PA, and CP are key to promoting SWM practices, in particular, reducing WC, recycling, and adopting cleaner production approaches (Balaei et al., 2020; Godinez-Madrigal et al., 2020; Sinha et al., 2023).

3.6 Considerations in the evaluations of the environmental impacts of the rural and urban water footprint

Assessing the EI of urban and rural WFs is necessary if the impact of human activities on water usage and availability is to be understood. Such assessments should focus on revealing the relationship between WFs and EI (Q. Huang et al., 2021; H. Huang et al., 2021; Jackson & Head, 2020; Paterson et al., 2015). In this context, assessments should be interdisciplinary and not focus on a single EI, considering in their calculation methodology the quantification and interaction between several EIs. These could include the impact of WF on WS, considering that excessive water use in RAs and UAs has the potential to deplete the local WR, endangering water supply for human consumption, agriculture, and wildlife (Meza-Herrera et al., 2022; Navarrete-Molina et al., 2020; 2019a; 2019b; Ornelas-Villarreal et al., 2022a, 2022b; Rios-Flores et al., 2018), resulting in devastating effects on local aquatic ecosystems and contributing to biodiversity loss (Barrios & Teixeira De Mello, 2022; Hogeboom et al., 2018; Jackson & Head, 2020). Another EI of overexploitation of an area's WR is the degradation of water quality, resulting from the discharge of polluted water into water bodies such as rivers and lakes. These pollutants can include chemicals, excess nutrients (e.g., agricultural fertilizers), and untreated WW (Kurnia et al., 2022; Wichelns, 2017; Zhang et al., 2019). This negatively affects drinking water quality and aquatic ecosystems, which may cause health problems and death of aquatic life (Hoekstra, 2015; Pecl et al., 2017; Ulayi et al., 2021; Veldkamp et al., 2017; Tortajada, 2016).

In the same vein, additional considerations in the evaluations of the EI of the WFs in the RAs and UAs should include the impact on surface water availability, considering that excessive water usage in RAs and UAs has the potential to reduce river flows and lake water levels, thus negatively affecting aquatic habitats and the species that depend on them (Anh et al., 2023; Daloğlu Çetinkaya et al., 2022; Leão et al., 2022; Garrick et al., 2019). In addition, excessive water abstraction from groundwater aquifers, along with lowering water tables, also contributes to the loss of wetlands, generating a negative impact on biodiversity and ecosystem services (Gebre & Gebremedhin, 2019; Martínez-Martínez et al., 2023; Pecl et al., 2017). All these considerations must be contextualized in the CC, considering that water usage in RAs and UAs is closely related to greenhouse gas emissions. In this regard, water treatment and distribution processes consume energy, which often comes from fossil fuel sources (Boretti & Rosa, 2019; Hirahara, 2020; Wen et al., 2019; Yan et al., 2021; Kirby & Mainuddin, 2022; UN, 2018). In addition, deforestation associated with the expansion of urban and agricultural areas can contribute to CC. In turn, CC impacts water availability and precipitation patterns, exacerbating WS problems (Degefu et al., 2018; Gesualdo et al., 2019; Li et al., 2018; Veldkamp et al., 2017; Wichelns, 2017).

To achieve this, the active participation of individuals, communities, governments, businesses, and above all, the scientific community, is necessary (Gómez-Llanos et al., 2020; Rouse, 2016). Research aimed at improving the SWM, reducing pollution, and conserving aquatic ecosystems is essential to minimize negative impacts and ensure a supportive use of WRs, additionally, to achieve greater sustainability (Balaei et al., 2020; Chen et al., 2022; Li et al., 2017; Long et al., 2022; Lee & Lin, 2020). In addition, research must possess a holistic view if it is to better understand these EIs and seek sustainable solutions (Meza-Herrera et al., 2022; Navarrete-Molina et al., 2020; 2019a; 2019b;

Ornelas-Villarreal 2022a; 2022b). However, the results of this research, will not have an adequate impact on society and the environment if they are not translated into the adoption of SPP aimed at promoting efficient water use practices, implementing WConser and reuse techniques, improving WW treatment systems, and promoting integrated WRs management (Balaei et al., 2020; Chapagain et al., 2022; Godinez-Madrigal et al., 2020; Sinha et al., 2023; Zarei et al., 2023). Finally, it is necessary to consider that the successful implementation of these SPP must be complemented with EE, PA, and CP on water care and associated EI to achieve real positive changes in human behavior (Aghayev & Pashayeva, 2021; Gómez-Llanos et al., 2020; Long et al., 2022).

3.7 Recommended actions to reduce the rural and urban water footprint

3.7.1 Recommendations to promote practices on sustainable water usage in rural areas

Considering the research conducted, it is necessary to promote SWM practices in the RAs, which is essential to ensure the availability of this vital resource in the long term. One of the main is to increase the EE, PA, and CP (Berger et al., 2021; Gómez-Llanos et al., 2020; Mishra et al., 2021; UN, 2023). This can be achieved by organizing workshops and trainings on the value of water and SWM practices, sharing with attendees at these events the consequences of inappropriate water use and ways to conserve water and considering that water harvesting in RAs is the main source of water available, including for UAs. Similarly, rainwater harvesting and storage in RAs should be promoted through the installation and operation of rainwater harvesting systems in households and rural communities. This will allow water to be used more efficiently and reduce dependence on other sources (Mahjabin et al., 2018; Manzardo et al., 2016; Pandit, 2016; Paterson et al., 2015; Sinha et al., 2023). However, it has been mentioned that agriculture is the main user of the available water. In this context, the promotion and installation of efficient irrigation techniques and systems for farmers, such as drip irrigation or sprinkler irrigation, will contribute to reducing the WFRAs. These techniques and systems minimize water wastage by providing water directly to the root zone of the plants and avoiding excessive evaporation. This can be complemented by soil conservation practices, including the promotion of minimum tillage techniques and no-tillage, which increase soil moisture retention and reduce the need for frequent irrigation (Barrera et al., 2021; Kassam et al., 2015; 2019; 2022; Martínez-Cruz et al., 2019; Montes de Oca Munguia et al., 2021; Speratti et al., 2015; Ruzzante & Bilton, 2018). In the same context, another recommendation is the planting and cultivation of drought-resistant plant varieties, considering that these crops will need less water to grow and can better adapt to local conditions, which additionally reduces the loss of diversity in RAs (Iglesias & Garrote, 2015; Kahil et al., 2015; Kassam et al., 2015; 2019; 2022; Long et al., 2016; Pulido-Velazquez & Ward, 2017; Rossi, 2022).

The above recommendations can be complemented by promoting water reuse and recycling by encouraging the use of water treatment and reuse systems in rural communities, considering that treated water can be used for irrigation or other nonpotable purposes like cleaning or livestock (Gómez-Llanos et al., 2020; He et al., 2021; Hogeboom et al., 2018; Q. Huang et al., 2021; H. Huang et al., 2021; Rouse, 2016). In addition, in RAs, it is necessary to educate inhabitants about the importance of protecting WSo, for example, local rivers, streams, and aquifers. Promoting the conservation of aquatic ecosystems and the prohibition of polluting activities near these sources (Berger et al., 2021; Gómez-Llanos et al., 2020; Mahjabin et al., 2018; Manzardo et al., 2016; Mishra et al., 2021; Pandit, 2016; UN, 2023). However, water monitoring systems need to be established to assess water quality and quantity in rural communities. This constant monitoring should generate data, which should be used to make informed decisions on SWM and implementation of WConser measures in RAs (Martínez-Martínez et al., 2023; Okumu et al., 2021; Wang et al., 2020; 2021; 2023). For this to be effective, it is necessary to create incentive programs for those inhabitants of rural communities who implement SWM practices. These programs could include discounts on water tariffs, along with financial support for the installation of water collection or treatment systems (Iglesias & Garrote, 2015; Kahil et al., 2015; Kassam et al., 2015, 2019, 2022; Long et al., 2016; Pulido-Velazquez & Ward, 2017; Rossi, 2022). It should not be overlooked that community collaboration and active participation in the planning and implementation of WConser measures should be encouraged. This will ensure that practices are appropriate to local needs and circumstances, increasing acceptance and adoption (Balaei et al., 2020; Godinez-Madrigal et al., 2020; Gómez-Llanos et al., 2020; Rouse, 2016; Sinha et al., 2023).

3.7.2 Recommendations to promote practices to reduce the water footprint in urban areas

Reducing WFUAs is crucial to conserve and sustainably use WRs. This should be based on SWM actions in these environments. These practices can include the installation of rainwater harvesting systems in buildings and urban structures. This collected water could be used for watering gardens, cleaning streets, and washing vehicles, thereby reducing dependence on potable WSo (Guma et al., 2023; Lu, 2016; Pandit, 2016; Silver & Wiig, 2023; Sinha et al., 2023). However, if their impact is to be maximized, the installation of efficient irrigation systems in parks, gardens, and urban green areas is recommended using drip irrigation techniques and automatic timers to minimize water wastage (Barrios & Teixeira De Mello, 2022; Boochani & Tabaei, 2022; Li et al., 2017; Valencia et al., 2022; Elkamel et al., 2023). In addition, it can be complemented by the reuse of treated WW, which can be used for nonpotable uses, such as irrigation of green areas or street cleaning. Decentralized WW treatment systems can also be used to reuse water in buildings and communities (Garrick et al., 2019; Hutchings et al., 2022; Lee & Lin, 2020).

Another recommendation that is currently gaining increasing attention is the use of advanced technology for water monitoring and efficient SWM, mainly in UAs. This includes the use of smart metering and control systems, leak detection, automated irrigation systems, and data analysis tools for informed decision making (Martínez-Martínez et al., 2023; Okumu et al., 2021; Wang et al., 2020; 2021; 2023). This may have a greater effect if complemented by the installation of green infrastructure, for example, green roofs, vertical gardens, and permeable areas. Such infrastructure helps to reduce rainwater runoff and recharge groundwater aquifers (Depietri, 2022; Li et al., 2018; Nouri et al., 2019; Sinha et al., 2023). However, there is a need to establish regulations and policies that encourage WConser in UAs. This could include water efficiency requirements in the construction of new buildings, fiscal incentives for those who install sustainable water technologies, and progressive water tariffs to promote responsible consumption (Balaei et al., 2020; Godinez-Madrigal et al., 2020; Sinha et al., 2023). In addition, as in the RAs, EE, PA. and CP campaigns should also be carried out in the UAs to raise awareness among the population about the operation of SWM practices. This must be complemented by continued commitment from governments if these practices are to be implemented and sustained in the long term (Aghayev & Pashayeva, 2021; Gómez-Llanos et al., 2020; Long et al., 2022; Rouse, 2016).

3.8 Conclusions

The study of WFRAs and WFUAs is essential to understand and address challenges related to water usage and management in different environments. Water is a vital resource and plays a fundamental role in human life, socioeconomic development, and ecosystem health. However, increasing pressure on WRs resulting from population growth, the CC, and uncontrolled urbanization has led to global WS. In this context, the determination of WFs in both areas provides a valuable tool to assess and understand water usage in different sectors and activities. This assessment should not only consider the direct use of water (domestic or agricultural consumption) but also how it is indirectly used in the production of goods and services. Therefore, the determination of WFs in RAs and UAs provides valuable information to identify consumption patterns, points of vulnerability in water availability, and opportunities for a more efficient and sustainable management of WRs.

In the same way, a comprehensive comparison between the WFRAs and WFUAs needs to be made, considering the published evidence. This has identified significant differences, in both areas, in SWM and WC. This is exacerbated by the CC, which affects rainfall patterns and freshwater availability, highlighting the importance of holistic and SWM to meet future challenges. Although UAs generate a higher WF, RAs have a high potential to minimize global WF, considering that the main productive activity in RAs is agriculture; therefore, it is crucial to adopt sustainable practices in these areas, for example, conservation agriculture, which promotes the minimization of soil disturbance and maximizes water retention. Another practice is the adoption of efficient irrigation systems, like drip or sprinkler irrigation, minimizing evaporation and wastage. Also, SWM measures include the implementation of agroforestry systems, which combine trees and crops in the same area; another recommended practice in RAs is integrated SWM in agriculture, through the collection and reuse of rainwater, combined with efficient irrigation techniques, as well as the use of crops more adapted to local conditions.

Considering that UAs generate, globally, the largest WFs, adopting all the practices recommended in the RAs would not be sufficient. In this sense, UA generates significant pressure on WR, caused by high population density, infrastructure, industrial and commercial demand, and lifestyles. This often results in overexploitation of WSo and the need to import water from surrounding RAs, with negative consequences for ecosystems and rural communities. Consequently, inefficient management of drinking water supply and sewerage systems can result in significant water losses and leakage, and urban water infrastructure plays an important role in this regard, as it has the potential to alter the natural water cycle and contribute to surface runoff. Therefore, advanced technologies, such as smart metering systems and data analysis tools, can play an important role in SWM, especially in UAs. In addition, in both areas (RAs and UAs), SWM policies should be promoted that address differences in water availability and demand and include the appropriation of technologies and practices that reduce WC in households, industries, and agriculture. Such consumption can include EE, PA, and CP strategies aimed at encouraging proper water usage and leak detection and repair. In addition, encouraging water reuse and recycling, especially in UAs, can reduce reliance on conventional WSo and minimize overall WFs.

Therefore, the assessment of EI in relation to water usage must be interdisciplinary and consider multiple factors, including water quality and quantity, aside from the impact on ecosystems. The results of EI assessments will generate valuable information for the adoption of policies and regulations that promote WConser, water efficiency, and reuse as fundamental pillars for achieving SWM in both areas. However, these results must be socialized through the promotion of a true and lasting EE, PA, and CP, to promote SWM practices and generate positive changes in human behavior. This can only be achieved through the active participation of individuals, communities, governments, businesses, and the scientific community. Finally, the SWM in the UAs and RAs is a global challenge that requires broad collaboration and a holistic approach to ensure the availability of this vital resource for the present and future generations, on top of, reducing the WFs of anthropogenic activities globally.

References

Adedeji, K. B., & Hamam, Y. (2020). Cyber-physical systems for water supply network management: Basics, challenges, and roadmap. *Sustainability*, *12*(22), 9555. Available from https://doi.org/10.3390/su12229555.

Aghayev, M., & Pashayeva, K. (2021). System for reducing water consumption at home. *International Research Journal of Innovations in Engineering and Technology*, *5*(7), 91–97. Available from https://doi.org/10.47001/irjiet/2021.507015.

Andersson, J. A., & D'Souza, S. (2014). From adoption claims to understanding farmers and contexts: A literature review of Conservation Agriculture (CA) adoption among smallholder farmers in southern Africa. *Agriculture, Ecosystems and Environment*, *187*, 116–132. Available from https://doi.org/10.1016/j.agee.2013.08.008, http://www.elsevier.com/inca/publications/store/5/0/3/2/9/8.

Anh, N. T., Can, L. D., Nhan, N. T., Schmalz, B., & Luu, T. L. (2023). Influences of key factors on river water quality in urban and rural areas: A review. *Case Studies in Chemical and Environmental Engineering*, *8*. Available from https://doi.org/10.1016/j.cscee.2023.100424, http://www.journals.elsevier.com/case-studies-in-chemical-and-environmental-engineering/.

Balaei, B., Wilkinson, S., Potangaroa, R., & McFarlane, P. (2020). Investigating the technical dimension of water supply resilience to disasters. *Sustainable Cities and Society*, *56*, 102077. Available from https://doi.org/10.1016/j.scs.2020.102077.

Baleta, H., Orr, S., & Chapagain, A. K. (2021). *Africa-wide trends in development and water resources through a climate change lens. Climate Change and Water Resources in Africa: Perspectives and Solutions Towards an Imminent Water Crisis* (pp. 13–28). South Africa: Springer International Publishing. Available from https://link.springer.com/book/10.1007/978-3-03-061225-2, https://doi.org/10.1007/978-3-030-61225-2_2.

Balland, P. A., Jara-Figueroa, C., Petralia, S. G., Steijn, M. P. A., Rigby, D. L., & Hidalgo, C. A. (2020). Complex economic activities concentrate in large cities. *Nature Human Behaviour*, *4*(3), 248–254. Available from https://doi.org/10.1038/s41562-019-0803-3, http://www.nature.com/nathumbehav/.

Barrera, V. H., Delgado, J. A., & Alwang, J. R. (2021). Conservation agriculture can help the South American Andean region achieve food security. *Agronomy Journal*, *113*(6), 4494–4509. Available from https://doi.org/10.1002/agj2.20879, https://acsess.onlinelibrary.wiley.com/journal/14350645.

Barrios, M., & Teixeira De Mello, F. (2022). Urbanization impacts water quality and the use of microhabitats by fish in subtropical agricultural streams. *Environmental Conservation*, *49*(3), 155–163. Available from https://doi.org/10.1017/S0376892922000200, https://www.cambridge.org/core/journals/environmental-conservation/all-issues.

Berger, M., Campos, J., Carolli, M., Dantas, I., Forin, S., Kosatica, E., Kramer, A., Mikosch, N., Nouri, H., Schlattmann, A., Schmidt, F., Schomberg, A., & Semmling, E. (2021). Advancing the water footprint into an instrument to support achieving the SDGs—Recommendations from the "water as a global resources" research initiative (GRoW). *Water Resources Management*, *35*(4), 1291–1298. Available from https://doi.org/10.1007/s11269-021-02784-9, http://www.wkap.nl/journalhome.htm/0920-4741.

Bhagat, S., Santra, A. K., Mishra, S., Khune, V. N., Bobade, M. D., Dubey, A., Yadav, A., Soni, A., Banjare, S., & Yadav, G. (2020). The water footprint of livestock production system and livestock products: A dark area: A review. *International Journal of Fauna and Biological Studies*, *7*, 83–88.

Boochani, M. H., & Tabaei, M. M. (2022). Analysis of urban gardens in urban natural system based on water footprint approach (A case study of Nazhvan area of Isfahan). *Iranian Journal of Ecohydrology*, *9*(3), 531–553. Available from https://doi.org/10.22059/IJE.2023.352735.1704.

Boretti, A., & Rosa, L. (2019). Reassessing the projections of the World Water Development Report. *npj Clean Water*, *2*(1). Available from https://doi.org/10.1038/s41545-019-0039-9, https://www.nature.com/npjcleanwater/.

Brown, B., Nuberg, I., & Llewellyn, R. (2017). Stepwise frameworks for understanding the utilisation of conservation agriculture in Africa. *Agricultural Systems*, *153*, 11–22. Available from https://doi.org/10.1016/j.agsy.2017.01.012, http://www.elsevier.com/inca/publications/store/4/0/5/8/5/1.

Cao, X., Bao, Y., Li, Y., Li, J., & Wu, M. (2023). Unravelling the effects of crop blue, green and grey virtual water flows on regional agricultural water footprint and scarcity. *Agricultural Water Management*, *278*, 108165. Available from https://doi.org/10.1016/j.agwat.2023.108165.

Chapagain, K., Aboelnga, H. T., Babel, M. S., Ribbe, L., Shinde, V. R., Sharma, D., & Dang, N. M. (2022). Urban water security: A comparative assessment and policy analysis of five cities in diverse developing countries of Asia. *Environmental Development*, *43*. Available from https://doi.org/10.1016/j.envdev.2022.100713, http://www.sciencedirect.com/science/journal/22114645.

Chen, S. S., Kimirei, I. A., Yu, C., Shen, Q., & Gao, Q. (2022). Assessment of urban river water pollution with urbanization in East Africa. *Environmental Science and Pollution Research, 29*(27), 40812–40825. Available from https://doi.org/10.1007/s11356-021-18082-1, https://link.springer.com/journal/11356.

Cooper, C. M., Troutman, J. P., Awal, R., Habibi, H., & Fares, A. (2022). Climate change-induced variations in blue and green water usage in U.S. urban agriculture. *Journal of Cleaner Production, 348*. Available from https://doi.org/10.1016/j.jclepro.2022.131326, https://www.journals.elsevier.com/journal-of-cleaner-production.

Cordingley, J. E., Snyder, K. A., Rosendahl, J., Kizito, F., & Bossio, D. (2015). Thinking outside the plot: Addressing low adoption of sustainable land management in sub-Saharan Africa. *Current Opinion in Environmental Sustainability, 15*, 35–40. Available from https://doi.org/10.1016/j.cosust.2015.07.010, http://www.elsevier.com/wps/find/journaldescription.cws_home/718675/description#description.

Daloğlu Çetinkaya, I., Yazar, M., Kılınç, S., & Güven, B. (2022). Urban climate resilience and water insecurity: Future scenarios of water supply and demand in Istanbul. *Urban Water Journal*, 1–12. Available from https://doi.org/10.1080/1573062x.2022.2066548.

Degefu, D. M., Weijun, H., Zaiyi, L., Liang, Y., Zhengwei, H., & Min, A. (2018). Mapping monthly water scarcity in global transboundary basins at country-basin mesh based spatial resolution. *Scientific Reports, 8*(1). Available from https://doi.org/10.1038/s41598-018-20032-w, http://www.nature.com/srep/index.html.

Depietri, Y. (2022). Planning for urban green infrastructure: Addressing tradeoffs and synergies. *Current Opinion in Environmental Sustainability, 54*, 101148. Available from https://doi.org/10.1016/j.cosust.2021.12.001.

Do, T. A. T., Do, A. N. T., & Tran, H. D. (2022). Quantifying the spatial pattern of urban expansion trends in the period 1987–2022 and identifying areas at risk of flooding due to the impact of urbanization in Lao Cai city. *Ecological Informatics, 72*. Available from https://doi.org/10.1016/j.ecoinf.2022.101912, http://www.elsevier.com/wps/find/journaldescription.cws_home/705192/description#description.

Durán-Sánchez, A., Álvarez-García, J., & del Río-Rama, M. (2018). Sustainable water resources management: A bibliometric overview. *Water, 10*(9), 1191. Available from https://doi.org/10.3390/w10091191.

Elkamel, M., Valencia, A., Zhang, W., Zheng, Q. P., & Chang, N. B. (2023). Multi-agent modeling for linking a green transportation system with an urban agriculture network in a food–energy–water nexus. *Sustainable Cities and Society, 89*. Available from https://doi.org/10.1016/j.scs.2022.104354, http://www.elsevier.com/wps/find/journaldescription.cws_home/724360/description#description.

Eshel, G., Volk, E., Maor, A., Argaman, E., & Levy, G. J. (2022). Degradation of Agricultural Lands in Israel. In P. Pereira, M. Muñoz-Rojas, I. Bogunovic, & W. Zhao (Eds.), *Impact of Agriculture on Soil Degradation I. The Handbook of Environmental Chemistry*. In: (120). Cham: Springer. Available from https://doi.org/10.1007/698_2022_931.

Fialkiewicz, W., Burszta-Adamiak, E., Kolonko-Wiercik, A., Manzardo, A., Loss, A., Mikovits, C., & Scipioni, A. (2018). Simplified direct water footprint model to support urban water management. *Water, 10*(5), 630. Available from https://doi.org/10.3390/w10050630.

Franci, F., Mandanici, E., & Bitelli, G. (2015). Remote sensing analysis for flood risk management in urban sprawl contexts. *Geomatics, Natural Hazards and Risk, 6*(5–7), 583–599. Available from https://doi.org/10.1080/19475705.2014.913695, http://www.tandfonline.com/toc/tgnh20/current.

Garrick, D., De Stefano, L., Yu, W., Jorgensen, I., O'Donnell, E., Turley, L., Aguilar-Barajas, I., Dai, X., de Souza Leão, R., Punjabi, B., Schreiner, B., Svensson, J., & Wight, C. (2019). Rural water for thirsty cities: A systematic review of water reallocation from rural to urban regions. *Environmental Research Letters, 14*(4), 043003. Available from https://doi.org/10.1088/1748-9326/ab0db7.

Gebre, T., & Gebremedhin, B. (2019). The mutual benefits of promoting rural-urban interdependence through linked ecosystem services. *Global Ecology and Conservation, 20*, e00707. Available from https://doi.org/10.1016/j.gecco.2019.e00707.

Gesualdo, G. C., Oliveira, P. T., Rodrigues, D. B. B., & Gupta, H. V. (2019). Assessing water security in the São Paulo metropolitan region under projected climate change. *Hydrology and Earth System Sciences, 23*(12), 4955–4968. Available from https://doi.org/10.5194/hess-23-4955-2019, http://www.hydrol-earth-syst-sci.net/volumes_and_issues.html.

Godinez-Madrigal, J., Van Cauwenbergh, N., & Van Der Zaag, P. (2020). Unraveling intractable water conflicts: The entanglement of science and politics in decision-making on large hydraulic infrastructure. *Hydrology and Earth System Sciences, 24*(10), 4903–4921. Available from https://doi.org/10.5194/hess-24-4903-2020, http://www.hydrol-earth-syst-sci.net/volumes_and_issues.html.

Gómez-Llanos, E., Durán-Barroso, P., & Robina-Ramírez, R. (2020). Analysis of consumer awareness of sustainable water consumption by the water footprint concept. *Science of the Total Environment, 721*, 137743. Available from https://doi.org/10.1016/j.scitotenv.2020.137743.

Gosling, S. N., & Arnell, N. W. (2016). A global assessment of the impact of climate change on water scarcity. *Climatic Change, 134*(3), 371–385. Available from https://doi.org/10.1007/s10584-013-0853-x, http://www.wkap.nl/journalhome.htm/0165-0009.

Guan, D., Wu, L., Cheng, L., Zhang, Y., & Zhou, L. (2022). How to measure the ecological compensation threshold in the upper Yangtze River basin, China? An approach for coupling InVEST and grey water footprint. *Frontiers in Earth Science, 10*. Available from https://doi.org/10.3389/feart.2022.988291, https://www.frontiersin.org/journals/earth-science.

Guma., Wiig, A., Ward, K., Enright, T., & Hodson, M. (2023). Infrastructuring urban futures: The politics of remaking cities. *Silver*, 199–208. Available from https://doi.org/10.2307/jj.3452814.15.

Haitsma Mulier, M. C. G., van de Ven, F. H. M., & Kirshen, P. (2022). Circularity in the urban water–energy–nutrients–food nexus. *Energy Nexus, 7*, 100081. Available from https://doi.org/10.1016/j.nexus.2022.100081.

Harris, F., Moss, C., Joy, E. J. M., Quinn, R., Scheelbeek, P. F. D., Dangour, A. D., & Green, R. (2020). The water footprint of diets: A global systematic review and meta-analysis. *Advances in Nutrition, 11*(2), 375–386. Available from https://doi.org/10.1093/advances/nmz091, http://advances.nutrition.org/.

He, G., Geng, C., Zhai, J., Zhao, Y., Wang, Q., Jiang, S., Zhu, Y., & Wang, L. (2021). Impact of food consumption patterns change on agricultural water requirements: An urban–rural comparison in China. *Agricultural Water Management, 243*. Available from https://doi.org/10.1016/j.agwat.2020.106504, http://www.journals.elsevier.com/agricultural-water-management/.

Hirahara, S. (2020). Regeneration of underused natural resources by collaboration between urban and rural residents: A case study in Fujiwara district, Japan. *International Journal of the Commons*, *14*(1), 173–190. Available from https://doi.org/10.5334/ijc.977, http://www.thecommonsjournal.org/articles/10.5334/ijc.977/galley/1033/download/.

Hoekstra, A.Y., Chapagain, A.K., Aldaya, M.M., & Mekonnen M.M. (2011). The water footprint assessment manual: Setting the global standard.

Hoekstra A.Y. (2003). Virtual water trade: Proceedings of the international expert meeting on virtual water trade, value of water research report series No. UNESCO-IHE. 12.

Hoekstra, A. Y. (2015). *The water footprint: The relation between human consumption and water use. The Water We Eat: Combining Virtual Water and Water Footprints* (pp. 35–48). Netherlands: Springer International Publishing. Available from http://doi.org/10.1007/978-3-319-16393-2_3.

Hogeboom, R. J., Knook, L., & Hoekstra, A. Y. (2018). The blue water footprint of the world's artificial reservoirs for hydroelectricity, irrigation, residential and industrial water supply, flood protection, fishing and recreation. *Advances in Water Resources*, *113*, 285–294. Available from https://doi.org/10.1016/j.advwatres.2018.01.028, http://www.elsevier.com/inca/publications/store/4/2/2/9/1/3/index.htt.

Hoekstra, A. Y. (2020). *The water footprint of modern consumer society*. (2nd, p. 294)Routledge. Available from https://www.routledge.com/The-Water-Footprint-of-Modern-Consumer-Society/Hoekstra/p/book/9781138354784.

Hogeboom, R. J. (2020). The water footprint concept and water's grand environmental challenges. *One Earth*, *2*(3), 218–222. Available from https://doi.org/10.1016/j.oneear.2020.02.010, http://www.cell.com/one-earth.

Holmatov, B., Hoekstra, A. Y., & Krol, M. S. (2019). Land, water and carbon footprints of circular bioenergy production systems. *Renewable and Sustainable Energy Reviews*, *111*, 224–235. Available from https://doi.org/10.1016/j.rser.2019.04.085, https://www.journals.elsevier.com/renewable-and-sustainable-energy-reviews.

Huang, Q., Zhang, H., van Vliet, J., Ren, Q., Wang, R. Y., Du, S., Liu, Z., & He, C. (2021). Patterns and distributions of urban expansion in global watersheds. *Earth's Future*, *9*(8). Available from https://doi.org/10.1029/2021EF002062, http://onlinelibrary.wiley.com/journal/10.1002/(ISSN)2328-4277.

Huang, H., Zhuo, L., Wang, R., Shang, K., Li, M., Yang, X., & Wu, P. (2021). Agricultural infrastructure: The forgotten key driving force of crop-related water footprints and virtual water flows in China. *Journal of Cleaner Production*, *309*, 127455. Available from https://doi.org/10.1016/j.jclepro.2021.127455.

Hutchings, P., Willcock, S., Lynch, K., Bundhoo, D., Brewer, T., Cooper, S., Keech, D., Mekala, S., Mishra, P. P., Parker, A., Shackleton, C. M., Venkatesh, K., Vicario, D. R., & Welivita, I. (2022). Understanding rural–urban transitions in the Global South through peri-urban turbulence. *Nature Sustainability*, *5*(11), 924–930. Available from https://doi.org/10.1038/s41893-022-00920-w, https://www.nature.com/natsustain/.

Ibidhi, R., & Ben Salem, H. (2020). Water footprint of livestock products and production systems: A review. *Animal Production Science*, *60*(11), 1369–1380. Available from https://doi.org/10.1071/AN17705, http://www.publish.csiro.au/nid/72.htm?nid = 73&aid = 65.

Iglesias, A., & Garrote, L. (2015). Adaptation strategies for agricultural water management under climate change in Europe. *Agricultural Water Management*, *155*, 113–124. Available from https://doi.org/10.1016/j.agwat.2015.03.014, http://www.journals.elsevier.com/agricultural-water-management/.

Jackson, S., & Head, L. (2020). Australia's mass fish kills as a crisis of modern water: Understanding hydrosocial change in the Murray–Darling Basin. *Geoforum; Journal of Physical, Human, and Regional Geosciences*, *109*, 44–56. Available from https://doi.org/10.1016/j.geoforum.2019.12.020, http://www.elsevier.com/inca/publications/store/3/4/4/index.

Kahil, M. T., Connor, J. D., & Albiac, J. (2015). Efficient water management policies for irrigation adaptation to climate change in Southern Europe. *Ecological Economics*, *120*, 226–233. Available from https://doi.org/10.1016/j.ecolecon.2015.11.004, http://www.elsevier.com/inca/publications/store/5/0/3/3/0/5.

Karwacka, M., Ciurzyńska, A., Lenart, A., & Janowicz, M. (2020). Sustainable development in the agri-food sector in terms of the carbon footprint: A review. *Sustainability*, *12*(16), 6463. Available from https://doi.org/10.3390/su12166463.

Kassam, A., Friedrich, T., & Derpsch, R. (2019). Global spread of Conservation Agriculture. *International Journal of Environmental Studies*, *76*(1), 29–51. Available from https://doi.org/10.1080/00207233.2018.1494927, http://www.tandf.co.uk/journals/titles/00207233.asp.

Kassam, A., Friedrich, T., Derpsch, R., & Kienzle, J. (2015). Overview of the worldwide spread of conservation agriculture. *Field Actions Science Report*, *8*. Available from http://factsreports.revues.org/pdf/3966.

Kassam, A., Friedrich, T., & Derpsch, R. (2022). Successful experiences and lessons from conservation agriculture worldwide. *Agronomy*, *12*(4), 769. Available from https://doi.org/10.3390/agronomy12040769.

Khatibi, S., & Arjjumend, H. (2019). Water crisis in making in Iran. *Grassroots Journal of Natural Resources*, *2*(3), 45–54. Available from https://doi.org/10.33002/nr2581.6853.02034.

Kirby, M., & Mainuddin, M. (2022). The impact of climate change, population growth and development on sustainable water security in Bangladesh to 2100. *Scientific Reports*, *12*(1). Available from https://doi.org/10.1038/s41598-022-26807-6, https://www.nature.com/srep/.

Kocalar, A.C. (2022). Evaluation of water management processes in terms of planning. 5.

Kurnia, A. A., Rustiadi, E., Fauzi, A., Pravitasari, A. E., Saizen, I., & Ženka, J. (2022). Understanding industrial land development on rural–urban land transformation of Jakarta megacity's outer suburb. *Land*, *11*(5). Available from https://doi.org/10.3390/land11050670, https://www.mdpi.com/2073-445X/11/5/670/pdf?version = 1651311077.

Langemeyer, J., Madrid-Lopez, C., Mendoza Beltran, A., & Villalba Mendez, G. (2021). Urban agriculture—A necessary pathway towards urban resilience and global sustainability? *Landscape and Urban Planning*, *210*, 104055. Available from https://doi.org/10.1016/j.landurbplan.2021.104055.

Leão, A. S., Sipert, S. A., Medeiros, D. L., & Cohim, E. B. (2022). Water footprint of drinking water: The consumptive and degradative use. *Journal of Cleaner Production*, *355*. Available from https://doi.org/10.1016/j.jclepro.2022.131731, https://www.journals.elsevier.com/journal-of-cleaner-production.

Li, F., Sutton, P., & Nouri, H. (2018). Planning green space for climate change adaptation and mitigation: A review of green space in the central city of Beijing. *Urban and Regional Planning*, *3*(2), 55. Available from https://doi.org/10.11648/j.urp.20180302.13.

Lee, Y. J., & Lin, S. Y. (2020). Vulnerability and ecological footprint: a comparison between urban Taipei and rural Yunlin, Taiwan. *Environmental Science and Pollution Research*, *27*(28), 34624–34637. Available from https://doi.org/10.1007/s11356-019-05251-6.

Li, J., Lin, M., & Feng, Y. (2023). Improved grey water footprint model based on uncertainty analysis. *Scientific Reports*, *13*(1). Available from https://doi.org/10.1038/s41598-023-34328-z, https://www.nature.com/srep/.

Li, Y., Tang, Z., Liu, C., & Kilic, A. (2017). Estimation and investigation of consumptive water use in residential area—Case cities in Nebraska, U.S. A. *Sustainable Cities and Society*, *35*, 637–644. Available from https://doi.org/10.1016/j.scs.2017.09.012, http://www.elsevier.com/wps/find/journaldescription.cws_home/724360/description#description.

Liu, H., Wei, L., Chen, C., & Wang, Z. (2022). The impact of consumption patterns and urbanization on the cross-regional water footprint in China: A decomposition analysis. *Frontiers in Environmental Science*, *9*. Available from https://doi.org/10.3389/fenvs.2021.792423, http://journal.frontiersin.org/journal/environmental-science.

Liu, J., Yang, H., Gosling, S. N., Kummu, M., Flörke, M., Pfister, S., Hanasaki, N., Wada, Y., Zhang, X., Zheng, C., Alcamo, J., & Oki, T. (2017). Water scarcity assessments in the past, present, and future. *Earth's Future*, *5*(6), 545–559. Available from https://doi.org/10.1002/2016EF000518, http://onlinelibrary.wiley.com/journal/10.1002/(ISSN)2328-4277.

Long, H., Shi, S., Tang, Z., & Zhang, S. (2022). Does living alone increase the consumption of social resources? *Environmental Science and Pollution Research*, *29*(47), 71911–71922. Available from https://doi.org/10.1007/s11356-022-20892-w, https://link.springer.com/journal/11356.

Long, T. B., Blok, V., & Coninx, I. (2016). Barriers to the adoption and diffusion of technological innovations for climate-smart agriculture in Europe: Evidence from the Netherlands, France, Switzerland and Italy. *Journal of Cleaner Production*, *112*, 9–21. Available from https://doi.org/10.1016/j.jclepro.2015.06.044.

Lu, S. (2016). Water infrastructure in China: the importance of full project life-cycle cost analysis in addressing water challenges. *Water Infrastructure. Routledge*. Available from https://doi.org/10.4324/9781315692173.

Lyu, H., Dong, Z., Roobavannan, M., Kandasamy, J., & Pande, S. (2019). Rural unemployment pushes migrants to urban areas in Jiangsu Province, China. *Palgrave Communications*, *5*(1). Available from https://doi.org/10.1057/s41599-019-0302-1, http://www.palgrave-journals.com/palcomms/.

Ma, W., Opp, C., & Yang, D. (2020). Past, present, and future of virtual water and water footprint. *Water*, *12*(11), 3068. Available from https://doi.org/10.3390/w12113068.

Ma, X., Zhai, Y., Zhang, R., Shen, X., Zhang, T., Ji, C., Yuan, X., & Hong, J. (2019). Energy and carbon coupled water footprint analysis for straw pulp paper production. *Journal of Cleaner Production*, *233*, 23–32. Available from https://doi.org/10.1016/j.jclepro.2019.06.069, https://www.journals.elsevier.com/journal-of-cleaner-production.

Mahjabin, T., Garcia, S., Grady, C., Mejia, A., & Creutzig, F. (2018). Large cities get more for less: Water footprint efficiency across the US. *PLoS One*, *13*(8), e0202301. Available from https://doi.org/10.1371/journal.pone.0202301.

Manzardo, A., Mazzi, A., Loss, A., Butler, M., Williamson, A., & Scipioni, A. (2016). Lessons learned from the application of different water footprint approaches to compare different food packaging alternatives. *Journal of Cleaner Production*, *112*, 4657–4666. Available from https://doi.org/10.1016/j.jclepro.2015.08.019.

Marenya, P. P., Usman, M. A., & Rahut, D. B. (2021). Community-embedded experiential learning and adoption of conservation farming practices in Eastern and Southern Africa. *Environmental Development*, *40*. Available from https://doi.org/10.1016/j.envdev.2021.100672, http://www.sciencedirect.com/science/journal/22114645.

Martínez-Cruz, T. E., Almekinders, C. J. M., & Camacho-Villa, T. C. (2019). Collaborative research on Conservation Agriculture in Bajío, Mexico: Continuities and discontinuities of partnerships. *International Journal of Agricultural Sustainability*, *17*(3), 243–256. Available from https://doi.org/10.1080/14735903.2019.1625593, http://www.tandfonline.com/toc/tags20/current.

Martínez-Martínez, Y., Dewulf, J., Aguayo, M., & Casas-Ledón, Y. (2023). Sustainable wind energy planning through ecosystem service impact valuation and exergy: A study case in south-central Chile. *Renewable and Sustainable Energy Reviews*, *178*, 113252. Available from https://doi.org/10.1016/j.rser.2023.113252.

Mekonnen, M. M., & Hoekstra, A. Y. (2020). Blue water footprint linked to national consumption and international trade is unsustainable. *Nature Food*, *1*(12), 792–800. Available from https://doi.org/10.1038/s43016-020-00198-1, https://www.nature.com/natfood/.

Meza-Herrera, C. A., Navarrete-Molina, C., Luna-García, L. A., Pérez-Marín, C., Altamirano-Cárdenas, J. R., Macías-Cruz, U., García de la Peña, C., & Abad-Zavaleta, J. (2022). Small ruminants and sustainability in Latin America & the Caribbean: Regionalization, main production systems, and a combined productive, socio-economic & ecological footprint quantification. *Small Ruminant Research*, *211*, 106676. Available from https://doi.org/10.1016/j.smallrumres.2022.106676.

Mirlas, V., Anker, Y., Aizenkod, A., & Goldshleger, N. (2020). Soil salinization risk assessment owing to poor water quality drip irrigation: A case study from an olive plantation at the arid to semi-arid Beit She'an Valley, Israel. *Geoscientific Model Development Discussions*, *10*, 1–31. Available from https://doi.org/10.5194/gmd-2020-231.

Mishra, B., Kumar, P., Saraswat, C., Chakraborty, S., & Gautam, A. (2021). Water security in a changing environment: Concept, challenges and solutions. *Water*, *13*(4), 490. Available from https://doi.org/10.3390/w13040490.

Mohan, G., Chapagain, S. K., Fukushi, K., Papong, S., Sudarma, I. M., Rimba, A. B., & Osawa, T. (2021). An extended input–output framework for evaluating industrial sectors and provincial-level water consumption in Indonesia. *Water Resources and Industry*, *25*. Available from https://doi.org/10.1016/j.wri.2021.100141, http://www.journals.elsevier.com/water-resources-and-industry/.

Montes de Oca Munguia, O., Pannell, D. J., Llewellyn, R., & Stahlmann-Brown, P. (2021). Adoption pathway analysis: Representing the dynamics and diversity of adoption for agricultural practices. *Agricultural Systems*, *191*, 103173. Available from https://doi.org/10.1016/j.agsy.2021.103173.

Mor-Mussery, A., Cohen, S., & Leu, S. (2021). Interrelation between harvester ant activity, soil fertility, and land management in the arid lands of the Negev Desert, Israel. *CATENA*, *207*105700. Available from https://doi.org/10.1016/j.catena.2021.105700.

Morris-Iveson, L., & Day, S. J. (2021). *Resilience of Water Supply in Practice: Experiences from the Frontline*. IWA Publishing.

Muller, M. (2017). Understanding the origins of Cape Town's water crisis. *Civil Engineering*, *5*, 11–16.

Navarrete-Molina, C., Meza-Herrera, C. A., Herrera-Machuca, M. A., Lopez-Villalobos, N., Lopez-Santos, A., & Veliz-Deras, F. G. (2019a). To beef or not to beef: Unveiling the economic environmental impact generated by the intensive beef cattle industry in an arid region. *Journal of Cleaner Production*, *231*, 1027–1035. Available from https://doi.org/10.1016/j.jclepro.2019.05.267.

Navarrete-Molina, C., Meza-Herrera, C. A., Ramirez-Flores, J. J., Herrera-Machuca, M. A., Lopez-Villalobos, N., Lopez-Santiago, M. A., & Veliz-Deras, F. G. (2019b). Economic evaluation of the environmental impact of a dairy cattle intensive production cluster under arid lands conditions. *Animal*, *13*(10), 2379–2387. Available from https://doi.org/10.1017/S175173111900048X.

Navarrete-Molina, C., Meza-Herrera, C. A., Herrera-Machuca, M. A., Macias-Cruz, U., & Veliz-Deras, F. G. (2020). Not all ruminants were created equal: Environmental and socio-economic sustainability of goats under a marginal-extensive production system. *Journal of Cleaner Production*, *255*, 120237. Available from https://doi.org/10.1016/j.jclepro.2020.120237.

Ngxumeshe, A. M., Ratsaka, M., Mtileni, B., & Nephawe, K. (2020). Sustainable application of livestock water footprints in different beef production systems of South Africa. *Sustainability*, *12*(23), 9921. Available from https://doi.org/10.3390/su12239921.

Nouri, H., Chavoshi Borujeni, S., & Hoekstra, A. Y. (2019). The blue water footprint of urban green spaces: An example for Adelaide, Australia. *Landscape and Urban Planning*, *190*. Available from https://doi.org/10.1016/j.landurbplan.2019.103613, http://www.elsevier.com/inca/publications/store/5/0/3/3/4/7.

Oh, S., & Lu, C. (2023). Vertical farming-smart urban agriculture for enhancing resilience and sustainability in food security. *The Journal of Horticultural Science and Biotechnology*, *98*(2), 133–140. Available from https://doi.org/10.1080/14620316.2022.2141666.

Okumu, B., Kehbila, A. G., & Osano, P. (2021). A review of water–forest–energy–food security nexus data and assessment of studies in East Africa. *Current Research in Environmental Sustainability*, *3*. Available from https://doi.org/10.1016/j.crsust.2021.100045, https://www.journals.elsevier.com/current-research-in-environmental-sustainability.

Ornelas-Villarreal, E. C., Navarrete-Molina, C., Meza-Herrera, C. A., Herrera-Machuca, M. A., Altamirano-Cardenas, J. R., Macias-Cruz, U., García-de la Peña, C., & Veliz-Deras, F. G. (2022a). Goat production and sustainability in Latin America & the Caribbean: A combined productive, socio-economic & ecological footprint approach. *Small Ruminant Research*, *211*. Available from https://doi.org/10.1016/j.smallrumres.2022.106677, http://www.elsevier.com/inca/publications/store/5/0/3/3/1/7/index.htt.

Ornelas-Villarreal, E. C., Navarrete-Molina, C., Meza-Herrera, C. A., Herrera-Machuca, M. A., Altamirano-Cardenas, J. R., Macias-Cruz, U., García-de la Peña, C., & Veliz-Deras, F. G. (2022b). Sheep production and sustainability in Latin America & the Caribbean: A combined productive, socio-economic & ecological footprint approach. *Small Ruminant Research*, *211*. Available from https://doi.org/10.1016/j.smallrumres.2022.106675, http://www.elsevier.com/inca/publications/store/5/0/3/3/1/7/index.htt.

Pandit. (2016). OPINION Environmental over enthusiasm. *Water infrastructure. Routledge*. Available from https://doi.org/10.4324/9781315692173.

Parra-Orobio, B. A., Soto-Paz, J., Ramos-Santos, A., Sanjuan-Quintero, K. F., Saldaña-Escorcia, R., Dominguez-Rivera, I. C., & Sánchez, A. (2023). Assessment of the water footprint in low-income urban neighborhoods from developing countries: Case study Fátima (Gamarra, Colombia). *Sustainability (Switzerland)*, *15*(9). Available from https://doi.org/10.3390/su15097115, http://www.mdpi.com/journal/sustainability/.

Paterson, W., Rushforth, R., Ruddell, B., Konar, M., Ahams, I., Gironás, J., Mijic, A., & Mejia, A. (2015). Water footprint of cities: A review and suggestions for future research. *Sustainability*, *7*(7), 8461–8490. Available from https://doi.org/10.3390/su7078461.

Pecl, G. T., Araújo, M. B., Bell, J. D., Blanchard, J., Bonebrake, T. C., Chen, I. C., Clark, T. D., Colwell, R. K., Danielsen, F., Evengård, B., Falconi, L., Ferrier, S., Frusher, S., Garcia, R. A., Griffis, R. B., Hobday, A. J., Janion-Scheepers, C., Jarzyna, M. A., Jennings, S., ... Williams, S. E. (2017). Biodiversity redistribution under climate change: Impacts on ecosystems and human well-being. *Science (New York, N.Y.)*, *355*(6332). Available from https://doi.org/10.1126/science.aai9214, http://science.sciencemag.org/content/sci/355/6332/eaai9214.full.pdf.

Pulido-Velazquez, M., & Ward, F. A. (2017). *Comparison of water management institutions and approaches in the United States and Europe—What can we learn from each other? Competition for Water Resources: Experiences and Management Approaches in the US and Europe* (pp. 423–441). Spain: Elsevier Inc. Available from http://www.sciencedirect.com/science/book/9780128032374, http://doi.org/10.1016/B978-0-12-803237-4.00024-0.

Regmi, P., Karl, M., & Chew, C. (2022). Digital twin: A path to efficient and intuitive water system operations. 95th Water Environment Federation Technical Exhibition and Conference, WEFTEC 2022. Water Environment Federation United States. 9781713870586 756-761.

Rios-Flores, J. L., Rios-Arredondo, B. E., Cantu-Brito, J. E., Rios-Arredondo, H. E., Armendariz-Erives, S., Chavez-Rivero, J. A., Navarrete-Molina, C., & Castro-Franco, R. (2018). Analisis de la eficiencia fisica, economica y social del agua en esparrago (Asparagus officinalis L.) y uva (Vitis vinifera) de mesa del DR-037 Altar-Pitiquito-Caborca, Sonora, Mexico 2014. *Revista de la Facultad de Ciencias Agrarias. Universidad Nacional de Cuyo*, *50*(1), 101–122.

Rossi, F. (2022). Method and practice for integrated water landscapes management: River contracts for resilient territories and communities facing climate change. *Urban Science*, *6*(4), 83. Available from https://doi.org/10.3390/urbansci6040083.

Rouse, M. (2016). The worldwide urban water and wastewater infrastructure challenge. *Water infrastructure*. Available from https://doi.org/10.4324/9781315692173, Routledge.

Ruíz-Pérez, M. R., Alba-Rodríguez, M. D., & Marrero, M. (2020). The water footprint of city naturalisation. Evaluation of the water balance of city gardens. *Ecological Modelling*, *424*, 109031. Available from https://doi.org/10.1016/j.ecolmodel.2020.109031.

Ruiz-Pérez, M. R., Alba-Rodríguez, M. D., & Marrero, M. (2022). Evaluation of water footprint of urban renewal projects. Case study in Seville, Andalusia. *Water Research*, *221*118715. Available from https://doi.org/10.1016/j.watres.2022.118715.

Ruzzante, S. W., & Bilton, A. M. (2018). *Agricultural technology in the developing world: A meta-analysis of the adoption literature*. World Development. Available from https://doi.org/10.1115/DETC2018-86343.

Shi, C., Wu, C., Zhang, J., Zhang, C., & Xiao, Q. (2022). Impact of urban and rural food consumption on water demand in China—From the perspective of water footprint. *Sustainable Production and Consumption, 34*, 148–162. Available from https://doi.org/10.1016/j.spc.2022.09.006, http://www.journals.elsevier.com/sustainable-production-and-consumption/.

Silver, J., & Wiig, A. (2023). Global infrastructure and urban futures: London's transforming royal albert dock. In A. Wiig, K. Ward, T. Enright, M. Hodson, H. Pearsall, & J. Silver (Eds.), Infrastructuring urban futures: The politics of remaking cities (pp. 164–187). Bristol University Press. Available from https://doi.org/10.2307/jj.3452814.13.

Sinha, S. K., Davis, C., Gardoni, P., Babbar-Sebens, M., Stuhr, M., Huston, D., Cauffman, S., Williams, W. D., Alanis, L. G., Anand, H., & Vishwakarma, A. (2023). Water sector infrastructure systems resilience. A social-ecological-technical system-of-systems and whole-life approach. *Cambridge Prisms: Water*, 1–50. Available from https://doi.org/10.1017/wat.2023.3.

Speratti, A., Turmel, M. S., Calegari, A., Araujo, C. F., Violic, A., Wall, P., & Govaerts, B. (2015). Conservation agriculture in Latin America. *Conservation Agriculture*, 391–415. Available from https://doi.org/10.1007/978-3-319-11620-4_16, http://www.dx.doi.org/10.1007/978-3-319-11620-4.

Stein, U., Bueb, B., Bouleau, G., & Rouillé-Kielo, G. (2023). Making urban water management tangible for the public by means of digital solutions. *Sustainability, 15*(2), 1280. Available from https://doi.org/10.3390/su15021280.

Sun, M., & Kato, T. (2022). The effect of urban agriculture on water security: a spatial approach. *Water, 14*(16), 2529. Available from https://doi.org/10.3390/w14162529.

Tompa, O., Kiss, A., Maillot, M., Sarkadi Nagy, E., Temesi, Á., & Lakner, Z. (2022). Sustainable diet optimization targeting dietary water footprint reduction—A country-specific study. *Sustainability, 14*(4), 2309. Available from https://doi.org/10.3390/su14042309.

Tortajada, C. (2016). Water infrastructure as an essential element for human development. In C. Tortajada (Ed.), *Water infrastructure* (1st, p. 12). Routledge. Available from https://doi.org/10.4324/9781315692173.

Uddin, M. G., Nash, S., & Olbert, A. I. (2021). A review of water quality index models and their use for assessing surface water quality. *Ecological Indicators, 122*107218. Available from https://doi.org/10.1016/j.ecolind.2020.107218.

Ulayi, A. I., Okpe, T. A., & Omang, T. N. (2021). Human population growth and environmental resources conservation in the Southern Senatorial District of Cross River State. *LWATI: A Journal of Contemporary Research, 18*(4), 184–199.

Ullrich, P. A., Xu, Z., Rhoades, A. M., Dettinger, M. D., Mount, J. F., Jones, A. D., & Vahmani, P. (2018). California's drought of the future: A mid-century recreation of the exceptional conditions of 2012–2017. *Earth's Future, 6*(11), 1568–1587. Available from https://doi.org/10.1029/2018EF001007, http://onlinelibrary.wiley.com/journal/10.1002/(ISSN)2328-4277.

UN. (2018). The United Nations world water development report 2018: Nature-based solutions for water.

UN. (2023). The sustainable development goals report 2023: Special edition—Towards a rescue plan for people and planet. 80.

Valencia, A., Qiu, J., & Chang, N.-B. (2022). Integrating sustainability indicators and governance structures via clustering analysis and multicriteria decision making for an urban agriculture network. *Ecological Indicators, 142*109237. Available from https://doi.org/10.1016/j.ecolind.2022.109237.

Van Ginkel, K. C. H., Hoekstra, A. Y., Buurman, J., & Hogeboom, R. J. (2018). Urban water security dashboard: Systems approach to characterizing the water security of cities. *Journal of Water Resources Planning and Management, 144*(12). Available from https://doi.org/10.1061/(ASCE)WR.1943-5452.0000997, https://ascelibrary.org/journal/jwrmd5.

Veldkamp, T. I. E., Wada, Y., Aerts, J. C. J. H., Döll, P., Gosling, S. N., Liu, J., Masaki, Y., Oki, T., Ostberg, S., Pokhrel, Y., Satoh, Y., Kim, H., & Ward, P. J. (2017). Water scarcity hotspots travel downstream due to human interventions in the 20th and 21st century. *Nature Communications, 8*(1). Available from https://doi.org/10.1038/ncomms15697.

Velpuri, N. M., & Senay, G. B. (2017). Partitioning evapotranspiration into green and blue water sources in the conterminous United States. *Scientific Reports, 7*(1). Available from https://doi.org/10.1038/s41598-017-06359-w, http://www.nature.com/srep/index.html.

Wang, R., Guo, Z., Cai, C., Zhang, J., Bian, F., Sun, S., & Wang, Q. (2021). Practices and roles of bamboo industry development for alleviating poverty in China. *Clean Technologies and Environmental Policy, 23*(6), 1687–1699. Available from https://doi.org/10.1007/s10098-021-02074-3, https://link.springer.com/journal/10098.

Wang, Z., Peng, D., Xu, D., Zhang, X., & Zhang, Y. (2020). Assessing the water footprint of afforestation in Inner Mongolia, China. *Journal of Arid Environments, 182,* 104257. Available from https://doi.org/10.1016/j.jaridenv.2020.104257.

Wang, Z., Xu, D., Peng, D., & Zhang, X. (2023). Future climate change would intensify the water resources supply-demand pressure of afforestation in inner Mongolia, China. *Journal of Cleaner Production, 407,* 137145. Available from https://doi.org/10.1016/j.jclepro.2023.137145.

Wen, Y., Liu, X., Bai, Y., Sun, Y., Yang, J., Lin, K., Pei, F., & Yan, Y. (2019). Determining the impacts of climate change and urban expansion on terrestrial net primary production in China. *Journal of Environmental Management, 240,* 75–83. Available from https://doi.org/10.1016/j.jenvman.2019.03.071, https://www.sciencedirect.com/journal/journal-of-environmental-management.

Wichelns, D. (2017). Volumetric water footprints, applied in a global context, do not provide insight regarding water scarcity or water quality degradation. *Ecological Indicators, 74*, 420–426. Available from https://doi.org/10.1016/j.ecolind.2016.12.008, http://www.elsevier.com/locate/ecolind.

Wiig, A., Ward, K., Enright, T., Hodson, M., Pearsall, H., Silver., Wiig, A., Ward, K., Enright, T., & Hodson, M. (2023). *Infrastructuring urban futures: The politics of remaking cities* (pp. 1–19). Bristol University Press. Available from https://doi.org/10.2307/jj.3452814.6.

Wiśniowska, E. (2023). Integrated water management — Directions of activities and policies. In M. Smol, M. N. V. Prasad, & I. Stefanakis (Eds.), Water in circular economy (pp. 21–30). Springer International Publishing. Available from https://doi.org/10.1007/978-3-031-18165-8_2.

Woltering, L., Alvarado, M. D. R. B., Stahl, J., Van Loon, J., Hernández, E. O., Brown, B., Gathala, M. K., & Thierfelder, C. (2022). Capacity development for scaling conservation agriculture in smallholder farming systems in Latin America, South Asia, and Southern Africa: exposing the hidden levels. *Knowledge Management for Development Journal*, 1–22. Available from https://km4djournal.org/index.php/km4dj/article/download/510/644.

Yan, Y., Wang, R., Chen, S., Zhang, Y., & Sun, Q. (2023). Three-dimensional agricultural water scarcity assessment based on water footprint: A study from a humid agricultural area in China. *Science of the Total Environment, 857,* 159407. Available from https://doi.org/10.1016/j.scitotenv.2022.159407.

Yan, Y., Wu, C., & Wen, Y. (2021). Determining the impacts of climate change and urban expansion on net primary productivity using the spatio-temporal fusion of remote sensing data. *Ecological Indicators, 127,* 107737. Available from https://doi.org/10.1016/j.ecolind.2021.107737.

Yang, L., Qian, F., Song, D. X., & Zheng, K. J. (2016). Research on urban heat-island effect. *Procedia Engineering, 169,* 11–18. Available from https://doi.org/10.1016/j.proeng.2016.10.002, Elsevier Ltd China 18777058, http://www.sciencedirect.com/science/journal/18777058.

Yue, C. (2020). The multi-level perspective in analysis of the irrigation innovations in Israel. The frontiers of society. *Science and Technology, 2*(18). Available from https://doi.org/10.25236/FSST.2020.021820.

Zarei, G., Ghazaii, F., & Zahra, A. (2023). Fitting the structural relationship between use of home appliance with sustainable water consumption based on attitude and environmental knowledge: The mediating role of behavioral intention.

Zhang, S., Taiebat, M., Liu, Y., Qu, S., Liang, S., & Xu, M. (2019). Regional water footprints and interregional virtual water transfers in China. *Journal of Cleaner Production, 228,* 1401–1412. Available from https://doi.org/10.1016/j.jclepro.2019.04.298, https://www.journals.elsevier.com/journal-of-cleaner-production.

Zhu, M., Wang, J., Zhang, J., & Xing, Z. (2022). The impact of virtual water trade on urban water scarcity: A nested MRIO analysis of Yangtze River Delta cities in China. *Journal of Cleaner Production, 381,* 135165. Available from https://doi.org/10.1016/j.jclepro.2022.135165.

Chapter 4

Drivers of the growing water footprint: a global scenario

Abdullah Kaviani Rad*

Department of Environmental Engineering and Natural Resources, College of Agriculture, Shiraz University, Shiraz, Iran

**Corresponding author. e-mail address: akaviani2020@yahoo.com.*

4.1 Introduction

Water-related challenges are presently viewed as the second most significant hazard to human civilization, following closely behind energy issues (Dong et al., 2013). The issue of limited access to freshwater is widely acknowledged as an enormous obstacle to the economy worldwide (Ewaid et al., 2019) and has emerged as a prominent environmental issue that is currently the subject of global discourse. Through the assessment of 405 river basins worldwide over the time frame of 1996–2005, it was determined that across 201 of these basins, which were inhabited by a population of about 2.67 billion individuals, a significant scarcity of water occurred for a minimum duration of 1 month per year (Hoekstra et al., 2012). In response to this concern, the notion of water footprint (WF) was established as a metric to mitigate the negative consequences of water utilization and has garnered attention from global enterprises (Manzardo et al., 2016). As a result, there has been a substantial growth in the number of WF assessments carried out over the recent decades (Fridman et al., 2021). The idea of WF presented for the first time in 2002 contains similarities to the ecological footprint (EF) approach that emerged in the 1990s. The EF is a measure of the amount of biological production space, measured in hectares, needed to support a certain population. Contrary to that, the WF quantifies the amount of fresh water, measured in cubic meters per year, that must be provided (Hoekstra, 2009). Arjen Y. Hoekstra, a Dutch scientist, made significant contributions to the field of water-related studies by developing an extensive methodological framework and terminology, thereby driving the establishment of a new community of scholars in this domain. He emphasized the involvement of nations in the global WF of virtual water production, consumption, and commerce when he introduced the concept of the "water footprint" on a global scale (Konar & Marston, 2020). At present, WF is widely employed as a prevalent metric for assessing the water usage linked to a particular service or product (Jorrat et al., 2018). The WF of a country is defined as the quantity of water necessary for the production of services and products that are used by the inhabitants of the society (Hoekstra & Chapagain, 2007b). WF can be defined as a metric of consumption that calculates the amount of fresh water used in both direct and indirect ways by an organization or country in the production of goods or delivery of services (Chenoweth et al., 2014). The utilization of subsurface and surface water is shown by the blue WF, while the utilization of rainfall is represented by the green WF (Harris et al., 2020). The concept of the grey WF pertains to the quantity of fresh water necessary for the dilution of contaminants to adhere to established water quality requirements (De Girolamo et al., 2019). The practice of WF calculation is a valuable tool in the field of water resource management, particularly in regions facing water scarcity (Roux et al., 2016). Within the expanding population, the discipline of water resources administration encounters the formidable task of ensuring the efficient and equitable utilization of water, particularly in the sectors of farming and food production, which account for the majority of its consumption (Novoa, Ahumada-Rudolph, Rojas, Munizaga et al., 2019; Novoa, Ahumada-Rudolph, Rojas, Sáez et al., 2019).

The agricultural sector is the primary receiver of freshwater resources, constituting around 70% of global water withdrawals (Lamastra et al., 2014). Lately, irrigated agriculture has faced a precarious predicament characterized by escalating production costs and mounting demands from governments to mitigate adverse environmental effects (García Morillo et al., 2015). The assessment of green, blue, and gray footprints at the national scale in Tunisia during the period of 1996–2005 revealed that the WF associated with crop production accounted for the majority share (87%) of

Water Footprints and Sustainable Development. DOI: https://doi.org/10.1016/B978-0-443-23631-0.00004-2

the overall national WF. On a national scale, tomatoes and potatoes emerged as the primary commodities, exhibiting comparatively high levels of economic productivity, whereas olives and barley had relatively lower levels of productivity. Concerning the economic productivity of the land, oranges exhibited the highest level of productivity, while barley demonstrated the lowest level of productivity. The WF attributed to agricultural production was 31% of the overall renewable water resources, indicating a notable occurrence of water scarcity in Tunisia. The WF attributed to groundwater utilization is 62% of the aggregate renewable groundwater resources, demonstrating a significant water scarcity issue associated with groundwater in the country (Chouchane et al., 2015). According to estimates, the annual WF of the Netherlands is approximately 2300 cubic meters per year. This WF is divided into three main categories: agricultural product consumption accounts for 67%, industrial product consumption accounts for 31%, and home water usage accounts for 2%. The WF of farming operations encompasses several sectors, with animal goods accounting for 46%, oil products for 17%, coffee, tea, cocoa, and tobacco for 12%, cereals for 8%, cotton products for 6%, fruits for 5%, and other crop-based goods for 6%. The residential WF in the Netherlands accounts for approximately 11% of the total, while the remaining 89% is attributed to external sources (van Oel et al., 2009). When countries engage in the importation of food, they indirectly import water as well. The data presented indicates that Morocco, being classified as an arid and semiarid region, relies on external water resources for around 14% of its total water supply. If Morocco decided to shift its current imports from the Netherlands toward its production, it would necessitate an annual water volume of 780 million cubic meters. Nonetheless, the imported goods originating from the Netherlands exhibited a yearly production output of merely 140 million cubic meters, resulting in a worldwide conservation of 640 million cubic meters annually (Hoekstra & Chapagain, 2007a). The evidence indicates that the United States possesses an average WF of 2480 cubic meters per capita annually, while China exhibits an average WF of 700 cubic meters per capita annually. On a global scale, the average WF per capita stands at 1240 cubic meters per year. It appears that four primary aspects directly influence the WF of a country. These factors include (1) the volume of consumption, which is closely associated with the gross national income; (2) the consumption pattern; (3) the climate; and (4) the efficiency of water consumption in agriculture (Hoekstra & Chapagain, 2007b). The findings of a bibliometric analysis conducted on WF surveys from 2006 to 2015 revealed that the United States, China, and the Netherlands contributed 24.1%, 19.2%, and 16% of the issued publications, respectively. The primary focus of scholarly investigation encompassed several subjects, such as the methodologies employed for calculating WFs, the interconnections between water, food, and energy, the factors that drive changes in WFs, and the environmental consequences associated with water utilization (Zhang et al., 2017).

Zhan-Ming and Chen (2013) assessed the virtual water profile of the globe in 2004 and the WF of 112 national regions in their study. The research revealed that, while the agricultural sector contributes to 69% of the total water withdrawal, agricultural products contribute to less than 35% of the worldwide virtual water demand. India, the United States, and China are the leading global consumers of virtual water on a national level. In the interim, the findings indicate that a significant portion, specifically 57%, of global virtual water transfers are encompassed within nonfood commerce. This underscores the necessity of taking into account not just food commodities but also nonfood commodities to comprehensively assess the water budget. Water and energy are intrinsically interconnected in the contemporary global economy. Water already serves a crucial function in the generation of electricity. Furthermore, as biofuels and electricity are projected to acquire a substantial portion of the transportation fuel market, they are expected to play a key role in the domain of energy for transportation as well. According to a study, the elimination of evaporative losses from power can lead to a range of outcomes for the WF, with an increase of 82% being the minimum and a decrease of 250% being the maximum (Scown et al., 2011).

The various footprints associated with the family of sustainability challenges, such as EFs, energy, carbon, and WFs, offer a comprehensive understanding of ecological dynamics. These footprints are particularly valuable for decision-makers, particularly when conducting national-level surveys (Fang et al., 2014). The most significant environmental concerns confronting China are water shortage and carbon emissions. According to a study conducted between 1990 and 2010, China had a net surplus in the exportation of virtual water and carbon. The sectors of large-scale manufacturing and transportation have the most substantial carbon footprints, whereas the domains of agriculture, fisheries, and light industries demonstrate the most significant WFs. According to a study, a significant proportion exceeding 90% of China's carbon emissions and water resources are encompassed within the context of trading (Wang & Ge, 2020). The region of North China is confronted with significant challenges related to water scarcity. A study conducted in 2007 evaluated the whole WF of Liaoning Province, which amounted to 7.3 billion cubic meters. The findings of this assessment indicate that the agricultural, food, and beverage production sectors have contributed to the status of the province as a net exporter of water resources (Dong et al., 2013). The quantification of the impact of socioeconomic development on water usage holds major significance for the Chinese government (Yang et al., 2016). Beijing is currently facing major difficulties in managing its water supplies as a result of fast economic growth and a growing population. The research revealed that the annual overall WF of Beijing amounts to 1524.5 $\times$ 10^6 cubic meters. Notably, 51% of this WF is attributed to the foreign WF, which is acquired through virtual water imports. Agriculture accounts for the highest proportion of the WF as it allocates 56% of external resources. Huebei, a water-scarce region, serves as the primary provider of virtual water to Beijing (Zhang et al., 2011). Beijing is situated in the water-scarce region of northern China, which has a historical record of implementing policies to restrict agricultural water consumption to cover the growing water needs of urban areas. As a result, there have been alterations in the production of crops in the region, which in turn has had implications for the importation of grains and vegetables from other regions within China. The aforementioned scenario in Beijing is likely prevalent in other metropolitan areas worldwide (Huang et al., 2014). The combination of swift economic and social progress, alongside a growing population density, is exerting considerable pressure on the availability and sustainability of regional water resources (Wang et al., 2013). The global avocado market has experienced rapid growth, with notable increases in consumption observed in North America and Europe over the past few decades. This surge can be attributed to a confluence of socioeconomic and advertising influences. Simultaneously, the cultivation of avocados is linked to substantial water disputes and adverse ecological and socioeconomic consequences for indigenous communities residing in the primary production regions. Hence, it is imperative to decrease importation from regions whose manufacturing processes contribute to water stress. Consumers have the power to actively engage in responsible consumption and environmental conservation, ultimately reducing detrimental impacts and enhancing the sustainability of production regions (Sommaruga & Eldridge, 2021).

Although there is a significant level of interest surrounding the advancement and utilization of WFs, many issues have been voiced regarding the conceptual framework and its practical value. In contrast to the carbon footprint, which offers a comprehensive assessment of the worldwide influence of human activities on the limited capacity for absorption of the atmosphere, the standard representation of WF solely measures a factor involved in production, neglecting to consider the diverse spatial and temporal effects of water usage (Chenoweth et al., 2014). In recent times, there has been a concerted attempt to establish an integrated footprint methodology to analyze the environmental impacts associated with both manufacturing and consumption. The study conducted by Galli et al. (2012) introduced the concept of the "footprint family" as a collection of indicators designed to measure human influence on the environment. The underlying premise of this approach is that no individual indicator can fully capture the extent of this impact. The relationship between man and the environment is not one of dominance but rather one in which indications should be used and analyzed in conjunction. Carbon footprint (CF), ecological footprint (EF), and WF are analytical instruments that facilitate the comprehension of the interplay between consumption patterns and environmental stressors in the Earth's atmosphere, bioproduction regions, and freshwater reservoirs. The acceptance of this measure among scientists and politicians is evident; yet, a lack of consensus exists between these two groups (Ewing et al., 2012). The international trade activities of a country contribute to a range of issues related to the environment on a global scale, resulting from the consumption of products and services. In the evaluation of environmental stressors, such as carbon footprint, land footprint, and WF, within the European Union, it was determined that the consumption patterns of an average European Union citizen in 2004 encompassed the emission of 13.3 tons of carbon dioxide equivalent greenhouse gases, utilization of 2.53 giga hectares of land, and consumption of 179 cubic meters of blue water. From an overall point of view, the European Union has been found to externalize all three categories of environmental stressors to other regions of the world by means of imported products. Poland, France, and Spain emerged as the primary contributors to environmental pressures in terms of net exports. Specifically, Poland was found to be the largest exporter of greenhouse gases, while France was identified as the leading exporter of land-related pressures. In addition, Spain was observed to have the highest net export of freshwater-related pressures. Numerous studies have been undertaken to examine the WF of various economies at global, regional, and national levels. However, a significant research gap exists regarding the WF of distinct economic sectors, which constitute a fundamental component of every economy (Li & Chen, 2014). The objective of this chapter is to review the three primary determinants of WF, which include agriculture, industry, and energy production, within a global context.

4.2 Drivers of water footprint

4.2.1 Agriculture

The consumption of a cup of coffee or tea entails the utilization of a significantly larger volume of water throughout the production process. The global population requires approximately 140 billion cubic meters of water annually for the consumption of coffee and tea, with a significant portion of this quantity being attributed to agricultural activities (Chapagain & Hoekstra, 2007). The agriculture industry has experienced dramatic growth since 1860. According to the research conducted by Duarte et al. (2014), there is substantial evidence indicating a significant strain on water resources due to the growth and expansion of the agricultural industry in Spain during the last 150 years. Agriculture is widely recognized as the primary user of water resources, exerting significant pressure on global water availability and contributing to water-related stress (Mekonnen & Hoekstra, 2020). Consequently, it plays a crucial role in influencing both food and water security. The assessment of long-term sustainability in irrigated farming relies on key parameters such as WF, water scarcity, and water productivity (Xu et al., 2019). WF is contingent upon various factors, such as the specific crop, prevailing climatic conditions, and the particular agricultural production technique (Gerbens-Leenes et al., 2009). The quantity of water essential to food production is contingent upon several parameters, including the number of people, conventional dietary habits, and food production efficiency in relation to water usage. These factors exhibit notable variations across different regions globally (Yang & Cui, 2014). The evaluation carried out by Mekonnen and Hoekstra (2020) revealed that a significant portion, specifically 70%, of the blue WF can be attributed to the cultivation of five specific crops. These crops include wheat, which accounts for 27% of the blue WF, rice at 17%, cotton at 10%, sugarcane at 8%, and fodder at 7% (Fig. 4.1). The regions predominantly characterized by a substantial share of blue WF about crop production are primarily located in the Middle East and Central Asia. According

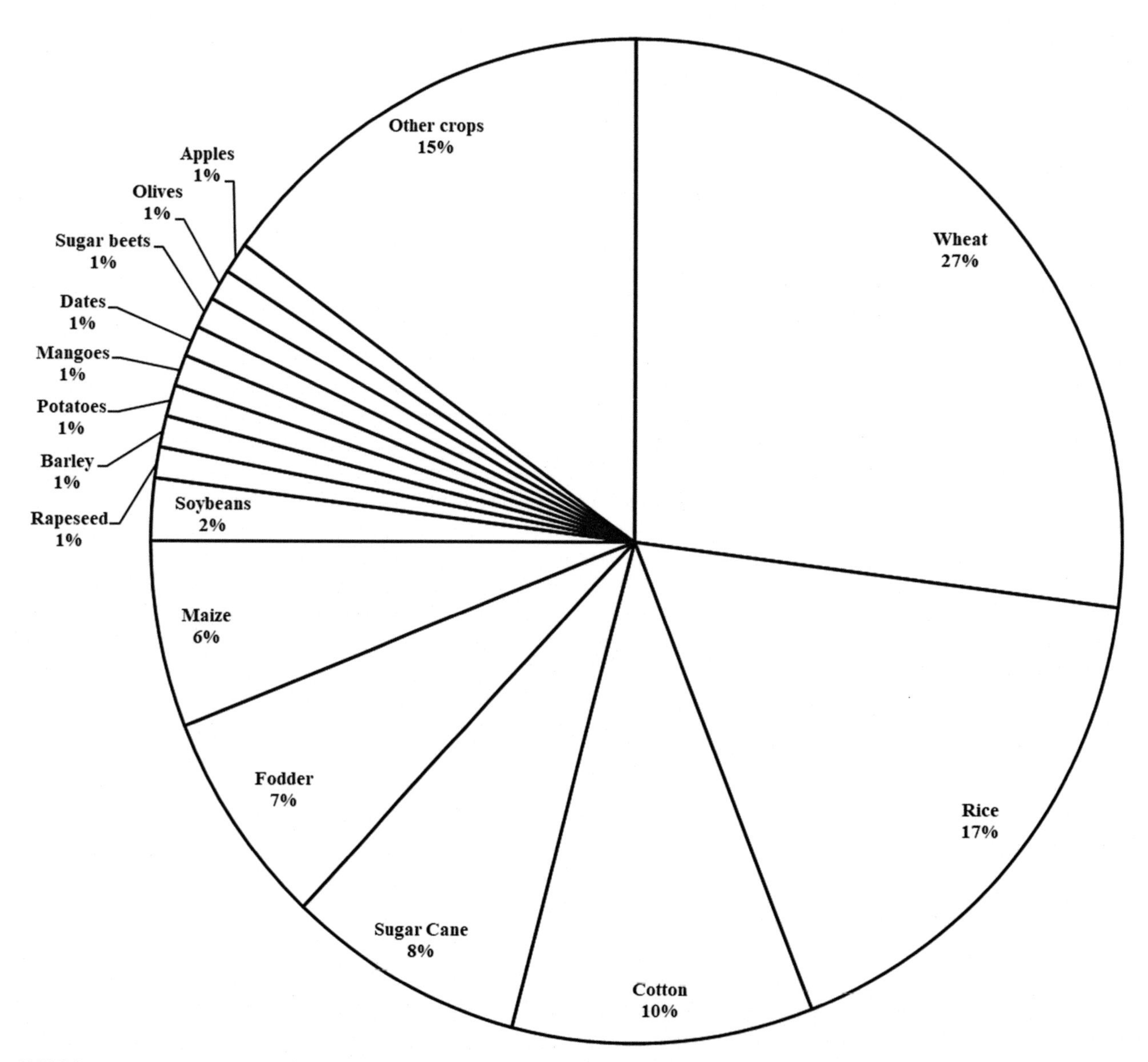

FIGURE 4.1 Share of various plant species in the global unsustainable blue water foootprint of crop production. *Adapted from Mekonnen and Gerbens-Leenes (2020).*

to the data provided, Qatar exhibits the highest proportion of its blue WF categorized as unstable, accounting for around 71%. Subsequently, Uzbekistan, Pakistan, and Turkmenistan followed suit with respective percentages of 68%, 68%, and 67%. Based on certain analyses, it has been determined that the annual worldwide WF associated with crop production from 1996 to 2005 amounted to approximately 7404 billion cubic meters. This WF can be further categorized into three components: 78% green, 12% blue, and 10% gray. The annual WF for wheat was determined to be 1087 billion cubic meters, whereas rice had a WF of 992 billion cubic meters per year and corn had a WF of 770 billion cubic meters per year. Wheat and rice possess the most substantial WFs, collectively constituting 45% of the global WF. At the national level, India had the highest overall WF of 1047 billion cubic meters per year, followed by China with 967 billion cubic meters per year, and the United States with 826 billion cubic meters per year (Mekonnen & Hoekstra, 2011). The estimation of the WF associated with national wheat cultivation was previously carried out on a worldwide basis in a study (Ababaei & Etedali, 2014). The annual global wheat production for the time frame from 1996 to 2005 amounted to approximately 108 billion cubic meters. The majority of the water, approximately 70%, can be classified as green water, while around 19% is categorized as blue water. The remaining 11% is known as gray water. Approximately 18% of the total WF can be attributed to the cultivation of wheat, specifically for export rather than domestic consumption (Mekonnen & Hoekstra, 2010).

The field of WF evaluation is experiencing substantial progress in research; however, there has been limited focus on addressing the uncertainties associated with this topic (Zhuo et al., 2014). The potential adverse effects of climate change on productivity in agriculture have been identified as an important concern within the agriculture sector. However, the specific implications of climate change on the WF of food production in various countries remain uncertain. The examination performed regarding the consequences of climate change between 1980 and 2010 on winter wheat production in Zimbabwe showed that the combined influence of various climate factors, such as temperature, humidity, wind speed, and solar radiation, resulted in a 4% increase in the blue WF of wheat production. In addition, the study recognized a decrease of 6.6% in wheat yield as a consequence of these overall climate-related effects. Consequently, it is plausible that climatic change could be related to the augmentation of the WF (Govere et al., 2020). The WF demonstrates considerable variations across geographic areas and is affected by multiple variables, including climate conditions and farming practices. In an investigation conducted in Thailand, researchers discovered that the WF of sugarcane varied between provinces, with one province exhibiting a WF of 202 cubic meters per ton, while another province had a higher WF of 252 cubic meters per ton (Kongboon & Sampattagul, 2012). The results of WF assessments are influenced by the timing of planting and prevailing meteorological circumstances (Roux et al., 2016). Zhuo et al. (2014) reported findings indicating that the overall WF of a crop is subject to a certain amount of uncertainty. This uncertainty arises from various factors, including reference evaporation and transpiration (ET_0), crop coefficient, precipitation, and crop calendar. The average level of uncertainty was determined to be approximately $\pm 30\%$ at a 95% confidence level. The sensitivity and uncertainty vary across various crop types, and overall, the WF of crops is particularly influenced by ET_0 and crop coefficient. The theory of virtual water content (VWC), alternatively referred to as crop WF, serves as a valuable instrument for examining the interplay between food and water resources in connection to climate, soil, and agricultural methods. The outcomes of the analysis of sensitivity to VWC for various crops indicated that the VWC of wheat exhibits the most sensitivity to the duration of the growing season. In contrast, rice has a greater sensitivity to ET_0. On the other hand, both corn and soybeans have a heightened sensitivity to the timing of crop planting. The sensitivity of VWC reflects variability not just between different crops but also within diverse harvest regions worldwide, even at the local level (Tuninetti et al., 2015). The research study executed by Vanham and Bidoglio (2014) in Milan, Italy, revealed that the city functions as a significant recipient of water resources, mostly due to the importation of agricultural commodities originating from regions outside Milan and Italy. This finding reflects a reliance on water resources from adjacent river basins. According to a WF study conducted in Morocco between 1996 and 2005, it was determined that the utilization of water for evaporation from storage tanks ranked as the second most significant water consumption category, following water crop cultivation. The majority of virtual water exports originating from Morocco are associated with the exportation of products that exhibit comparatively low water financial efficiency (Schyns et al., 2014).

Given that agriculture is the predominant user of water, ensuring food security necessitates the acquisition of comprehensive knowledge regarding the precise water requirements for crop irrigation (Ewaid et al., 2021). The development and adoption of strategies to address water scarcity in arid regions is an essential policy issue, particularly within the context of agriculture (Adetoro et al., 2020). Assessing the impact of water usage in relation to resource availability might assist producers in selecting the optimal timing and strategy for crop cultivation (Deepa et al., 2022). Egypt is situated in dry geographical areas, demanding the implementation of water resource management strategies to effectively preserve limited water supplies and attain a desirable degree of socioeconomic progress and food security. According to the research conducted by El-Marsafawy and Mohamed (2021), the proportion of green water in the overall EF of Egyptian goods was determined to be relatively lower in comparison to blue water. The mean combined WF, encompassing both green and blue water, of Egyptian products was approximately 680 cubic meters per metric ton. Accordingly, the agricultural products in Egypt exhibited lower WF values in comparison to the global average WF. The geographical location of Iran includes arid and semiarid regions, making agricultural water management a matter of significant relevance within the country. The assessment of the green WF in Iran between 2006 and 2012 demonstrated that the cultivation of wheat, barley, and corn accounted for 47%, 42%, and 2% of the overall WF, respectively. Hence, it may be inferred that producers of wheat and barley act as important users of green water resources. This proof emphasizes that there exists a multitude of prospects for enhancing green water production by means of augmenting crop yield, particularly in rainfed areas (Ababaei & Ramezani Etedali, 2017). The augmentation of agricultural output in rainfed regions leads to a decrease in the demand for production in irrigated lands, resulting in a reduction in water usage (Ababaei & Etedali, 2014).

According to the available data, the annual WF attributed to agricultural activities in 207 counties located in North China has shown an upward trend, rising from 53 billion cubic meters in 1986 to 78 billion cubic meters in 2010. Despite acute water scarcity, 173 cities produced large harvests with high WFs (Xu et al., 2019). The period from 1999 to 2014 witnessed an escalation in water scarcity within China's agriculture sector (Xinchun et al., 2017). In the case of Taiwan, it has been observed that the WF associated with rice agriculture is approximately 13 times greater than that of sweet potatoes, and between 8.8 and 10.4 times higher than that of sugarcane. This discrepancy highlights the relatively inefficient utilization of resources in rice production as an agricultural product (Su et al., 2015). An investigation performed in Australia examined the WF of nine different crops, revealing that almonds had the highest WF, measuring 6672 cubic meters per ton, while tomatoes displayed the lowest WF, measuring 212.4 cubic meters per ton. Therefore, Australia remains capable of cultivating tomatoes, carrots, turnips, potatoes, and apples, while it is not advisable to cultivate stone fruits like almonds (Hossain et al., 2021). In 2017, the WF of paddy rice cultivation was estimated in seven provinces of Iraq. The research revealed that the WF of rice is 3072 cubic meters per ton, surpassing the global average of 1325 cubic meters per ton. Hence, it is plausible to substitute rice with other crops, such as vegetables, due to their enhanced economic advantages and reduced water requirements (Ewaid et al., 2021). The province of Nineveh in Iraq had the maximum quantity of wheat WF, with subsequent levels found in the provinces of Al-Muthani, Anbar, and Basra. In the aforementioned provinces, it is possible to substitute wheat cultivation with alternative crops that require lower water consumption and yield greater economic advantages. The rise in production levels in rainfed provinces has led to a decrease in the demand for output from irrigated regions, resulting in a reduction in the utilization of aquifer water (Ewaid et al., 2019). One effective strategy for mitigating the WF is the selection of suitable irrigation systems. The study pointed out that crop cultivation with rain irrigation generally exhibits the highest WF consumption, followed by furrow, drip, and subsurface drip irrigation methods (Chukalla et al., 2015). A study that was carried out in the Malelane region of South Africa revealed that the utilization of irrigation water for sugarcane cultivation, when implemented with a thick mulch cover, exhibited a notable decrease compared to the implementation of a light mulch cover. Accordingly, by the installation of enhanced irrigation systems, the utilization of a substantial mulch cover to mitigate evaporation, and the adoption of efficient irrigation planning, it would be conceivable to enhance the effectiveness of water usage in the cultivation of sugarcane and consequently minimize the associated WF (Adetoro et al., 2020).

Approximately 33% of the WF in agriculture can be attributed to livestock-derived goods. Three primary components contribute to the WF of meat production: feed conversion efficiency, feed composition, and feed origin. The enhancement in productivity is observed as a progression from grazing systems to mixed systems and finally to industrial systems. This can be explained by the fact that animals in manufacturing environments are provided with more concentrated feed, exhibit less mobility, and experience accelerated growth rates. As a result, these factors collectively lead to a decrease in WF, as reported in previous studies. Generally, beef displays a greater total WF compared to pork, whereas pork has a higher WF than chicken. However, the average worldwide blue and gray WFs for these three meat products are similar (Gerbens-Leenes et al., 2013). Assessing the water demands of various consumer items, particularly those that have a significant WF, proves to be beneficial for food manufacturers, dealers, and those who purchase these products (Ercin et al., 2011). The need to mitigate the impact of water usage in the cultivation of irrigated crops has become increasingly crucial in light of the escalating competition for freshwater resources (Chukalla et al., 2015). The execution of WF assessments can provide valuable insights for water resource managers and legislators in establishing appropriate pricing strategies for products. By considering aspects such as plantation area, yield, and total water consumption, this approach enables the identification of optimal pricing levels that can effectively incentivize reduced water usage (Deepa et al., 2022). WF offers a comprehensive perspective on the utilization of freshwater across the supply chain, hence aiding in the formulation of informed decisions (Ewaid et al., 2019).

4.2.2 Industry

The application of the WF idea has been extensively observed in the farming sector, while comparatively less emphasis has been placed on manufacturing products (Nezamoleslami & Mahdi Hosseinian, 2020). The ongoing execution of heavy manufacturing operations has been found to result in water pollution, potentially leading to a water crisis (Wang et al., 2021). The concept of WF encompasses the quantification of freshwater utilization, considering both direct and indirect water consumption by individuals or entities involved in production processes. The practical implementation of the WF concept for commercial enterprises offers helpful insights into the sustainability of their production processes. This study examined the range of water usage, expressed as a WF, associated with the production of 1 kilogram of pasta, which was found to vary between 1.336 and 2.847 liters of water (Ruini et al., 2013). The findings of a case study conducted in Iran revealed that the WF connected with steel production is significantly higher when compared to other industrial goods (Nezamoleslami & Mahdi Hosseinian, 2020). According to a study performed between 2005 and 2014, it was determined that hydraulic fracturing of unconventional shale gas and oil in the United States resulted in a total water consumption of 708 billion liters for unconventional gas extraction and 232 billion liters for shale oil extraction (Kondash & Vengosh, 2015). The concept of the gray WF pertains to the quantification of water volume necessary for assimilating contaminated water, hence serving as an indicator of the extent of water pollution resulting from human actions (Liu et al., 2017). The concept of the gray WF, which quantifies water pollution by assessing the volume of water required to dilute contaminants, has garnered significant interest in light of the imperative to address concerns related to water pollution and scarcity (Zhang et al., 2019). The research project presented by Liu et al. (2020) aimed to assess the environmental sustainability of blue, green, and gray WFs throughout 31 Chinese provinces during 2002, 2007, and 2012. The study revealed a notable rise of almost 30% in the overall WF for the period from 2002 to 2012. The inadequate levels of WF sustainability can be attributed to the existence of water scarcity and pollution, which further intensify the deterioration of rivers and local habitats.

There is a growing recognition among industries of their role in contributing to both water scarcity and contamination. Hence, a growing number of enterprises are progressively recognizing the need to assess their WFs and seek strategies for improving water management (Hoekstra, 2015). According to the study conducted by van Oel and Hoekstra (2012), around 100 liters of water were used in the production of the printed iteration of their research paper. The WF associated with the manufacturing of paper with wood resources is estimated to range from 300 to 2600 cubic meters per metric ton (equivalent to 13.2 liters for each sheet of paper). Therefore, the utilization of recycled paper has the potential to be efficacious in mitigating the WF. The investigation carried out by Hossain and Khan (2020) examined the WF of the textile industry in Bangladesh from 2012 to 2016. The findings of the study revealed that the yearly WF of the sector amounted to 1.8 billion cubic meters. The substantial WF and water pollution associated with the situation have the potential to diminish groundwater levels and pose health risks to the surrounding population. The aggregate WF for prepared-to-wear products was calculated to be around 27.56 billion cubic meters. When appropriate water purifying infrastructure and recycling methods are implemented, it is possible to decrease the volume of gray water to approximately 1.26 billion cubic meters. Miglietta et al. (2015) figured out that, from a freshwater resource viewpoint, it is more effective to derive protein from worms than from other conventionally farmed animals. While the energy efficacy of the Internet, information and communication technologies, and data centers (DC) has been extensively explored, their WF has received comparatively less interest. The study conducted by Ristic et al. (2015) aimed to determine the WF of DC, yielding a range of 1047 to 151,061 m^3/TJ. The WF associated with DC output data traffic ranges from 1 to 205 liters per GB. Liu et al. (2017) highlighted the necessity of establishing consistent water quality criteria for evaluating the gray WF, taking into account various aquatic habitats and the existence of numerous contaminants. To address the issues associated with groundwater depletion in China, various policy measures have been suggested. These include the enhancement of discharge requirements for wastewater treatment plants and the promotion of wastewater recycling (Zhang et al., 2019). To mitigate China's WF, it is essential to enhance the efficiency of wastewater and sludge treatment processes (Shao & Chen, 2013). Morera et al. (2016) revealed that the implementation of secondary treatment and chemical phosphorus removal resulted in a significant reduction of the WF, with reductions of 51.5% and 72.4%, respectively. Therefore, the execution of wastewater treatment methods yields a substantial decrease in the gray WF when compared to the absence of any treatment.

4.2.3 Energy

Water and energy are two vital resources that are intricately connected and cannot be separated from one another. Indeed, the process of energy production necessitates a substantial amount of water, as exemplified by the utilization of

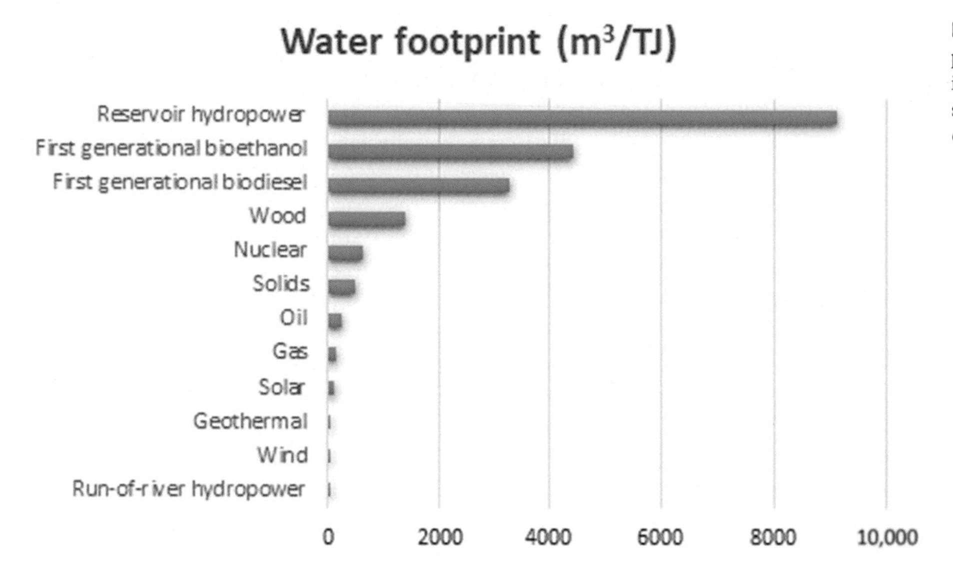

FIGURE 4.2 The average water footprint associated with energy generation in the European Union (EU), excluding soil moisture. *Source: European Commission, EU Science Hub (2019).*

water for cooling in power plants (Okadera et al., 2014). The use of water in hydraulic fracturing has witnessed a substantial surge in key regions of shale oil and gas production within the United States. Notably, the water consumption per well has had a remarkable growth of 770% between 2011 and 2017. The upward trend in the WF of hydraulic fracturing over time indicates that future unconventional operations in the oil and gas industry will demand a greater quantity of water, resulting in an amplified WF (Kondash et al., 2018). The WF associated with energy production in Thailand experienced a noticeable rise from 1986 to 2010, with the latter year exhibiting a footprint approximately nine times greater than the former (Okadera et al., 2014). On a global scale, electricity production is linked to a significant WF (Fig. 4.2). Hydropower constitutes around 16% of global electricity generation. There exists a discourse surrounding the potential water consumption associated with power production. Through the assessment of 35 locations, it has been ascertained that the collective WF of the chosen hydroelectric facilities amounts to approximately 90 billion cubic meters annually. This figure corresponds to 10% of the global production's WF in 2000. As a result, the generation of hydropower entails substantial water consumption. Simultaneously, considerable variations in WFs exist among hydropower plants owing to disparities in the climatic conditions of their respective sites (Mekonnen & Hoekstra, 2012a, 2012b). Hydropower, as a significant kind of renewable energy, exhibits the potential to exert a substantial demand on water resources owing to the evaporation occurring from the surface of the reservoir. This process, in turn, has the potential to contribute to the issue of water scarcity (Scherer & Pfister, 2016).

A research study conducted by Mekonnen et al. (2015) provided an estimation of the annual worldwide WF use for both heating and electricity production, which amounted to 378 billion cubic meters. Europe accounts for the highest share of the world's WF, comprising around 22% of the total. China follows closely behind with a contribution of 15%, while Latin America contributes 14%. The United States and Canada collectively account for 12% of the global WF, while India contributes 9%. The global WF of power and heat experienced a 1.8-fold increase between 2000 and 2012. The application of wind energy, photovoltaics, and geothermal energy has the potential to decrease the carbon footprint associated with power and heat generation. Carbon capture and sequestration (CCS) devices typically include significant water usage in the process of capturing energy. The WF analysis conducted on four CCS technologies, including postcombustion CCS, precombustion CCS, direct air CCS, and bioenergy, demonstrated a range of 0.74–575 cubic meters per metric ton of CO_2. As a consequence, bioenergy technology indicates the most substantial WF, mostly attributable to its significant transpiration-related water demands. The deployment of CCS on a large scale to achieve the 1.5°C climate target would result in an almost twofold increase in the anthropogenic WF. Climate change mitigation scenarios relying on a varied range of CCS technologies have reduced effects on water resources compared to scenarios that predominantly rely on a single CCS technology. Evaluating the WF of CCS technologies is a crucial aspect to consider when assessing their viability. In places where water scarcity is a concern, it is advisable to prioritize CCS technologies that are specifically designed to minimize water usage (Rosa et al., 2021). Furthermore, it is crucial to take into account the potential for relocating the manufacturing of each product to regions that offer the highest level of efficiency (Nyambo & Wakindiki, 2015).

4.2.4 Concluding remarks

The most pressing environmental issues of the 21st century are the widespread scarcity of water, contamination of water sources, and the depletion of freshwater reservoirs. The primary emphasis of these concerns pertains to the excessive utilization of water by human beings. However, for what purposes do we use this huge water resource? The utilization of the idea of WF can provide insights toward addressing this inquiry (Hogeboom, 2020). WF is a recently developed metric that quantifies the overall amount of fresh water used in the context of production processes (Pellicer-Martínez and Martínez-Paz, 2016). Present developments in gaining knowledge of the WF concept have demonstrated its potential as a valuable tool for attaining the Sustainable Development Goals (SDGs; Berger et al., 2021). At the same time, it has been determined that SDG 6, while covering the matter of water usage efficiency and sustainability, does not encompass any objectives pertaining to the enhanced utilization of green water and the equitable distribution of water resources (Hoekstra et al., 2017). One of the primary obstacles in the field of water management involves the assessment of how the present water utilization practices can influence the future availability of this resource for subsequent generations, hence affecting its long-term viability (Pellicer-Martínez and Martínez-Paz, 2016).

Agriculture is widely acknowledged on a global scale as a significant source of greenhouse gas emissions, encompassing CO_2, CH_4, and N_2O. Farming is the primary user of fresh water, constituting around 70% of global water withdrawals. Hence, it is critically important to move toward sustainable agricultural and food production systems at the national, regional, and global scales. This shift should involve the adoption of methods that significantly reduce greenhouse gas emissions while simultaneously ensuring the preservation of yield and quality (Pellegrini et al., 2016). According to the findings of Harris et al. (2020), the consumption of animal-derived foods significantly contributes to the overall green WF of dietary patterns. Grains, fruits, nuts, and oils significantly contribute to the WF of diets. The research findings indicate that adopting healthier dietary patterns can lead to a decrease in overall water usage in the agricultural sector. However, it is important to note that such modifications do not have a significant impact on water consumption. The necessity for enhanced comprehension of the quantity and nature of water used in food production is accentuated by the swift fluctuations in diets and escalating apprehensions about water security. This imperative arises to facilitate well-informed policy determinations. The escalation in the consumption of animal-derived goods is anticipated to exert additional strain on global freshwater supplies. Approximately 33% of the global agricultural WF is attributed to the manufacturing of animal-based commodities. The WF associated with the production of animal-based food products exceeds that of crop-based food products with comparable nutritional content. The WF per calorie for beef is significantly larger, almost 20 times compared to that of cereals and starchy roots. The WF per gram of protein for milk, eggs, and chicken meat exhibits a 1.5-fold increase compared to that of legumes. Animal products derived from factories tend to exhibit a higher utilization of groundwater and surface water resources compared to animal products sourced from grazing systems. This disparity is mostly attributed to the increased blue and gray WFs associated with industrial systems. The anticipated rise in worldwide consumption of meat and the heightened intensity of animal production systems are expected to exert additional strain on freshwater resources in the forthcoming decades (Mekonnen & Hoekstra, 2012a, 2012b). Souissi et al. (2019) employed the Water Footprint Network methodology to examine the WF associated with food intake across households in Tunisia. The findings revealed a notable surge of 31% in the average WF for the primary food groups consumed, rising from 1208 cubic meters per person per year in recent decades. The per capita annual consumption of cubic meters in 1985 increased to 1586 in 2010. The WF of urban meals exceeds that of rural ones. To promote food security and water management approaches, policymakers should implement several measures pertaining to price policies, consumer awareness, and modifications in the agricultural system. The WF of tourism is significantly influenced by food intake, wherein meals high in calories are about five times more water intensive for foreign tourists compared to vegetable-based diets. According to reports, tourism in Jing-jin-jiby, China, is projected to contribute a substantial portion of the overall water usage in the region by 2020. Tourism patterns play an important role in influencing water use (Li, 2018). Furthermore, according to available data, tourism contributes to over 10% of Spain's GDP, whereas the WF associated with foreign tourists in Spain is estimated to be approximately 3.7 cubic kilometers (Cazcarro et al., 2014).

The comprehension of water usage holds significant importance in the realm of sustainable water resource management. The demand for government initiatives that guarantee efficiency and justice in water resources is expected to increase in response to climate change projections that anticipate significant fluctuations in water availability (Novoa, Ahumada-Rudolph, Rojas, Munizaga et al., 2019; Novoa, Ahumada-Rudolph, Rojas, Sáez et al., 2019). The transition toward renewable energy is being motivated by growing apprehensions regarding the security of energy and climate change. It is anticipated that the transportation sector will undergo a transition from its reliance on fossil fuels to the utilization of fuel mixes containing a greater proportion of biofuels, specifically bioethanol and biodiesel (Gerbens-Leenes et al., 2012). Over the course of the upcoming decades, there is projected to be a rise in the global requirement for freshwater resources to cater to the escalating needs for food and biofuel commodities (Mekonnen & Hoekstra, 2014). The manufacturing of biofuels necessitates the utilization of water, a resource that is limited in availability. Based on the announced pledges scenario (APS) by international energy agency (IEA), it is projected that the worldwide production of biofuels will experience a growth of over 10 times between 2005 and 2030. The collective contributions of the United States, China, and Brazil will account for 50% of the worldwide biofuel WF. Biofuel WFs provide a substantial contribution to water scarcity in numerous countries (Gerbens-Leenes et al., 2012). Consequently, conducting a WF evaluation serves as the initial stage in the implementation of a nationwide initiative aimed at adopting ecologically sustainable biofuels (Dominguez-Faus et al., 2009).

The adoption of equitable water use is a critical necessity that needs collaborative political efforts from all nations concerned (Aldaya et al., 2010). The sustainable management of water resources is an essential requirement for the advancement of societal progress (Hoekstra et al., 2017). And the establishment of precise water stress monitoring is a crucial requirement for effective water management (Wang et al., 2021). In Europe, education and awareness initiatives, as well as legislation, have traditionally been centered on the goal of reducing the amount of water that is wasted in both home and manufacturing environments. In Europe, education and awareness initiatives, as well as legislation, have traditionally been centered on the goal of reducing the amount of water that is wasted in both home and manufacturing environments. Simultaneously, the implementation of measures to mitigate food waste and the adoption of altered dietary patterns among consumers in Europe can significantly contribute to water conservation in agricultural production operations (Vanham & Bidoglio, 2013). The achievement of a more environmentally friendly, effective, and equitable utilization of resources in the commodity supply chain necessitates the implementation of collective and coordinated efforts across many levels and throughout all stages of the process (Hoekstra et al., 2019). It is imperative to compute the WF of the whole production within the agricultural sector to facilitate sustainable production and effective water management. Moreover, it is crucial to incorporate this information into the formulation of enduring strategies for the farming sector (Yoo et al., 2014). In light of the escalating phenomenon of climate change and the imperative to enhance agricultural productivity, the implementation of suitable management solutions in agricultural areas becomes highly important in mitigating water scarcity issues (Shamshiri et al., 2022). Failure to implement appropriate strategies for sustainable agricultural production and the maintenance of the supply-and-demand equilibrium may lead to a critical situation in terms of public health and food security (Zarei & Rad, 2020). Freshwater scarcity has emerged as a pressing worldwide issue in numerous regions, giving rise to apprehensions regarding global food security and the potential destruction of freshwater ecosystems. The anticipated outcome of this scenario is projected to escalate, as indicated by the estimation made by the FAO that there is a need for a twofold increase in world food production by 2050. Consequently, there is a pressing need for food systems to enhance their water usage efficiency (Ridoutt et al., 2010). It is feasible to achieve sustainable levels of humanity's WF, even in the face of population growth, as long as there are modifications in consumption patterns and other influential factors such as economic growth, production, and trade patterns (Ercin & Hoekstra, 2014).

References

Ababaei, B., & Etedali, H. R. (2014). Estimation of water footprint components of Iran's wheat production: Comparison of global and national scale estimates. *Environmental Processes*, *1*(3), 193–205. Available from https://doi.org/10.1007/s40710-014-0017-7, http://www.springer.com/earth-+sciences+and+geography/environmental+science+%26+engineering/journal/40710.

Ababaei, B., & Ramezani Etedali, H. (2017). Water footprint assessment of main cereals in Iran. *Agricultural Water Management*, *179*, 401–411. Available from https://doi.org/10.1016/j.agwat.2016.07.016, http://www.journals.elsevier.com/agricultural-water-management/.

Adetoro, A. A., Abraham, S., Paraskevopoulos, A. L., Owusu-Sekyere, E., Jordaan, H., & Orimoloye, I. R. (2020). Alleviating water shortages by decreasing water footprint in sugarcane production: The impacts of different soil mulching and irrigation systems in South Africa. *Groundwater for Sustainable Development*, *11*. Available from https://doi.org/10.1016/j.gsd.2020.100464, http://www.journals.elsevier.com/groundwater-for-sustainable-development/.

Aldaya, M., Muñoz, G., & Hoekstra, A. (2010). Water footprint of cotton, wheat and rice production in Central Asia: UNESCO-IHE Delft. *The Netherlands*. Available from https://www.researchgate.net/publication/254884133_Water_footprint_of_cotton_wheat_and_rice_production_in_Central_Asia.

Berger, M., Campos, J., Carolli, M., Dantas, I., Forin, S., Kosatica, E., Kramer, A., Mikosch, N., Nouri, H., Schlattmann, A., Schmidt, F., Schomberg, A., & Semmling, E. (2021). Advancing the water footprint into an instrument to support achieving the SDGs—Recommendations from the "Water as a Global Resources" Research Initiative (GRoW). *Water Resources Management*, *35*(4), 1291–1298. Available from https://doi.org/10.1007/s11269-021-02784-9, http://www.wkap.nl/journalhome.htm/0920-4741.

Cazcarro, I., Hoekstra, A. Y., & Sánchez Chóliz, J. (2014). The water footprint of tourism in Spain. *Tourism Management*, *40*, 90–101. Available from https://doi.org/10.1016/j.tourman.2013.05.010, http://www.elsevier.com/inca/publications/store/3/0/4/7/2/.

Chapagain, A. K., & Hoekstra, A. Y. (2007). The water footprint of coffee and tea consumption in the Netherlands. *Ecological Economics*, *64*(1), 109–118. Available from https://doi.org/10.1016/j.ecolecon.2007.02.022.

Chenoweth, J., Hadjikakou, M., & Zoumides, C. (2014). Quantifying the human impact on water resources: A critical review of the water footprint concept. *Hydrology and Earth System Sciences*, *18*(6), 2325–2342. Available from https://doi.org/10.5194/hess-18-2325-2014, http://www.hydrol-earth-syst-sci.net/volumes_and_issues.html.

Chouchane, H., Hoekstra, A. Y., Krol, M. S., & Mekonnen, M. M. (2015). The water footprint of Tunisia from an economic perspective. *Ecological Indicators*, *52*, 311–319. Available from https://doi.org/10.1016/j.ecolind.2014.12.015, http://www.elsevier.com/locate/ecolind.

Chukalla, A. D., Krol, M. S., & Hoekstra, A. Y. (2015). Green and blue water footprint reduction in irrigated agriculture: Effect of irrigation techniques, irrigation strategies and mulching. *Hydrology and Earth System Sciences*, *19*(12), 4877–4891. Available from https://doi.org/10.5194/hess-19-4877-2015, http://www.hydrol-earth-syst-sci.net/volumes_and_issues.html.

Deepa, R., Anandhi, A., Bailey, N. O., Grace, J. M., Betiku, O. C., & Muchovej, J. J. (2022). Potential environmental impacts of peanut using water footprint assessment: A case study in Georgia. *Agronomy*, *12*(4), 20734395. Available from https://doi.org/10.3390/agronomy12040930, https://www.mdpi.com/2073-4395/12/4/930/pdf.

Dominguez-Faus, R., Powers, S. E., Burken, J. G., & Alvarez, P. J. (2009). The water footprint of biofuels: A drink or drive issue? *Environmental Science and Technology*, *43*(9), 3005–3010. Available from https://doi.org/10.1021/es802162x, http://pubs.acs.org/doi/pdfplus/10.1021/es802162x, United States.

Dong, H., Geng, Y., Sarkis, J., Fujita, T., Okadera, T., & Xue, B. (2013). Regional water footprint evaluation in China: A case of Liaoning. *Science of the Total Environment*, *442*, 215–224. Available from https://doi.org/10.1016/j.scitotenv.2012.10.049.

Duarte, R., Pinilla, V., & Serrano, A. (2014). The water footprint of the Spanish agricultural sector: 1860–2010. *Ecological Economics*, *108*, 200–207. Available from https://doi.org/10.1016/j.ecolecon.2014.10.020, http://www.elsevier.com/inca/publications/store/5/0/3/3/0/5.

El-Marsafawy, S. M., & Mohamed, A. I. (2021). Water footprint of Egyptian crops and its economics. *Alexandria Engineering Journal*, *60*(5), 4711–4721. Available from https://doi.org/10.1016/j.aej.2021.03.019, http://www.elsevier.com/wps/find/journaldescription.cws_home/724292/description#description.

Ercin, A. E., Aldaya, M. M., & Hoekstra, A. Y. (2011). Corporate water footprint accounting and impact assessment: The case of the water footprint of a sugar-containing carbonated beverage. *Water Resources Management*, *25*(2), 721–741. Available from https://doi.org/10.1007/s11269-010-9723-8, http://www.wkap.nl/journalhome.htm/0920-4741.

Ercin, A. E., & Hoekstra, A. Y. (2014). Water footprint scenarios for 2050: A global analysis. *Environment International*, *64*, 71–82. Available from https://doi.org/10.1016/j.envint.2013.11.019, http://www.elsevier.com/locate/envint.

Ewaid, S., Abed, S., & Al-Ansari, N. (2019). Water footprint of wheat in Iraq. *Water*, *11*(3), 535. Available from https://doi.org/10.3390/w11030535.

Ewaid, S. H., Abed, S. A., Chabuk, A., & Al-Ansari, N. (2021). Water footprint of rice in Iraq. *IOP Conference Series: Earth and Environmental Science*, *722*(1), 012008. Available from https://doi.org/10.1088/1755-1315/722/1/012008.

Ewing, B. R., Hawkins, T. R., Wiedmann, T. O., Galli, A., Ertug Ercin, A., Weinzettel, J., & Steen-Olsen, K. (2012). Integrating ecological and water footprint accounting in a multi-regional input–output framework. *Ecological Indicators*, *23*, 1–8. Available from https://doi.org/10.1016/j.ecolind.2012.02.025.

Fang, K., Heijungs, R., & De Snoo, G. R. (2014). Theoretical exploration for the combination of the ecological, energy, carbon, and water footprints: Overview of a footprint family. *Ecological Indicators*, *36*, 508–518. Available from https://doi.org/10.1016/j.ecolind.2013.08.017, http://www.elsevier.com/locate/ecolind.

Fridman, D., Biran, N., & Kissinger, M. (2021). Beyond blue: An extended framework of blue water footprint accounting. *Science of the Total Environment*, *777*, 146010. Available from https://doi.org/10.1016/j.scitotenv.2021.146010.

Galli, A., Wiedmann, T., Ercin, E., Knoblauch, D., Ewing, B., & Giljum, S. (2012). Integrating ecological, carbon and water footprint into a "footprint family" of indicators: Definition and role in tracking human pressure on the planet. *Ecological Indicators*, *16*, 100–112. Available from https://doi.org/10.1016/j.ecolind.2011.06.017.

García Morillo, J., Rodríguez Díaz, J. A., Camacho, E., & Montesinos, P. (2015). Linking water footprint accounting with irrigation management in high value crops. *Journal of Cleaner Production*, *87*(1), 594–602. Available from https://doi.org/10.1016/j.jclepro.2014.09.043.

Gerbens-Leenes, P. W., Hoekstra, A. Y., & van der Meer, T. (2009). The water footprint of energy from biomass: A quantitative assessment and consequences of an increasing share of bio-energy in energy supply. *Ecological Economics*, *68*(4), 1052–1060. Available from https://doi.org/10.1016/j.ecolecon.2008.07.013.

Gerbens-Leenes, P. W., Mekonnen, M. M., & Hoekstra, A. Y. (2013). The water footprint of poultry, pork and beef: A comparative study in different countries and production systems. *Water Resources and Industry*, *1–2*, 25–36. Available from https://doi.org/10.1016/j.wri.2013.03.001.

Gerbens-Leenes, P. W., van Lienden, A. R., Hoekstra, A. Y., & van der Meer, Th. H. (2012). Biofuel scenarios in a water perspective: The global blue and green water footprint of road transport in 2030. *Global Environmental Change*, *22*(3), 764–775. Available from https://doi.org/10.1016/j.gloenvcha.2012.04.001.

De Girolamo, A. M., Miscioscia, P., Politi, T., & Barca, E. (2019). Improving grey water footprint assessment: Accounting for uncertainty. *Ecological Indicators*, *102*, 822–833. Available from https://doi.org/10.1016/j.ecolind.2019.03.040, http://www.elsevier.com/locate/ecolind.

Govere, S., Nyamangara, J., & Nyakatawa, E. Z. (2020). Climate change signals in the historical water footprint of wheat production in Zimbabwe. *Science of The Total Environment*, *742*140473. Available from https://doi.org/10.1016/j.scitotenv.2020.140473.

Harris, F., Moss, C., Joy, E. J. M., Quinn, R., Scheelbeek, P. F. D., Dangour, A. D., & Green, R. (2020). The water footprint of diets: A global systematic review and meta-analysis. *Advances in Nutrition*, *11*(2), 375–386. Available from https://doi.org/10.1093/advances/nmz091, http://advances.nutrition.org/.

Hoekstra, A. Y. (2009). Human appropriation of natural capital: A comparison of ecological footprint and water footprint analysis. *Ecological Economics*, *68*(7), 1963–1974. Available from https://doi.org/10.1016/j.ecolecon.2008.06.021.

Hoekstra, A. Y., & Chapagain, A. K. (2007a). The water footprints of Morocco and the Netherlands: Global water use as a result of domestic consumption of agricultural commodities. *Ecological Economics*, *64*(1), 143–151. Available from https://doi.org/10.1016/j.ecolecon.2007.02.023.

Hoekstra, A. Y., & Chapagain, A. K. (2007b). Water footprints of nations: Water use by people as a function of their consumption pattern. *Water Resources Management (Netherlands)*, *21*, 35–48. Available from https://doi.org/10.1007/s11269-006-9039-x, January 1, 2007.

Hoekstra, A. Y., Chapagain, A. K., & van Oel, P. R. (2019). Progress in water footprint assessment: Towards collective action in water governance. *Water (Switzerland)*, *11*(5). Available from https://doi.org/10.3390/w11051070, https://res.mdpi.com/water/water-11-01070/article_deploy/water-11-01070.pdf?filename = &attachment = 1.

Hoekstra, A., Chapagain, A., & van Oel, P. (2017). Advancing water footprint assessment research: Challenges in monitoring progress towards sustainable development goal 6. *Water*, *9*(6), 438. Available from https://doi.org/10.3390/w9060438.

Hoekstra, A. Y., Mekonnen, M. M., Chapagain, A. K., Mathews, R. E., & Richter, B. D. (2012). Global monthly water scarcity: Blue water footprints versus blue water availability. *PLoS One*, *7*(2), 19326203, http://www.plosone.org/article/fetchObjectAttachment.action?uri = info%3Adoi%2F10.1371%2Fjournal.pone.0032688&representation = PDF. doi: 10.1371/journal.pone.0032688.

Hoekstra, A. Y. (2015). *The water footprint of industry. Assessing and Measuring Environmental Impact and Sustainability* (pp. 221–254). Netherlands: Elsevier Inc. Available from http://www.sciencedirect.com/science/book/9780127999685, http://doi.org/10.1016/B978-0-12-799968-5.00007-5.

Hogeboom, R. J. (2020). The water footprint concept and water's grand environmental challenges. *One Earth*, *2*(3), 218–222. Available from https://doi.org/10.1016/j.oneear.2020.02.010, http://www.cell.com/one-earth.

Hossain, I., Imteaz, M. A., & Khastagir, A. (2021). Water footprint: Applying the water footprint assessment method to Australian agriculture. *Journal of the Science of Food and Agriculture*, *101*(10), 4090–4098. Available from https://doi.org/10.1002/jsfa.11044, http://onlinelibrary.wiley.com/journal/10.1002/(ISSN)1097-0010.

Hossain, L., & Khan, M. S. (2020). Water footprint management for sustainable growth in the Bangladesh apparel sector. *Water*, *12*(10), 2760. Available from https://doi.org/10.3390/w12102760.

Huang, J., Ridoutt, B. G., Zhang, H., Xu, C., & Chen, F. (2014). Water footprint of cereals and vegetables for the Beijing market. *Journal of Industrial Ecology*, *18*(1), 40–48. Available from https://doi.org/10.1111/jiec.12037.

Jorrat, Md. M., Araujo, P. Z., & Mele, F. D. (2018). Sugarcane water footprint in the province of Tucumán, Argentina. Comparison between different management practices. *Journal of Cleaner Production*, *188*, 521–529. Available from https://doi.org/10.1016/j.jclepro.2018.03.242.

Konar, M., & Marston, L. (2020). The water footprint of the United States. *Water*, *12*(11), 3286. Available from https://doi.org/10.3390/w12113286.

Kondash, A. J., Lauer, N. E., & Vengosh, A. (2018). The intensification of the water footprint of hydraulic fracturing. *Science Advances*, *4*(8). Available from https://doi.org/10.1126/sciadv.aar5982, http://advances.sciencemag.org/content/4/8/eaar5982.

Kondash, A., & Vengosh, A. (2015). Water footprint of hydraulic fracturing. *Environmental Science and Technology Letters*, *2*(10), 276–280. Available from https://doi.org/10.1021/acs.estlett.5b00211, http://pubs.acs.org/page/estlcu/about.html.

Kongboon, R., & Sampattagul, S. (2012). The water footprint of sugarcane and cassava in northern Thailand. *Procedia—Social and Behavioral Sciences*, *40*, 451–460. Available from https://doi.org/10.1016/j.sbspro.2012.03.215.

Lamastra, L., Suciu, N. A., Novelli, E., & Trevisan, M. (2014). A new approach to assessing the water footprint of wine: An Italian case study. *Science of the Total Environment*, *490*, 748–756. Available from https://doi.org/10.1016/j.scitotenv.2014.05.063.

Li, J. (2018). Scenario analysis of tourism's water footprint for China's Beijing–Tianjin–Hebei region in 2020: Implications for water policy. *Journal of Sustainable Tourism*, *26*(1), 127–145. Available from https://doi.org/10.1080/09669582.2017.1326926, http://www.tandfonline.com/toc/rsus20/current.

Li, J. S., & Chen, G. Q. (2014). Water footprint assessment for service sector: A case study of gaming industry in water scarce Macao. *Ecological Indicators*, *47*, 164–170. Available from https://doi.org/10.1016/j.ecolind.2014.01.034, http://www.elsevier.com/locate/ecolind.

Liu, W., Antonelli, M., Liu, X., & Yang, H. (2017). Towards improvement of grey water footprint assessment: With an illustration for global maize cultivation. *Journal of Cleaner Production*, *147*, 1–9. Available from https://doi.org/10.1016/j.jclepro.2017.01.072, https://www.journals.elsevier.com/journal-of-cleaner-production.

Liu, J., Zhao, D., Mao, G., Cui, W., Chen, H., & Yang, H. (2020). Environmental sustainability of water footprint in Mainland China. *Geography and Sustainability*, *1*(1), 8–17. Available from https://doi.org/10.1016/j.geosus.2020.02.002, http://www.journals.elsevier.com/geography-and-sustainability.

Manzardo, A., Mazzi, A., Loss, A., Butler, M., Williamson, A., & Scipioni, A. (2016). Lessons learned from the application of different water footprint approaches to compare different food packaging alternatives. *Journal of Cleaner Production*, *112*, 4657–4666. Available from https://doi.org/10.1016/j.jclepro.2015.08.019.

Mekonnen, M. M., & Gerbens-Leenes, W. (2020). The water footprint of global food production. *Water*, *12*(10), 2696. Available from https://doi.org/10.3390/w12102696.

Mekonnen, M. M., Gerbens-Leenes, P. W., & Hoekstra, A. Y. (2015). The consumptive water footprint of electricity and heat: A global assessment. *Environmental Science: Water Research and Technology*, *1*(3), 285–297. Available from https://doi.org/10.1039/c5ew00026b, http://pubs.rsc.org/en/journals/journalissues/ew#!recentarticles&adv.

Mekonnen, M. M., & Hoekstra, A. Y. (2020). Sustainability of the blue water footprint of crops. *Advances in Water Resources, 143*. Available from https://doi.org/10.1016/j.advwatres.2020.103679, http://www.elsevier.com/inca/publications/store/4/2/2/9/1/3/index.htt.

Mekonnen, M. M., & Hoekstra, A. Y. (2011). The green, blue and grey water footprint of crops and derived crop products. *Hydrology and Earth System Sciences, 15*(5), 1577–1600. Available from https://doi.org/10.5194/hess-15-1577-2011.

Mekonnen, M. M., & Hoekstra, A. Y. (2014). Water footprint benchmarks for crop production: A first global assessment. *Ecological Indicators, 46*, 214–223. Available from https://doi.org/10.1016/j.ecolind.2014.06.013, http://www.elsevier.com/locate/ecolind.

Mekonnen, M. M., & Hoekstra, A. Y. (2012a). The blue water footprint of electricity from hydropower. *Hydrology and Earth System Sciences, 16*(1), 179–187. Available from https://doi.org/10.5194/hess-16-179-2012.

Mekonnen, M. M., & Hoekstra, A. Y. (2012b). A global assessment of the water footprint of farm animal products. *Ecosystems, 15*(3), 401–415. Available from https://doi.org/10.1007/s10021-011-9517-8.

Mekonnen, M. M., & Hoekstra, A. Y. (2010). A global and high-resolution assessment of the green, blue and grey water footprint of wheat. *Hydrology and Earth System Sciences, 14*(7), 1259–1276. Available from https://doi.org/10.5194/hess-14-1259-2010.

Miglietta, P., De Leo, F., Ruberti, M., & Massari, S. (2015). Mealworms for food: A water footprint perspective. *Water, 7*(11), 6190–6203. Available from https://doi.org/10.3390/w7116190.

Morera, S., Corominas, L., Poch, M., Aldaya, M. M., & Comas, J. (2016). Water footprint assessment in wastewater treatment plants. *Journal of Cleaner Production, 112*, 4741–4748. Available from https://doi.org/10.1016/j.jclepro.2015.05.102.

Nezamoleslami, R., & Mahdi Hosseinian, S. (2020). An improved water footprint model of steel production concerning virtual water of personnel: The case of Iran. *Journal of Environmental Management, 260*, 110065. Available from https://doi.org/10.1016/j.jenvman.2020.110065.

Novoa, V., Ahumada-Rudolph, R., Rojas, O., Munizaga, J., Sáez, K., & Arumí, J. L. (2019a). Sustainability assessment of the agricultural water footprint in the Cachapoal River basin, Chile. *Ecological Indicators, 98*, 19–28. Available from https://doi.org/10.1016/j.ecolind.2018.10.048, http://www.elsevier.com/locate/ecolind.

Novoa, V., Ahumada-Rudolph, R., Rojas, O., Sáez, K., de la Barrera, F., & Arumí, J. L. (2019b). Understanding agricultural water footprint variability to improve water management in Chile. *Science of the Total Environment, 670*, 188–199. Available from https://doi.org/10.1016/j.scitotenv.2019.03.127, http://www.elsevier.com/locate/scitotenv.

Nyambo, P., & Wakindiki, I. I. C. (2015). Water footprint of growing vegetables in selected smallholder irrigation schemes in South Africa. *Water SA, 41*(4), 571. Available from https://doi.org/10.4314/wsa.v41i4.17.

Okadera, T., Chontanawat, J., & Gheewala, S. H. (2014). Water footprint for energy production and supply in Thailand. *Energy, 77*, 49–56. Available from https://doi.org/10.1016/j.energy.2014.03.113, https://www.journals.elsevier.com/energy.

Pellegrini, G., Ingrao, C., Camposeo, S., Tricase, C., Contò, F., & Huisingh, D. (2016). Application of water footprint to olive growing systems in the Apulia region: A comparative assessment. *Journal of Cleaner Production, 112*, 2407–2418. Available from https://doi.org/10.1016/j.jclepro.2015.10.088, https://www.journals.elsevier.com/journal-of-cleaner-production.

Pellicer-Martínez, F., & Martínez-Paz, J. M. (2016). The Water Footprint as an indicator of environmental sustainability in water use at the river basin level. *Science of the Total Environment, 571*, 561–574. Available from https://doi.org/10.1016/j.scitotenv.2016.07.022, http://www.elsevier.com/locate/scitotenv.

Ridoutt, B. G., Juliano, P., Sanguansri, P., & Sellahewa, J. (2010). The water footprint of food waste: Case study of fresh mango in Australia. *Journal of Cleaner Production, 18*(16–17), 1714–1721. Available from https://doi.org/10.1016/j.jclepro.2010.07.011.

Ristic, B., Madani, K., & Makuch, Z. (2015). The water footprint of data centers. *Sustainability, 7*(8), 11260–11284. Available from https://doi.org/10.3390/su70811260.

Rosa, L., Sanchez, D. L., Realmonte, G., Baldocchi, D., & D'Odorico, P. (2021). The water footprint of carbon capture and storage technologies. *Renewable and Sustainable Energy Reviews, 138*, 18790690. Available from https://doi.org/10.1016/j.rser.2020.110511, https://www.journals.elsevier.com/renewable-and-sustainable-energy-reviews.

Roux, B., van der Laan, M., Vahrmeijer, T., Annandale, J., & Bristow, K. (2016). Estimating water footprints of vegetable crops: Influence of growing season, solar radiation data and functional unit. *Water, 8*(10), 473. Available from https://doi.org/10.3390/w8100473.

Ruini, L., Marino, M., Pignatelli, S., Laio, F., & Ridolfi, L. (2013). Water footprint of a large-sized food company: The case of Barilla pasta production. *Water Resources and Industry, 1–2*, 7–24. Available from https://doi.org/10.1016/j.wri.2013.04.002.

Scherer, L., & Pfister, S. (2016). Global water footprint assessment of hydropower. *Renewable Energy, 99*, 711–720. Available from https://doi.org/10.1016/j.renene.2016.07.021, http://www.journals.elsevier.com/renewable-and-sustainable-energy-reviews/.

Schyns, J. F., Hoekstra, A. Y., & Magar, V. (2014). The added value of water footprint assessment for national water policy: A case study for Morocco. *PLoS ONE. 1932–6203, 9*(6), e99705. Available from https://doi.org/10.1371/journal.pone.0099705.

Scown, C. D., Horvath, A., & McKone, T. E. (2011). Water footprint of U.S. transportation fuels. *Environmental Science and Technology, 45*(7), 2541–2553. Available from https://doi.org/10.1021/es102633h.

Shao, L., & Chen, G. Q. (2013). Water footprint assessment for wastewater treatment: Method, indicator, and application. *Environmental Science and Technology, 47*(14), 7787–7794. Available from https://doi.org/10.1021/es402013t, http://pubs.acs.org/journal/esthag.

Shamshiri, R.R., Balasundram, S.K., Rad, A.K., Sultan, M., & Hameed, I.A. (2022). An overview of soil moisture and salinity sensors for digital agriculture applications. IntechOpen. doi: 10.5772/intechopen.103898.

Sommaruga, R., & Eldridge, H. M. (2021). Avocado production: Water footprint and socio-economic implications. *EuroChoices, 20*(2), 48–53. Available from https://doi.org/10.1111/1746-692X.12289, http://onlinelibrary.wiley.com/journal/10.1111/(ISSN)1746-692X.

Souissi, A., Mtimet, N., Thabet, C., Stambouli, T., & Chebil, A. (2019). Impact of food consumption on water footprint and food security in Tunisia. *Food Security*, *11*(5), 989–1008. Available from https://doi.org/10.1007/s12571-019-00966-3, http://www.springer.com/life + sci/agriculture/journal/12571.

Su, M. H., Huang, C. H., Li, W. Y., Tso, C. T., & Lur, H. S. (2015). Water footprint analysis of bioethanol energy crops in Taiwan. *Journal of Cleaner Production*, *88*, 132–138. Available from https://doi.org/10.1016/j.jclepro.2014.06.020, https://www.journals.elsevier.com/journal-of-cleaner-production.

Tuninetti, M., Tamea, S., D'Odorico, P., Laio, F., & Ridolfi, L. (2015). Global sensitivity of high-resolution estimates of crop water footprint. *Water Resources Research*, *51*(10), 8257–8272. Available from https://doi.org/10.1002/2015WR017148, http://onlinelibrary.wiley.com/journal/10.1002/(ISSN)1944-7973.

van Oel, P. R., & Hoekstra, A. Y. (2012). Towards quantification of the water footprint of paper: A first estimate of its consumptive component. *Water Resources Management*, *26*(3), 733–749. Available from https://doi.org/10.1007/s11269-011-9942-7.

van Oel, P. R., Mekonnen, M. M., & Hoekstra, A. Y. (2009). The external water footprint of the Netherlands: Geographically-explicit quantification and impact assessment. *Ecological Economics*, *69*(1), 82–92. Available from https://doi.org/10.1016/j.ecolecon.2009.07.014.

Vanham, D., & Bidoglio, G. (2013). A review on the indicator water footprint for the EU28. *Ecological Indicators*, *26*, 61–75. Available from https://doi.org/10.1016/j.ecolind.2012.10.021.

Vanham, D., & Bidoglio, G. (2014). The water footprint of Milan. *Water Science and Technology*, *69*(4), 789–795. Available from https://doi.org/10.2166/wst.2013.759Italy, http://www.iwaponline.com/wst/06904/0789/069040789.pdf.

Wang, Q., & Ge, S. (2020). Carbon footprint and water footprint in China: Similarities and differences. *Science of the Total Environment*, *739*, 140070. Available from https://doi.org/10.1016/j.scitotenv.2020.140070.

Wang, Z., Huang, K., Yang, S., & Yu, Y. (2013). An input–output approach to evaluate the water footprint and virtual water trade of Beijing, China. *Journal of Cleaner Production*, *42*, 172–179. Available from https://doi.org/10.1016/j.jclepro.2012.11.007.

Wang, D., Hubacek, K., Shan, Y., Gerbens-Leenes, W., & Liu, J. (2021). A review of water stress and water footprint accounting. *Water*, *13*(2), 201. Available from https://doi.org/10.3390/w13020201.

Xinchun, C., Mengyang, W., Xiangping, G., Yalian, Z., Yan, G., Nan, W., & Weiguang, W. (2017). Assessing water scarcity in agricultural production system based on the generalized water resources and water footprint framework. *Science of the Total Environment*, *609*, 587–597. Available from https://doi.org/10.1016/j.scitotenv.2017.07.191, http://www.elsevier.com/locate/scitotenv.

Xu, Z., Chen, X., Wu, S. R., Gong, M., Du, Y., Wang, J., Li, Y., & Liu, J. (2019). Spatial-temporal assessment of water footprint, water scarcity and crop water productivity in a major crop production region. *Journal of Cleaner Production*, *224*, 375–383. Available from https://doi.org/10.1016/j.jclepro.2019.03.108, https://www.journals.elsevier.com/journal-of-cleaner-production.

Yang, C., & Cui, X. (2014). Global changes and drivers of the water footprint of food consumption: A historical analysis. *Water*, *6*(5), 1435–1452. Available from https://doi.org/10.3390/w6051435.

Yang, Z., Liu, H., Xu, X., & Yang, T. (2016). Applying the water footprint and dynamic structural decomposition analysis on the growing water use in china during 1997–2007. *Ecological Indicators*, *60*, 634–643. Available from https://doi.org/10.1016/j.ecolind.2015.08.010, http://www.elsevier.com/locate/ecolind.

Yoo, S. H., Choi, J. Y., Lee, S. H., & Kim, T. (2014). Estimating water footprint of paddy rice in Korea. *Paddy and Water Environment*, *12*(1), 43–54. Available from https://doi.org/10.1007/s10333-013-0358-2, http://www.springerlink.com/content/1611-2490.

Zarei, M., & Rad, A. K. (2020). Covid-19, challenges and recommendations in agriculture. *Journal of Botanical Research*, *2*(1), 12–15. Available from https://doi.org/10.30564/jrb.v2i1.1841.

Zhan-Ming, C., & Chen, G. Q. (2013). Virtual water accounting for the globalized world economy: National water footprint and international virtual water trade. *Ecological Indicators*, *28*, 142–149. Available from https://doi.org/10.1016/j.ecolind.2012.07.024.

Zhang, L., Dong, H., Geng, Y., & Francisco, M. J. (2019). China's provincial grey water footprint characteristic and driving forces. *Science of the Total Environment*, *677*, 427–435. Available from https://doi.org/10.1016/j.scitotenv.2019.04.318, http://www.elsevier.com/locate/scitotenv.

Zhang, Y., Huang, K., Yu, Y., & Yang, B. (2017). Mapping of water footprint research: A bibliometric analysis during 2006–2015. *Journal of Cleaner Production*, *149*, 70–79. Available from https://doi.org/10.1016/j.jclepro.2017.02.067.

Zhang, Z., Yang, H., & Shi, M. (2011). Analyses of water footprint of Beijing in an interregional input–output framework. *Ecological Economics*, *70*(12), 2494–2502. Available from https://doi.org/10.1016/j.ecolecon.2011.08.011.

Zhuo, L., Mekonnen, M. M., & Hoekstra, A. Y. (2014). Sensitivity and uncertainty in crop water footprint accounting: A case study for the Yellow River basin. *Hydrology and Earth System Sciences*, *18*(6), 2219–2234. Available from https://doi.org/10.5194/hess-18-2219-2014, http://www.hydrol-earth-syst-sci.net/volumes_and_issues.html.

Chapter 5

Global scenario of water footprints in smart cities

Ravinder Kumar[1,*], **Ranbir Singh**[2], **Anubhav Agrawal**[3] **and Ankur**[1]
[1]*Mechanical Engineering Department, School of Engineering & Technology, BML Munjal University, Gurugram, Haryana, India,* [2]*School of Mechanical Engineering, Lovely Professional University, Phagwara, Punjab, India,* [3]*Electronics Engineering Department, School of Engineering & Technology, BML Munjal University, Gurugram, Haryana, India*
**Corresponding author. e-mail address: rav.chauhan@yahoo.co.in.*

5.1 Background of water footprint

The present upward trajectory of the world's urban population with a growth rate of 55% is expected to rise to 66% by 2050 (Mekonnen & Hoekstra, 2011a). The number of megacities with over 10 million inhabitants is expected to reach to a number of 40 by 2030 (UNDESA, 2010) with a substantial residential water demand (Cosgrove & Cosgrove, 2012). In the 2009 statistics of residential water consumption, 60%–80% of Europe (Collins et al., 2009) and 58% of the United States. consumption was urban consumption. Such concentrated water demand will significantly impact the demand and supply of freshwater (McDonald, Douglas et al., 2011).

There can be an exponential rise in the population facing water shortage (McDonald, Green et al., 2011). Increasing demand for fresh water in urban areas can underpin the actual thought of sustainability. Although researchers and governments have focused on the development of strategies like management of water demand, water-saving technologies, economic policies and regulations, and education for the last three decades (Gleick et al., 2003; Inman & Jeffrey, 2006). Even then, the implementation and effectiveness are still a challenge (Jorgensen et al., 2009). The studies on water planning and management systems with a primary focus on water consumption patterns and socioeconomic and climatic drivers have shown insufficient outcomes (House-Peters & Chang, 2011). Smart meters have enabled the public administration to collect data for the development of individual water consumption profiles (Hilty et al., 2014; Laniak et al., 2013). The challenges associated with water supply, distribution, and conservation will increase in the days to come. The ideation of "smart cities" can address the problem of water management in the growing urban world. Data collection and analysis in real time to support decision making is the major characteristics of the conceptualized smart cities. The data-driven approach can control water quality, and distribution networks and can provide real-time information. Integrated sensor networks, Internet of Things (IoT) devices, and smart meters are key enablers.

The concept of "water footprints" has been the center of attraction for the last two decades. It can be regarded as a comprehensive framework for understanding and quantifying the utilization of water and its associated environmental impacts. The term "water footprints" outlines a holistic view of water consumption across any region across the world. The evolution of "water footprint," with its significance, methodologies, and applications across different sectors needs to be analyzed in detail.

5.2 Concept of "water footprint"

Arjen Hoekstra introduced the concept of water footprints in 2002 and defined "water footprint" as the total volume of freshwater used directly or indirectly to produce a product, process, or service. Thereafter, some classified "water footprints" like blue, green, and gray "water footprints" have evolved to represent surface and groundwater, rainwater, and pollution aspects, respectively (Hoekstra, 2003). "Water footprints" provide a quantitative framework to assess the usage of water in the context of sustainable development, resource management, and environmental conservation and

Water Footprints and Sustainable Development. DOI: https://doi.org/10.1016/B978-0-443-23631-0.00005-4

offer significant insights into water-related challenges like water scarcity, pollution, and environmental impact (Ridoutt & Pfister, 2010). "Water footprint" assessments offer a comprehensive evaluation of water usage throughout the entire supply chain with different types of water resources and pollution. "Water footprint network" quantifies water consumption based on the volume of water required to produce specific goods or services and is the most common method to calculate the "water footprints" (Chapagain & Hoekstra, 2008).

The concept of "water footprints" has been applied in different sectors. By applying "water footprints," one can analyze the water usage efficiency in agricultural practices and environmental impact in food production, assess virtual water trades, and aid the decision-making for sustainable water management and policy development (Mekonnen and Hoekstra, 2011b). Gender-based "water footprint analysis" highlights the unequal distribution of water-related work among men and women (Zwarteveen & van der Kooij, 2006). The United Nations' Sustainable Development Goals have water-specific sustainability targets. "Water footprints analysis" assists in achieving the sustainability goals of the United Nations by tracking water usage consumption in different sectors (United Nations, 2015). "Water footprint analysis" can identify possible ecological impacts (Tilman et al., 2001).

5.3 Assessment of sector-specific water footprints

Virtual water trade is a critical application of water footprints, representing the amount of water embedded in traded goods to guide water-scarce regions to import water-intensive products, thereby reducing local water stress (Hoekstra & Hung, 2005). "Water footprints analysis" has been utilized in local and **global businesses** to assess and manage water-related risks and their impacts to identify opportunities for efficient utilization of water, reduce water pollution, and support water sustainability initiatives (Pfister et al. 2011). "Water footprint assessments" also provides critical information for policymakers and government agencies to develop effective water management strategies, policies, regulations, and sustainability initiatives, specific to household, agriculture, and industry (Yang & Li, 2009). "Water footprints" of agriculture and land use are found to impact ecosystems and biodiversity. Large-scale agricultural practices are the reason behind habitat loss, water pollution, and reduced biodiversity (Tilman et al. 2009). Food waste is a significant contributor to water waste. "Water footprints" of food waste can analyze the amount of water wasted along the food supply chain. Because urban areas are more water intensive, urban planning significantly impacts water footprints (Trowsdale & O'Sullivan, 2003). Understanding "water footprints" and their impacts on the ecosystem is crucial for preserving vital resources like clean water, pollination, and climate regulation to support the existence of mankind on the earth (Costanza et al., 1997). "Water footprint analysis" across different sectors outlined that climate change is rapidly altering patterns of water availability. The increasing frequency of extreme weather events is found to directly impact water use efficiency and management strategies (Fig. 5.1; IPCC, 2019).

The fashion industry has significant water consumption footprints right from cotton cultivation to dyeing processes. "Water footprint analysis" of the fashion industry promotes sustainable practices of efficient textile production and sourcing of materials (Fletcher & Tham, 2015). Like "circular economy," the concept of the "circular water economy" focuses on minimization of water waste and maximizing water recycling and reuse. "Water footprint analysis" helps in understanding the role of "circular water practices" to reduce water footprints and contribute to sustainability (Brock & Smith, 2020). "Water footprints" can assess the water-saving potential of circular practices such as recycling food waste and reducing food losses (Food & Agriculture Organization, 2019). "Water footprints" align itself with the principles of Integrated Water Resources Management (IWRM) to promote the coordinated development and management of water,

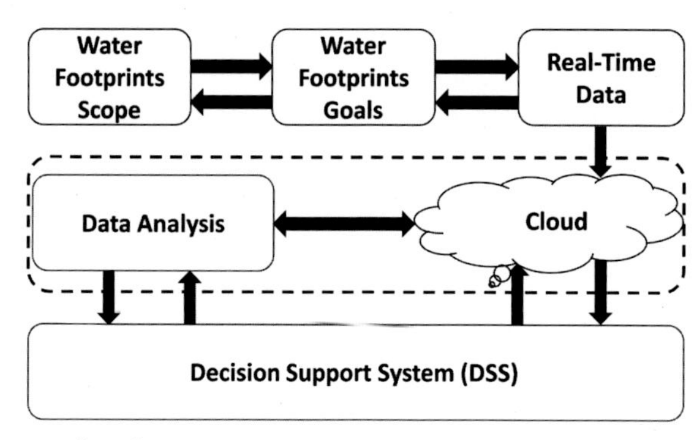

FIGURE 5.1 Assessment of smart city water footprints management system.

land, and related resources. A proper understanding of the water footprints of different activities helps develop and promote IWRM strategies to ensure water sustainability (Global Water Partnership, 2000). Different governments and international organizations are increasingly adding "water footprints" into their policies and regulations. "Water footprint assessments" can influence national water management plans, agricultural policies, and strategies to achieve Sustainable Development Goal 6 (Clean Water and Sanitation; UN-Water, 2015). Companies are including water footprint data in their sustainability reports to highlight their efforts to reduce water consumption as their standard practice in corporate social responsibility initiatives (Global Reporting Initiative, 2016).

Advanced geographic information systems and remote sensing technologies are enabling the development of spatially explicit water footprints to provide insights about water consumption at regional, national, and local levels. "Water footprints" incorporated with indigenous knowledge and perspectives is an emerging trend to recognize the cultural and ecological significance of water in different regions and to encourage inclusive and sustainable water management practices. "Water footprint assessments" are increasingly getting integrated into life cycle assessment studies to analyze the water impact of products or processes for a more comprehensive understanding of environmental sustainability (ISO, 2006). "Water footprint assessments" are also engaging the stakeholders to influence water usage decisions to ensure sustainable and equitable water management (Reed, 2011).

Water footprint assessments when applied to regions facing water scarcity can help to identify water-efficient strategies and prioritize water use to address local and regional water stress (Allan, 1998). "Water footprint assessments are also linked to climate adaptation strategies to understand the water impact of various sectors to aid the development of resilient adaptation measures to cope with changing climate conditions (IPCC, 2014). "Water footprints" are also part of sustainable and green finance initiatives to promote environment-friendly investment choices to support water-efficient projects (UNEP, 2018). Food—energy—water nexus also integrates "water footprints" to study the interdependencies and trade-offs among the critical resources to make informed decisions for resource management (Bazilian et al., 2011). Ecological economics incorporates "water footprint analysis" to explore the sustainability of economic systems about natural resources to evaluate the long-term ecological and social impacts of economic activities (Costanza, 1991). Understanding "water footprints" helps plan for recovery from water-related disasters, such as floods and droughts (UNDRR, 2019).

Water footprints guide cross-border water management through international agreements and cooperation in shared water basins (United Nations Economic Commission for Europe, 1992). "Water footprint assessments" have become a fundamental component of academic research (The Sustainability Tracking, Assessment, & Rating System, 2021) Water footprint data has also been integrated into environmental education programs to enable the student to understand the environmental and social implications of water usage (UNESCO, 2019). Data analytics and machine learning are enhancing the accuracy and efficiency of water footprint assessments by enabling more detailed and real-time data collection, making water management decisions more data-driven (Chui et al., 2016). Different sectors have unique water consumption patterns and environmental impacts. "Water footprints" analysis in agriculture can identify water-efficient farming practices to reduce water usage (Vanham et al., 2018).

5.4 Concept of "water footprints" management system in smart cities: a "water smart city"

Urban planners and designers should integrate "water footprint assessments" into urban development projects. Sustainable urban design includes considerations for water-efficient infrastructure, green spaces, water recycling systems, public health, travel practices, social impact assessments, green buildings, water-positive buildings, smart city water management, climate-adaptive building design, and artificial intelligence (AI)-enhanced healthcare (Fig. 5.2).

The IoT is used for real-time monitoring of water consumption in smart cities and industries. IoT devices and sensors are instrumental in collecting data for water footprint assessments and facilitating water-saving strategies (Atzori et al., 2010).

5.5 Critiques and challenges

"Water footprints" help to understand local, regional, national as well and international water dependencies and water management challenges (Gleick, 2014). Usage and access to water have been reported as gender-differentiated because particularly in rural areas of developing countries women bear the responsibility for water collection and management (Zwarteveen & van der Kooij, 2006). Although "water footprints" enable a comprehensive review of water consumption and environmental impacts, there are a few critics and challenges associated with it. Researchers argue that this concept of water footprints oversimplifies complex hydrological systems without due consideration to local context, and the availability of global data with the required level of consistency is still a challenge (Allan, 2011). The virtual water

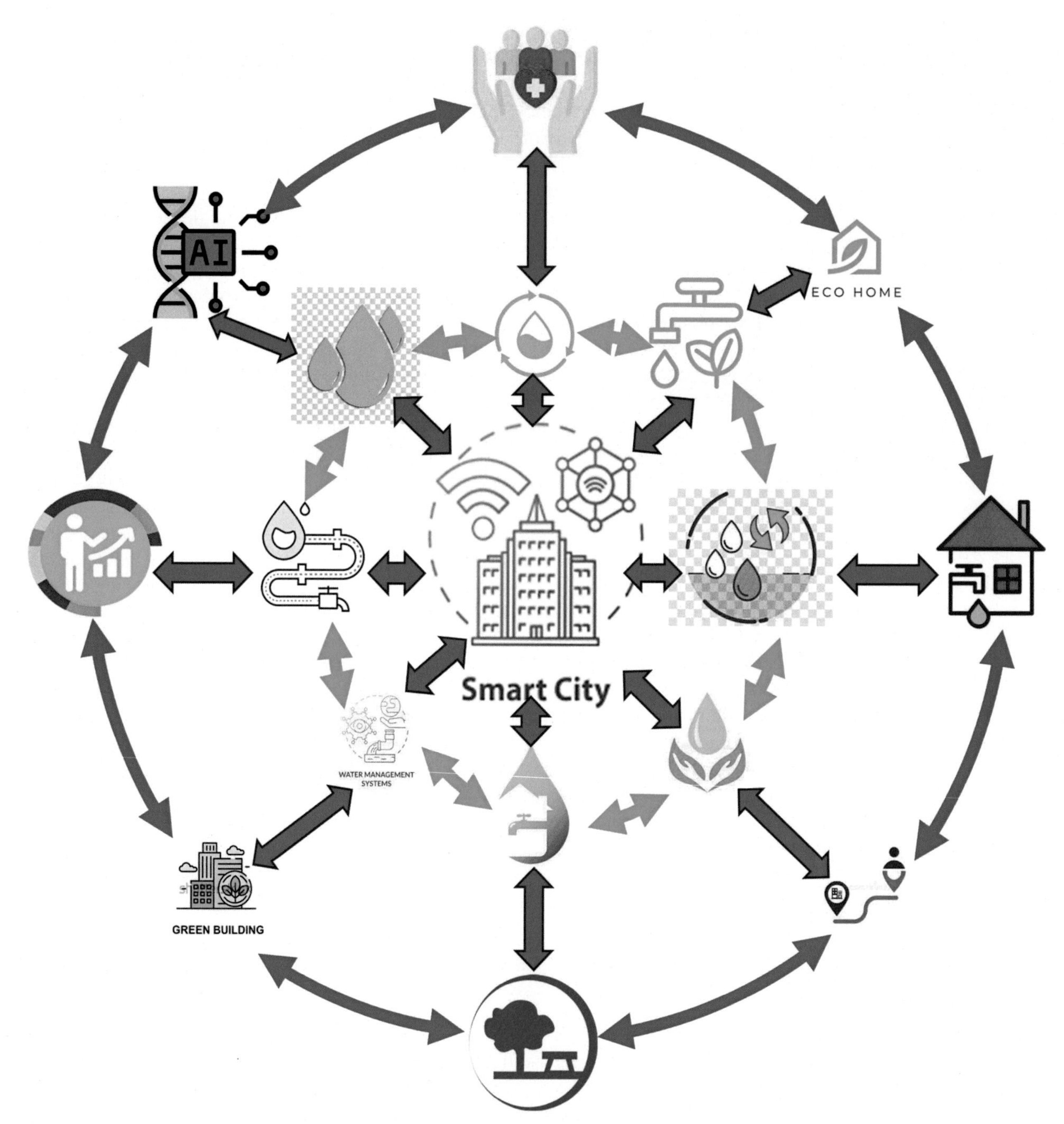

FIGURE 5.2 Smart city with smart water management system.

trade influences global food security and resource allocation (Hoekstra & Hung, 2005). Open data initiatives and the promotion of water footprint databases can enable researchers, policymakers, and the public to access water footprint information for several different purposes. Data transparency is crucial for informed decision-making and global collaboration on water sustainability.

5.6 Future directions

The field of "water footprints" will continue to grow in the coming time, with an ongoing research focus on refining and developing new methodologies, and data accuracy techniques, broadening the scope of application, enhancement of the spatial and temporal resolution of assessments and integrating water footprint analysis into broader sustainability frameworks

(Aldaya et al., 2010). With technological advancements in data collection and analysis techniques, the field of "water footprint assessment" will continually evolve to improve the accuracy and reliability of assessments by incorporating local context, climate change impacts, and emerging technologies in its analysis. Because the principles of circular economy are pivoted around recycling, reusing, and remanufacturing processes to emphasize resource efficiency and waste reduction, "water footprint analysis" can be applied to assess the water implications of circular economy practices, (Kirchherr, 2017).

There is a need for gender-inclusive water policies (Zwarteveen & van der Kooij, 2006). Strategies need to be developed to reduce water waste with waste food (Mekonnen & Hoekstra, 2011a). Considering water availability and quality, urban planning should account for minimizing water consumption, reduce runoff, and promote integrated water management (Trowsdale & O'Sullivan, 2003). Debates and discussions on equitable access to water resources raising ethical and moral questions at national and international forums need to be organized and the responsibilities of high-water-consuming nations need to be highlighted across the globe (Brown & Merz, 2018). Public education and outreach programs about "water footprint" can encourage responsible water usage across the world. "Water footprint information" can raise public awareness about water conservation and its environmental consequences.

5.7 Conclusions

"Water footprint" analysis is applied to different sectors, including agriculture, industry, and households. It offers a comprehensive and quantitative approach to understanding water usage and its environmental implications. These footprints influence policy and corporate decisions. It also extends beyond the traditional analysis to cover a wide range of social, ecological, and ethical dimensions. As this field continues to evolve, these trends and aspects will grow accordingly with increased relevance and versatility from education to environmental management and public health and will continue to address complex global challenges related to climate change and sustainability. Application of "water footprints" in fields like energy, humanitarian aid, cultural heritage preservation, climate resilience, gender-responsive water management, circular economies, tourism, cybersecurity, high-performance computing, AI development, and smart waste management **opens new scope for future developments**. The expanding scope of water footprint assessments in addressing diverse global challenges is a valid tool for fostering sustainability, sustainable urban planning, climate-adaptive building design, smart city water management, and many more such fields and sectors will continue to evolve in the time to come. The science will witness a realistic smart city with full water footprint assessments soon.

References

Aldaya, M. M., Garrido, A., Llamas, M. R., Varela-Ortega, C., Novo, P., & Casado, R. R. (2010). Water footprint and virtual water trade in Spain. In A. Garrido, & M. R. Llamas (Eds.), *Water Policy in Spain* (pp. 49–59). Leiden: CRC Press.

Aldaya, M. M., Martínez-Santos, P., & Llamas, M. R. (2010). Incorporating the Water Footprint and Virtual Water into Policy: Reflections from the Mancha Occidental Region, Spain. *Water Recourses Management*, *24*, 941–958. Available from https://doi.org/10.1007/s11269-009-9480-8.

Allan, J. A. (1998). Virtual water: A long-term solution for water short Middle Eastern economies? *Water International*, *23*(1), 3–13.

Allan, J. A. (2011). Virtual water: The water, food, and trade nexus. Useful concept or misleading metaphor? *Water International*, *35*(1), 4–12.

Atzori, L., Iera, A., & Morabito, G. (2010). The internet of things: A survey. *Computer networks*, *54*(15), 2787–2805.

Bazilian, M., Rogner, H., Howells, M., Hermann, S., Arent, D., Gielen, D., Steduto, P., Mueller, A., Komor, P., Tol, R. S., & Yumkella, K. K. (2011). Considering the energy, water and food nexus: Towards an integrated modelling approac. *Energy Policy*, *39*(12), 7896–7906.

Brock, A., & Smith, S. (2020). *The Circular Water Economy: A blueprint for progress*. The World Bank.

Brown, C., & Merz, B. (2018). The ethics of water governance. *Water International*, *43*(5), 587–604.

Chapagain, A. K., & Hoekstra, A. Y. (2008). The global component of freshwater demand and supply: An assessment of virtual water flows between nations as a result of trade in agricultural and industrial products. *Water International*, *33*(1), 19–32.

Chui, M., Manyika, J., & Miremadi, M. (2016). Where machines could replace humans-and where they can't (yet). *The McKinsey Quarterly*, 1–12.

Collins, R., Kristensen, P., & Thyssen, N. (2009). Water resources across Europe-confronting water scarcity and drought (Doctoral dissertation, Univerza v Mariboru, Fakulteta za kmetijstvo in biosistemske vede).

Cosgrove, C. E., & Cosgrove, W. J. (2012). The United Nations World Water Development Report–N 4–The Dynamics of Global Water Futures: Driving Forces 2011–2050 (Vol. 2). UNESCO.

Costanza, R. (1991). Ecological economics: The science and management of sustainability. *Ecological Economics*, *1*(1), 1–7.

Costanza, R., d'Arge, R., De Groot, R., Farber, S., Grasso, M., Hannon, B., Limburg, K., Naeem, S., O'neill, R. V., Paruelo, J., & Van Den Belt, M. (1997). The value of the world's ecosystem services and natural capital. *Nature*, *387*(6630), 253–260.

Fletcher, K., & Tham, M. (2015). *The Sustainable Fashion Handbook*. Thames & Hudson.

Food and Agriculture Organization (FAO). (2019), Sustainable Food Systems: Concept and Framework.

Gleick, P. H. (2014). Water, drought, climate change, and conflict in Syria. *Weather, Climate, and Society*, *7*(3), 331–340.

Gleick, P. H., Wolff, G. H., & Cushing, K. K. (2003). Waste not, want not: The potential for urban water conservation in California.

Global Reporting Initiative (GRI). (2016), GRI Sustainability Reporting Standards.

Global Water Partnership (2000), Integrated Water Resources Management".

Hilty, L. M., Aebischer, B., & Rizzoli, A. E. (2014). Modeling and evaluating the sustainability of smart solutions. *Environmental Modelling & Software*, *56*, 1–5.

House-Peters, L., & Chang, H. (2011). Urban water demand modeling: review of concepts, methods, and organizing principles. *Water Resources Research*, *47*, W05401.

Hoekstra, A. (2003), Virtual water trade: Proceedings of the International Expert Meeting on Virtual Water Trade, Value of Water Research Report Series No. 12, UNESCO-IHE, Delft, the Netherlands.

Hoekstra, A. Y., & Hung, P. Q. (2005). Globalisation of water resources: International virtual water flows in relation to crop trade. *Global Environmental Change*, *15*(1), 45–56.

Inman, D., & Jeffrey, P. (2006). A review of residential water conservation tool performance and influences on implementation effectiveness. *Urban Water Journal*, *3*(3), 127–143.

IPCC (2019), Special Report on Climate Change and Land.

IPCC (2014), Climate Change 2014: Impacts, Adaptation, and Vulnerability. Part A: Global and Sectoral Aspects, Contribution of Working Group II to the Fifth Assessment Report of the Intergovernmental Panel on Climate Change.

ISO (2006). ISO 14040: Environmental management—life cycle assessment—principles and framework.

Jorgensen, B., Graymore, M., & O'Toole, K. (2009). Household water use behavior: An integrated model. *Journal of Environmental Management*, *91* (1), 227–236.

Kirchherr, J., Reike, D., & Hekkert, M. (2017). Conceptualizing the circular economy: An analysis of 114 definitions. *Resources, conservation and recycling*, *127*, 221–232.

Laniak, G. F., Olchin, G., Goodall, J., Voinov, A., Hill, M., Glynn, P., Whelan, G., Geller, G., Quinn, N., Blind, M., Peckham, S., Reaney, S., Gaber, N., Kennedy, R., & Hughes, A. (2013). Integrated environmental modeling: a vision and roadmap for the future. *Environ. Model. Softw*, *39*, 3–23.

McDonald, R., Douglas, I., Grimm, N., Hale, R., Revenga, C., Gronwall, J., & Fekete, B. (2011). Implications of fast urban growth for freshwater provision. *Ambio*, *40*, 437.

McDonald, R., Green, P., Balk, D., Fekete, B., Revenga, C., Todd, M., & Montgomery, M. (2011b). Urban growth, climate change, and freshwater availability. *Proc. Natl. Acad. Sci*, *108*, 6312–6317.

Mekonnen, M. M., & Hoekstra, A. Y. (2011a). The water footprint of humanity. *Proceedings of the National Academy of Sciences*, *109*(9), 3232–3237.

Mekonnen, M. M., & Hoekstra, A. Y. (2011b). The green, blue and grey water footprint of crops and derived crop products. *Hydrology and Earth System Sciences*, *15*(5), 1577–1600.

Pfister, S., Bayer, P., Koehler, A., & Hellweg, S. (2011). Environmental impacts of water use in global crop production: hotspots and trade-offs with land use. *Environmental Science & Technology*, *45*(13), 5761–5768.

Reed, M. S., Buenemann, M., Atlhopheng, J., Akhtar-Schuster, M., Bachmann, F., Bastin, G., Bigas, H., Chanda, R., Dougill, A. J., Essahli, W., Evely, A. C., Fleskens, L., Geeson, N., Glass, J. H., Hessel, R., Holden, J., Ioris, A. A. R., Kruger, B., Liniger, H. P., Mphinyane, W., Nainggolan, D., Perkins, J., Raymond, C. M., Ritsema, C. J., Schwilch, G., Sebego, R., Seely, M., Stringer, L. C., Thomas, R., Twomlow, S., & Verzandvoort, S. (2011). Cross-scale monitoring and assessment of land degradation and sustainable land management: A methodological framework for knowledge management. *Land Degradation & Development*, *22*(2), 261–271.

Ridoutt, B. G., & Pfister, S. (2010). A revised approach to water footprinting to make transparent the impacts of consumption and production on global freshwater scarcity. *Global Environmental Change*, *20*(1), 113–120.

The Sustainability Tracking, Assessment & Rating System (STARS). (2021), AASHE's Sustainability Tracking, Assessment & Rating System (STARS).

Tilman, D., Socolow, R., Foley, J. A., Hill, J., Larson, E., Lynd, L., Pacala, S., Reilly, J., Searchinger, T., Somerville, C., & Williams, R. (2009). Beneficial biofuels—the food, energy, and environment trilemma. *Science (New York, N.Y.)*, *325*(5938), 270–271.

Trowsdale, S., & O'Sullivan, A. D. (2003). Integrated urban water management. *Water Science and Technology*, *48*(12), 15–22.

UNEP (2018), The Global Status of Sustainable Finance.

UNDRR (2019), Global Assessment Report on Disaster Risk Reduction.

UNESCO (2019), Education for Sustainable Development Goals: Learning Objectives.

United Nations Economic Commission for Europe (UNECE) (1992), Convention on the Protection and Use of Transboundary Watercourses and International Lakes.

UNDESA (2010). World urbanization prospects: the 2014 revision. Highlights (ST/ESA/SER.A/352). United Nations Department of Economic and Social Affairs Population Division.

United Nations (2015). Transforming our world: The 2030 Agenda for Sustainable Development.

UN-Water (2015). Global Analysis and Assessment of Sanitation and Drinking-Water (GLAAS) 2014/2015: Investing in Water and Sanitation: Increasing Access, Reducing Inequalities.

Vanham, D., Comero, S., Gawlik, B. M., & Bidoglio, G. (2018). The water footprint of different diets within European sub-national geographical entities, Nature. *Sustainability*, *1*(9), 518–525.

Yang, H., & Li, L. (2009). The water footprint of Shanghai and its implications for water policy. *Environmental Science & Technology*, *43*(7), 1927–1932.

Zwarteveen, M., & van der Kooij, S. (2006). Water is life: Women's human rights to water and sanitation in poor urban areas. *Environment and Urbanization*, *18*(2), 375–394.

Chapter 6

Assessing water footprints for global sustainability: a comprehensive review

Chinmaya Kumar Swain* **and Lipsa Tripathy**
ICAR-National Rice Research Institute, Cuttack, Odisha, India
**Corresponding author. e-mail address: chinu.swain@gmail.com.*

6.1 Introduction

Rapid growth of population, economic growth, demand for agricultural products for both food and nonfood use, and a shift in consumption patterns toward meat and sugar-based products, competition for freshwater resources has been increasing for decades (de Fraiture & Wichelns, 2010). Due to rising demand and freshwater scarcity, supply and pollution will likely worsen in the future. It is assumed that our water needs will increase, threatening food security and environmental sustainability (Rosegrant et al., 2009). Worldwide water withdrawal will increase from 4500 billion cubic meters to 6900 billion cubic meters by 2030 (McKinsey, 2009). Water is a vital natural resource that is essential for agricultural production and human societies. Approximately 40 countries have a lower per capita availability of renewable internal freshwater resources than the global average (World Bank 2016). The nations situated along the Belt and Road initiative are currently encountering the challenge of insufficient water supply.

According to Hoekstra and Mekonnen (2012) and Willaarts et al. (2014), agriculture is constantly growing and plays a crucial role in providing food for both humans and livestock. The two primary sources of water used in this production type rainfall and irrigation account for approximately 86% of the world's total consumption (Aldaya et al., 2010). Commodity production also leads to biodiversity loss, deforestation, surface and groundwater degradation, and pollution. Loss of water for other uses, like household water usage, occurs as a result in nations that produce and export water (Hoekstra and Mekonnen, 2012; Aldaya et al., 2010). The vast majority of the water used in Latin America comes from agricultural output, specifically crop production (71%). The majority of this water is consumed in rainfed conditions, also known as green water, accounting for 90 to 94% of the total water utilised (Hoekstra and Mekonnen, 2012; Mekonnen et al., 2015). In contrast, during the aforementioned, about 2500 Gm^3yr^{-1} of the world's water was withdrawn for agricultural uses, namely irrigation. If we include in the amount of rainwater that crops use, the estimated annual global agricultural water use is 5400 Gm^3yr^{-1}. Worldwide, people use an estimated 1200 Gm^3 of water per year for various uses in their homes and businesses. Instead of being used for local use, over 15% of the world's water that is designated for human use is used virtually for export.

The water footprint (WF) idea is frequently used as a method to increase water usage productivity and sustainability while also better managing water resources within watersheds (Lamastra et al., 2014). As water shortage affects water productivity, the WF concept is used to investigate the economic productivity of water components (Garrido et al., 2010; Lambooy, 2011). The amount of freshwater used in the production of a crop, as defined by Aldaya and Hoekstra (2010), is measured at the location where the crop is produced. The WF is then broken down into a total volume of either used or contaminated direct or indirect freshwater from directors. The projected increase in population necessitates an augmented water supply to facilitate the production of the anticipated 60% surplus food demand by the year 2050, as reported by the Food and Agriculture Organization of the United Nations in 2017 (FAO, 2017). The utilization of water in agriculture plays a crucial role in mitigating worldwide water scarcity and redistributing global water resources. The issue at hand is not solely a technological one but rather necessitates a more comprehensive and macroscopic approach. This entails taking into account not only visible water but also virtual water, not only at a regional level but on a global scale.

Water Footprints and Sustainable Development. DOI: https://doi.org/10.1016/B978-0-443-23631-0.00006-6

As the field of WF accounting advanced, researchers developed into the connection between WF and the promotion of environmental sustainability. Hoekstra's (2009) analysis was one of the initial investigations into the concept of water footprinting from a sustainability standpoint, wherein WF was juxtaposed against freshwater availability. Hoekstra et al. (2011) proposed a methodology consisting of four steps to evaluate the sustainability of WF. The first step involves identifying the sustainability criteria, followed by identifying the WF sustainability hotspots. The third step involves quantifying the primary impacts of the hotspots, and finally, in the fourth step, the secondary impacts of the hotspots are determined. The study of Mekonnen and Hoekstra (2012) evaluated the ecological viability of blue water footprint (BWF) across the primary river basins worldwide. Their findings revealed that a staggering four billion individuals experience acute water scarcity for a minimum of 1 month annually. Several scholarly investigations have evaluated the ecological sustainability of gray WF by contrasting it with the assimilative potential of aquatic systems (Aldaya et al., 2010; Karandish, 2019; Liu et al., 2013; Mekonnen & Hoekstra, 2012).

Water footprint is categorized into three groups: blue, green and gray WF. Green water footprints (GWFs) are defined as rainwater that does not run off or recharge groundwater whereas BWFs are defined as irrigation water extracted from the ground or surface water, and gray water footprints as the amount of freshwater required to dissolve the load of pollutants (Chapagain & Hoekstra, 2011). As a measure of the total amount of water consumed in a given area, the water footprint also functions as a tool for evaluating the sustainability of water resource management. A river basin's sustainability assessment is frequently carried out (Zeng et al., 2012), serving as a spatial setting for the process of water planning. The maximum values (i.e., water withdrawal from the ground surface) for each of the WF components (blue, green, and gray) are then used to compare them in a river basin. Therefore, water resource management is deemed sustainable if water footprints within the basin support the environmental flow requirements (EFRs) and are lower than the withdrawal volume. Sustainability evaluation is gaining importance in water management and provides policymakers with knowledgeable guidelines.

The objective of this study is to analyze the WF or water usage globally, and how it affects environmental sustainability. This includes an analysis of the current trade situation of different WF, the various types of water involved, and the structures of trade. This novel viewpoint offers valuable insights that can inform policymakers in their efforts to effectively manage water resources and adapt trade frameworks.

6.2 Human impact on water footprint

Water footprint is an ecological metric that quantifies the amount of freshwater required to manufacture the commodities and amenities. This facilitates the determination of the extent of the impact produced by human actions and the acquisition of impartial information. Subsequently, the implementation of more sustainable choices can be facilitated to diminish water consumption and enhance water efficacy. The WF concept was introduced in 2002, which aimed to establish a consumption-based metric for water usage. This metric was intended to supplement the conventional production sector–based indicators of water usage (Hung, 2002). The water footprint of a nation is a quantification of the aggregate amount of freshwater used in the production of goods and services that are consumed by the populace of the nation.

The computation of a WF can comprehensively evaluate the sustainability of a given commodity, encompassing its consumption of blue, green, and gray water. The notion of virtual water is employed to monitor the exchange of virtual water as a means of assessing the advantages of water transfer throughout the value chain. The concept of the water footprint was established by drawing an analogy to the ecological footprint, which was originally proposed in the 1990s by Rees (1992) and further developed by Wackernagel and Rees (1996) and Wackernagel et al. (1997). The term "ecological footprint" pertains to the total expanse of productive land and aquatic ecosystems necessary to generate the resources used and absorb the waste generated by a specific population at a designated material standard of living, regardless of the location of that land on the planet. This concept serves to measure the amount of land necessary to support human habitation, while the "water footprint" quantifies the amount of water needed to sustain a given population.

6.2.1 Understanding the differences between green, blue, and gray water footprint

Water footprint is a metric that quantifies the amount of freshwater used for consumption and degradation purposes. The WF of consumption comprises two distinct components: a green component, which pertains to the consumption of rainwater, and a blue component, which pertains to the consumption of surface water or groundwater. The degradative WF, commonly referred to as the gray WF, quantifies the amount of water necessary for the dilution of pollutants that are introduced into freshwater systems (Hoekstra et al., 2011). There are various categories of WF. Upon further examination, it is evident that there exist three distinct categories of WF, which are determined by the source of the water in question. The ecological impact of precipitation on the environment is commonly referred to as the "green footprint," The aforementioned water refers to the superficial storage of water in soil or on plants, including precipitation in the form of rain. The GWF is a metric that quantifies the amount of water used in the production of agricultural, horticultural, and forestry goods. The BWF, which refers to the amount of fresh water used in the production of a product is a significant metric in water resource management. The aforementioned refers to the accumulation of water in substantial quantities, originating from either surface or subsurface sources, including rivers, lakes, and underground water-bearing layers. The BWF is a metric that quantifies the amount of water used in irrigated agriculture, industrial processes, and domestic activities.

The gray water footprint, which pertains to water quality, is necessary to mitigate the concentration of contaminants resulting from manufacturing operations until it falls below the threshold established by prevailing regulatory standards. However, due to limitations in the applied models, it was presented as a combined value without explicit differentiation between the two components (Hoekstra & Hung 2002). Chapagain and Hoekstra (2008) introduced a comprehensive framework that unified the GWF, BWF, and gray water footprints. Hoekstra et al. (2011) proposed a refinement to the definition of the gray water footprint. This was achieved by incorporating the presence of naturally occurring substances in water bodies, which in turn reduced the ability of these bodies to absorb additional loads resulting from human activities while adhering to the maximum allowable concentrations. While initial gray water footprint studies were restricted to examining nitrogen pollution, contemporary research has expanded to encompass a diverse range of water quality indicators, such as nutrients, dissolved solids, metals, and pesticides. Several studies have differentiated between various categories of BWFs based on the origin of the water, such as surface water or renewable sources. The assessment of water footprint has emerged as a novel research field. It involves the evaluation of various sources of water, such as groundwater, fossil groundwater, or capillary rise. As data availability increases, it is anticipated that the assessment of different shades of BWF will become more prevalent. This is because the potential implications of these various shades of BWF may differ.

6.2.2 Different perspectives of water footprint toward future projections

In the construction of WF scenarios, reliance is placed on past global scenario exercises to the greatest extent feasible. There are two primary benefits associated with this approach. Firstl it allows us to construct future projections (FP) based on thoroughly researched and documented potential future outcomes. Second, it enables readers to quickly orient themselves to the various storylines. The 2 × 2 matrix system of scenarios developed by the Intergovernmental Panel on Climate Change (IPCC; Nakicenovic et al., 2000) was used as organized based on two axes that signify two fundamental aspects of ambiguity: the dichotomy between globalization and regional self-reliance, and the dichotomy between economy-centric progress and progress driven by social and environmental goals. The four quadrants are formed by the intersection of two axes, with each quadrant denoting a specific scenario (Ercin & Hoekstra, 2014). These projections include globalization (FP-I), regional-level sustainability (FP-II), environmental sustainability (FP-III), and economic development (FP-IV; Fig. 6.1).

The narrative structures that have been developed bear a resemblance to the scenarios presented by the IPCC concerning factors such as population growth, economic growth, technological advancement, and governance. To conduct the analysis, a majority of the intricate assumptions for the scenarios were formulated independently. However, these assumptions were derived from the narrative frameworks of the preexisting IPCC scenarios. The scenarios exhibit consistency and provide dependable narratives regarding potential future occurrences. It is imperative to comprehend that the scenarios presented do not serve as prognostications of forthcoming events, but rather offer alternative viewpoints on the potential evolution of water footprints leading up to 2050.

The incorporation of green water usage into the WF metric was significant and intentional, influenced by Falkenmark's (2000) research, which introduced the green–blue water lexicon to expand the scope of water management beyond the traditional emphasis on blue water. Chapagain et al. (2006) conducted the initial evaluation of a crop's GWF and BWF independently. Water pollution poses a challenge to water consumption. In 2008, Hoekstra and Chapagain introduced a comprehensive framework that integrated the concepts of GWF, BWF, and gray water footprints. In their study, Hoekstra et al. (2011) proposed a refinement to the definition of the gray water footprint. This refinement involved the consideration of naturally occurring concentrations of substances in water bodies, which in turn reduced the ability of these bodies to absorb additional loads resulting from human activities, given the maximum allowable concentrations. While initial gray water footprint studies were restricted to examining nitrogen pollution,

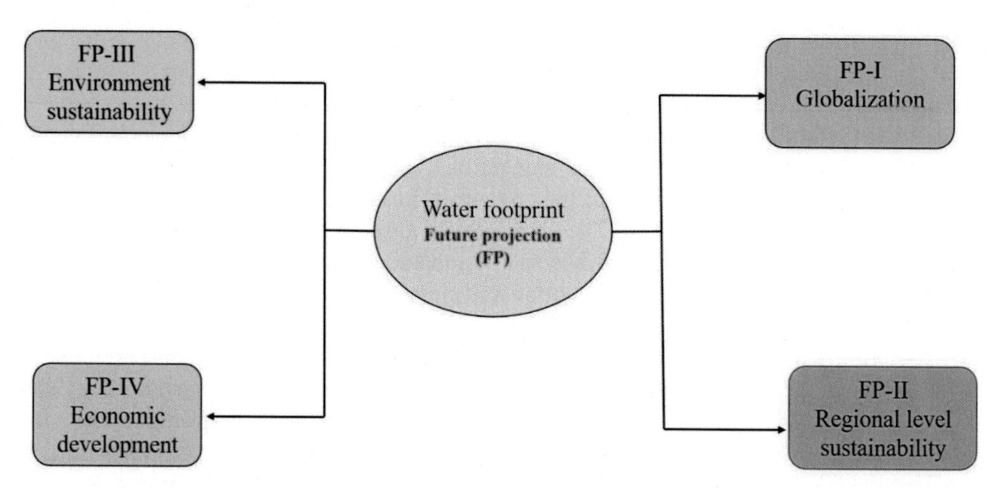

FIGURE 6.1 Different scenarios for constructing water footprint. *Courtesy*: Ercin, A. E., & Hoekstra, A. Y. (2014). Water footprint scenarios for 2050: A global analysis. Environment International. 64, *71–82, 18736750. https://doi.org/10.1016/j.envint.2013.11.019. Elsevier Ltd, Netherlands. http://www.elsevier.com/locate/envint.*

contemporary research has expanded to encompass a diverse range of water quality indicators, such as nutrients, dissolved solids, metals, and pesticides.

Initially, a baseline was established for 2050, predicated on the assumption that present circumstances will persist unabated. The four scenarios were developed by using varying alternatives for the drivers of change, building upon the baseline. The baseline projection formulated for 2050 is predicated on the assumption that both per capita food consumption and nonfood crop demand will remain consistent with the levels observed in 2000. This year was taken into account about technology, production, and trade. The population growth rate is derived from the medium-fertility population projection provided by the United Nations (UN, 2011). The projected economic growth aligns with the FP-II projection outlined by the IPCC. The consideration of climate change is not being taken into account. Hence, alterations in both food and nonfood consumption patterns, as well as modifications in the water footprint of agriculture and domestic water supply, are solely contingent upon population expansion (Oestigaard, 2012).

The FP-I scenario about the global market is derived from the fourth by the IPCC. The situation is distinguished by a notable upsurge in economic expansion and the relaxation of restrictions on global commerce. Individual consumption and material well-being are the driving forces behind the global economy. Environmental policies worldwide are predominantly reliant on economic instruments, with long-term sustainability not being prioritized in policy agendas. Animal-based protein sources, such as meat and dairy products, are considered crucial components of human dietary intake. Anticipation exists for the swift advancement of novel and effective technologies. Fossil fuels are the primary source of energy. It is anticipated that there will be a decrease in both fertility and mortality rates.

The FP-II projection is additionally propelled by economic expansion, albeit with a greater emphasis on delineations of regional and national scope. The level of regional self-sufficiency is on the rise. Like the findings of FP-I, environmental concerns do not appear to significantly influence decision-making processes. Instead, there is a notable trend toward the swift development and implementation of innovative and effective technologies. Fossil fuels currently hold a dominant position, there is a projected marginal rise in the utilization of biofuels. The scenario under consideration exhibits the highest rate of population growth.

The third scenario denoted as FP-III, about global sustainability, bears resemblance to IPCC. The current situation is distinguished by heightened social and environmental considerations that have been incorporated into worldwide trade regulations. The pace of economic expansion is comparatively sluggish in the current period as compared to the preceding two quarters, and there is a heightened emphasis on ensuring social fairness and impartiality. Efficient and environmentally friendly technologies are being developed with a focus on resource conservation. Due to the emphasis on environmental concerns, there has been a reduction in the consumption of meat and dairy products. The process of globalization and liberalization has led to an increase in international trade. It is anticipated that there will be a decrease in the utilization of agrochemicals and a shift toward more environmentally friendly industrial practices.

The FP-IV scenario, which pertains to local sustainability and economic development, was developed by the IPCC and characterized by prominent national or regional values. Prioritization of self-sufficiency, equity, and environmental

TABLE 6.1 Comparison of existing global water demand scenarios with the current study.

Study	Disaggregation	Drivers	References
Domain of time: 2025 Scope: blue water withdrawal	Agriculture Industry Domestic	Population growth Economic growth Technology change	Vörösmarty, Fekete, et al. (2000), Vörösmarty, Green, et al. (2000), Alcamo et al. (2003), Shen et al. (2008)
Domain of time: 2050 Scope: green and blue water consumption	Agriculture Industry Domestic	Population growth Economic growth Technology change Consumption patterns—diet	de Fraiture and Wichelns (2010)
Domain of time: 2050 Scope: green and blue water consumption, pollution as gray water footprint	Agriculture Industry Domestic	Population growth Economic growth Technology change Consumption patterns(diets) Consumption patterns(biofuel)	Ercin and Hoekstra (2014)

Courtsey: Ercin, A. E., & Hoekstra, A. Y. (2014). Water footprint scenarios for 2050: A global analysis. *Environment International. 64*, 71–82, 18736750. https://doi.org/10.1016/j.envint.2013.11.019. Elsevier Ltd, Netherlands. http://www.elsevier.com/locate/envint.

sustainability has emerged as a key focus of policymakers. It is anticipated that there will be a deceleration in economic growth over an extended period. The levels of pollution in the agricultural and industrial domains have been reduced. The utilization of biofuels as a source of energy has experienced a significant expansion.

The aforementioned scenarios have been formulated for a total of distinct global regions, with a focus on 2050 (Calzadilla et al., 2011). The study encompasses a range of regions, including the United States of America, Canada, Japan and South Korea, Western Europe, Australia and New Zealand, Eastern Europe, the Former Soviet Union, the Middle East, Central America, South America, South Asia, South East Asia, China, North Africa, Sub-Saharan Africa, and the remainder of the world.

Table 6.1 presents a comparative analysis of the scope of other recent studies on water demand scenarios. Vörösmarty, Fekete, et al. (2000), Vörösmarty, Green, et al. (2000) conducted an estimation of water withdrawal in the agricultural, industrial, and domestic sectors for 2025. The study differentiated various trajectories for population growth, economic development, and alterations in water usage efficiency. In 2000, Shiklomanov evaluated water withdrawals and consumption across global regions, with a focus on the projected figures for 2025. Rosegrant et al. (2003) conducted a study on the global water scenario, focusing on the issue of water and food security at a global level for 2025. In contrast to other recent research endeavors, the research incorporates the most comprehensive compilation of factors that drive change, encompassing population expansion, urbanization, economic development, technological advancements, policies, and water availability limitations. The research conducted by Alcamo et al. (2003) examined the alterations in water withdrawals that may occur in 2025, assuming that present patterns in population, economy, and technology persist, under future business-as-usual conditions. Alcamo et al. (2007) conducted a revised evaluation that differentiated between two distinct developmental pathways for population and economic expansion. A similar focus by Shen et al. (2008) examined alterations in water withdrawals across the agricultural, industrial, and domestic sectors during 2020, 2050, and 2070. Furthermore, the interconnections among patterns of consumption, trade, and social and economic progress have not been comprehensively incorporated. Water footprint scenarios for 2050 may change by considering various factors that may influence change, such as population growth, economic growth, production and trade patterns, consumption patterns (including dietary changes and bioenergy usage), and technological advancements (Zhuo et al., 2016). Blue and green water consumption, rather than solely blue water withdrawal volumes and then water pollution by using the gray water footprint. Lastly, it evaluates water consumption in the agricultural, domestic, and industrial sectors and breaks down consumption by major commodity groups. At last, it integrates all significant drivers of change within a unified and consistent framework. The integration of all essential factors is of utmost importance in establishing effective policies for prudent water management. This approach can assist policymakers in comprehending the enduring implications of their decisions, even beyond political and administrative borders. The present study has opted to examine WF scenarios, as opposed to water withdrawal scenarios that have been the focus of previous research endeavors.

6.2.3 Idea to implement field study

The primary focus of the developmental phase was the measurement and assessment of water footprints of various crops, VWT associated with crop production, and WF concerning national consumption (Hoekstra & Hung, 2002). The fundamental framework used for the estimation of the national water footprint was the accounting scheme. Hoekstra and Chapagain (2007, 2008) enhanced the national WF accounts by incorporating various types of consumption and trade, encompassing animal and industrial products, as well as municipal water usage. Subsequently, the scope expanded to encompass the production perspective, which garnered greater attention due to the increase interest from corporations that became aware of the applicability of the WF concept . One of the driving factors to examine and consolidate the water footprints of production activities in specific geographical regions, to contextualize them within the constraints of limited water resources in those areas. The progress made has led to the expansion of a more comprehensive conceptual framework. This framework facilitates the measurement of water footprints at various levels, including the WF of individual processes or activities, the WF of products, the WF of consumption at the individual or community level, the WF of production in a specific region, and the operational and supply-chain WF of companies (Fig. 6.2).

The terminology used to describe water consumption per unit of product has evolved. This allows for consistency when aggregating the WF of individual products on a larger scale, such as the WF of a basket of products, or even further, the WF of a consumption pattern or diet (Hoekstra et al., 2011).

6.2.4 Water Footprints: Understanding Virtual Water Concepts

A philosophical discourse has arisen regarding the ontological status of Virtual World Technology and the coherence of World Farms). As per the studies conducted by Merett (2003) and Wichelns and Qadir (2015), it can be inferred that nations tend to import food rather than virtual water. The fundamental criticism posits that the concepts of Virtual Water Trade (VWT) and WF exhibit redundancy, thereby lacking the potential to augment comprehension. According to Wichelns and Qadir (2015), the notion that countries can conserve water by importing virtual water is untenable. According to certain economists, trade pertains exclusively to tangible goods and not intangible or digital items, while the term "saving" denotes a particular type of economic efficiency enhancement. The adoption of a rigid neoclassical economic framework may impede the recognition of the concepts of virtual work teams and work flexibility. However, it is precisely in this context that their potential benefits can be fully appreciated. Jordan, a nation facing severe water scarcity, has outsourced 86% of its water footprint. The country is a significant importer of virtual water, with a net gain, and has saved approximately 7 billion cubic meters of water annually through trade. This volume of water is equivalent to the amount that would have been necessary if Jordan had produced all of the imported goods domestically (Schyns et al., 2015).

The validity of the VWT concept has been subject to scrutiny, particularly in light of the "virtual water hypothesis," which posits that nations experiencing water scarcity ought to import products that require high amounts of water from countries with ample water resources (Merett, 2003). The hypothesis, which is attributed to Allen et al. (2011), is

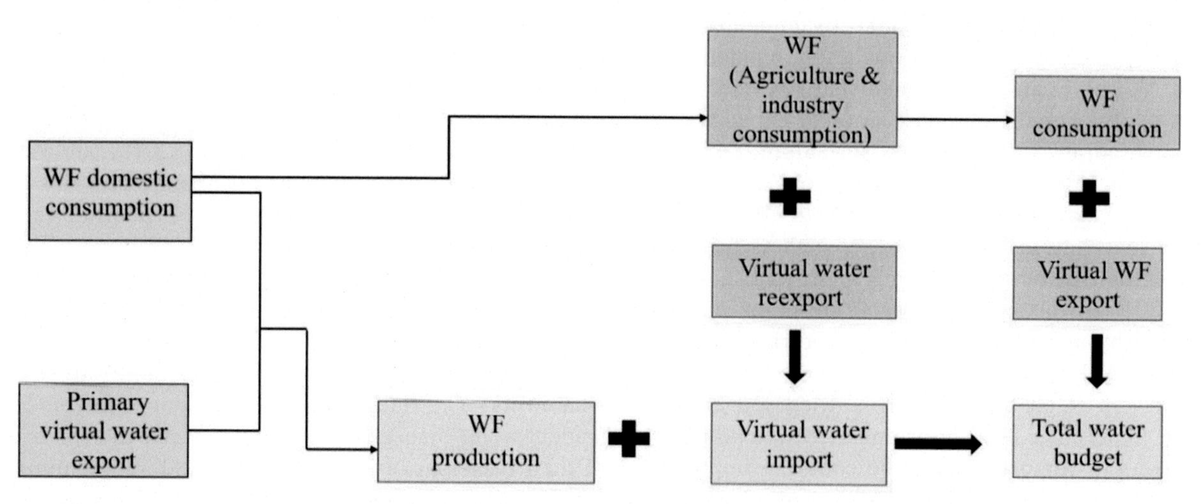

FIGURE 6.2 Relation between the water footprints (of production and consumption) and virtual water trade. *Courtsey: Hoekstra, A. Y., Chapagain, A. K., Aldaya, M. M., & Mekonnen, M. M. (2011). The water footprint assessment manual: Setting the global standard.*

deemed nonexistent and is founded on a misinterpretation. The significance of incorporating VWT in the discourse on national water security and global interdependencies is highlighted. Numerous scholars, such as Allen et al. (2011), have proposed exploring the possibility of augmenting net virtual water import in nations facing water scarcity. However, it is crucial to note that this recommendation differs fundamentally from the notion that these countries ought to escalate their importation efforts. The interpretation of VWT as a trade policy solution for the global water crisis should be avoided (Horlemann & Neubert, 2007).

6.3 Virtual water unveiled: a comprehensive exploration

The concepts of virtual water and WF were developed as useful mechanisms for tackling the issue of water scarcity. The notion of "virtual water," as introduced by Tony Allan in 1993, pertains to the water used in the manufacturing of any given commodity (Hoekstra, 2003). The three types of WF comprise the green, blue, and gray categories (Mekonnen & Hoekstra, 2012). Virtual water and water footprint are two methods used to approximate the amount of water present in a given product or service. According to Velázquez et al. (2011), the distinction between virtual water and WF lies in their respective definitions, with virtual water being defined in terms of production and WF being defined in terms of consumption. The notion of virtual water is commonly employed in discussions about commerce (Chapagain et al., 2006; Dalin et al., 2012; Zeitoun et al., 2010). According to Zeitoun et al. (2010), the water used in the production of a traded commodity can be perceived as "traded" alongside the product itself. According to Novo et al. (2009), the exchange of products in international trade results in the occurrence of virtual water "flow." Virtual water trade occurs through the transfer of water resources that are embedded in products traded across international borders (Duarte et al., 2014).

The notion of "virtual water" has been in circulation since the 1990s, positing that the entirety of the water used in the production of a given commodity remains integrated within said commodity as it traverses the value chain. The notion of virtual water allows us to assert that a bowl of cereal devoid of any moisture embodies the entirety of the water resources that were required to cultivate and reap the crops, purify the grain, eliminate extraneous substances, refine the ultimate product, and fabricate the packaging material. It is a measure of the indirect water consumption that is associated with the production of a particular product or service. The virtual water content of a product or service takes into account the water that is used in all stages of its production, including the extraction of raw materials, processing, packaging, and transportation.

Water is consumed by individuals not solely through direct ingestion or bathing. In 1993, Professor John Allan, who was awarded the Stockholm Water Prize in 2008, introduced the concept of "virtual water." This concept quantifies the amount of water that is used in the manufacturing and exchange of food and consumer goods. In developed countries, it is estimated that the widely consumed hamburger necessitates approximately 2400 liters of water. On a per capita basis, the virtual water consumption of Americans is approximately 6800 liters per day, which is more than three times the amount consumed by an individual in China. The concept of virtual water has significant implications for global trade policy and research, particularly in regions where water resources are limited. It has also transformed the conversation surrounding water policy and management. The virtual water concept, as presented by Hoekstra and Hung (2002), has facilitated more efficient water usage by elucidating the mechanisms and rationales behind the annual exportation of billions of liters of water by nations such as the United States, Argentina, and Brazil, and the corresponding importation of billions of liters by nations such as Japan, Egypt, and Italy. Global virtual water trade was very much higher for certain period in different uses (Hoekstra and Hung, 2002; Chapagain & Hoekstra, 2003). The majority of this trade, approximately 67%, was related to the international trade of crops, while 23% was attributed to the trade of livestock and livestock products and the remaining 10% was associated with the trade of industrial products (Fig. 6.3).

Similar estimates have been reported in other global studies on global virtual water trade (Zimmer & Renault, 2003). There exists an inequitable distribution of global virtual water trade among nations. The primary nations that export virtual water in significant quantities are the United States, Canada, Australia, Argentina, and Thailand. Japan, Sri Lanka, Italy, South Korea, and the Netherlands are nations that exhibit a substantial net import of virtual water. The virtual trading of blue water reveals notable disparities emerging in China, Pakistan, India, and the Middle East due to the diminishing availability of water for irrigation and shifting demographics over the course of the century. By 2100, China is projected to be a significant contributor to the global virtual water export market, primarily through the exchange of wheat and rice commodities (Zhuo et al., 2016). It is noteworthy that China is projected to transition from an importing nation to an exporting nation in the coming years, due to a decrease in growth rate that will ultimately lead to a decline in population after 2030. The reduction of domestic demands enables the allocation of surplus production toward fulfilling global agricultural requirements. According to Graham et al. (2020), the regions in Africa are characterized by an inverse relationship between population growth and domestic production, leading to a surge in demand that remains unfulfilled.

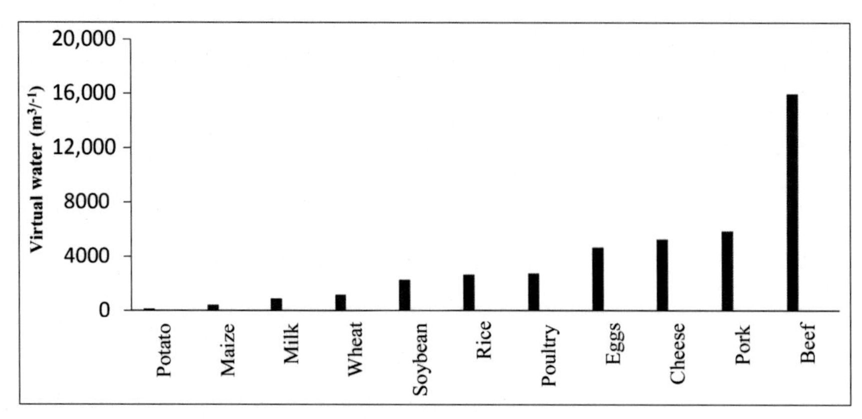

FIGURE 6.3 Use of virtual water content in different products (*Courtsey*: Chapagain & Hoekstra, 2003).

6.3.1 The global water footprint

Global estimates of consumptive water usage for several crops across multiple nations were originally calculated by Hung (2002), although they did not differentiate between green, blue, and gray water at that time. Since then, a plethora of international research has been conducted (Chapagain & Hoekstra, 2011; Chapagain et al., 2006; Mekonnen & Hoekstra, 2012). Chapagain and Hoekstra (2004) report that between 1997 and 2001, a total of 6390 Gm^3/year of water was used in the production of crops around the world. Because the virtual water content of animal products also includes water used in feed production, the former tends to have a lower virtual water content than the latter. In the latter, a total of 7404 Gm^3/year of water was used in crop production and 9087 Gm^3/year of water in all total production including industrial, domestic, and animal products (Enne et al., 2006; Gerbens-Leenes et al., 2013; Hoekstra, 2003; Hung, 2002; Mekonnen & Hoekstra, 2012; Oki et al., 2003; Palhares & Pezzopane, 2015; Table 6.2).

The freshwater usage of 13 major rice-producing countries (90% of global production, average yield 4.49 t/ha) was evaluated by Chapagain and Hoekstra (2011). In contrast to earlier studies that omitted gray WF and to that by Chapagain and Hoekstra (2004) that added an average constant percolation loss (300 mm) to green and blue water. The virtual water content of products was low in the northern and western European Union (EU), while the southern and eastern EU had the highest (Table 6.3). According to Mekonnen and Hoekstra (2012), most of the former literature values on a worldwide scale were not representative of global variability and could not be used to draw clear conclusions. From a statistical analysis emerged that some former literature data was too much approximated and that WF had a lower impact than what was attributed.

There were about 1325 m^3/t of water in rice (48% green, 44% blue, and 8% gray, on average). Fig. 6.4 shows the subdivision among cereals (e.g., maize, rice, wheat) according to their presence in studies in which these crops were analyzed.

The virtual water content of rice (broken) that a

consumer buys in the shop is about 3420 m^3/t. This is larger than the virtual water content of paddy rice as harvested from the field because of the weight loss if paddy rice is processed into broken rice. The virtual water content of some selected crop and livestock products for several selected countries is presented in Table 6.3. In general, livestock products have a higher virtual water content than crop products. This is because a live animal consumes a lot of feed crops, drinking water and service water in its lifetime before it produces some output. The higher we go up in the product chain, the higher will be the virtual water content of the product. However, the virtual water content of products strongly varies from place to place, depending upon the climate, technology adopted for farming. and corresponding yields (Fig. 6.5).

6.3.2 Impact of crop production on water resources

The amount of water used in crop production is known as its "Water footprint". The methodology given in the WF manual created by the WF Network (Hoekstra et al., 2011; WF Network, 2013) was followed to perform WF on crop production. Fig. 6.6 shows the graphical summary of WF of crop production.

TABLE 6.2 Water footprint of agricultural productions.

Global WF of production (Gm^3/year) (1996–2005)			
sectors	**Green**	**Blue**	**Gray**
Agricultural crop production	5771	899	733
Industrial production	–	38	362
Domestic water supply	–	42	282
Pasture and animal raising	913	46	–

Courtsey: Mekonnen and Hoekstra (2011).

TABLE 6.3 Use of water footprints in different area for different products.

Area	Products	Component	Author
China	Maize, wheat, vegetables, melons, oil plants, and cotton	Green, blue, gray	Xu et al. (2015)
China	Maize and wheat	Green,blue	Sun et al. (2013)
World (China, Indonesia, United States, European Union)	Multiple (wheat, maize, potatoes, beets, cane, vegetables, citrus, fruits, groundnuts, etc.)	Green, blue	Bruinsma (2003)
China, India, Indonesia, Bangladesh, Vietnam, Thailand, Myanmar, Philippines, Brazil, Japan, United States, Pakistan, Korea	Rice	Green, blue, gray	Chapagain and Hoekstra (2010); Chapagain and Hoekstra (2011)
Japan	Rice, wheat, soybean, maize, barley, beef, pork, chicken, and egg	Green, blue	Oki et al. (2003)
Indonesia	Rice, maize, cassava, soybeans, groundnuts, coconuts, oil palm, bananas, coffee, and cocoa	Green, blue, gray	Bulsink et al. (2010)
China	Wheat, rice, maize, soybeans, and potatoes	Green, blue	Cao et al. (2015)
World	Temperate cereals, rice, maize, tropical cereals, pulses, temperate roots, tropical roots, sunflower, soybeans, groundnuts, and rapeseed	Green, blue	Fader et al. (2011)

Courtsey: Lovarelli, D., Bacenetti, J., & Fiala M. (2016). Water footprint of crop productions: A review. *Science of the Total Environment. 548–549,* 236–251, https://doi.org/10.1016/j.scitotenv.2016.01.022. 18791026. Elsevier B.V., Italy. http://www.elsevier.com/locate/scitotenv.

6.3.3 The water footprint: A look on past, present and tomorrow

The resources of water, energy, and food (WEF) are essential for sustaining human life and facilitating socioeconomic progress. The escalating population growth, urbanization, and alterations in dietary patterns have led to a surge in the focus on the utilization of WEF, as well as their interdependencies in contemporary times. The WEF nexus framework represents a fresh viewpoint aimed at tackling intricate interrelationships and discerning potential synergies and trade-offs among the aforementioned sectors. The prominence of the issue has grown among policymakers, in part due to its connection to the Sustainable Development Goals outlined in the post-2015 agenda (Hák et al., 2016). Currently, the majority of research on water scarcity, efficiency, and productivity solely focuses on blue water, while neglecting the role of green water. As previously stated, in the context of worldwide food production, green water is the primary

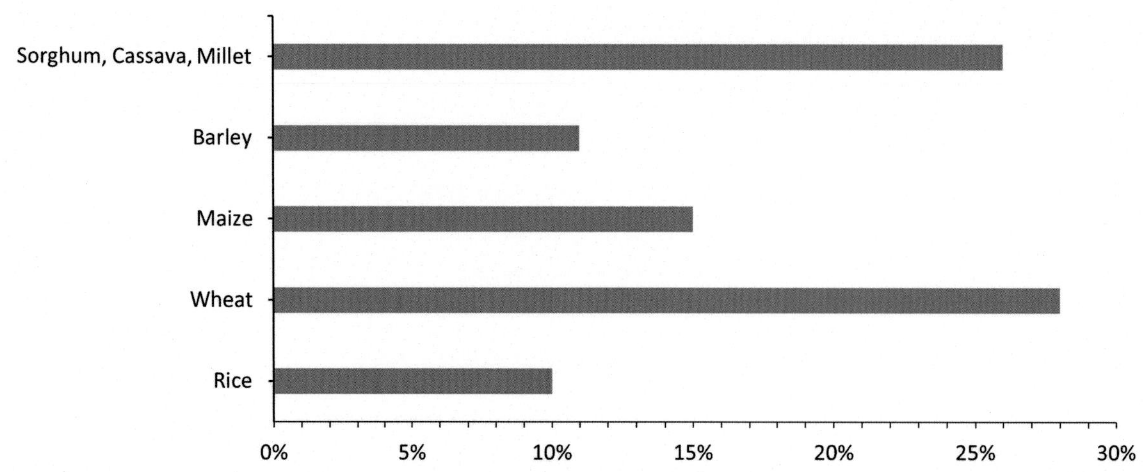

FIGURE 6.4 .Categorization of cereals such as maize, rice, wheat, etc. *Courtsey: Lovarelli, D., Bacenetti, J., & Fiala M. (2016). Water footprint of crop productions: A review.* Science of the Total Environment. 548–549, *236–251, https://doi.org/10.1016/j.scitotenv.2016.01.022.18791026. Elsevier B.V., Italy. http://www.elsevier.com/locate/scitotenv.*

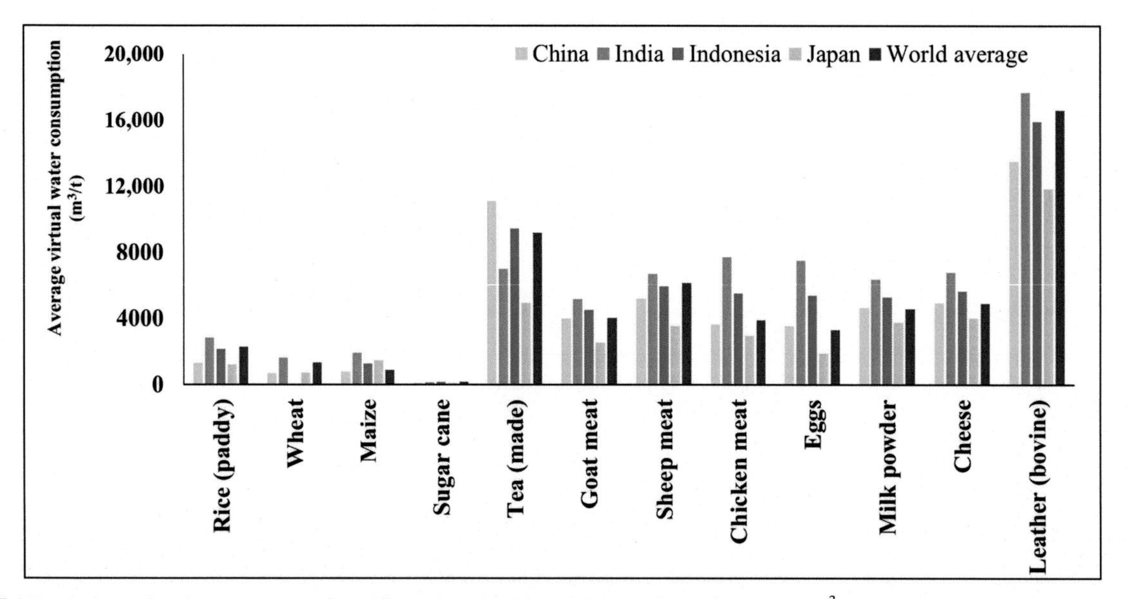

FIGURE 6.5 Average virtual water content of specific products in Asian countries with world average (m^3/t). *Courtsey: Hoekstra et al. (2007).*

contributor and assumes a more significant role than blue water. In the context of climate change, global warming is expected to impact regional water resources and agricultural patterns by altering rainfall patterns and their spatial distribution. This, in turn, is anticipated to affect food production and energy consumption.

6.3.3.1 Concept of material metabolism is a significant aspect in the field of sustainability

The relationship between human activities and ecosystems within the specified region is analogous to a metabolic process characterized by the exchange of materials and the transfer of energy. Material flow analysis is a widely employed approach for studying material metabolism. This method entails examining the origin, trajectory, and destination of material circulation. The field of material metabolism research examines the inflow of natural resources into the socioeconomic system and the outflow of pollutants into the ecoenvironmental system using material flow analysis. This enables an assessment of the interplay between human activities and the natural environment. The material flow account is a structured system of accounting that is used to quantify the utilization of human material and its ecological repercussions. The application of this method has been observed at various scales including global, national, and city levels, and has resulted in the development of a comprehensive theoretical framework.

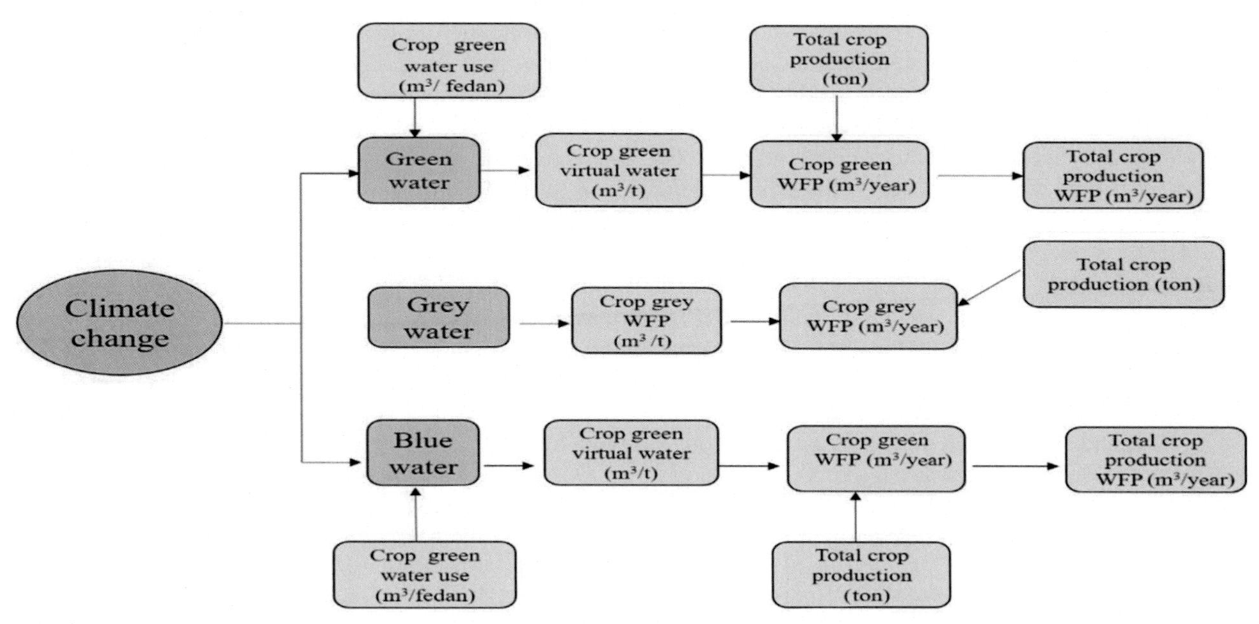

FIGURE 6.6 Model structure of WFP of crop production sector. *Courtsey: El-Gafy, I. (2014). System dynamic model for crop production, water footprint, and virtual water nexus.* Water Resources Management. 28*(13), 4467–4490, 15731650. https://doi.org/10.1007/s11269-014-0667-2. Kluwer Academic Publishers, Egypt. http://www.wkap.nl/journalhome.htm/0920-4741.*

6.3.3.2 *Evaluation of Telecoupling's Potential Future Impacts*

In the 21st century, sustainable development is considered to be the most significant challenge that humanity must address. In the realm of water usage, it is crucial to ensure not only the sufficient allocation of water resources to promote social equity and facilitate economic efficiency, but also to monitor and control water pollution within predetermined limits. In contemporary times, remote areas are interconnected and mutually impacted through various means, while the allocation of resources is determined by the market economy's supply and demand dynamics. The interdependence of the "social-economic-environmental" system within a particular region and that of other regions is established through trade. The ongoing trade dispute between China and the United States has resulted in increased consumption costs for both nations. Furthermore, it has had, or is projected to have, significant ramifications on the agricultural framework, the revenue of farmers, and the provision of ecosystem services in various other countries. The framework of telecoupling was proposed by a group of scholars and is used to assess the interrelatedness of social, economic, and environmental factors between geographically separated human–natural systems. This approach is regarded as a novel viewpoint for addressing multifaceted obstacles confronting worldwide sustainable development.

6.4 Assessing water safety: comparing blue and green water footprint models

The Water Footprint concept, as proposed by Hoekstra and Hung (2002) and further developed by Hoekstra et al. (2011), serves to establish a correlation between the depletion of water resources and the growth of the human population. BWF refers to the amount of water consumed by humans from blue water resources. This quantity can be measured by determining the volume of surface and groundwater used during the production of a particular good or service, such as domestic, industrial, power production, or irrigation activities. The term "Green water footprint" pertains to the utilization of green water resources, such as the process of evapotranspiration from agricultural and forested regions (Hoekstra et al., 2011; Rodrigues & De la Riva, 2014). The quantification of water security, specifically water scarcity and vulnerability, can be achieved through the calculation of the ratio between water consumed and water available. This metric is widely recognized as a global standard (Hoekstra et al., 2011; Rodrigues & De la Riva, 2014) for identifying and mapping areas of high water stress. The BWF is positively correlated with human water consumption, while the GWF is subject to indirect influence from human activities, such as agriculture. Consequently, the BWFs and GWFs are interconnected with the utilization of water by humans. The analysis of water footprint in spatiotemporal terms examines the potential connection between climatic factors, specifically water supply, and anthropogenic factors, specifically water demand, and how these connections change over time.

Hoekstra and Mekonnen (2012), incorporation of green water consumption in the analysis can render WF pertinent to the metric of environmental sustainability. According to Liu et al. (2013), a significant proportion of the GWF, amounting to almost 80%, is linked to worldwide agricultural production. This encompasses various products such as wheat (Mekonnen & Hoekstra, 2012), animal-based items (Mekonnen, & Hoekstra, 2010b, 2012), cotton (Chapagain et al., 2006), and bioenergy (Gerbens-Leenes et al., 2009). The ecosystem of a stream can be negatively impacted by the overuse of blue water, thus necessitating the implementation of the EFR principle to ensure the preservation of a robust river system (Honrado et al., 2013). The utilization of the presumptive standard method (Richter, 2010; Richter et al., 2012) has the potential to facilitate the analysis of EFR, as well as the assessment of the blue water accessibility within a given stream. Based on the presumptive standard, it can be inferred that the extraction of stream flow exceeding 20% is likely to result in a decline in ecological well-being and environmental equilibrium. Several techniques have been devised to compute EFR, as documented by Tharme (2003), Hoekstra et al. (2011), and Rodrigues and De la Riva (2014). These methods have also been employed to conduct water security assessments. Several methodologies have been suggested for the examination of BWFs and GWFs. CROPWAT (Allen et al., 1998), Aqua Crop model (Heng et al., 2009), and the Soil and Water Assessment Tool (SWAT; Arnold et al., 1998) are widely used models for the assessment of blue and green water resources. Abbaspour et al. (2015) conducted an estimation of green and blue water for continental Europe. Zhang et al. (2012) employed the SWAT model to assess the availability of blue and green water in Africa and the Heihe River Basin in northeast China, respectively.

6.5 Methods for assessing environmental sustainability

The significance of sustainability in business practices has led to the emergence of sustainable supply chain management as a crucial subsequent measure for the operations function (Dey et al., 2011; Hassini et al., 2012). The connection between accounting or measurement methods and their implementation in production and supply chains/networks to evaluate sustainability performance is a significant component of environmentally sustainable supply chain management. Schaltegger and Wagner (2006) propose that sustainability performance measurement aims to encompass the social, environmental, and economic dimensions of corporate management, with a specific focus on sustainability management. The pursuit of sustainability has resulted in a growing emphasis on adopting a lifecycle perspective. In the context of environmental sustainability, the overarching framework of lifecycle assessment serves as the foundation for gauging environmental performance at the product level (Heijungs et al., 2010). This framework can also be extended to the industry level, as observed by Joshi (1999). The notion of lifecycle thinking is advocated on the basis that it encompasses all supply chain activities involved in the production and consumption processes, even at the industrial level. The framework encompasses all supply chain activities, such as raw materials extraction, processing, and transportation, as well as value-adding activities, that are involved in the production of goods and services. According to Koh et al. (2013), implementing a bottom-up performance measurement approach at the product level of the value chain can provide advantages such as the evaluation of carbon hot spots and the mapping of the supply chain. The effectiveness of sectorial policies is subject to certain limitations, such as the truncation error (Feng et al., 2014) and the challenge of scaling up from the product level. Thus, it suggests mitigating some of these constraints by conducting environmental performance evaluations from an industrial standpoint. A performance measurement approach that operates from a top-down perspective, such as at the industrial level, can offer a comprehensive perspective of value chains. This can include an assessment of the overall performance of particular industries, which may reveal opportunities to evaluate the efficacy of policies that have been implemented in those industries. Furthermore, a particular sector within a nation can evaluate its efficacy in analogous industries in disparate regions and nations. Conducting environmental performance assessments at the industrial level and connecting the results to sustainable supply chain tactics is of significant importance. This is because industrial systems, as highlighted by Azapagic and Perdan (2000), play a crucial role in the human economy by regulating the movement of materials and energy, thereby contributing to environmental deterioration and resource exhaustion. Hence, the industry must assume a pivotal position in recognizing and executing sustainable alternatives.

In general, the environmental performance measurements are twofold. Initially, the aim is to establish a connection between ecological systems and corporate activities, as well as in the context of ecological sustainability reporting (Clarkson et al., 2011; Melnyk et al., 2003). The second fold is to establish a correlation between environmental management and business as well as competitive strategy (Hart & Milstein, 2003; Porter & Kramer, 2006; Wagner & Schaltegger, 2003). Using suitable analytical frameworks for model development and the outcomes produced as a mechanism for advising the administration of supply chains is an essential aspect of sustainable supply chain design and management. This perspective has been recently reaffirmed by several authors. Taticchi et al. (2013) observe that the field of sustainable supply chain performance measurements is still in its nascent stages, but is rapidly expanding. Conversely, Schaltegger and Burritt (2014) note that there is a dearth of research on sustainability performance issues, including available methodologies and potential strategies for measuring and overseeing sustainable supply chains.

6.6 Balancing Act: Optimizing water allocation for a sustainable future

The field of Water Footprint Assessment (WFA) is a methodological approach that examines the interconnection between human consumption and the utilization of freshwater resources for the production of goods and services that are consumed. Water footprint is a comprehensive metric that quantifies the amount of freshwater used by a producer or consumer, either directly or indirectly. The WF associated with consumption can be divided into two components: the GWF, which pertains to the consumption of rainfall, and the GWF, which pertains to the consumption of surface and groundwater. On the other hand, the water footprint associated with pollution is commonly referred to as the gray water footprint, which represents the volume of water required to assimilate pollutants that enter freshwater bodies (Hoekstra et al., 2011). The WF of individuals residing in a particular region comprises an intrinsic component, which pertains to the consumption and contamination of freshwater resources within the region, and an extrinsic component, which pertains to the freshwater consumption and contamination associated with the production of goods imported and consumed within the region under consideration. The exchange of water-intensive commodities results in a conservation of water resources for the importing region, as it obviates the necessity of using its internal water resources for local production. The concept of maximum sustainable BWF within a river basin pertains to the quantification of the amount of renewable freshwater that can be used for consumptive purposes annually. This evaluation takes into account the available runoff, storage capacity, and environmental flow requirements. It is possible to differentiate between the maximum sustainable surface water footprint and the maximum sustainable groundwater footprint, taking into account factors such as the recharge of groundwater, the acceptable level of decline in groundwater, and the necessary base flow for the river. The determination of the maximum sustainable gray water footprint in a river basin is contingent upon the quantity of available runoff that can effectively assimilate pollutants. The examination of water use equity may be conducted through the comparative analysis of WF among various communities, as well as through the comparison of individual WFs against equitable allocations, given the constraints of limited availability.

6.7 Policy Imperativeness

Water Footprint and VWT have raised water awareness, but their policy importance is unclear (Chenoweth et al., 2014). Two items may have the same WF but differing environmental impacts, making it risky to use the WF to influence environmental policy. For the same reason, there are concerns about disclosing the WF on product labels or using it in product site certification (Postle 2011). The objection is based on the idea that the WF measure should give a complete, immediate response. Wichelns (2015) believes that the WF measure is inadequate for tracking company, consumer, or country water sustainability due to such expectations. However, expecting an indicator to tell you what to do is simplistic.

The different Governmental policies and corporate strategies can be informed based on a full WFA, not just based on one number but with different assumptions. The Dutch Environmental Agency notes that instead of revealing their overall WF in their sustainability reports, companies would do better to report progress made in reducing the separate components of their WF in unsustainable hotspots. The strength of this approach would be the involvement of distant consumers, producers, retailers, and investors, in addition to local stakeholders and authorities, in addressing water problems in hotspot areas (Witmer & Cleij, 2012). Over the past few years, an increasing number of companies and governments have found or started to explore the relevance of WFA. A good governmental example is the United Kingdom's Environmental Agency which carried out a detailed WFA for the Hertfordshire and North London Area to assist water resources and water quality regulators in managing the quantity and quality of water resources sustainably. Other examples are the Spanish government adopting a regulation that requires WFA as part of the process of developing river basin plans (Aldaya et al., 2010; Zhang et al., 2014) and the Indian government including the goal of WF reduction in its draft national water framework bill (GoI, 2016). WFA is a partial analysis, as any other analysis, and should always be integrated into or combined with other analyses for developing water or other policies.

6.8 Conclusions

Water footprint discipline innovates water management by incorporating new perspectives. By uncovering indirect sources of local water problems, it prepares to research what may be done "elsewhere" to enhance water sustainability and equality. WFA enhanced the analysis to include more relevant participants by bringing supply-chain thinking into water management. Future resources include water, which is needed to produce food and electricity. To advance WFA, we must understand the many stakeholders' roles in water governance models that balance ecological preservation, social fairness, economic efficiency, and supply stability. Sustainable supply chain management principles are aligning. Measuring the efficiency of eco-friendly supply chains is difficult. This is due to supply network complexity, many indicators to evaluate their efficacy, and methodological limitations. This study suggests that the global supply chain should design, monitor, and analyse industry-level environmental sustainability models due to value-added activities.

References

Abbaspour, K. C., Rouholahnejad, E., Vaghefi, S., Srinivasan, R., Yang, H., & Kløve, B. (2015). A continental-scale hydrology and water quality model for Europe: Calibration and uncertainty of a high-resolution large-scale SWAT model. *Journal of Hydrology, 524*, 733–752. Available from https://doi.org/10.1016/j.jhydrol.2015.03.027, 00221694. Elsevier, Switzerland, http://www.elsevier.com/inca/publications/store/5/0/3/3/4/3.

Akita, F., & Matsumoto, N. (2004). Hydrological responses induced by the Tokachi-oki earthquake in 2003 at hot spring wells in Hokkaido, Japan. *Geophysical Research Letters, 31*(16). Available from https://doi.org/10.1029/2004GL020433, L16603-4.

Alcamo, J., Döll, P., Henrichs, T., Kaspar, F., Lehner, B., Rösch, T., & Siebert, S. (2003). Global estimates of water withdrawals and availability under current and future "business-as-usual" conditions. *Hydrological Sciences Journal, 48*(3), 339–348. Available from https://doi.org/10.1623/hysj.48.3.339.45278.

Alcamo, J., Flörke, M., & Märker, M. (2007). Future long-term changes in global water resources driven by socio-economic and climatic changes. *Hydrological Sciences Journal, 52*(2), 247–275. Available from https://doi.org/10.1623/hysj.52.2.247, 02626667.

Aldaya, M. M., Allan, J. A., & Hoekstra, A. Y. (2010). Strategic importance of green water in international crop trade. *Ecological Economics, 69*(4), 887–894. Available from https://doi.org/10.1016/j.ecolecon.2009.11.001, 09218009.

Aldaya, M. M., & Hoekstra, A. Y. (2010). The water needed for Italians to eat pasta and pizza. *Agricultural Systems, 103*(6), 351–360. Available from https://doi.org/10.1016/j.agsy.2010.03.004, 0308521X.

Allen, D. E., Singh, B. P., & Dalal, R. C. (2011). *Soil health indicators under climate change: A review of current knowledge* (pp. 25–45). Springer Science and Business Media LLC. https://doi.org/10.1007/978-3-642-20256-8_2.

Allen, R. G., Pereira, L. S., Raes, D., & Smith, M. (1998). Crop evapotranspiration—Guidelines for computing crop water requirements—FAO Irrigation and Drainage Paper 56. 300.

Arnold, J. G., Srinivasan, R., Muttiah, R. S., & Williams, J. R. (1998). Large area hydrologic modeling and assessment part I: Model development. *Journal of the American Water Resources Association, 34*(1), 73–89. Available from https://doi.org/10.1111/j.1752-1688.1998.tb05961.x. 1093474X. Blackwell Publishing Inc., United States. http://www.blackwellpublishing.com.

Azapagic, A., & Perdan, S. (2000). Indicators of sustainable development for industry: A general framework. *Process Safety and Environmental Protection, 78*(4), 243–261. Available from https://doi.org/10.1205/095758200530763. 09575820. Institution of Chemical Engineers, United Kingdom. http://www.elsevier.com/wps/find/journaldescription.cws_home/713889/description#description.

Brandenburg, M., Govindan, K., Sarkis, J., & Seuring, S. (2014). Quantitative models for sustainable supply chain management: Developments and directions. *European Journal of Operational Research, 233*(2), 299–312. Available from https://doi.org/10.1016/j.ejor.2013.09.032. 03772217. Elsevier B.V., Germany. https://www.journals.elsevier.com/european-journal-of-operational-research/.

Bruinsma, J. (2003). World agriculture: Towards 2015/2030: An FAO perspective. Earthscan.

Bulsink, F., Hoekstra, A. Y., & Booij, M. J. (2010). The water footprint of Indonesian provinces related to the consumption of crop products. *Hydrology and Earth System Sciences, 14*(1), 119–128. Available from https://doi.org/10.5194/hess-14-119-2010. 16077938. Copernicus GmbH, Netherlands. http://www.hydrol-earth-syst-sci.net/volumes_and_issues.html.

Calzadilla, A., Rehdanz, K., & Tol, R. S. J. (2011). Water scarcity and the impact of improved irrigation management: A computable general equilibrium analysis. *Agricultural Economics, 42*(3), 305–323. Available from https://doi.org/10.1111/j.1574-0862.2010.00516.x, 15740862.

Cao, Q., Yu, D., Georgescu, M., Han, Z., & Wu, J. (2015). Impacts of land use and land cover change on regional climate: A case study in the agro-pastoral transitional zone of China. *Environmental Research Letters, 10*(12), 124025. Available from https://doi.org/10.1088/1748-9326/10/12/124025, 1748-9326.

Chapagain, A. K., Hoekstra, A. Y., Savenije, H. H. G., & Gautam, R. (2006). The water footprint of cotton consumption: An assessment of the impact of worldwide consumption of cotton products on the water resources in the cotton producing countries. *Ecological Economics, 60*(1), 186–203. Available from https://doi.org/10.1016/j.ecolecon.2005.11.027, 09218009.

Chapagain, A. K., & Hoekstra, A. Y. (2011). The blue, green and grey water footprint of rice from production and consumption perspectives. *Ecological Economics, 70*(4), 749–758. Available from https://doi.org/10.1016/j.ecolecon.2010.11.012, 09218009.

Chapagain, A. K., & Hoekstra, A. Y. (2008). The global component of freshwater demand and supply: An assessment of virtual water flows between nations as a result of trade in agricultural and industrial products. *Water International, 33*(1), 19–32. Available from https://doi.org/10.1080/02508060801927812, 02508060.

Chapagain, A. K., & Hoekstra, A. Y. (2003). Virtual water flows between nations in relation to trade in livestock and livestock products. *UNESCO-IHE, 13*.

Chapagain, A. K., & Hoekstra, A. Y. (2004). Water footprints of nations.

Chapagain, A., & Hoekstra, A. Y. (2010). The blue, green and grey water footprint of rice from both a production and consumption perspective. (Value of water research report 40; No. 40). Unesco-IHE Institute for Water Education.

Chenoweth, J., Hadjikakou, M., & Zoumides, C. (2014). Quantifying the human impact on water resources: A critical review of the water footprint concept. *Hydrology and Earth System Sciences*, *18*(6), 2325–2342. Available from https://doi.org/10.5194/hess-18-2325-2014. 16077938. Copernicus GmbH, United Kingdom. http://www.hydrol-earth-syst-sci.net/volumes_and_issues.html.

Clarkson, P. M., Li, Y., Richardson, G. D., & Vasvari, F. P. (2011). Does it really pay to be green? Determinants and consequences of proactive environmental strategies. *Journal of Accounting and Public Policy*, *30*(2), 122–144. Available from https://doi.org/10.1016/j.jaccpubpol.2010.09.013, 02784254.

Dalin, C., Konar, M., Hanasaki, N., Rinaldo, A., & Rodriguez-Iturbe, I. (2012). Evolution of the global virtual water trade network. *Proceedings of the National Academy of Sciences*, *109*(16), 5989–5994. Available from https://doi.org/10.1073/pnas.1203176109, 0027-8424.

de Fraiture, C., & Wichelns, D. (2010). Satisfying future water demands for agriculture. *Agricultural Water Management*, *97*(4), 502–511. Available from https://doi.org/10.1016/j.agwat.2009.08.008, 03783774.

Dey, A., LaGuardia, P., & Srinivasan, M. (2011). Building sustainability in logistics operations: A research agenda. *Management Research Review*, *34* (11), 1237–1259. Available from https://doi.org/10.1108/01409171111178774, 20408269.

Duarte, T. M., Carinhas, N., Barreiro, L. C., Carrondo, M. J. T., Alves, P. M., & Teixeira, A. P. (2014). Metabolic responses of CHO cells to limitation of key amino acids. *Biotechnology and Bioengineering*, *111*(10), 2095–2106. Available from https://doi.org/10.1002/bit.25266. 10970290. John Wiley and Sons Inc., Portugal. http://www.interscience.wiley.com/jpages/0006-3592.

El-Gafy, I. (2014). System dynamic model for crop production, water footprint, and virtual water nexus. *Water Resources Management*, *28*(13), 4467–4490. Available from https://doi.org/10.1007/s11269-014-0667-2. 15731650. Kluwer Academic Publishers, Egypt. http://www.wkap.nl/journalhome.htm/0920-4741.

Enne, V. I., Delsol, A. A., Roe, J. M., & Bennett, P. M. (2006). Evidence of antibiotic resistance gene silencing in Escherichia coli. *Antimicrobial Agents and Chemotherapy*, *50*(9), 3003–3010. Available from https://doi.org/10.1128/AAC.00137-06, 00664804.

Ercin, A. E., & Hoekstra, A. Y. (2014). Water footprint scenarios for 2050: A global analysis. *Environment International*, *64*, 71–82. Available from https://doi.org/10.1016/j.envint.2013.11.019. 18736750. Elsevier Ltd, Netherlands. http://www.elsevier.com/locate/envint.

FAO. (2017). The future of food and agriculture – Trends and challenges. Rome. https://www.fao.org/3/i6583e/i6583e.pdf

Fader, M., Gerten, D., Thammer, M., Heinke, J., Lotze-Campen, H., Lucht, W., & Cramer, W. (2011). Internal and external green–blue agricultural water footprints of nations, and related water and land savings through trade. *Hydrology and Earth System Sciences*, *15*(5), 1641–1660. Available from https://doi.org/10.5194/hess-15-1641-2011, 16077938.

Falkenmark, M. (2000). Competing freshwater and ecological services in the river basin perspective: An expanded conceptual framework. *Water International*, *25*(2), 172–177. Available from https://doi.org/10.1080/02508060008686815, 02508060.

Feng, K., Hubacek, K., Pfister, S., Yu, Y., & Sun, L. (2014). Virtual scarce water in China. *Environmental Science and Technology*, *48*(14), 7704–7713. Available from https://doi.org/10.1021/es500502q. 15205851. American Chemical Society, United States. http://pubs.acs.org/journal/esthag.

Food and Agriculture Organization. (2007). AQUASTAT—FAO's information system on water and agriculture.

Garrido, N., Marinho, D. A., Barbosa, T. M., Costa, A. M., Silva, A. J., Pérez Turpin, J. A., & Marques, M. C. (2010). Relationships between dry land strength, power variables and short sprint performance in young competitive swimmers. *Journal of Human Sport and Exercise*, *5*(2), 240–249. Available from https://doi.org/10.4100/jhse.2010.52.12, 19885202.

Gerbens-Leenes, P. W., Hoekstra, A. Y., & van der Meer, T. (2009). The water footprint of energy from biomass: A quantitative assessment and consequences of an increasing share of bio-energy in energy supply. *Ecological Economics*, *68*(4), 1052–1060. Available from https://doi.org/10.1016/j.ecolecon.2008.07.013, 09218009.

Gerbens-Leenes, P. W., Mekonnen, M. M., & Hoekstra, A. Y. (2013). The water footprint of poultry, pork and beef: A comparative study in different countries and production systems. *Water Resources and Industry*, *1–2*, 25–36. Available from https://doi.org/10.1016/j.wri.2013.03.001, 22123717.

GoI, 2016: Annual report of government of India. https://dea.gov.in/sites/default/files/Annual%20Report-2016-17-E.pdf

Graham, J. L., Dubrovsky, N. M., Foster, G. M., King, L. R., Loftin, K. A., Rosen, B. H., & Stelzer, E. A. (2020). Cyanotoxin occurrence in large rivers of the United States. *Inland Waters*, *10*(1), 109–117. Available from https://doi.org/10.1080/20442041.2019.1700749. 2044205X. Taylor and Francis Ltd., United States. https://www.tandfonline.com/loi/tinw20.

Hák, T., Janoušková, S., & Moldan, B. (2016). Sustainable development goals: A need for relevant indicators. *Ecological Indicators*, *60*, 565–573. Available from https://doi.org/10.1016/j.ecolind.2015.08.003, 1470160X. Elsevier B.V., Czech Republic, http://www.elsevier.com/locate/ecolind.

Hart, S. L., & Milstein, M. B. (2003). Creating sustainable value. *Academy of Management Perspectives*, *17*(2), 56–67. Available from https://doi.org/10.5465/ame.2003.10025194, 1558-9080.

Hassini, E., Surti, C., & Searcy, C. (2012). A literature review and a case study of sustainable supply chains with a focus on metrics. *International Journal of Production Economics*, *140*(1), 69–82. Available from https://doi.org/10.1016/j.ijpe.2012.01.042, 09255273.

Heijungs, R., Huppes, G., & Guinée, J. B. (2010). Life cycle assessment and sustainability analysis of products, materials and technologies. Toward a scientific framework for sustainability life cycle analysis. *Polymer Degradation and Stability*, *95*(3), 422–428. Available from https://doi.org/10.1016/j.polymdegradstab.2009.11.010, 01413910.

Heng, D. Y. C., Xie, W., Regan, M. M., Warren, M. A., Golshayan, A. R., Sahi, C., Eigl, B. J., Ruether, J. D., Cheng, T., North, S., Peter, V., Knox, J. J., Chi, K. N., Kollmannsberger, C., McDermott, D. F., Oh, W. K., Atkins, M. B., Bukowski, R. M., Rini, B. I., & Choueiri, T. K. (2009). Prognostic factors for overall survival in patients with metastatic renal cell carcinoma treated with vascular endothelial growth factor-targeted agents: Results from a large, multicenter study. *Journal of Clinical Oncology*, *27*(34), 5794–5799. Available from https://doi.org/10.1200/JCO.2008.21.4809. 0732183X. http://jco.ascopubs.org/cgi/reprint/27/34/5794.

Hoekstra, A. K., & Chapagain, A. Y. (2007). Water footprints of nations: Water use by people as a function of their consumption pattern. *Water Resources Management*, *21*, 35–48. Available from https://doi.org/10.1007/s11269-006-9039-x.

Hoekstra, A. Y., Chapagain, A. K., Aldaya, M. M., & Mekonnen, M. M. (2011). *The water footprint assessment manual: Setting the global standard.*

Hoekstra, A. Y. (2003). Virtual water: An introduction. *Virtual water trade*, *13*.

Hoekstra, A. Y. (2017). Water footprint assessment: Evolvement of a new research field. *Water Resources Management*, *31*(10), 3061–3081. Available from https://doi.org/10.1007/s11269-017-1618-5, 15731650. Springer Netherlands, Netherlands, http://www.wkap.nl/journalhome.htm/0920-4741.

Hoekstra, A. Y., & Mekonnen, M. M. (2012). The water footprint of humanity. *Proceedings of the National Academy of Sciences*, *109*(9), 3232–3237. Available from https://doi.org/10.1073/pnas.1109936109, 0027-8424.

Hoekstra, A. Y., & Hung, P. Q. (2002). *Virtual Water Trade: A Quantification of Virtual Water Flows between Nations in Relation to Crop Trade.* Delft, The Netherlands: IHE.

Hoekstra, A. Y., & Hung, P. Q. (2005). Globalisation of water resources: international virtual water flows in relation to crop trade. *Global environmental change*, *15*(1), 45–56.

Hoekstra, M. (2009). The Effect of Attending the Flagship State University on Earnings: A Discontinuity-Based Approach. *Review of Economics and Statistics*, *91*(4), 717–724.

Honrado, J. P., Vieira, C., Soares, C., Monteiro, M. B., Marcos, B., Pereira, H. M., & Partidário, M. R. (2013). Can we infer about ecosystem services from EIA and SEA practice? A framework for analysis and examples from Portugal. *Environmental Impact Assessment Review*, *40*(1), 14–24. Available from https://doi.org/10.1016/j.eiar.2012.12.002, 01959255. Elsevier Inc., Portugal, http://www.elsevier.com/inca/publications/store/5/0/5/7/1/8.

Horlemann, L., & Neubert, S. (2007). Virtual water trade. A realistic concept for resolving the water crisis.

Hull, V., & Liu, J. (2018). Telecoupling: A new frontier for global sustainability. *Ecology and Society*, *23*(4). Available from https://doi.org/10.5751/ES-10494-230441. 17083087. Resilience Alliance, United States. https://www.ecologyandsociety.org/vol23/iss4/art41/ES-2018-10494.pdf.

Hung, A.H. P. (2002).Virtual water trade a quantification of virtual water flows between nations in relation to international crop trade.

Joshi, S. (1999). Product environmental life-cycle assessment using input-output techniques. *Journal of industrial ecology*, *3*(2-3), 95–120.

Karandish, F. (2019). Applying grey water footprint assessment to achieve environmental sustainability within a nation under intensive agriculture: A high-resolution assessment for common agrochemicals and crops. *Environmental Earth Sciences*, *659*, 807–820.

Koh, K. T., Park, C. J., Ryu, G. S., Park, J. J., Kim, D. G., & Lee, J. H. (2013). An experimental investigation on minimum compressive strength of early age concrete to prevent frost damage for nuclear power plant structures in cold climates. *Nuclear Engineering and Technology*, *45*(3), 393–400. Available from https://doi.org/10.5516/NET.09.2012.046, 2234358X. Korean Nuclear Society, South Korea, http://www.kns.org/jknsfile/v45/11-12-46.pdf.

Lambooy, T. (2011). Corporate social responsibility: Sustainable water use. *Journal of Cleaner Production*, *19*(8), 852–866. Available from https://doi.org/10.1016/j.jclepro.2010.09.009, 09596526.

Lamastra, L., Suciu, N. A., Novelli, E., & Trevisan, M. (2014). A new approach to assessing the water footprint of wine: An Italian case study. *Science of the total Environment*, *490*, 748–756.

Liu, J., Hull, V., Batistella, M., deFries, R., Dietz, T., Fu, F., Hertel, T. W., Izaurralde, R. C., Lambin, E. F., Li, S., Martinelli, L. A., McConnell, W. J., Moran, E. F., Naylor, R., Ouyang, Z., Polenske, K. R., Reenberg, A., Rocha, Gd. M., Simmons, C. S., ... Zhu, C. (2013). Framing sustainability in a telecoupled world. *Ecology and Society*, *18*(2). Available from https://doi.org/10.5751/ES-05873-180226. 17083087. Resilience Alliance, United States. http://www.ecologyandsociety.org/vol18/iss2/art26/ES-2013-5873.pdf.

Liu, J., Zhao, D., Mao, G., Cui, W., Chen, H., & Yang, H. (2020). Environmental sustainability of water footprint in Mainland China. *Geography and Sustainability*, *1*(1), 8–17. Available from https://doi.org/10.1016/j.geosus.2020.02.002. 26666839. Beijing Normal University Press, China. http://www.journals.elsevier.com/geography-and-sustainability.

Lodhia, S., & Hess, N. (2014). Sustainability accounting and reporting in the mining industry: Current literature and directions for future research. *Journal of Cleaner Production*, *84*(1), 43–50. Available from https://doi.org/10.1016/j.jclepro.2014.08.094, 09596526.

Lovarelli, D., Bacenetti, J., & Fiala, M. (2016). Water footprint of crop productions: A review. *Science of the Total Environment*, *548-549*, 236–251. Available from https://doi.org/10.1016/j.scitotenv.2016.01.022. 18791026. Elsevier B.V., Italy. http://www.elsevier.com/locate/scitotenv.

Marano, R. P., & Filippi, R. A. (2015). Water footprint in paddy rice systems. Its determination in the provinces of Santa Fe and Entre Ríos, Argentina. *Ecological Indicators*, *56*, 229–236. Available from https://doi.org/10.1016/j.ecolind.2015.03.027. 1470160X. Elsevier, Argentina. http://www.elsevier.com/locate/ecolind.

Mekonnen, M., Keesstra, S. D., Stroosnijder, L., Baartman, J. E. M., & Maroulis, J. (2015). Soil conservation through sediment trapping: A review. *Land Degradation and Development*, *26*(6), 544–556. Available from https://doi.org/10.1002/ldr.2308. 1099145X. John Wiley and Sons Ltd, Netherlands. http://onlinelibrary.wiley.com/journal/10.1002/(ISSN)1099-145X.

Mekonnen, M. M., & Hoekstra, A. Y. (2012). A global assessment of the water footprint of farm animal products. *Ecosystems*, *15*(3), 401–415. Available from https://doi.org/10.1007/s10021-011-9517-8, 14350629.

Mekonnen, M. M., & Hoekstra, A. Y. (2011). The green, blue and grey water footprint of crops and derived crop products. *Hydrology and Earth System Sciences*, *15*(5), 1577–1600.

Melnyk, S. A., Sroufe, R. P., & Calantone, R. (2003). Assessing the impact of environmental management systems on corporate and environmental performance. *Journal of Operations Management*, *21*(3), 329–351. Available from https://doi.org/10.1016/S0272-6963(02)00109-2.

Merett, S. (2003). Virtual water and Occam's razor. *Water International*, *28*(1), 103–105. Available from https://doi.org/10.1080/02508060.2003.9724811, 0250-8060.

McKinsey. (2009). *Charting our water future: Economic frameworks to inform decision-making*. Munich: McKinsey Company, 2030 Water Resource Group.

Nakicenovic, N., Alcamo, J., Davis, G., Vries, B.D., Fenhann, J., Gaffin, S., Gregory, K., Grubler, A., Jung, T.Y., Kram, T. and La Rovere, E.L., (2000). Special report on emissions scenarios. https://escholarship.org/uc/item/9sz5p22f

Narayanamoorthy, A. (2003). Averting water crisis by drip method of irrigation: A study of two water-intensive crops. *Indian Journal of Agricultural Economics*, *58*(3), 427–437.

Novo, P., Garrido, A., & Varela-Ortega, C. (2009). Are virtual water "flows" in Spanish grain trade consistent with relative water scarcity? *Ecological Economics*, *68*(5), 1454–1464. Available from https://doi.org/10.1016/j.ecolecon.2008.10.013, 09218009.

Oestigaard, T. (2012). Water Scarcity and Food Security along the Nile: Politics, population increase and climate change. Nordiska Afrika institutet.

Oki, T., Sato, M., Kawamura, A., Miyake, M., Kanae, S., & Musiake, K. (2003). Virtual water trade to Japan and in the world. In A. Y. Hoekstra (Ed.), *Virtual Water Trade: Proceedings of the International Expert Meeting on Virtual Water Trade, Research Report Series No. 12*. The Netherlands: IHE Delft.

Palhares, J. C. P., & Pezzopane, J. R. M. (2015). Water footprint accounting and scarcity indicators of conventional and organic dairy production systems. *Journal of Cleaner Production*, *93*, 299–307. Available from https://doi.org/10.1016/j.jclepro.2015.01.035. 09596526. Elsevier Ltd, Brazil. https://www.journals.elsevier.com/journal-of-cleaner-production.

Park, A. Y., Ince, R. A. A., Schyns, P. G., Thut, G., & Gross, J. (2015). Frontal top-down signals increase coupling of auditory low-frequency oscillations to continuous speech in human listeners. *Current Biology*, *25*(12), 1649–1653. Available from https://doi.org/10.1016/j.cub.2015.04.049. 09609822. Cell Press, United Kingdom. http://www.elsevier.com/journals/current-biology/0960-9822.

Porter, M. E., & Kramer, M. R. (2006). Strategy & society: The link between competitive advantage and corporate social responsibility. *Harvard Business Review*, *84*(12), 78–92, 00178012.

Postle, B. R. (2011). What underlies the ability to guide action with spatial information that is no longer present in the environment. *Spatial working memory*, 897–901.

Rees, W. E. (1992). Ecological footprints and appropriated carrying capacity: what urban economics leaves out. *Environment and Urbanization*, *4*(2), 121–130. Available from https://doi.org/10.1177/095624789200400212, 0956-2478.

Richter, A., Glunz, S. W., Werner, F., Schmidt, J., & Cuevas, A. (2012). Improved quantitative description of Auger recombination in crystalline silicon. *Physical review B*, *86*(16), 165202.

Richter, B. D. (2010). Re-thinking environmental flows: from allocations and reserves to sustainability boundaries. *River Research and Applications*, *26*(8), 1052–1063.

Rodrigues, M., & De la Riva, J. (2014). An insight into machine-learning algorithms to model human-caused wildfire occurrence. *Environmental Modelling and Software*, *57*, 192–201. Available from https://doi.org/10.1016/j.envsoft.2014.03.003. 13648152. Elsevier Ltd, Spain. http://www.elsevier.com/inca/publications/store/4/2/2/9/2/1.

Rosegrant, M. W., Ringler, C., & Zhu, T. (2009). Water for agriculture: Maintaining food security under growing scarcity. *Annual Review of Environment and Resources*, *34*, 205–222. Available from https://doi.org/10.1146/annurev.environ.030308.090351, 15435938.

Rosegrant, M. W., & Cline, S. A. (2003). Global food security: challenges and policies. *Science (New York, N.Y.)*, *302*, 1917–1919. Available from https://doi.org/10.1126/science.1092958.

Santini, P., Maiolino, R., Magnelli, B., Lutz, D., Lamastra, A., Li Causi, G., Eales, S., Andreani, P., Berta, S., Buat, V., Cooray, A., Cresci, G., Daddi, E., Farrah, D., Fontana, A., Franceschini, A., Genzel, R., Granato, G., Grazian, A., ... Xu, K. (2014). The evolution of the dust and gas content in galaxies. *Astronomy & Astrophysics*, *562*, A30. Available from https://doi.org/10.1051/0004-6361/201322835, 0004-6361.

Schaltegger, S., & Burritt, R. (2014). Measuring and managing sustainability performance of supply chains: Review and sustainability supply chain management framework. *Supply Chain Management*, *19*(3), 232–241. Available from https://doi.org/10.1108/SCM-02-2014-0061. 13598546. Emerald Group Holdings Ltd., Germany. http://www.emeraldinsight.com/info/journals/scm/scm.jsp.

Schaltegger, S., & Wagner, M. (2006). Integrative management of sustainability performance, measurement and reporting. *International Journal of Accounting, Auditing and Performance Evaluation*, *3*(1), 1–19. Available from https://doi.org/10.1504/IJAAPE.2006.010098. 17408016. Inderscience Publishers, Germany. http://www.inderscience.com/ijaape.

Schyns, J. F., Hamaideh, A., Hoekstra, A. Y., Mekonnen, M. M., & Schyns, M. (2015). Mitigating the risk of extreme water scarcity and dependency: The case of Jordan. *Water*, *7*(10), 5705–5730.

Shen, Z., Arimoto, R., Cao, J., Zhang, R., Li, X., Du, N., Okuda, T., Nakao, S., & Tanaka, S. (2008). Seasonal variations and evidence for the effectiveness of pollution controls on water-soluble inorganic species in total suspended particulates and fine particulate matter from Xi'an, China. *Journal of the Air and Waste Management Association*, *58*(12), 1560–1570. Available from https://doi.org/10.3155/1047-3289.58.12.1560. 21622906. Taylor and Francis Inc., China. http://www.tandfonline.com/loi/uawm20?.

Shiklomanov, I. A. (2000). Appraisal and assessment of world water resources. *Water International*, *25*(1), 11–32. Available from https://doi.org/10.1080/02508060008686794, 02508060.

Song, D., Zhou, X., Peng, Q., Chen, Y., Zhang, F., Huang, T., Zhang, T., Li, A., Huang, D., Wu, Q., He, H., & Tang, Y. (2015). Newly emerged porcine deltacoronavirus associated with diarrhoea in swine in China: Identification, prevalence and full-length genome sequence analysis. *Transboundary and Emerging Diseases*, *62*(6), 575–580. Available from https://doi.org/10.1111/tbed.12399. 18651682. http://onlinelibrary.wiley.com/journal/10.1111/(ISSN)1865-1682.

Sun, Y., Wang, Z., Fu, P., Jiang, Q., Yang, T., Li, J., & Ge, X. (2013). The impact of relative humidity on aerosol composition and evolution processes during wintertime in Beijing, China. *Atmospheric Environment*, *77*, 927–934. Available from https://doi.org/10.1016/j.atmosenv.2013.06.019, 18732844.

Taticchi, P., Tonelli, F., & Pasqualino, R. (2013). Performance measurement of sustainable supply chains: A literature review and a research agenda. *International Journal of Productivity and Performance Management*, *62*(8), 782–804. Available from https://doi.org/10.1108/IJPPM-03-2013-0037, 17410401.

Tharme, R. E. (2003). A global perspective on environmental flow assessment: Emerging trends in the development and application of environmental flow methodologies for rivers. *River Research and Applications*, *19*(5–6), 397–441. Available from https://doi.org/10.1002/rra.736.

The World Bank Annual Report 2016: Washington, DC, hdl.handle.net/10986/24985

United Nations (Committee on Civil and Political Rights). General Comment No. 34, Article 19: Freedoms of Opinion and Expression. 2011

Velázquez, E., Madrid, C., & Beltrán, M. J. (2011). Rethinking the concepts of virtual water and water footprint in relation to the production–consumption binomial and the water–energy nexus. *Water Resources Management*, *25*(2), 743–761. Available from https://doi.org/10.1007/s11269-010-9724-7. 09204741. Kluwer Academic Publishers, Spain. http://www.wkap.nl/journalhome.htm/0920-4741.

Vörösmarty, C. J., Fekete, B. M., Meybeck, M., & Lammers, R. B. (2000). Global system of rivers: Its role in organizing continental land mass and defining land-to-ocean linkages. *Global Biogeochemical Cycles*, *14*(2), 599–621. Available from https://doi.org/10.1029/1999gb900092, 0886-6236.

Vörösmarty, C. J., Green, P., Salisbury, J., & Lammers, R. B. (2000). Global water resources: Vulnerability from climate change and population growth. *Science*, *289*(5477), 284–288. Available from https://doi.org/10.1126/science.289.5477.284.

Wagner, M., & Schaltegger, S. (2003). Introduction: How does sustainability performance relate to business competitiveness? *Greener Management International*, *2003*(44), 5–16. Available from https://doi.org/10.9774/GLEAF.3062.2003.wi.00003, 09669671.

Wackernagel, M., & Rees, W. (1996). *Our Ecological Footprint: Reducing Human Impact on the Earth*. Philadelphia: New Society Publishers.

Wackernagel, M., Onisto, L., Callejas Linares, A., López Falfán, I. S., Méndez Garcia, J., Suárez Guerrero, A. I., & Suárez Guerrero, M. G. (1997). *Ecological Footprints of Nations: How Much Nature Do They Useł How Much Nature Do They Havel? Commissioned by the Earth Council for the Rio+5 Forum*. Toronto: International Council for Local Environmental Initiatives.

Water footprint Network (2013). http://www.waterfootprint.org/?page=files/NationalWaterAccountingFramework, (Sited on May 2013).

Wichelns, D., & Qadir, M. (2015). Achieving sustainable irrigation requires effective management of salts, soil salinity, and shallow groundwater. *Agricultural Water Management*, *157*, 31–38. Available from https://doi.org/10.1016/j.agwat.2014.08.016. 18732283. Elsevier B.V., United States. http://www.journals.elsevier.com/agricultural-water-management/.

Willaarts, B. A., Garrido, A., & Llamas, M. R. (2014). Water for food security and well-being in Latin America and the Caribbean: Social and environmental implications for a globalized economy. *Water for Food Security and Well-Being in Latin America and the Caribbean: Social and Environmental Implications for a Globalized Economy* (pp. 1–432). Spain: Taylor and Francis, Spain Taylor and Francis. Available from http://www.taylorandfrancis.com/books/details/9781315883137/, https://doi.org/10.4324/9781315883137.

Witmer, M. C. H., & Cleij, P. (2012). Water footprint: Useful for sustainability policies. PBL Netherlands Environmental Assessment Agency.

Xu, K., Ba, J. L., Kiros, R., Cho, K., Courville, A., Salakhutdinov, R., Zemel, R. S., & Bengio, Y. (2015). Show, attend and tell: Neural image caption generation with visual attention. Proceedings of the 32nd International Conference on Machine Learning, ICML 2015. International Machine Learning Society (IMLS) Canada. January 2015. 3. 9781510810587 2048-2057.

Yoo, H. Y., & Bruckenstein, S. (2013). A novel quartz crystal microbalance gas sensor based on porous film coatings. A high sensitivity porous poly (methylmethacrylate) water vapor sensor. *Analytica Chimica Acta*, *785*, 98–103. Available from https://doi.org/10.1016/j.aca.2013.04.052. 18734324. Elsevier B.V., United States. http://www.journals.elsevier.com/analytica-chimica-acta/.

Zeitoun, M., Allan, J. A., & Mohieldeen, Y. (2010). Virtual water 'flows' of the Nile Basin, 1998–2004: A first approximation and implications for water security. *Global Environmental Change*, *20*(2), 229–242. Available from https://doi.org/10.1016/j.gloenvcha.2009.11.003, 09593780.

Zeng, W. C., Zhang, Z., Gao, H., Jia, L. R., & Chen, W. Y. (2012). Characterization of antioxidant polysaccharides from Auricularia auricular using microwave-assisted extraction. *Carbohydrate Polymers*, *89*(2), 694–700. Available from https://doi.org/10.1016/j.carbpol.2012.03.078, 01448617.

Zhang, B., Song, X., Zhang, Y., Han, D., Tang, C., Yu, Y., & Ma, Y. (2012). Hydrochemical characteristics and water quality assessment of surface water and groundwater in Songnen plain, Northeast China. *Water Research*, *46*(8), 2737–2748. Available from https://doi.org/10.1016/j.watres.2012.02.033. 18792448. Elsevier Ltd, China. http://www.elsevier.com/locate/watres.

Zhang, S., Sadras, V., Chen, X., & Zhang, F. (2014). Water use efficiency of dryland maize in the Loess Plateau of China in response to crop management. *Field Crops Research*, *163*, 55–63. Available from https://doi.org/10.1016/j.fcr.2014.04.003. 03784290. Elsevier, China. http://www.elsevier.com/inca/publications/store/5/0/3/3/0/8.

Zhuo, L., Mekonnen, M. M., & Hoekstra, A. Y. (2016). Consumptive water footprint and virtual water trade scenarios for China—With a focus on crop production, consumption and trade. *Environment International*, *94*, 211–223. Available from https://doi.org/10.1016/j.envint.2016.05.019. 18736750. Elsevier Ltd, Netherlands. http://www.elsevier.com/locate/envint.

Zimmer D., & Renault D. (2003).Virtual water in food production and global trade: Review of methodological issues and preliminary results. Virtual water trade: Proceedings of the International Expert Meeting on Virtual Water Trade. Value of Water Research Report Series.

Chapter 7

Industrial water conservation by water footprint and Sustainable Development Goals

Ashish Kumar[1,2,*] **and Abhinay Thakur**[1,2]

[1]*Department of Science, Technology and Technical Education, Nalanda College of Engineering, Bihar Engineering University, Patna, Bihar, India,*
[2]*Division of Research and Development, Lovely Professional University, Punjab, India*
**Corresponding author. e-mail address: drashishchemlpu@gmail.com.*

7.1 Introduction

Water is a finite and essential resource that plays a critical role in sustaining life and supporting ecosystems. It is essential for drinking, sanitation, agriculture, energy production, and industrial activities. However, the increasing water demand, coupled with various environmental challenges, has led to widespread concerns about water scarcity and its impact on socioeconomic development (Bosire et al., 2015; Deb et al., 2019a). Industries, in particular, have been identified as major contributors to water consumption and pollution, making their sustainable water management practices imperative. Water scarcity is a pressing global issue that affects both developed and developing countries. Population growth, urbanization, climate change, and inefficient water use practices have exacerbated the problem, leading to water stress in many regions. According to the United Nations (UN), approximately 2.2 billion people lack access to safely managed drinking water services, and 4.2 billion people lack access to safely managed sanitation services (Ahmed et al., 2015; Boretti & Rosa, 2019). Moreover, water scarcity has severe consequences for ecosystems, biodiversity, and food security. Industries are major consumers of water, using it for various purposes such as processing, cooling, cleaning, and sanitation. The industrial sector accounts for a significant portion of global water consumption, often competing with other sectors and communities for limited water resources. Industrial activities also generate wastewater and pollutants that can have detrimental effects on water quality and ecosystems if not properly managed.

Given these challenges, the concept of industrial water conservation has gained significant attention in recent years as a means to address the growing water-related issues (Chowdhary et al., 2019). It involves adopting strategies and practices aimed at reducing water consumption, improving efficiency, and minimizing the negative environmental impacts associated with industrial processes. Industrial water conservation is not only an environmental imperative but also an economic necessity, as it can lead to cost savings, improved resource efficiency, and enhanced reputation. The water footprint provides a comprehensive framework for quantifying the amount of freshwater consumed, directly and indirectly, throughout the life cycle of a product, process, or organization. It takes into account water used in the production of raw materials, manufacturing, transportation, and disposal or recycling. By analyzing the water footprint, industries can identify hotspots of water consumption, evaluate their efficiency, and implement targeted conservation measures. The integration of water footprint assessment into industrial practices offers a holistic approach to water management that aligns with the principles of sustainable development. The UN Sustainable Development Goals (SDGs) have emerged as a global framework for addressing pressing social, economic, and environmental challenges (Aladuwaka & Momsen, 2010; Carr et al., 2012). Goal 6 specifically focuses on ensuring the availability and sustainable management of water and sanitation for all. By linking the water footprint concept with the SDGs, industries can contribute to multiple goals simultaneously, including responsible consumption and production (Goal 12), climate action (Goal 13), and life below water (Goal 14; Bradley et al., 2001). The integration of water footprint assessment into SDG implementation strategies offers a powerful mechanism for achieving sustainable industrial water management.

Water Footprints and Sustainable Development. DOI: https://doi.org/10.1016/B978-0-443-23631-0.00007-8

This book chapter aims to explore the intersection of industrial water conservation, water footprint assessment, and the pursuit of SDGs. It provides a comprehensive overview of the challenges posed by water scarcity and industrial water consumption. This chapter will delve into the concept of water footprint, examining its methodology, calculation, and application in industrial contexts. Moreover, it will highlight the relevance of water footprint to the SDGs, showcasing the potential synergies and alignment between these two frameworks. Practical approaches to industrial water conservation will be explored, including technological innovations, policy interventions, and stakeholder engagement. Real-life case studies from various industries will be examined to showcase the successful implementation of water-saving initiatives and the benefits achieved. This chapter will also outline future directions and challenges, emphasizing the need for integrating water footprint assessment into sustainability assessments, promoting transparent reporting practices, and fostering collaboration between stakeholders (Chapagain & Orr, 2009; D'Odorico et al., 2012).

By addressing the gaps in knowledge and practice, this chapter aims to contribute to the understanding of industrial water conservation and its relationship with water footprint assessment and SDGs. It provides insights into the significance of adopting holistic and integrated approaches to water management in industrial sectors, highlighting the potential for improved resource efficiency, reduced environmental impact, and long-term sustainability. By leveraging the concepts of water footprint and SDGs, industries can become proactive agents of change, leading the way towards a more water-secure and sustainable future.

7.2 Water scarcity and industrial activities

7.2.1 Global water scarcity challenges

Water scarcity is a pressing global issue that poses significant challenges to sustainable development and the well-being of communities worldwide. The growing population, urbanization, climate change, and inefficient water management practices have contributed to the worsening water scarcity crisis (Bhatti et al., 2011). This chapter will delve into the global water scarcity challenges, exploring its causes, consequences, and potential solutions. Water scarcity refers to the lack of sufficient water resources to meet the needs of a particular region or population. It occurs when the water demand exceeds the available supply, either due to physical scarcity (insufficient water availability) or economic scarcity (lack of infrastructure and resources to access water). Water scarcity can manifest in various forms, including limited access to clean drinking water, inadequate sanitation facilities, and reduced availability of water for agriculture, industry, and ecosystems (Baumgartner, 2019; Wang et al., 2020). One of the primary causes of water scarcity is population growth. As the global population continues to rise, so does the water demand. More people require water for drinking, sanitation, and food production, placing strain on available water resources. Urbanization also exacerbates water scarcity, as cities require significant amounts of water to support their inhabitants and industries. The rapid expansion of urban areas often outpaces the development of infrastructure and water management systems, leading to increased water stress.

Climate change is another critical factor contributing to global water scarcity challenges. Changing weather patterns, including altered rainfall distribution and increased frequency of droughts and floods, have profound implications for water availability (Murata & Sakamoto, 2019). Climate change impacts both surface water and groundwater sources, affecting river flows, groundwater recharge rates, and the overall water cycle. Rising temperatures also increase evaporation rates, further depleting water resources. Inefficient water management practices aggravate water scarcity. Poor infrastructure, inadequate water storage and distribution systems, and lack of wastewater treatment facilities contribute to water losses and inefficient use. In many regions, water is wasted due to leaky pipes, outdated irrigation techniques, and unsustainable agricultural practices. Moreover, water pollution and contamination further reduce the available freshwater resources, rendering them unfit for human consumption and agricultural use. The consequences of water scarcity are far-reaching and impact various sectors of society. The most immediate and severe consequence is the lack of access to clean drinking water and sanitation facilities (Deb et al., 2018; Galli et al., 2012). Millions of people, particularly in developing countries, face the daily struggle of obtaining safe water for their basic needs. This leads to increased vulnerability to waterborne diseases, malnutrition, and poverty. Agriculture, which accounts for the largest share of global water consumption, is profoundly affected by water scarcity. Insufficient water for irrigation limits crop yields affects food production and threatens food security. Farmers may be forced to cultivate less water-intensive crops or abandon agriculture altogether, exacerbating rural poverty and migration to urban areas.

Industries also face significant challenges due to water scarcity. Water is a vital input for many industrial processes, including manufacturing, energy production, and mining. Limited water availability can disrupt production, increase costs, and hinder economic growth. Industries may be compelled to implement water conservation measures, invest in water-efficient technologies, or relocate to areas with more abundant water resources (Gerbens-Leenes & Hoekstra, 2012). The environmental impacts of water scarcity are equally concerning. Reduced water availability affects ecosystems, rivers, lakes, and wetlands, leading to habitat loss, species extinction, and imbalanced ecosystems. Declining water levels in rivers and lakes can affect aquatic biodiversity, disrupt migration patterns of fish, and harm dependent communities. Addressing global water scarcity challenges requires comprehensive and integrated solutions. One crucial approach is improving water governance and management. This includes developing robust water policies, strengthening institutional capacities, and promoting stakeholder engagement. Integrated water resources management (IWRM) provides a holistic framework for managing water sustainably, considering social, economic, and environmental aspects (Fang et al., 2014; Fox et al., 1991).

Investments in water infrastructure are crucial to ensure a reliable and efficient water supply. This includes expanding water storage capacity, upgrading distribution networks, and promoting rainwater harvesting and wastewater reuse. Efficient irrigation practices, such as drip irrigation and precision agriculture, can optimize water use in agriculture. Water pricing mechanisms that reflect the true value of water can incentivize conservation and promote efficient use. Promoting water conservation and awareness among individuals and communities is essential. Educating people about the importance of water, encouraging responsible water use, and implementing water-saving measures in households, schools, and businesses can contribute significantly to reducing water demand (Ebenstein, 2012; Evans & Miguel, 2007). Climate change mitigation and adaptation strategies are integral to addressing water scarcity challenges. Reducing greenhouse gas emissions and transitioning to renewable energy sources can help mitigate the impacts of climate change on water resources. Developing climate-resilient infrastructure and implementing drought preparedness plans can enhance water security in vulnerable regions. International cooperation and partnerships are vital to addressing global water scarcity challenges. Collaborative efforts among governments, organizations, and stakeholders can facilitate knowledge sharing, technology transfer, and financial support for water-related projects (Eriksen et al., 2014). Initiatives such as the UN Water Action Decade and the SDGs provide a global platform for collective action and progress toward water security.

7.2.2 Impact of industrial water consumption

Industrial water consumption has a significant impact on water resources, ecosystems, and communities. Industries require water for various purposes, including manufacturing, processing, cooling, cleaning, and sanitation. However, the scale and intensity of industrial water use, combined with inefficient practices and inadequate wastewater management, have led to a range of negative environmental and socioeconomic consequences (Chapagain & Hoekstra, 2007; Chukalla et al., 2015). One of the primary impacts of industrial water consumption is the depletion of water sources. Industries often extract large volumes of water from rivers, lakes, and groundwater aquifers, leading to reduced water levels and the drying up of water bodies. This not only affects the availability of water for other users but also disrupts ecosystems and habitats that rely on these water sources. Furthermore, the high demand for water by industries can create competition and conflicts with other sectors, such as agriculture and domestic use. In regions where water resources are already scarce, industrial water consumption can exacerbate water stress and limit access to water for local communities, particularly in developing countries (Cantin et al., 2005). This can have severe social and economic consequences, as communities depend on water for their livelihoods, agriculture, and basic needs. Industrial water consumption also contributes to water pollution and contamination. Industries generate wastewater that contains various pollutants, including heavy metals, chemicals, and organic substances. Without proper treatment, these pollutants can find their way into water bodies, causing water pollution and posing risks to human health and ecosystems. The discharge of heated water from industrial processes can also disrupt aquatic ecosystems and affect the survival of aquatic species.

Moreover, industrial water use can have detrimental effects on biodiversity and ecosystems. Water withdrawals from natural sources can disrupt the natural flow regimes of rivers and streams, affecting aquatic habitats and species. Reduced water flow can lead to habitat degradation, loss of wetlands, and negative impacts on fish populations and migratory routes (Berger et al., 2015; Bjornlund et al., 2007). This, in turn, can have cascading effects on the overall health and resilience of ecosystems. Another significant impact of industrial water consumption is the energy requirement for water extraction, treatment, and distribution. Energy-intensive industries, such as manufacturing and mining, consume substantial amounts of water, resulting in a high energy footprint. This contributes to greenhouse gas emissions and exacerbates climate change, further impacting water resources through altered precipitation patterns, increased evaporation, and reduced snowpack. The socioeconomic consequences of industrial water consumption are multifaceted. Water scarcity caused by industrial activities can hinder agricultural productivity, leading to food insecurity and rural livelihood challenges. Industries heavily reliant on water may face disruptions in production and increased costs due to water shortages or increased water prices. Small-scale industries and local businesses that cannot afford water-intensive processes may struggle to compete or even face closure. Communities living near industrial facilities often bear the brunt of the negative impacts of industrial water consumption. They may experience reduced access to clean water due to contamination or water diversion for industrial purposes. Pollution from industrial activities can harm human health, leading to waterborne diseases, respiratory problems, and other health issues. Displaced communities, particularly in the case of large-scale water-intensive projects like dams or mining operations, can face social and economic upheaval, loss of traditional livelihoods, and displacement.

Addressing the impact of industrial water consumption requires a comprehensive approach that focuses on water conservation, sustainable practices, and effective management. Industries can adopt water-efficient technologies and practices, such as recycling and reuse of water, optimizing production processes, and implementing leak detection and repair programs. Water footprint assessments can help industries identify hotspots of water use and implement targeted conservation measures (Aldaya & Hoekstra, 2010; Chapagain & Orr, 2009; Licker et al., 2019). Government regulations and policies play a crucial role in promoting sustainable industrial water management. Implementing water permits, setting limits on water withdrawals, and enforcing wastewater treatment standards can ensure responsible water use by industries. Economic instruments, such as water pricing mechanisms, can encourage efficient water use and incentivize industries to invest in water-saving technologies. Collaboration between industries, governments, and communities is essential for sustainable water management. Stakeholder engagement and participatory approaches can facilitate dialogue, knowledge sharing, and joint decision-making. Partnerships between industries and local communities can promote water stewardship initiatives, community-based water management, and equitable access to water resources.

7.2.3 Need for industrial water conservation

The need for industrial water conservation has become increasingly urgent due to the growing water scarcity crisis and the environmental, economic, and social challenges it presents. Industries are major consumers of water, using it for various processes and activities, and their unsustainable water practices contribute to water stress, environmental degradation, and social inequities (Cao et al., 2018; Carr et al., 2015). This section will explore the need for industrial water conservation, highlighting its benefits and the key drivers behind its implementation. One of the primary reasons for industrial water conservation is the finite nature of water resources. Freshwater is a limited and essential resource for sustaining life, ecosystems, and economic activities. However, the water demand is continuously increasing due to population growth, urbanization, and industrial development. This growing demand, coupled with the effects of climate change, has led to water scarcity in many regions, making it imperative to conserve water for present and future generations. Industries have a significant impact on water resources due to their large-scale water consumption. They use water for processes such as manufacturing, cooling, cleaning, and transportation. The inefficient usage of water in these operations leads to excessive withdrawals from freshwater sources, putting a strain on already limited water supplies. By implementing water conservation measures, industries can reduce their water footprint, alleviate pressure on water resources, and contribute to water security.

Industrial water conservation is not only essential from an environmental perspective but also has economic benefits. Water is often a significant cost factor for industries, and inefficient water usage can lead to financial losses. By adopting water-saving technologies, optimizing processes, and implementing water management strategies, industries can reduce water consumption and associated costs (Bhatnagar et al., 2014). Water conservation measures can also enhance resource efficiency, improve operational performance, and support sustainable business practices. Furthermore, industrial water conservation plays a crucial role in promoting environmental sustainability. Excessive water consumption by industries can deplete water sources, disrupt ecosystems, and harm biodiversity. Reduced water availability can lead to habitat loss, impact fish populations, and degrade water quality. By conserving water, industries contribute to the preservation of ecosystems, protect sensitive habitats, and promote the overall health and resilience of natural systems. Another key driver for industrial water conservation is the growing recognition of corporate social responsibility (CSR) and SDGs (Miao et al., 2015; Zhou & Etzkowitz, 2021). Many industries are increasingly adopting CSR initiatives and integrating sustainability principles into their operations. Water conservation aligns with these objectives by demonstrating a commitment to environmental stewardship, resource efficiency, and the well-being of communities. It enhances the reputation of industries, and fosters trust among stakeholders, and supports their social license to operate.

Moreover, industrial water conservation can contribute to achieving the SDGs, particularly Goal 6, which aims to ensure the availability and sustainable management of water and sanitation for all. By implementing water-saving practices, industries directly contribute to water-related targets under Goal 6, such as improving water-usage efficiency, reducing pollution, and supporting integrated water resources management. Industrial water conservation also has indirect positive impacts on other SDGs, such as promoting responsible consumption and production (Goal 12), combating climate change (Goal 13), and protecting life below water (Goal 14; Mhlanga, 2021). To achieve industrial water conservation, a multifaceted approach is necessary. Industries can implement a range of strategies and technologies to optimize water use, such as adopting water-efficient equipment, implementing water recycling and reuse systems, and optimizing production processes to minimize water consumption. Conducting water audits and assessments can help identify inefficiencies and prioritize conservation measures. In addition, raising awareness among employees and stakeholders about the importance of water conservation can foster a culture of responsible water usage within industrial settings. Government policies and regulations also play a crucial role in driving industrial water conservation. Governments can establish water management frameworks, set water efficiency targets, and provide incentives for industries to adopt water-saving practices. They can also implement pricing mechanisms that reflect the true value of water and encourage industries to invest in water-efficient technologies.

Collaboration and partnerships between industries, governments, and other stakeholders are vital for promoting industrial water conservation. Public–private partnerships can facilitate knowledge exchange, technology transfer, and financial support for implementing water-saving measures. Engaging with local communities, nongovernmental organizations (NGOs), and water management authorities can foster dialogue, build trust, and ensure that conservation efforts are socially inclusive and environmentally sustainable.

7.3 Understanding water footprint

7.3.1 Definition and concept of water footprint

The concept of water footprint has emerged as a valuable tool for understanding and managing water use in various contexts. The water footprint represents the total volume of freshwater used, directly and indirectly, to produce goods and services consumed by individuals, communities, or organizations. It provides insights into the water resources required to meet human needs, as well as the environmental impacts associated with water consumption. This section will delve into the definition and concept of water footprint, its components, calculation methods, and its significance for sustainable water management. The water footprint encompasses three main components: the blue water footprint, the green water footprint, and the gray water footprint. The blue water footprint refers to the volume of surface and groundwater consumed during the production process. It includes water withdrawn from freshwater sources, such as rivers and aquifers, for irrigation, industrial processes, and domestic use (Ma et al., 2018; Papadopoulou et al., 2022). The green water footprint represents the volume of rainwater used in crop growth and is primarily related to agricultural activities. It accounts for the amount of water evaporated from the soil and transpired by plants. The gray water footprint refers to the volume of water required to dilute and assimilate pollutants generated during production processes. It provides an indicator of the potential environmental impact associated with water pollution. Calculating the water footprint involves assessing the direct and indirect water usage throughout the entire supply chain of a product or service. The direct water footprint involves measuring the water used within a specific process or activity, such as agricultural irrigation or industrial production. It can be quantified using various methods, including metering, flow measurement, or modeling techniques. The indirect water footprint, also known as virtual water, considers the water used in the production of intermediate goods and services that contribute to the final product. It accounts for the water embedded in the entire supply chain, including raw material extraction, processing, and transportation.

Several methodologies and tools have been developed to calculate water footprints at different scales, ranging from individual products to entire countries. The Water Footprint Network (WFN) has developed the most widely used framework, which includes standardized guidelines and indicators for water footprint assessment (Croese et al., 2020; Lin et al., 2018). This framework considers the blue, green, and gray components of water footprint and allows for comparisons between different products or sectors. Other methods, such as life cycle assessment (LCA), also incorporate water footprint assessment as part of a broader environmental impact analysis (Luján-Ornelas et al., 2020). Understanding the water footprint of products and services is essential for sustainable water management and decision-making. It provides valuable information about the water requirements and impacts associated with various consumption patterns and production processes. By quantifying the water footprint, policymakers, businesses, and consumers can identify hotspots of water usage, assess the environmental risks, and develop strategies to optimize water consumption and minimize negative impacts. The water footprint concept has several significant implications for sustainable water management. First, it highlights the interconnectedness between water resources, consumption patterns, and environmental sustainability (Fonseca et al., 2020; Gunkel et al., 2007). The water footprint of a product or activity often extends beyond geographical boundaries, as it considers the virtual water embedded in traded goods. This emphasizes the need for integrated water management, considering both local water resources and global water flows.

Second, water footprint provides a comprehensive perspective on water usage, going beyond traditional measures such as water withdrawal or consumption. It considers the entire life cycle of a product, including indirect water usage, which can account for a significant portion of the overall water footprint. This holistic approach enables decision-makers to identify potential water-saving opportunities across the supply chain, from raw material extraction to waste management. Third, water footprint allows for the assessment of the environmental impacts associated with water usage, particularly through the gray water footprint component. Quantifying the volume of water required to dilute and assimilate pollutants, provides insights into the potential risks of water pollution and the need for effective wastewater treatment and management.

Moreover, the water footprint concept can help inform policy development and water-related strategies. It can guide the formulation of water-usage efficiency targets, water allocation schemes, and water pricing mechanisms. It also supports the identification of water-intensive sectors, allowing policymakers to prioritize water conservation measures and promote sustainable practices. For businesses, understanding and managing their water footprint can lead to operational and financial benefits. By identifying water-efficient technologies, optimizing processes, and implementing water-saving measures, businesses can reduce water consumption, improve resource efficiency, and lower operational costs. Water footprint assessment can also enhance corporate sustainability strategies, CSR initiatives, and stakeholder engagement (Hoekstra, 2015). However, it is important to note that water footprint has some limitations and challenges. Calculating water footprints can be complex, requiring accurate data on water usage and availability, which may not always be readily available, particularly in data-scarce regions. The variability of water footprints across different geographical locations and climatic conditions adds another layer of complexity. Moreover, water footprint does not capture all aspects of water management, such as water quality and the social and cultural dimensions of water usage.

7.3.2 Methodology and calculation

The methodology and calculation of water footprint play a crucial role in understanding and assessing water usage in different sectors, products, or activities. It involves a systematic approach to quantify the direct and indirect water consumption associated with a particular process or supply chain. This section will explore the methodology and calculation methods used for determining water footprint, including key considerations, data requirements, and challenges.

7.3.2.1 System boundary definition

The first step in calculating water footprint is to define the system boundary. This involves identifying the specific product, sector, or activity under consideration and delineating the boundaries of the analysis (Hsu et al., 2006). The system boundary determines what processes and inputs will be included in the assessment and helps establish the scope and scale of the analysis.

7.3.2.2 Identification of water consumption

The next step is to identify and quantify the direct and indirect water consumption within the system boundary. Direct water consumption refers to the water used within the specific process or activity being assessed. It can be measured through water meters, flow measurements, or other monitoring techniques (Gleeson et al., 2012). Indirect water consumption, also known as virtual water, refers to the water used in the production of intermediate goods and services that contribute to the final product. This requires tracing the water usage along the entire supply chain.

7.3.2.3 Blue, green, and gray water footprints

Water footprint consists of three components: blue, green, and gray. The blue water footprint represents the volume of surface and groundwater consumed within the system boundary. It includes water withdrawals from freshwater sources such as rivers, lakes, or aquifers for various purposes like irrigation, industrial processes, and domestic usage (Guo et al., 2021; Zhao et al., 2022). The green water footprint represents the volume of rainwater used in crop growth and is mainly associated with agricultural activities. It accounts for the water evaporated from the soil and transpired by plants. The gray water footprint refers to the volume of water required to dilute and assimilate pollutants generated within the system boundary. It provides an indicator of the potential environmental impact associated with water pollution. Utilizing water footprint matrices to improve the advancement of SDG targets is depicted in Fig. 7.1 (Hoekstra et al., 2017).

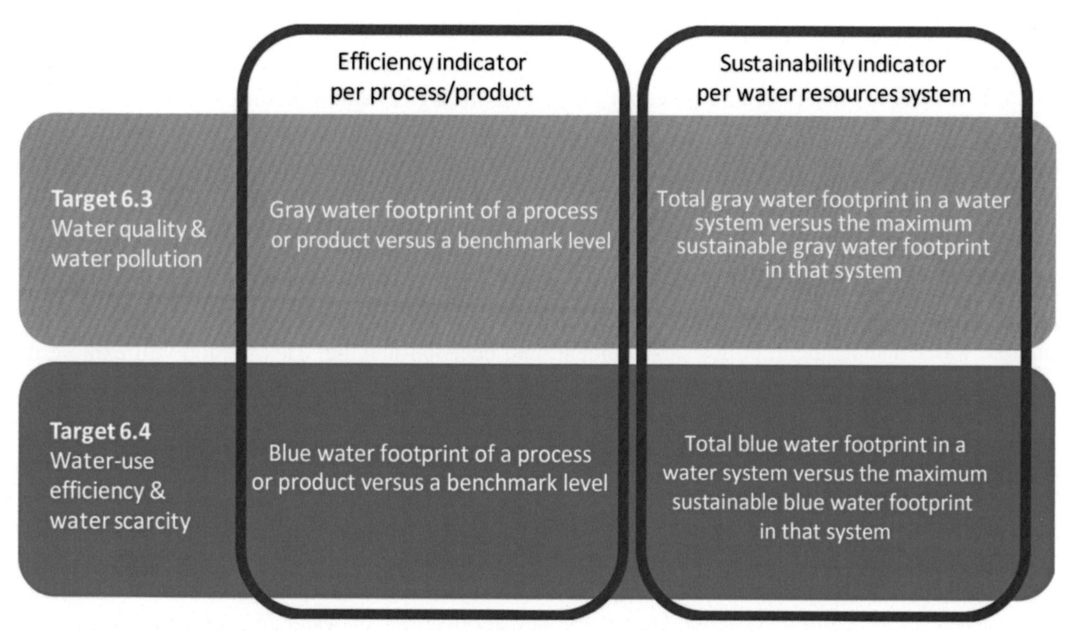

FIGURE 7.1 Utilization of water footprint matrices to enhance progress for SDG targets . *From Hoekstra, A. Y., Chapagain, A. K. & van Oel, P. R. (2017). Advancing water footprint assessment research: challenges in monitoring progress towards sustainable development goal 6.* Water (Switzerland), *9(6). doi: 10.3390/w9060438.*

7.3.2.4 Calculation methods

Several methods and tools have been developed to calculate water footprint, each with its strengths and limitations. The WFN has developed a widely used framework for water footprint assessment (Hoekstra, 2017). This framework provides standardized guidelines and indicators for calculating the water footprint at different scales, ranging from individual products to national or global levels. It considers the blue, green, and gray components and allows for comparisons between different products or sectors.

The WFN framework follows a process-based approach that involves assessing water usage along the supply chain using input–output analysis. It requires data on water usage at each stage of production, including agricultural inputs, manufacturing processes, and service provision. Water footprint is calculated by multiplying the water usage at each stage by the corresponding water scarcity or pollution factor. These factors reflect the local water availability and the environmental impact of water usage. LCA is another widely used method that incorporates water footprint assessment as part of a broader environmental impact analysis. It considers water footprint in conjunction with other environmental indicators, such as energy usage, greenhouse gas emissions, and land usage (Huang et al., 2012). It provides a comprehensive assessment of the environmental performance of a product or process throughout its life cycle.

7.3.2.5 Data requirements and limitations

Accurate and reliable data is essential for calculating water footprint. Data requirements include information on water usage, water quality, local hydrological conditions, and regional water scarcity or pollution factors. Data sources can vary, including government statistics, industry reports, research studies, and expert knowledge (Chen et al., 2015; Lu et al., 2015). However, data availability and quality can pose significant challenges, particularly in data-scarce regions or for complex supply chains. Uncertainty and variability are inherent in water footprint calculations due to various factors, such as spatial and temporal heterogeneity of water resources, data limitations, and modeling assumptions. Sensitivity analyses and scenario modeling can help address some of these uncertainties and provide a range of possible outcomes.

7.3.2.6 Advancements and emerging trends

The methodology and calculation of water footprint are continuously evolving, with advancements in data collection, modeling techniques, and assessment frameworks. Efforts are being made to improve the accuracy and robustness of water footprint calculations, including the incorporation of local-scale data, remote sensing technologies, and advanced modeling approaches. Furthermore, there is an increasing focus on sector-specific water footprint assessments, such as for agriculture, manufacturing, or the textile industry. These sector-specific assessments provide more detailed insights into the water usage patterns, hotspots, and potential mitigation strategies within specific industries. Integration of the water footprint with other sustainability assessment tools, such as carbon footprinting or social impact assessments, is also gaining traction (Li, Luo, et al., 2017). This integrated approach provides a more comprehensive understanding of the environmental, economic, and social dimensions of water usage and allows for more informed decision-making.

7.3.3 Application of water footprint assessment

The application of water footprint assessment has gained significant traction in various sectors and contexts, contributing to sustainable water management, policy development, and decision-making. This section will explore the application of water footprint assessment in different areas, including agriculture, industry, households, and policy formulation, highlighting its benefits, challenges, and potential future developments.

7.3.3.1 Agriculture

Agriculture is one of the most water-intensive sectors, accounting for a significant portion of global water consumption. As the global population continues to grow and the demand for food increases, there is a pressing need for sustainable water management practices in agriculture. Water footprint assessment provides a valuable tool for understanding and managing water usage in agricultural systems. One of the key applications of water footprint assessment in agriculture is the identification of water-intensive crops. Different crops have varying water requirements, and some are more water intensive than others (Miller and O'Callaghan, 2015; Wu et al., 2021). By quantifying the water footprint of different crops, farmers can make informed decisions about which crops to grow, taking into consideration the availability of water resources in their region. This knowledge helps farmers allocate water resources effectively and choose crops that are more suited to their local water conditions. Water footprint assessment also helps identify regions facing water scarcity in the context of agricultural production. By assessing the water footprint at a regional level, policymakers and agricultural stakeholders can pinpoint areas where water resources are under pressure. This information is crucial for developing targeted strategies and interventions to address water scarcity, such as implementing water-efficient irrigation systems, promoting crop rotation, or introducing drought-resistant crop varieties.

Furthermore, water footprint assessment enables the evaluation of different water-saving strategies and technologies in agriculture. Farmers can assess the effectiveness of various irrigation methods, such as drip irrigation or precision agriculture, in reducing water consumption and increasing water usage efficiency. They can also explore the potential of implementing water-saving technologies, such as rainwater harvesting, water recycling, or using precision farming techniques. By quantifying the water footprint before and after implementing these strategies, farmers can assess their impact on water usage and make informed decisions about their adoption. Water footprint assessment supports the development of sustainable agricultural policies and practices. Policymakers can use the insights from water footprint assessments to design effective water management strategies, such as water pricing mechanisms that incentivize water-saving practices and discourage excessive water use. In addition, land-use planning can be informed by water footprint assessments to ensure that water-intensive crops are not cultivated in water-scarce regions, reducing the strain on water resources. The implementation of water footprint assessment in agriculture, however, comes with its challenges. Data availability and quality are key concerns, as water footprint assessments require accurate and reliable data on crop water requirements, irrigation practices, and local water availability. Collaboration between researchers, farmers, and policymakers is essential to collect and share this data effectively.

7.3.3.2 Industry

The industrial sector is a significant consumer of water, and its operations can have a significant impact on water resources and the environment. In this context, water footprint assessment has emerged as a valuable tool for evaluating and optimizing water usage in industries. Water-intensive industries, such as textiles, food processing, and mining, can benefit from water footprint assessment by identifying opportunities to reduce water consumption and improve water efficiency (Mekonnen & Hoekstra, 2011). By quantifying the water footprint of their products or processes, companies can gain insights into the water requirements at different stages of their operations. This knowledge allows them to identify water-intensive processes and prioritize areas for improvement. For example, in the textile industry, water footprint assessment can help identify stages of production that consume excessive amounts of water, such as dyeing or finishing processes, and implement measures to reduce water use in those areas. Water footprint assessment also helps industries identify opportunities for water reuse and recycling. By understanding the sources and destinations of water within their operations, companies can implement strategies to treat and reuse wastewater, reducing their reliance on freshwater sources. This not only conserves water resources but also reduces the discharge of wastewater into the environment, minimizing pollution and the potential for negative ecological impacts.

In addition to water conservation, water footprint assessment supports the adoption of cleaner production techniques in industries. By evaluating the environmental impact associated with water usage, companies can identify areas where improvements can be made. For example, they can explore the use of alternative materials or technologies that require less water or have lower environmental footprints. This can include the implementation of water-saving technologies, such as water-efficient equipment or closed-loop systems that minimize water losses. Water footprint assessment also plays a crucial role in helping industries meet CSR goals and enhance their sustainability performance. Many companies are increasingly recognizing the importance of sustainable water management and are incorporating water-related metrics into their CSR reporting (Akenji & Bengtsson, 2014; Weerasooriya et al., 2021). Water footprint assessment provides a quantitative measure that allows companies to track their water consumption, set targets for improvement, and demonstrate their commitment to sustainable practices. Moreover, water footprint assessment helps industries respond to consumer demands for sustainable products. As consumers become more environmentally conscious, there is a growing interest in products that have a reduced water footprint or are manufactured using water-efficient processes. By quantifying and disclosing the water footprint of their products, companies can provide transparency and allow consumers to make informed choices. This can create a market advantage for companies that prioritize sustainable water management and can lead to increased demand for their products.

However, implementing water footprint assessment in industries is not without challenges. Data availability and quality are crucial considerations, as companies need accurate and reliable information on water consumption at various stages of their operations. Collaboration with suppliers, stakeholders, and relevant industry associations can help gather the necessary data and ensure consistency in reporting.

7.3.3.3 Household consumption

Household consumption is a significant contributor to overall water usage, and understanding and managing water consumption at the household level is crucial for sustainable water management. Water footprint assessment offers a valuable approach to measuring and managing household water consumption. Water footprint assessment provides insights into the water requirements of different activities within a household, including bathing, laundry, cooking, gardening, and other domestic uses (le Roux et al., 2017; Mekonnen & Hoekstra, 2014). By quantifying the water footprint associated with each activity, households can identify areas of high water consumption and prioritize actions to reduce their water footprint. For example, they may realize that their gardening practices consume a substantial amount of water and can explore water-efficient landscaping techniques or the use of native, drought-tolerant plants. By understanding their water footprint, households can adopt water-saving behaviors and practices. This can include the use of efficient appliances, such as low-flow showerheads, faucets, and toilets, which reduce water consumption without sacrificing performance. Water footprint assessment can also raise awareness among household members about the water intensity of different activities, promoting behavior changes such as shorter showers or collecting and reusing rainwater for nonpotable uses. Water footprint assessment at the household level can also inform the development of water pricing structures and incentives for water conservation. By understanding the water footprint of households in a community or city, policymakers can implement progressive water pricing mechanisms that charge higher rates for excessive water consumption. This encourages households to reduce their water footprint and promotes more responsible water usage. Incentive programs, such as rebates or subsidies for water-saving appliances or landscaping, can also be designed based on water footprint assessments to motivate households to adopt water-saving measures.

Moreover, water footprint assessment in households can drive the development of water-saving technologies. Quantifying the water footprint associated with various activities provides valuable insights for innovators and manufacturers to develop and improve water-saving technologies (Oki & Kanae, 2006; Trubetskaya et al., 2021). For example, knowing the water footprint of laundry can lead to the development of washing machines that use less water or incorporate water recycling systems. Water footprint assessment also facilitates benchmarking and comparisons between households, communities, or cities. By assessing the water footprints of different households, communities can identify best practices and successful strategies for water conservation. This information can be shared through awareness campaigns, educational programs, or community initiatives, promoting the adoption of water-saving behaviors and practices. However, implementing water footprint assessments in households may face certain challenges. Data collection and accuracy are essential, as water consumption in households can vary significantly based on factors such as household size, lifestyle, and geographic location. Raising awareness and educating households about the importance of water footprint assessment and the benefits of water-saving behaviors is also crucial for driving behavior change.

7.3.3.4 Policy formulation

Water footprint assessment plays a crucial role in policy formulation related to water resources management. Policymakers and government agencies can use water footprint assessment to gain a comprehensive understanding of water usage patterns, identify areas of water stress, and develop effective strategies to manage water resources sustainably (Mekonnen & Hoekstra, 2012). At a regional or national scale, water footprint assessment helps policymakers identify water stress hotspots. By quantifying the water footprint of different sectors and activities, policymakers can identify regions or areas that are experiencing high levels of water demand or facing water scarcity. This information enables them to prioritize water allocation and management efforts in these regions, ensuring that water resources are distributed efficiently and equitably. Water footprint assessment also aids in the development of water management plans. By analyzing the water footprints of various sectors, policymakers can identify opportunities for water conservation, efficiency improvements, and sustainable water usage practices. For example, they can develop targets for water-usage efficiency in different sectors and establish guidelines for water-saving technologies and practices. Water footprint assessment contributes to the formulation of water pricing mechanisms and economic instruments (McIntyre et al., 2013). By understanding the water footprint of different sectors and activities, policymakers can design water pricing structures that encourage efficient water usage and discourage wasteful consumption. Pricing mechanisms can be tailored to incentivize water-saving practices and promote water-usage efficiency. In addition, economic instruments such as water tariffs, fees, or subsidies can be developed based on the water footprint assessment to influence consumer behavior and promote sustainable water usage.

Regulations and standards for water-intensive industries can be informed by water footprint assessment. By quantifying the water footprint of industrial processes and products, policymakers can establish regulations and benchmarks for water-intensive industries to reduce their water consumption and environmental impact. This can include implementing water reuse and recycling requirements, promoting the use of water-efficient technologies, and setting limits on water pollution. Water footprint assessment also supports the integration of water considerations into broader policy frameworks. Policymakers can align water footprint assessment with SDGs to ensure that water-related targets are effectively addressed (Kongboon & Sampattagul, 2012; Landrigan et al., 2018; le Roux et al., 2017; Trubetskaya et al., 2021; Zhang et al., 2020). Water footprint assessment can also inform policy formulation in the context of climate change adaptation, as it helps assess the vulnerability of water resources to changing climate conditions and supports the development of adaptation strategies. Furthermore, water footprint assessment provides insights into the impact of water use on ecosystems, facilitating the development of policies for the protection and restoration of aquatic ecosystems. Challenges in policy formulation using water footprint assessment include data availability and quality, as well as the need for interdisciplinary collaboration. Reliable data on water consumption and associated impacts across sectors and regions are essential for accurate assessment and informed policy decisions. Collaboration between policymakers, researchers, and stakeholders is crucial to ensure that water footprint assessments reflect the diverse perspectives and needs of different sectors and communities (Hoekstra, 2014).

7.3.3.5 Corporate sustainability reporting

Corporate sustainability reporting has evolved to encompass a wide range of environmental, social, and governance (ESG) metrics, and water footprint assessment has emerged as a crucial component of this reporting framework (Deb et al., 2019b). Companies are recognizing the importance of water as a finite resource and are incorporating water-related metrics into their sustainability reports to demonstrate their commitment to sustainable water management. Water footprint assessment provides a comprehensive and standardized approach for companies to measure and report their water-related impacts. It allows companies to quantify their water consumption, identify areas of high water usage, and assess the efficiency of their water management practices. By conducting water footprint assessments, companies gain insights into the environmental implications of their water usage, including potential water scarcity issues, pollution risks, and impacts on local ecosystems. Incorporating water footprint assessment into corporate sustainability reporting enhances transparency and accountability. By disclosing their water footprint data, companies provide stakeholders with valuable information about their environmental performance and water management practices. This transparency builds trust and credibility among stakeholders, including investors, customers, employees, communities, and regulatory bodies. Stakeholders can assess a company's water-related risks and opportunities, evaluate its water stewardship efforts, and make informed decisions based on this information.

Water footprint assessment in corporate sustainability reporting also enables companies to track their progress over time. By establishing baseline measurements and setting targets for water use reduction or efficiency improvements, companies can monitor their performance and measure the impact of their water conservation initiatives (Cheng et al., 2003). This tracking mechanism allows companies to identify areas for further improvement and evaluate the effectiveness of their water management strategies. Furthermore, water footprint assessment facilitates stakeholder engagement. Companies can actively involve stakeholders in their water footprint assessment processes by seeking input, sharing information, and inviting feedback. This engagement fosters collaboration, enhances understanding of local water challenges, and supports the development of context-specific water management strategies. Companies can work with local communities, NGOs, and other stakeholders to implement water-saving initiatives, support water-related projects, and contribute to water stewardship efforts. Including water footprint assessment in corporate sustainability, reporting aligns with the growing demand for responsible business practices. Investors and consumers are increasingly concerned about the environmental impacts of corporate activities, including water use and management. By disclosing their water footprint and related sustainability metrics, companies can address these concerns and showcase their commitment to sustainable water stewardship. This information allows investors and consumers to make informed decisions, favouring companies with robust water management practices and responsible water usage.

However, implementing water footprint assessment in corporate sustainability reporting may present challenges. Collecting accurate and reliable data on water use across different operations, supply chains, and geographic locations can be complex. Companies need to establish robust data collection systems, engage with suppliers, and ensure consistent measurement methodologies to ensure data accuracy. Furthermore, aligning water footprint assessment with other sustainability frameworks and reporting standards requires careful integration and coordination.

7.3.3.6 Consumer awareness and labeling

Consumer awareness and labeling initiatives play a vital role in promoting sustainable consumption patterns and encouraging responsible choices. Water footprint assessment can contribute to these initiatives by providing information about the water impact of products and enabling the development of water footprint labels or certifications. Water footprint labeling involves providing consumers with information about the water footprint associated with a product throughout its lifecycle (Ercin et al., 2012; Wang et al., 2017). This includes the water used in raw material extraction, manufacturing processes, packaging, distribution, and disposal. By quantifying and communicating the water footprint of a product, consumers can make more informed decisions, considering the environmental implications of their purchases. Water footprint labeling creates awareness among consumers about the water intensity of different products and helps them understand the connection between their consumption choices and water resources. It allows consumers to prioritize products with lower water footprints, encouraging demand for water-efficient and sustainable alternatives. This increased demand can incentivize companies to improve their water management practices, adopt water-saving technologies, and reduce their water footprints to remain competitive in the market. Moreover, water footprint labeling creates a market for water-saving technologies and innovations. As consumers become more aware of the water footprints associated with products, there is a growing demand for water-efficient alternatives. This drives companies to develop and promote technologies and practices that reduce water consumption and improve water efficiency throughout the supply chain. By creating a market for such technologies, water footprint labeling encourages innovation and supports the transition to a more sustainable and water-conscious economy.

Water footprint labels or certifications can be developed based on rigorous water footprint assessment methodologies (Babel et al., 2019; Ene et al., 2013). These labels can be standardized and displayed on product packaging, providing clear and concise information about the water footprint of the product. In addition, digital platforms and mobile applications can be used to provide more detailed information to consumers, allowing them to compare the water footprints of different products and make well-informed choices. Water footprint labeling initiatives may face challenges in implementation. Data collection and transparency across supply chains can be complex, requiring collaboration and cooperation among companies, suppliers, and certification bodies. Standardizing measurement methodologies and ensuring the accuracy of water footprint data is crucial to maintaining the credibility and integrity of labeling initiatives (Chapagain & Hoekstra, 2011; Geller et al., 1983). Consumer education and awareness campaigns are also essential to ensure that consumers understand the significance of water footprint labels and how to interpret and use the information provided. While the application of water footprint assessment offers numerous benefits, several challenges exist. Data availability and quality pose significant challenges, particularly in data-scarce regions or complex supply chains. There is also a need for standardized methodologies and harmonized indicators to ensure consistency and comparability of water footprint assessments across different sectors and regions. Addressing these challenges requires collaboration between stakeholders, data sharing, and capacity-building efforts.

In the future, the application of water footprint assessment is expected to evolve further. Advancements in data collection, remote sensing technologies, and modeling techniques will improve the accuracy and efficiency of water footprint calculations. Integration with other sustainability assessment tools, such as carbon footprinting and social impact assessments, will provide a more comprehensive understanding of the environmental, economic, and social dimensions of water usage. The development of sector-specific water footprint guidelines and the promotion of international cooperation and knowledge exchange will support the widespread adoption of water footprint assessment and enhance its effectiveness in achieving sustainable water management goals.

7.4 Integrating water footprint and Sustainable Development Goals

7.4.1 Overview of Sustainable Development Goals

**The Sustainable Development Goals (SDGs) are a set of global objectives adopted by the UN in 2015 as a framework for addressing the world's most pressing social, economic, and environmental challenges. Consisting of 17 goals and 169 targets, the SDGs aim to guide governments, organizations, and individuals towards a more sustainable and inclusive future. The SDGs build upon the Millennium Development Goals and provide a more comprehensive and integrated approach to sustainable development (Florea et al., 2005). They address a wide range of interconnected issues, including poverty, hunger, health, education, gender equality, clean water and sanitation, affordable and clean energy, sustainable cities and communities, responsible consumption and production, climate action, and biodiversity conservation, among others. The overarching objective of the SDGs is to achieve sustainable development, which is defined as development that meets the needs of the present without compromising the ability of future generations to meet their own needs. This includes addressing social, economic, and environmental dimensions of development in an integrated and balanced manner. The 17 SDGs are as follows and shown in Fig. 7.2 (Ho & Goethals, 2019):

- **No poverty:** End poverty in all its forms and dimensions, ensuring social protection for all.
- **Zero hunger:** End hunger, achieve food security and improved nutrition, and promote sustainable agriculture.
- **Good health and well-being:** Ensure healthy lives and promote well-being for all at all ages.
- **Quality education:** Ensure inclusive and equitable quality education and promote lifelong learning opportunities for all.

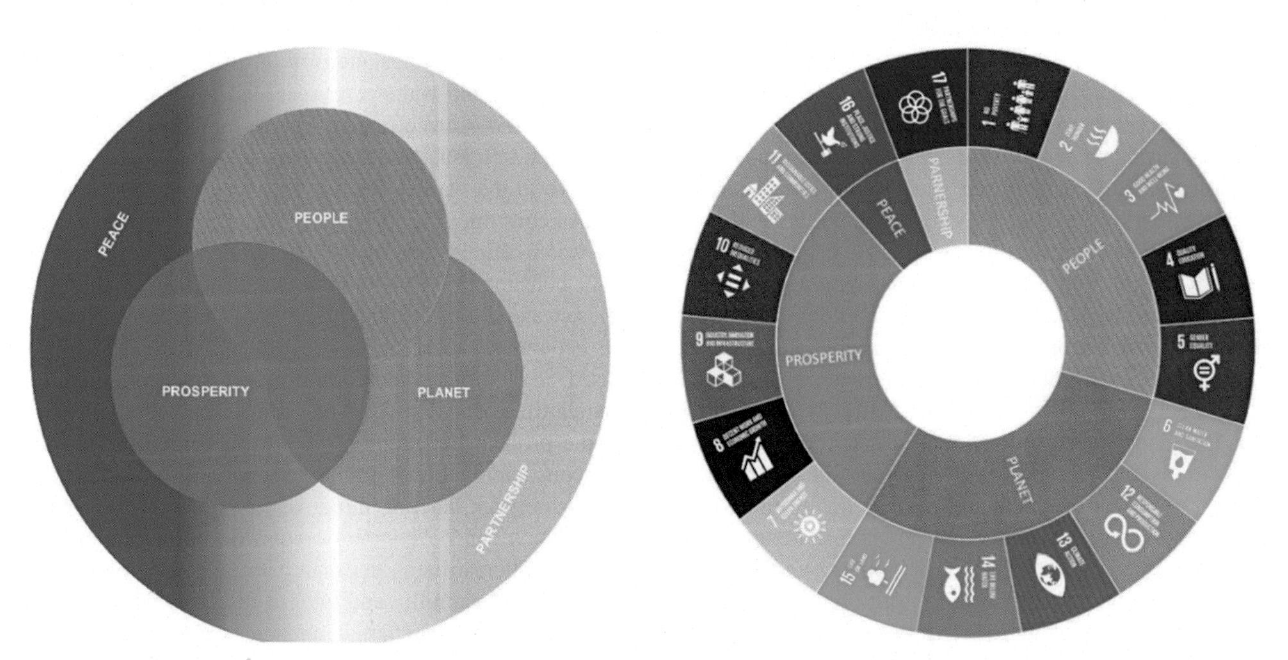

FIGURE 7.2 **The concept of the 5Ps within the framework of the 2030 Agenda for Sustainable Development**. *From Ho, L. T. & Goethals, P. L. M. (2019). Opportunities and challenges for the sustainability of lakes and reservoirs in relation to the Sustainable Development Goals (SDGs).* Water (Switzerland), 11*(7). doi: 10.3390/w11071462.*

- **Gender equality:** Achieve gender equality and empower all women and girls.
- **Clean water and sanitation:** Ensure availability and sustainable management of water and sanitation for all.
- **Affordable and clean energy:** Ensure access to affordable, reliable, sustainable, and modern energy for all.
- **Decent work and economic growth:** Promote sustained, inclusive, and sustainable economic growth, full and productive employment, and decent work for all.
- **Industry, innovation, and infrastructure:** Build resilient infrastructure, promote inclusive and sustainable industrialization, and foster innovation.
- **Reduced inequalities:** Reduce inequality within and among countries.
- **Sustainable cities and communities:** Make cities and human settlements inclusive, safe, resilient, and sustainable.
- **Responsible consumption and production:** Ensure sustainable consumption and production patterns.
- **Climate action:** Take urgent action to combat climate change and its impacts.
- **Life below water:** Conserve and sustainably use the oceans, seas, and marine resources for sustainable development.
- **Life on land:** Protect, restore, and promote sustainable use of terrestrial ecosystems, sustainably manage forests, combat desertification, halt and reverse land degradation, and halt biodiversity loss.
- **Peace, justice, and strong institutions:** Promote peaceful and inclusive societies for sustainable development, provide access to justice for all, and build effective, accountable, and inclusive institutions at all levels.
- **Partnerships for the goals:** Strengthen the means of implementation and revitalize the global partnership for sustainable development.

Each SDG consists of specific targets that provide clear milestones for measuring progress. The targets are interrelated and mutually reinforcing, recognizing the interconnectedness of various sustainable development issues. Achieving the SDGs requires collaboration and partnerships among governments, businesses, civil society organizations, and individuals. The SDGs provide a shared vision and framework for action, guiding countries and stakeholders in their efforts to achieve sustainable development. They encourage integrated and holistic approaches, recognizing that progress in one area often depends on progress in others. The SDGs also emphasize the importance of leaving no one behind, ensuring that the benefits of development reach all segments of society, particularly the most vulnerable and marginalized. Monitoring progress towards the SDGs is a crucial aspect of implementation. National governments, international organizations, and other stakeholders track and report on indicators related to each goal to assess progress and identify areas requiring further attention. This monitoring and reporting process helps guide policies, mobilize resources, and promote accountability.

7.4.2 Relevance of water footprint to Sustainable Development Goals

The concept of water footprint has gained significant relevance in the context of the SDGs due to its close alignment with several goals and targets related to water, environment, sustainable development, and responsible consumption. The water footprint provides a valuable framework for assessing and managing water use and promoting sustainable water management practices (Fonseca et al., 2020; Hoekstra, 2015). In this section, we will explore the relevance of the water footprint to the SDGs and how it contributes to achieving the goals.

- **SDG 6—Clean water and sanitation:** SDG 6 aims to ensure the availability and sustainable management of water and sanitation for all. The water footprint is directly linked to this goal as it provides a comprehensive assessment of water usage throughout the supply chain and helps identify opportunities for reducing water consumption, improving efficiency, and managing water resources effectively. By applying water footprint assessment, stakeholders can make informed decisions about water management strategies, water-saving technologies, and responsible water usage practices, contributing to achieving SDG 6.
- **SDG 12—Responsible consumption and production:** SDG 12 focuses on promoting sustainable consumption and production patterns. The water footprint is a valuable tool for measuring and managing the water impact associated with products and processes. By quantifying the water footprint of goods and services, stakeholders can identify water-intensive products, promote water-efficient production practices, and encourage responsible consumption choices. Water footprint assessment supports the implementation of sustainable production and consumption strategies, aligning with SDG 12 objectives.
- **SDG 13—Climate action:** SDG 13 calls for urgent action to combat climate change and its impacts. Water footprint assessment contributes to this goal by highlighting the link between water usage and greenhouse gas emissions. By reducing water consumption and adopting more efficient water management practices, stakeholders can also reduce energy usage and associated carbon emissions. The water footprint provides insights into the water–energy nexus and supports integrated approaches to address climate change and promote climate resilience.

- **SDG 15—Life on land**: SDG 15 aims to protect, restore, and promote sustainable use of terrestrial ecosystems. The water footprint is relevant to this goal as it helps assess the environmental impact of water usage on land ecosystems. By evaluating the water footprint of agricultural practices, land-use changes, and other activities, stakeholders can identify measures to mitigate negative impacts, promote sustainable land management, and preserve biodiversity. Water footprint assessment facilitates the integration of water considerations into land-use planning and conservation efforts.
- **SDG 17—Partnerships for the goals:** SDG 17 highlights the importance of partnerships and collaboration for achieving the SDGs. Water footprint assessment promotes collaboration among stakeholders, including governments, businesses, civil society organizations, and research institutions. By sharing knowledge, data, and best practices related to water management, stakeholders can work together to address water-related challenges, foster innovation, and support the implementation of the SDGs. Water footprint assessment provides a common language and framework for collaboration, contributing to SDG 17 objectives.

Overall, the water footprint is highly relevant to the SDGs as it provides a holistic and quantitative approach to understanding and managing water usage in the context of sustainable development. By promoting water-efficient practices, responsible consumption, and integrated water management, the water footprint contributes to achieving multiple goals, including SDG 6, SDG 12, SDG 13, SDG 15, and SDG 17 (Keesstra et al., 2018; Sopper, 1992). It helps stakeholders make informed decisions, track progress, and mobilize resources toward sustainable and responsible water usage, thereby advancing the broader agenda of the SDGs.

7.4.3 Synergies and alignment between water footprint and SDGs

The water footprint concept and the SDGs are closely aligned and share synergies in their objectives and approaches. The water footprint provides a comprehensive framework for assessing and managing water usage and pollution, while the SDGs offer a holistic agenda for sustainable development across various dimensions. This section will explore the synergies and alignment between the water footprint and the SDGs, highlighting how they can reinforce each other in achieving sustainable water management and broader development objectives.

- **Goal 6—Clean water and sanitation:** SDG 6 aims to ensure the availability and sustainable management of water and sanitation for all (Lian et al., 2022). The water footprint provides a quantitative assessment of water usage and pollution, helping to identify areas of high water stress, inefficiencies, and pollution hotspots. By understanding water footprints, policymakers, businesses, and communities can develop strategies and practices to improve water-usage efficiency, reduce pollution, and ensure sustainable water management, thereby contributing to SDG 6.
- **Goal 12—Responsible consumption and production:** SDG 12 promotes sustainable consumption and production patterns. The water footprint assessment is a valuable tool for evaluating the environmental impact of products and processes, including water consumption. By quantifying the water footprint of goods and services, businesses can identify opportunities for reducing water usage, optimizing production processes, and promoting responsible consumption. This aligns with SDG 12 objective of promoting sustainable resource use and reducing waste generation.
- **Goal 13—Climate action:** SDG 13 focuses on taking urgent action to combat climate change and its impacts. The water footprint assessment contributes to climate action by addressing the water–energy nexus. It helps identify water-related energy consumption and greenhouse gas emissions, enabling stakeholders to develop strategies for reducing carbon footprints (Gare, 2001; Hoekstra et al., 2017). By managing water footprints, such as through water-efficient technologies and practices, countries and businesses can contribute to SDG 13 targets of mitigating climate change and increasing resilience to its impacts.
- **Goal 14—Life below water and Goal 15—Life on land:** SDGs 14 and 15 emphasize the conservation and sustainable use of marine and terrestrial ecosystems. The water footprint assessment provides insights into the impact of water usage on ecosystems, including freshwater ecosystems, oceans, and forests. By managing water footprints, stakeholders can reduce the ecological footprint associated with water usage, mitigate habitat degradation, and promote sustainable ecosystem management, supporting the objectives of SDGs 14 and 15.
- **Goal 9—Industry, innovation, and infrastructure:** SDG 9 aims to build resilient infrastructure, promote inclusive and sustainable industrialization, and foster innovation. The water footprint assessment can support this goal by identifying opportunities for reducing water consumption, improving water efficiency, and minimizing the environmental impact of industrial processes. By implementing water-saving technologies and practices, businesses can enhance their sustainability performance, reduce their water footprints, and contribute to SDG 9 targets (Hoekstra, 2012).

- **Goal 17—Partnerships for the goals:** SDG 17 emphasizes the importance of strengthening partnerships and means of implementation for achieving the SDGs. The water footprint assessment promotes collaboration among governments, businesses, civil society organizations, and other stakeholders. By sharing information, best practices, and lessons learned, stakeholders can work together to address water-related challenges, align their efforts with the SDGs, and mobilize resources for sustainable water management. The water footprint serves as a common language for stakeholders to communicate and cooperate on water-related issues.

Moreover, the water footprint assessment has broader implications for achieving other SDGs indirectly. For example, by promoting water-usage efficiency and sustainable water management, it contributes to **poverty reduction (SDG 1), improved health and well-being (SDG 3), sustainable agriculture (SDG 2), and sustainable cities and communities (SDG 11), among others. The alignment between the water footprint and the SDGs provides a powerful framework for integrated and sustainable development. The water footprint assessment helps to identify priorities, set targets, and monitor progress toward achieving the SDGs. It enables stakeholders to make informed decisions, implement effective strategies, and mobilize resources for sustainable water management. By recognizing the synergies and aligning efforts between the water footprint and the SDGs, countries, businesses, and communities can foster a more integrated and holistic approach to water management, leading to a more sustainable and resilient future.

7.5 Practical approaches to industrial water conservation

7.5.1 Technological innovations and best practices

Technological innovations and best practices play a crucial role in addressing water-related challenges and achieving sustainable water management. As water resources become increasingly scarce and water quality deteriorates, innovative solutions are needed to improve water efficiency, reduce pollution, and enhance water availability. This section will explore the importance of technological innovations and best practices in the context of water management, highlighting their potential benefits and providing examples of successful implementation.

7.5.1.1 Water efficiency technologies

Technological innovations can significantly improve water efficiency across various sectors. For example, in agriculture, precision irrigation systems, such as drip irrigation and sensor-based irrigation, enable farmers to deliver water directly to plant roots, minimizing water wastage. Similarly, in industrial processes, advanced water treatment technologies, such as reverse osmosis and membrane filtration, enable efficient water reuse and recycling. These technologies reduce water consumption, enhance productivity, and contribute to sustainable water management. Israel is a leading example of efficient agricultural water usage. Through the adoption of precision irrigation systems, including drip irrigation, the country has achieved remarkable results in water conservation. Israel's agricultural sector has significantly reduced water consumption per unit of production while maintaining high crop yields. The success of these technologies has allowed Israel to transform arid regions into productive agricultural areas. In India, the state of Gujarat has implemented the Sardar Sarovar Narmada Canal Irrigation Project, which uses drip and sprinkler irrigation techniques (Hoekstra & Chapagain, 2007). This project has helped conserve water by providing precise amounts of water directly to plant roots, resulting in increased crop productivity and reduced water wastage. Singapore has emerged as a global leader in water reuse and recycling technologies. The city-state has implemented advanced water treatment processes, including reverse osmosis and membrane filtration, to treat wastewater and transform it into high-quality drinking water. The "NEWater" initiative has enabled Singapore to achieve water self-sufficiency and reduce its reliance on imported water sources.

In the textile industry, Levi Strauss & Co. implemented innovative water efficiency measures in its production processes. By adopting advanced water treatment technologies and recycling systems, the company significantly reduced water consumption and improved water quality. These measures demonstrate how industrial water efficiency technologies can contribute to sustainable production practices. Water efficiency technologies are particularly crucial in regions facing severe water scarcity (Hoekstra & Mekonnen, 2012). In the Middle East, countries like Qatar and the United Arab Emirates have implemented advanced irrigation systems, including drip irrigation, to cultivate crops in arid desert conditions. These technologies allow efficient water use and support agricultural production in water-stressed environments. In California, the implementation of precision irrigation techniques, such as sensor-based irrigation, has helped agricultural producers cope with drought conditions. These technologies enable farmers to apply water precisely when and where it is needed, minimizing water wastage and ensuring optimal crop growth.

7.5.1.2 Smart water management systems

Smart water management systems have revolutionized the way water resources are monitored, managed, and distributed. These systems leverage real-time data, remote sensing, and analytics to optimize water distribution networks, detect leaks, and improve water efficiency. By collecting data from sensors and meters, utilities can gain valuable insights into water usage patterns and make informed decisions regarding water allocation. Advanced analytics and modeling tools enable utilities to forecast water demand, detect anomalies, and optimize water allocation (Fang & Wu, 2020). In addition, these systems promote water conservation by providing consumers with real-time feedback and recommendations for water-saving behaviors. By identifying and addressing leaks in distribution networks, smart water management systems help minimize water loss and improve overall system efficiency. Moreover, these systems contribute to enhanced resilience in emergencies, as real-time monitoring and early warning systems provide timely information to emergency responders. Several cities worldwide have embraced smart water management systems. For example, in Barcelona, Spain, the installation of smart meters and a comprehensive water management platform has resulted in significant water savings and improved leak detection. The Smart Water Network project in Miami-Dade County, USA, uses a network of sensors and analytics to optimize water distribution, reduce water loss, and improve water quality (Sheng et al., 2020; Zhuo et al., 2020).

7.5.1.2.1 Rainwater harvesting and graywater recycling

Rainwater harvesting and graywater recycling systems are effective strategies for sustainable water management. These systems offer a range of benefits, including water conservation, reduced reliance on freshwater sources, and decreased strain on municipal water supplies. Here, we delve into the details of rainwater harvesting and graywater recycling.

7.5.1.2.2 Rainwater harvesting

Rainwater harvesting involves the collection and storage of rainwater for subsequent use. It can be as simple as capturing rainwater from rooftops and directing it to storage tanks or more advanced with underground cisterns, filters, and pumps. The collected rainwater can be used for various nonpotable purposes, such as landscape irrigation, toilet flushing, and industrial processes. Rainwater harvesting systems provide several advantages. First, they reduce the demand for freshwater resources, particularly in regions with limited water availability or during periods of drought. By using rainwater for nonpotable purposes, pressure on municipal water supplies is alleviated, ensuring a more sustainable water balance. In addition, rainwater harvesting helps prevent stormwater runoff, which can contribute to urban flooding and water pollution.

7.5.1.2.3 Graywater recycling

Graywater recycling systems treat and reuse wastewater from sinks, showers, baths, and laundry for nonpotable purposes. Graywater is relatively clean and can be effectively treated to remove contaminants, making it suitable for irrigation, toilet flushing, and other similar applications. Graywater recycling helps reduce freshwater demand and promotes efficient water usage. Greywater recycling systems typically involve collection, treatment, and storage processes. The collected graywater undergoes treatment, which can include filtration, disinfection, and nutrient removal, depending on the desired application. After treatment, the recycled graywater is stored and distributed for non-potable uses, reducing the need for fresh water supplies. Implementing rainwater harvesting and graywater recycling systems offers numerous benefits. These include:

Water conservation: By using rainwater and recycled graywater for nonpotable purposes, overall water consumption is reduced, conserving freshwater resources. This is particularly important in water-scarce regions or areas experiencing high water stress.

Cost savings: Rainwater harvesting and graywater recycling systems can lead to significant cost savings on water bills, especially for large consumers such as households, commercial buildings, and industrial facilities.

Environmental impact: By reducing reliance on freshwater sources, rainwater harvesting and graywater recycling contribute to a more sustainable water cycle. They help preserve natural water bodies, minimize the energy required for water treatment and distribution, and decrease the carbon footprint associated with water usage.

Resilience: Rainwater harvesting and graywater recycling systems provide an alternative water source during periods of water scarcity or emergencies. This enhances water supply resilience and reduces the vulnerability of communities to water shortages.

Educational and awareness opportunities: Implementing rainwater harvesting and graywater recycling systems offers educational and awareness-building opportunities for individuals, communities, and organizations. It promotes a greater understanding of water resources, encourages responsible water usage, and fosters a culture of sustainability.

Many regions around the world have successfully implemented rainwater harvesting and graywater recycling systems. For example, in Australia, where water scarcity is a prevalent issue, rainwater tanks are commonly installed in households and commercial buildings to capture and use rainwater for various purposes. Singapore has also implemented a comprehensive approach to water management, including the collection, treatment, and recycling of graywater for nonpotable applications.

7.5.1.3 Integrated water resource management

IWRM is a comprehensive and holistic approach to managing water resources sustainably. It recognizes that water is a finite and essential resource that is interconnected with various sectors, ecosystems, and communities. IWRM aims to balance the social, economic, and environmental dimensions of water management by integrating different disciplines, stakeholders, and sectors involved in water governance. The core principle of IWRM is the recognition that water resources are interconnected and that actions in one part of the water cycle can have impacts on other components. It emphasizes the need for an integrated and coordinated approach to water management, taking into account the diverse needs and priorities of various stakeholders. One of the key aspects of IWRM is the promotion of collaborative decision-making processes. It encourages the active participation of stakeholders, including government agencies, communities, NGOs, and the private sector, in the planning and management of water resources. By involving multiple perspectives and interests, IWRM seeks to ensure that decisions are transparent, inclusive, and responsive to the needs of all stakeholders (Mamun et al., 2022; Mpofu, 2022). IWRM also emphasizes the integration of water considerations into land-use planning and policy frameworks. It recognizes the interconnectedness between water resources and land management practices, such as agriculture, urban development, and industrial activities. By integrating water into land-use planning, IWRM aims to minimize conflicts and optimize the use of water resources across different sectors. This integration also takes into account the potential impacts of land-use activities on water quality and ecosystems.

Furthermore, IWRM promotes the application of an adaptive management approach. This involves monitoring, assessing, and adjusting water management strategies based on new information, changing conditions, and emerging challenges. It recognizes the dynamic nature of water resources and the need for flexibility in responding to uncertainties and evolving conditions. Adaptive management allows for the continuous improvement of water management practices and the ability to adapt to future challenges, such as climate change and population growth. The implementation of IWRM requires the use of various tools and approaches to support decision-making and planning processes. These include:

Water governance: IWRM emphasizes the need for effective and accountable water governance structures. This involves the development of clear policies, legal frameworks, and institutional arrangements that enable integrated decision-making and coordination among different stakeholders. Good governance practices, such as transparency, participation, and accountability, are essential for the successful implementation of IWRM.

Water allocation and management: IWRM promotes the efficient and equitable allocation of water resources. It involves assessing water availability, demands, and priorities across different sectors and users. Tools such as water rights systems, water markets, and water pricing mechanisms can be used to allocate water resources fairly and sustainably.

Water conservation and efficiency: IWRM emphasizes the importance of water conservation and efficiency measures to optimize water usage. This includes promoting water-saving technologies, improving water management practices, and raising awareness about the value of water. Water footprint assessment, as discussed earlier, can also play a significant role in identifying opportunities for water conservation and efficiency.

Water quality management: IWRM recognizes the importance of maintaining and improving water quality. It involves monitoring and managing water pollution sources, implementing treatment technologies, and promoting sustainable practices to protect water ecosystems and ensure the availability of safe drinking water. IWRM recognizes the importance of maintaining and improving water quality. It involves monitoring and managing water pollution sources, implementing treatment technologies, and promoting sustainable practices to protect water ecosystems and ensure the availability of safe drinking water.

Ecosystem protection: IWRM acknowledges the critical role of ecosystems in maintaining water resources. It emphasizes the need to protect and restore water-related ecosystems, such as wetlands, rivers, and groundwater recharge areas. By maintaining healthy ecosystems, IWRM ensures the resilience of water resources and supports biodiversity conservation.

The implementation of IWRM is not without challenges. It requires strong political will, institutional capacity, and financial resources. Stakeholder coordination and collaboration can be complex, particularly when there are competing interests and limited resources. In addition, IWRM needs to consider the social and economic dimensions of water management, taking into account issues of equity, affordability, and access to water services. However, despite these challenges, IWRM offers significant benefits. It provides a framework for addressing water-related challenges comprehensively and sustainably. By promoting integration, collaboration, and adaptive management, IWRM helps optimize the use of water resources, minimize conflicts, and enhance water security. It contributes to the achievement of the SDGs, particularly Goal 6, which focuses on ensuring the availability and sustainable management of water and sanitation for all.

7.5.1.4 Nature-based solutions

Nature-based solutions (NBS) are innovative approaches that leverage the power of nature to address water-related challenges. They involve the use of natural processes, ecosystems, and biodiversity to provide sustainable and cost-effective solutions for water management. NBS focus on the conservation, restoration, and sustainable management of natural habitats and ecosystems to enhance water availability, improve water quality, and reduce the impacts of water-related hazards. One example of an NBS is the restoration and conservation of wetlands. Wetlands act as natural sponges, absorbing and storing water during periods of heavy rainfall and slowly releasing it during drier periods (Hoekstra et al., 2017). They help regulate water flows, reduce the risk of floods, and recharge groundwater aquifers. Wetlands also provide important habitats for diverse plant and animal species, contributing to biodiversity conservation. Reforestation is another NBS that has significant water management benefits. Forests play a crucial role in regulating water cycles by intercepting rainfall, reducing soil erosion, and promoting infiltration. They act as natural water filters, improving water quality by trapping sediments and pollutants. Forested areas also serve as important water catchments, replenishing surface and groundwater resources. By restoring and protecting forests, water availability and quality can be enhanced, benefiting both humans and ecosystems. Green infrastructure is another NBS approach that involves the use of natural elements, such as green roofs, permeable pavements, and urban parks, to manage water. These green features absorb rainwater, reduce stormwater runoff, and promote infiltration into the ground. Green infrastructure not only helps manage urban water challenges but also provides additional benefits, such as urban heat island mitigation, air purification, and recreational spaces for communities. NBS offer a range of benefits, making them highly attractive for water management:

Water availability: By restoring and conserving natural habitats and ecosystems, NBS enhance water availability. They promote groundwater recharge, maintain stream flows, and contribute to the overall water balance in a region. This is particularly crucial in areas facing water scarcity or seasonal water availability.

Water quality: NBS contribute to improved water quality by acting as natural filters. Wetlands, forests, and other natural habitats remove sediments, nutrients, and pollutants from water, improving its quality and reducing the need for costly water treatment infrastructure.

Biodiversity conservation: NBS support biodiversity conservation by providing habitats for a wide range of plant and animal species. They enhance ecological resilience and contribute to the preservation of ecosystem services that are essential for water and food security.

Climate change mitigation and adaptation: NBS play a vital role in mitigating climate change by sequestering carbon dioxide from the atmosphere. Forests, in particular, are significant carbon sinks, helping to reduce greenhouse gas emissions. In addition, NBS enhance community resilience by reducing the impacts of climate-related hazards, such as floods and droughts.

Cost-effectiveness: NBS often provide cost-effective solutions compared to traditional gray infrastructure. They require lower capital investments, have lower maintenance costs, and provide multiple cobenefits, such as improved air quality, recreation opportunities, and aesthetic value.

To effectively implement NBS, a combination of scientific knowledge, policy support, and community engagement is required. Governments, water managers, and stakeholders need to recognize the value of NBS and integrate them into water management plans and policies. This includes providing incentives for NBS implementation, developing guidelines and standards, and raising awareness about the benefits of nature-based approaches. Global examples demonstrate the successful application of NBS in water management. In the Netherlands, for instance, the Room for the River program implemented a combination of measures, including the restoration of floodplains, the creation of retention areas, and the construction of flood-resistant infrastructure (Ho & Goethals, 2019; Li, Lu, et al., 2017). This approach not only increased flood safety but also improved ecological conditions and created recreational opportunities.

7.5.1.5 Best practices in water governance and policy

Effective water governance and policy frameworks are critical for achieving sustainable water management and addressing the complex challenges associated with water resources. Best practices in water governance encompass a range of principles and approaches that promote transparency, participation, accountability, and integration across different sectors and levels of governance. In addition, the implementation of robust water regulations, policy instruments, and enforcement mechanisms is necessary to protect water resources, ensure equitable access, and promote sustainable water usage. One key element of effective water governance is transparent decision-making processes (Octavia et al., 2022). Transparent governance ensures that decisions related to water allocation, infrastructure development, and policy formulation are made openly and inclusively. This involves providing access to information, engaging stakeholders in decision-making processes, and fostering public participation. Transparent governance helps build trust among stakeholders, enhances accountability, and improves the quality of water management decisions. Stakeholder participation is another essential aspect of water governance. Involving a diverse range of stakeholders, including water users, local communities, civil society organizations, and indigenous groups, in decision-making processes enhances the legitimacy and effectiveness of water management. By considering multiple perspectives, knowledge, and interests, stakeholder participation helps identify innovative solutions, improve the implementation of water policies, and ensure that the needs and priorities of all stakeholders are taken into account.

Integration of local knowledge and traditional practices is crucial for sustainable water governance. Local communities often have valuable insights into the local hydrological conditions, water usage practices, and traditional water management systems. Integrating local knowledge into decision-making processes can lead to more context-specific and culturally appropriate solutions, ensuring the sustainability and resilience of water management initiatives. Water regulations and enforcement mechanisms are vital for protecting water resources and ensuring compliance with sustainable water management practices. Robust regulatory frameworks define water rights, establish water quality standards, and set limits on water abstraction to prevent overexploitation and pollution. Effective enforcement mechanisms, such as monitoring systems, inspections, and penalties for noncompliance, help ensure that regulations are followed and water resources are effectively managed and protected. Policy instruments play a crucial role in incentivizing efficient water use and promoting water conservation. Water pricing mechanisms, for example, can provide economic incentives for users to adopt water-saving practices and technologies. Implementing progressive water tariffs, where higher prices are charged for excessive water consumption, encourages users to be more conscious of their water usage and promotes water conservation. Similarly, establishing water rights systems, where water allocations are based on equitable and environmentally sustainable principles, ensures efficient water allocation and protects vulnerable water users. Water markets, where water rights can be bought and sold, can also contribute to efficient water allocation. Water markets allow for the transfer of water from low-value to high-value uses, enabling water to be allocated to its most productive and economically efficient uses. However, it is important to design water markets carefully to prevent negative social and environmental impacts, ensure equitable access, and protect the rights of marginalized communities.

Promoting capacity building and knowledge sharing among water governance institutions and stakeholders is crucial for implementing best practices in water governance. This includes providing training programs, fostering collaboration and learning networks, and facilitating the exchange of experiences and lessons learned. Capacity building helps strengthen the technical, institutional, and governance capacities needed to address complex water challenges effectively. Several countries have implemented best practices in water governance and policy, providing valuable examples for others to learn from. Australia, for instance, has adopted an integrated water management approach that promotes collaboration between government agencies, water utilities, communities, and businesses. The Australian experience highlights the importance of stakeholder engagement, water pricing mechanisms, and water planning frameworks in achieving sustainable water management outcomes. In the Netherlands, the Delta Program is a successful example of integrated water governance. It adopts a long-term, adaptive approach to address flood risks, water scarcity, and water quality issues. The program brings together various stakeholders and experts to develop strategies, invest in infrastructure, and implement measures to enhance water resilience in the face of climate change.

7.5.1.6 Capacity building and knowledge sharing

Capacity building and knowledge sharing are crucial components of sustainable water management efforts. They enable stakeholders to acquire the necessary skills, knowledge, and resources to implement innovative technologies and best practices effectively. Capacity-building initiatives provide training programs, workshops, and educational resources to enhance the technical, institutional, and governance capacities of water professionals, policymakers, and local communities. Knowledge-sharing platforms facilitate the exchange of experiences, lessons learned, and best practices among different stakeholders at local, regional, and international levels. Capacity-building initiatives in the water sector focus on enhancing technical skills and knowledge related to water management, conservation, and efficiency. These programs provide training on the operation and maintenance of water infrastructure, water treatment technologies, water quality monitoring, and data analysis (Bailey et al., 2022). They also promote the adoption of best practices in water governance, policy formulation, and stakeholder engagement. Capacity-building efforts aim to equip water professionals with the skills and knowledge needed to address the evolving challenges in water management and make informed decisions. Workshops and seminars are valuable tools for capacity building as they provide a platform for experts, practitioners, and stakeholders to share their experiences, research findings, and innovative approaches. These events foster dialogue, promote networking, and encourage collaboration among water professionals from different sectors and regions. Workshops often include practical exercises, case studies, and interactive sessions to facilitate hands-on learning and the application of knowledge in real-world situations.

Knowledge-sharing platforms, such as online databases, webinars, and discussion forums, enable stakeholders to access and exchange information on sustainable water management practices. These platforms facilitate the dissemination of research findings, policy documents, technical guidelines, and success stories, allowing stakeholders to learn from each other's experiences and replicate successful approaches. International organizations, research institutions, and governmental agencies play a crucial role in curating and maintaining these knowledge-sharing platforms, ensuring the availability and accessibility of up-to-date information and resources. International cooperation and partnerships are key drivers of capacity building and knowledge sharing in the water sector. Collaborative initiatives between developed and developing countries promote the transfer of knowledge, expertise, and resources to support the implementation of sustainable water management practices. These partnerships can take the form of technical assistance programs, twinning arrangements, and peer-to-peer exchanges. Developed countries often provide financial and technical support to assist developing countries in building their capacity and implementing sustainable water management strategies. One example of international cooperation in capacity building is the United Nations Development Programme Water Governance Facility. The facility supports capacity development efforts in the water sector by providing technical assistance, training programs, and knowledge-sharing platforms. It promotes the exchange of experiences and best practices among countries, facilitates policy dialogue, and assists in the development of water governance frameworks. Capacity-building and knowledge-sharing efforts should be tailored to the specific needs and contexts of different regions and communities. Local knowledge and traditional practices should be integrated into capacity-building programs to ensure their relevance and effectiveness. Participatory approaches that involve local communities in the design and implementation of capacity-building initiatives are essential for sustainable outcomes. By empowering local stakeholders and fostering ownership, capacity-building efforts are more likely to lead to long-term positive impacts on water management practices. Examples of successful implementation of technological innovations and best practices can be found worldwide. In Singapore, the "NEWater" initiative has implemented advanced water treatment technologies to recycle wastewater and turn it into high-quality drinking water (Moglia et al., 2018). This has reduced Singapore's reliance on imported water and enhanced water security. In Australia, the implementation of water-sensitive urban design approaches, such as green roofs, permeable pavements, and rain gardens, has improved stormwater management and enhanced urban water sustainability (Esetlili et al., 2022).

7.5.2 Policy and regulatory interventions

Policy and regulatory interventions play a crucial role in driving water conservation initiatives and ensuring sustainable water management practices. Effective policies and regulations provide a framework for guiding water usage, promoting efficiency, and protecting water resources. They create incentives for conservation, establish standards for water quality, and allocate water resources equitably. In this section, we will discuss the importance of policy and regulatory interventions in water conservation and explore various approaches and strategies that have been successfully implemented. One key aspect of policy and regulatory interventions is the establishment of water rights and allocation systems. These systems define the legal rights and responsibilities associated with water resources and ensure their fair and sustainable distribution among competing users. Water rights can be based on principles of equity, efficiency, and environmental sustainability. Implementing clear and transparent water rights systems helps prevent overextraction, encourages efficient water usage, and fosters responsible stewardship of water resources. Water pricing mechanisms are another effective policy tool for promoting water conservation. By accurately reflecting the true value of water, pricing mechanisms encourage users to adopt more efficient practices and technologies. Incentivizing water conservation through tiered pricing structures, where higher prices are charged for excessive water usage, can effectively discourage wasteful consumption and promote conservation behaviors. In addition, implementing progressive pricing schemes that consider the socioeconomic context can ensure that water remains affordable for vulnerable populations while discouraging excessive usage. Regulations and standards for water quality also play a vital role in water conservation efforts. These regulations aim to prevent pollution and ensure that water resources remain safe for human consumption and ecosystem health. Implementing strict regulations on industrial discharge, agricultural runoff, and wastewater treatment helps protect water sources from contamination. Incentives and penalties can be incorporated into regulations to encourage industries and agricultural operations to adopt cleaner production practices and treat their wastewater before discharge.

IWRM approaches, which consider the interactions between different sectors and stakeholders, are gaining prominence in policy and regulatory frameworks (Dolganova et al., 2019; Mingkhwan & Worakhunpiset, 2018). IWRM emphasizes collaborative decision-making, stakeholder participation, and the integration of social, economic, and environmental considerations in water management. By involving diverse stakeholders in the decision-making process, IWRM promotes a better understanding of the water challenges and facilitates the identification of shared solutions. It also encourages the adoption of holistic and long-term approaches to water management, considering the interdependencies between water resources, ecosystems, and human activities. Institutional capacity building is crucial for effective policy implementation and enforcement. Governments and water management authorities need to have the necessary expertise, resources, and infrastructure to develop, implement, and monitor water conservation policies. Strengthening institutional capacity involves investing in training programs, knowledge-sharing platforms, and technical assistance to support the development and implementation of robust policies and regulations. It also involves enhancing coordination and collaboration among different government agencies, water utilities, and stakeholders to ensure effective policy implementation and enforcement.

7.5.3 Stakeholder engagement and collaboration

Stakeholder engagement and collaboration are fundamental to the success of water conservation initiatives. Effective engagement ensures that diverse perspectives, knowledge, and interests are considered in decision-making processes, leading to more inclusive and sustainable outcomes. Stakeholders in water management include government agencies, water utilities, industries, farmers, local communities, NGOs, and the general public. Engaging these stakeholders promotes ownership, builds consensus, and fosters cooperation in implementing water conservation measures. One of the key benefits of stakeholder engagement is the access to local knowledge and traditional practices. Local communities often possess valuable insights into water resources, ecological systems, and sustainable water management practices. Engaging with local communities and indigenous groups ensures that their knowledge and perspectives are incorporated into decision-making processes, leading to more context-specific and effective water conservation measures. It also strengthens the sense of ownership and responsibility among local communities, enhancing the long-term sustainability of conservation efforts. Stakeholder engagement processes should be inclusive, transparent, and participatory. Meaningful engagement involves creating opportunities for stakeholders to voice their concerns, provide input, and contribute to decision-making processes. Public consultations, workshops, and participatory planning exercises can facilitate dialogue and collaboration among stakeholders. These processes enable stakeholders to better understand the challenges and opportunities associated with water conservation, build trust among different parties, and identify common goals and shared solutions. Collaboration among stakeholders is essential for the implementation of water conservation initiatives. Collaborative efforts often involve multistakeholder partnerships, where different actors come together to collectively address water challenges. These partnerships can include government agencies, private sector entities, civil society organizations, research institutions, and community-based organizations. Collaborative initiatives enable the pooling of resources, expertise, and knowledge, leading to innovative solutions and shared responsibilities.

Water users, such as industries and agriculture, can actively participate in water conservation through voluntary initiatives and partnerships. For example, industry-led initiatives that promote water efficiency and sustainable production practices can significantly reduce water consumption and environmental impact. Collaborative platforms that bring together different sectors, such as the agriculture, energy, and water sectors, foster integrated approaches to water management, considering the interdependencies between these sectors and promoting more sustainable outcomes. International cooperation and knowledge sharing also play a crucial role in stakeholder engagement and collaboration. Global platforms, such as the World Water Forum and international conferences, provide opportunities for stakeholders from different countries and regions to exchange experiences, share best practices, and learn from each other's successes and challenges. International cooperation enables developing countries to access technical expertise, financial resources, and capacity-building support to implement effective water conservation measures.

7.5.4 Case studies: successful implementation of water conservation initiatives

Case studies provide valuable insights into the successful implementation of water conservation initiatives across different sectors and regions. Let's explore a few examples that highlight effective strategies and practices in water conservation.

7.5.4.1 *Singapore's integrated water management approach*

Singapore, a small island city-state facing water scarcity and dependence on imported water, has implemented a comprehensive and integrated approach to water management. The country has successfully reduced water demand, maximized water efficiency, and achieved self-sufficiency in water supply through a combination of policies, technologies, and public engagement. One key aspect of Singapore's approach is water conservation. The city-state has implemented various measures to promote water-saving behaviors and technologies. For instance, water-efficient appliances and fixtures are encouraged through incentives and labeling schemes. The widespread use of water-efficient toilets, showers, and taps has significantly reduced per capita water consumption. In addition, Singapore has implemented a dual-piping system that separates potable and nonpotable water sources, enabling the use of recycled water for nonpotable applications such as toilet flushing and industrial processes. Singapore's efforts in water conservation are complemented by its investment in advanced water treatment technologies. The country has developed the NEWater system, which treats wastewater to ultra-high-quality standards for indirect potable reuse (Aladuwaka & Momsen, 2010; Bhatti et al., 2011; Boretti & Rosa, 2019; Deb et al., 2019a; Shi, 2022). NEWater is used for industrial processes, cooling systems, and even blended with raw water supplies. The country has also embraced desalination technologies to diversify its water sources. These advancements in water treatment and reuse have significantly reduced the reliance on imported water and ensured a sustainable water supply for the population. Furthermore, Singapore has prioritized public engagement and education to raise awareness about the importance of water conservation. The "ABC Waters" program, for example, focuses on transforming water bodies into beautiful and functional spaces while educating the public about water resources. The program encourages community involvement in water conservation initiatives and highlights the interconnectedness between water and the environment.

7.5.4.2 *The Los Angeles water conservation program*

Los Angeles, located in a region known for its arid climate and recurring droughts, has implemented a successful water conservation program to address water scarcity challenges. The city's approach includes a combination of regulatory measures, financial incentives, and public awareness campaigns. Regulatory measures play a crucial role in promoting water conservation in Los Angeles. The city has implemented water restrictions, such as limits on outdoor watering and restrictions on the usage of potable water for nonessential purposes. These measures help regulate water consumption and raise awareness about the need for efficient water use. To incentivize water-saving behaviors, Los Angeles offers financial incentives and rebates for water-efficient appliances, fixtures, and landscaping. Residents and businesses can take advantage of these programs to reduce their water consumption and receive financial benefits.

Public awareness campaigns are another essential component of Los Angeles' water conservation efforts. The city conducts educational campaigns to inform residents about water-saving practices, such as installing low-flow fixtures, adopting drought-tolerant landscaping, and implementing rainwater harvesting systems. These campaigns empower the community to actively participate in water conservation efforts and contribute to the overall water sustainability of the region. Through the implementation of these measures, Los Angeles has successfully reduced per capita water consumption and improved water usage efficiency. During drought periods, the city's efforts in water conservation have minimized the strain on water resources and ensured a reliable water supply for its residents.

7.5.4.3 *The water efficient maize for Africa Project*

The Water Efficient Maize for Africa (WEMA) project is an excellent example of the successful implementation of water conservation initiatives in the agricultural sector, specifically in Sub-Saharan Africa. The project focuses on developing and promoting drought-tolerant maize varieties to address the challenges of water scarcity and climate change. Maize is a staple crop in many African countries, but its production is highly vulnerable to drought conditions. The WEMA project aims to improve food security and enhance farmers' resilience by developing genetically modified maize varieties that require less water and exhibit greater tolerance to drought. Through collaboration between public and private sectors, the WEMA project has successfully developed and deployed drought-tolerant maize varieties in several African countries (Ahmed et al., 2015; Aladuwaka & Momsen, 2010; D'Odorico et al., 2012; Murata & Sakamoto, 2019). These varieties have shown remarkable resilience in water-limited conditions, enabling farmers to achieve higher yields with reduced water inputs. In addition to the development of drought-tolerant varieties, the WEMA project emphasizes capacity building and knowledge sharing. It provides training and resources to farmers on sustainable agricultural practices, including efficient irrigation techniques and water management strategies. By empowering farmers with the necessary skills and knowledge, the project enhances their ability to conserve water resources while maintaining agricultural productivity.

The success of the WEMA project demonstrates the potential of technological innovations and collaborative approaches to address water scarcity in the agricultural sector. By promoting water-efficient crop varieties and supporting farmers with the necessary tools and knowledge, the project contributes to sustainable water management and food security in Sub-Saharan Africa.

7.5.4.4 Coca-Cola's water stewardship initiatives

Coca-Cola, a multinational beverage company, has implemented a range of water stewardship initiatives to address water conservation challenges in the communities where it operates. Recognizing the importance of water resources for its business and the communities, Coca-Cola has committed to replenishing the water it uses and minimizing its water footprint. One of Coca-Cola's key initiatives is its focus on water efficiency within its manufacturing plants. The company has implemented innovative technologies and practices to reduce water consumption during the production process. For instance, advanced water treatment systems, such as membrane filtration and reverse osmosis, are employed to treat and recycle process water. In addition, water-efficient cleaning methods and equipment are used to minimize water waste. These initiatives have significantly improved water usage efficiency and reduced the overall water consumption of Coca-Cola's operations. To address water scarcity and promote water conservation in communities, Coca-Cola engages in watershed restoration projects. The company collaborates with local organizations and communities to restore and protect watersheds, including reforestation efforts, conservation programs, and the promotion of sustainable land management practices. These initiatives help improve water availability, regulate water flows, and enhance ecosystem health.

Furthermore, Coca-Cola actively engages with local communities to promote water conservation and access to clean water. The "Every Drop Matters" program implemented in India focuses on rainwater harvesting, groundwater recharge, and water conservation awareness campaigns. Through partnerships with local governments, NGOs, and communities, Coca-Cola has implemented rainwater harvesting structures, such as check dams and rooftop rainwater collection systems, to recharge groundwater and enhance water availability. These initiatives have had a positive impact on water resources and have empowered communities to actively participate in water conservation efforts. By integrating water stewardship into its business practices and engaging with stakeholders, Coca-Cola has made significant progress in reducing its water consumption and replenishing water sources in water-stressed areas. The company's initiatives demonstrate the importance of corporate responsibility and collaboration in addressing water conservation challenges.

7.5.4.5 Israel's drip irrigation technology

Israel, a country known for its arid climate and limited water resources, has become a global leader in water conservation through the widespread adoption of drip irrigation technology. Drip irrigation delivers water directly to the plant roots, minimizing evaporation and water wastage. This technology has revolutionized agriculture in arid regions and has significantly contributed to water conservation efforts. Israel's success in implementing drip irrigation can be attributed to various factors. First, the government has actively supported and incentivized the adoption of drip irrigation systems. Financial incentives, subsidies, and technical assistance are provided to farmers to facilitate the installation and maintenance of drip irrigation infrastructure. These measures have encouraged widespread adoption across the agricultural sector. Second, Israel has invested in research and development to improve drip irrigation technologies and optimize their performance. Continuous innovation has led to the development of advanced drip irrigation systems that are highly efficient and precise in water delivery (Guo et al., 2019; Kumar et al., 2021). These systems allow farmers to tailor irrigation schedules to specific crop needs, further minimizing water waste. Lastly, Israel has implemented water pricing mechanisms that encourage efficient water usage in agriculture. The country employs a tiered pricing structure where the cost of water increases as consumption levels rise. This pricing system incentivizes farmers to use water efficiently and motivates them to invest in water-saving technologies like drip irrigation. The successful implementation of drip irrigation in Israel has resulted in significant water savings, increased agricultural productivity, and improved water usage efficiency. It has transformed arid lands into productive agricultural areas and reduced the pressure on freshwater resources. The experience of Israel serves as a valuable example for other water-stressed regions seeking to enhance water conservation in agriculture. These case studies highlight the successful implementation of water conservation initiatives across different contexts and sectors. They demonstrate the importance of a combination of policies, technologies, stakeholder engagement, and public awareness in achieving sustainable water management. By learning from these examples and promoting best practices, we can effectively address water scarcity challenges and ensure the availability of clean water resources for future generations.

7.6 Future directions and challenges

Industrial water conservation, driven by the water footprint concept and aligned with the SDGs, is crucial for ensuring the sustainable usage and management of water resources (Flores & Ghisi, 2022; Hossain & Khan, 2020; Peng et al., 2015). As we look toward the future, several key directions and challenges need to be addressed to further advance industrial water conservation efforts.

7.6.1 Integration of water footprint assessment into industrial practices

One future direction is the widespread integration of water footprint assessment into industrial practices. While there has been progress in implementing water footprint assessment in some industries, it is important to expand its application across a broader range of sectors. This requires raising awareness about the benefits of water footprint assessment, providing guidance and tools for implementation, and encouraging companies to adopt and use the assessment as a standard practice. Integration of water footprint assessment into industrial practices will enable a better understanding of water usage patterns, identification of hotspots, and implementation of targeted water conservation measures.

7.6.1.1 Technological innovations and efficiency improvements

Advancements in technology will play a crucial role in driving industrial water conservation efforts. There is a need for continuous innovation and development of technologies that improve water efficiency, reduce water consumption, and minimize pollution in industrial processes. This includes the adoption of water-saving technologies, such as water-efficient equipment, recycling systems, and advanced treatment processes. In addition, the integration of digital technologies, data analytics, and automation can enable real-time monitoring and optimization of water usage in industrial operations. Continued research and development in these areas will contribute to more efficient and sustainable industrial water management.

7.6.1.2 Collaboration and partnerships

Addressing the complex challenges of industrial water conservation requires collaboration and partnerships among various stakeholders, including industries, governments, NGOs, and research institutions. Collaboration can foster knowledge sharing, resource pooling, and joint problem-solving approaches. It can also facilitate the development and implementation of collective initiatives, such as water stewardship programs, industry-wide standards, and best practice sharing platforms. Strengthening collaboration and partnerships will enhance the effectiveness of industrial water conservation efforts and ensure the alignment of actions with broader sustainability goals.

7.6.1.3 Policy and regulatory frameworks

Effective policy and regulatory frameworks are essential for driving industrial water conservation. Governments play a critical role in developing and implementing policies that incentivize and enforce water-efficient practices in industries. This includes the establishment of water pricing mechanisms that reflect the true value of water, the development of water usage permits and regulations, and the promotion of water-saving technologies through subsidies and incentives. Policy frameworks should also consider the integration of water footprint assessment and sustainable development goals into industrial regulations and reporting requirements. Continual evaluation and revision of policies to address emerging challenges and promote best practices are necessary for effective industrial water conservation (Aslam et al., 2021; Cook & Webber, 2016).

7.6.1.4 Education and awareness

Enhancing education and awareness about the importance of industrial water conservation is a key challenge that needs to be addressed. Companies, industry associations, and educational institutions should actively promote water stewardship and provide training and resources to employees, managers, and decision-makers. Building awareness about the water footprint concept, sustainable development goals, and best practices in industrial water conservation can drive behavioral change and encourage the adoption of water-saving measures. Engaging the broader public through awareness campaigns, media, and community outreach programs is also crucial for fostering a collective understanding and commitment to water conservation.

7.6.1.5 Climate change resilience

Climate change poses additional challenges to industrial water conservation efforts. Increasing temperatures, changing rainfall patterns, and extreme weather events can disrupt water availability and increase water stress in many regions. Therefore, industrial water conservation strategies should incorporate climate change resilience measures. This includes the implementation of adaptive technologies, diversification of water sources, and the development of contingency plans to mitigate the impact of climate-related water shortages (Abdelzaher & Awad, 2022; Ignjatović et al., 2022; Knuth et al., 2020; Juanga-Labayen et al., 2022). Integrating climate change considerations into water footprint assessments and sustainable development goals will help industries adapt to the changing water availability and ensure long-term resilience.

7.7 Conclusions

Industrial water conservation, guided by the principles of water footprint assessment and sustainable development goals, is of paramount importance for ensuring the sustainable use and management of water resources. Throughout this chapter, we have explored the concept of industrial water conservation, its relevance to water footprint assessment and sustainable development goals, and the various approaches, challenges, and future directions associated with this crucial endeavor. Industrial water consumption has significant implications for water scarcity, ecosystem health, and socioeconomic development. By adopting strategies and practices aimed at reducing water consumption, improving efficiency, and minimizing environmental impacts, industries can contribute to sustainable water management. The water footprint assessment serves as a powerful tool for quantifying and managing water usage in industrial contexts, enabling companies to identify areas of improvement, implement water-saving measures, and monitor progress over time.

The alignment between industrial water conservation and theSDGs is evident. The SDGs provide a comprehensive framework for addressing global challenges, including water scarcity, poverty, hunger, and climate change. Industrial water conservation directly contributes to several SDGs, including SDG 6 (Clean Water and Sanitation), SDG 9 (Industry, Innovation, and Infrastructure), and SDG 12 (Responsible Consumption and Production). By embracing industrial water conservation practices, companies can actively contribute to achieving these goals and contribute to sustainable development on a global scale. Several key conclusions emerge from our discussion. First, integrating water footprint assessment into industrial practices is essential for better understanding water usage patterns, identifying hotspots, and implementing targeted water conservation measures. It requires concerted efforts to raise awareness, provide guidance, and encourage adoption across industries. Second, technological innovations play a crucial role in improving water efficiency across sectors. Advanced water treatment technologies, precision irrigation systems, and smart water management systems enable industries to optimize water usage, reduce consumption, and minimize pollution. Continued investment in research and development, as well as the adoption of innovative technologies, is necessary to drive further progress in industrial water conservation.

Third, collaboration and partnerships are fundamental to address the complex challenges of industrial water conservation. Engaging various stakeholders, including industries, governments, NGOs, and research institutions, foster knowledge sharing, resource pooling, and joint problem-solving approaches. Collaborative initiatives, such as water stewardship programs and industry-wide standards, can significantly enhance the effectiveness of industrial water conservation efforts. Fourth, policy and regulatory frameworks are essential for driving industrial water conservation. Governments should establish supportive policies, including water pricing mechanisms, regulations, and incentives, to incentivize water-efficient practices in industries. Regular evaluation and revision of policies are necessary to adapt to emerging challenges and promote best practices. Fifth, education and awareness are crucial for fostering a culture of water stewardship. Companies, industry associations, and educational institutions should actively promote water conservation and provide training and resources to employees, managers, and decision-makers. Building awareness about the water footprint concept, SDGs, and best practices in industrial water conservation can drive behavioral change and encourage the adoption of water-saving measures. Lastly, addressing climate change resilience is an imperative component of industrial water conservation. Climate change poses additional challenges to water availability, and industries need to integrate climate change considerations into their water management strategies. By implementing adaptive technologies, diversifying water sources, and developing contingency plans, industries can mitigate the impact of climate-related water shortages and ensure long-term resilience.

(Abdelzaher & Awad, 2022; Ahmed et al., 2015; Akenji & Bengtsson, 2014; Aladuwaka & Momsen, 2010; Aldaya & Hoekstra, 2010; Aslam et al., 2021; Babel et al., 2019; Bailey et al., 2022; Baumgartner, 2019; Berger et al., 2015; Bhatnagar et al., 2014; Bhatti et al., 2011; Bjornlund et al., 2007; Boretti & Rosa, 2019; Bosire et al., 2015; Bradley et al., 2001; Cantin et al., 2005; Cao et al., 2018; Carr et al., 2012, 2015, Chapagain & Hoekstra, 2007, 2011; Chapagain & Orr, 2009; Chen et al., 2015; Cheng et al., 2003; Chowdhary et al., 2019; Chukalla et al., 2015; Cook & Webber, 2016; Croese et al., 2020; Deb et al., 2018, 2019b, 2019a; D'Odorico et al., 2012; Dolganova et al., 2019; Ebenstein, 2012; Ene et al., 2013; Ercin et al., 2012; Eriksen et al., 2014; Esetlili et al., 2022; Evans & Miguel, 2007; Fang et al., 2014; Fang & Wu, 2020; Florea et al., 2005; Flores & Ghisi, 2022; Fonseca et al., 2020; Fox et al., 1991; Galli et al., 2012; Gare, 2001; Geller et al., 1983; Gerbens-Leenes & Hoekstra, 2012; Gleeson et al., 2012; Gunkel et al., 2007; Guo et al., 2019, 2021; Ho & Goethals, 2019; Hoekstra, 2012, 2014, 2015, 2017; Hoekstra et al., 2017; Hoekstra & Chapagain, 2007; Hoekstra & Mekonnen, 2012; Hossain & Khan, 2020; Hsu et al., 2006; Huang et al., 2012; Ignjatović et al., 2022; Juanga-Labayen et al., 2022; Keesstra et al., 2018; Knuth et al., 2020; Kongboon & Sampattagul, 2012; Kumar et al., 2021; Landrigan et al., 2018; le Roux et al., 2017; Li et al., 2017a, 2017b; Lian et al., 2022; Licker et al., 2019; Lin et al., 2018; Lu et al., 2015; Luján-Ornelas et al., 2020; Ma et al., 2018; Mamun et al., 2022; McIntyre et al., 2013; Mekonnen & Hoekstra, 2011, 2012, 2014; Mhlanga, 2021; Miao et al., 2015; Miller & O'Callaghan, 2015; Mingkhwan & Worakhunpiset, 2018; Moglia et al., 2018; Mpofu, 2022; Murata & Sakamoto, 2019; Octavia et al., 2022; Oki & Kanae, 2006; Papadopoulou et al., 2022; Peng et al., 2015; Sheng et al., 2020; Shi, 2022; Sopper, 1992; Trubetskaya et al., 2021; Wang et al., 2017, 2020; Weerasooriya et al., 2021; Wu et al., 2021; Zhang et al., 2020; Zhao et al., 2022; Zhou & Etzkowitz, 2021; Zhuo et al., 2020).

References

Abdelzaher, M. A., & Awad, M. M. (2022). Sustainable development goals for the circular economy and the water-food nexus: full implementation of new drip irrigation technologies in Upper Egypt. *Sustainability (Switzerland)*, *14*(21). Available from https://doi.org/10.3390/su142113883, MDPI, Egypt, http://www.mdpi.com/journal/sustainability/.

Ahmed, G., Anawar, H. M., Takuwa, D. T., Chibua, I. T., Singh, G. S., & Sichilongo, K. (2015). Environmental assessment of fate, transport and persistent behavior of dichlorodiphenyltrichloroethanes and hexachlorocyclohexanes in land and water ecosystems. *International Journal of Environmental Science and Technology*, *12*(8), 2741–2756. Available from https://doi.org/10.1007/s13762-015-0792-3, Center for Environmental and Energy Research and Studies, Botswana, http://www.springerlink.com/content/1735-1472.

Akenji, L., & Bengtsson, M. (2014). Making sustainable consumption and production the core of sustainable development goals. *Sustainability*, *6*(2), 513–529. Available from https://doi.org/10.3390/su6020513.

Aladuwaka, S., & Momsen, J. (2010). Sustainable development, water resources management and women's empowerment: the Wanaraniya Water Project in Sri Lanka. *Gender & Development*, *18*(1), 43–58. Available from https://doi.org/10.1080/13552071003600026.

Aldaya, M. M., & Hoekstra, A. Y. (2010). The water needed for Italians to eat pasta and pizza. *Agricultural Systems*, *103*(6), 351–360. Available from https://doi.org/10.1016/j.agsy.2010.030.004.

Aslam, S., Aftab, H., Martins, J. M., Mata, M. N., Qureshi, H. A., Adriano, A. M., & Mata, P. N. (2021). Sustainable model: Recommendations for water conservation strategies in a developing country through a psychosocial wellness program. *Water (Switzerland)*, *13*(14). Available from https://doi.org/10.3390/w13141984, MDPI AG, Pakistan, https://www.mdpi.com/2073-4441/13/14/1984/pdf.

Babel, M. S., Deb, P., & Soni, P. (2019). Performance evaluation of AquaCrop and DSSAT-CERES for maize under different irrigation and manure application rates in the Himalayan region of India. *Agricultural Research*, *8*(2), 207–217. Available from https://doi.org/10.1007/s40003-018-0366-y, Springer, Thailand., http://www.springer.com/life + sciences/cell + biology/journal/40003.

Bailey, K., Basu, A., & Sharma, S. (2022). The environmental impacts of fast fashion on water quality: a systematic review. *Water (Switzerland)*, *14* (7). Available from https://doi.org/10.3390/w14071073, MDPI, Canada, https://www.mdpi.com/2073-4441/14/7/1073/pdf.

Baumgartner, R. J. (2019). Sustainable development goals and the forest sector—a complex relationship. *Forests*, *10*(2). Available from https://doi.org/10.3390/f10020152, MDPI AG, Austria, https://www.mdpi.com/1999-4907/10/2/152/pdf.

Berger, M., Pfister, S., Bach, V., & Finkbeiner, M. (2015). Saving the planet's climate or water resources? The trade-off between carbon and water footprints of European biofuels. *Sustainability*, *7*(6), 6665–6683. Available from https://doi.org/10.3390/su7066665.

Bhatnagar, A., Kaczala, F., Hogland, W., Marques, M., Paraskeva, C. A., Papadakis, V. G., & Sillanpää, M. (2014). Valorization of solid waste products from olive oil industry as potential adsorbents for water pollution control—a review. *Environmental Science and Pollution Research*, *21*(1), 268–298. Available from https://doi.org/10.1007/s11356-013-2135-6.

Bhatti, Z. A., Mahmood, Q., Raja, I. A., Malik, A. H., Khan, M. S., & Wu, D. (2011). Chemical oxidation of carwash industry wastewater as an effort to decrease water pollution. *Physics and Chemistry of the Earth*, *36*(9–11), 465–469. Available from https://doi.org/10.1016/j.pce.2010.030.022.

Bjornlund, H., Nicol, L., & Klein, K. K. (2007). Challenges in implementing economic instruments to manage irrigation water on farms in southern Alberta. *Agricultural Water Management*, *92*(3), 131–141. Available from https://doi.org/10.1016/j.agwat.2007.050.018.

Boretti, A., & Rosa, L. (2019). Reassessing the projections of the World Water Development Report. *npj Clean Water*, *2*(1). Available from https://doi.org/10.1038/s41545-019-0039-9.

Bosire, C. K., Ogutu, J. O., Said, M. Y., Krol, M. S., Leeuw, J. D., & Hoekstra, A. Y. (2015). Trends and spatial variation in water and land footprints of meat and milk production systems in Kenya. *Agriculture, Ecosystems and Environment*, *205*, 36–47. Available from https://doi.org/10.1016/j.agee.2015.020.015, Elsevier, Netherlands. http://www.elsevier.com/inca/publications/store/5/0/3/2/9/8.

Bradley, R. H., Corwyn, R. F., McAdoo, H. P., & García Coll, C. (2001). The home environments of children in the United States part I: variations by age, ethnicity, and poverty status. *Child Development*, *72*(6), 1844–1867. Available from https://doi.org/10.1111/1467-8624.t01-1-00382, Blackwell Publishing Inc., United States. http://onlinelibrary.wiley.com/journal/10.1111/(ISSN)1467-8624/issues.

Cantin, B., Shrubsole, D., & Aït-Ouyahia, M. (2005). Using economic instruments for water demand management: introduction. *Canadian Water Resources Journal*, *30*(1), 1–10. Available from https://doi.org/10.4296/cwrj30011.

Cao, X., Huang, X., Huang, H., Liu, J., Guo, X., Wang, W., & She, D. (2018). Changes and driving mechanism of water footprint scarcity in crop production: a study of Jiangsu Province, China. *Ecological Indicators*, *95*, 444–454. Available from https://doi.org/10.1016/j.ecolind.2018.070.059.

Carr, J. A., D'Odorico, P., Laio, F., & Ridolfi, L. (2012). On the temporal variability of the virtual water network. *Geophysical Research Letters*, *39* (6). Available from https://doi.org/10.1029/2012GL051247, Blackwell Publishing Ltd, United States. http://onlinelibrary.wiley.com/journal/10.1002/(ISSN)1944-8007/issues?year = 2012.

Carr, J. A., Seekell, D. A., & D'Odorico, P. (2015). Inequality or injustice in water use for food? *Environmental Research Letters*, *10*(2). Available from https://doi.org/10.1088/1748-9326/10/2/024013.

Chapagain, A. K., & Hoekstra, A. Y. (2007). The water footprint of coffee and tea consumption in the Netherlands. *Ecological Economics*, *64*(1), 109–118. Available from https://doi.org/10.1016/j.ecolecon.2007.020.022.

Chapagain, A. K., & Hoekstra, A. Y. (2011). The blue, green and grey water footprint of rice from production and consumption perspectives. *Ecological Economics*, *70*(4), 749–758. Available from https://doi.org/10.1016/j.ecolecon.2010.110.012.

Chapagain, A. K., & Orr, S. (2009). An improved water footprint methodology linking global consumption to local water resources: a case of Spanish tomatoes. *Journal of Environmental Management*, *90*(2), 1219–1228. Available from https://doi.org/10.1016/j.jenvman.2008.060.006.

Chen, W., Bai, Y., Zhang, W., Lyu, S., & Jiao, W. (2015). Perceptions of different stakeholders on reclaimed water reuse: the case of Beijing, China. *Sustainability*, *7*(7), 9696–9710. Available from https://doi.org/10.3390/su7079696.

Cheng, H., Yang, Z., & Chan, C. W. (2003). An expert system for decision support of municipal water pollution control. *Engineering Applications of Artificial Intelligence*, *16*(2), 159–166. Available from https://doi.org/10.1016/S0952-1976(03)00055-1.

Chowdhary, P., Bharagava, R.N., Mishra, S., & Khan, N. (2019). Role of industries in water scarcity and its adverse effects on environment and human health. Springer Science and Business Media LLC, pp. 235–256, doi: 10.1007/978-981-13-5889-0_12.

Chukalla, A. D., Krol, M. S., & Hoekstra, A. Y. (2015). Green and blue water footprint reduction in irrigated agriculture: effect of irrigation techniques, irrigation strategies and mulching. *Hydrology and Earth System Sciences*, *19*(12), 4877–4891. Available from https://doi.org/10.5194/hess-19-4877-2015.

Cook, M., & Webber, M. (2016). Food, fracking, and freshwater: the potential for markets and cross-sectoral investments to enable water conservation. *Water (Switzerland)*, *8*(2). Available from https://doi.org/10.3390/w8020045, MDPI AG, United States. http://www.mdpi.com/2073-4441/8/2/45/pdf.

Croese, S., Green, C., & Morgan, G. (2020). Localizing the sustainable development goals through the lens of urban resilience: lessons and learnings from 100 resilient cities and Cape Town. *Sustainability*, *12*(2). Available from https://doi.org/10.3390/su12020550.

Deb, P., Babel, M. S., & Denis, A. F. (2018). Multi-GCMs approach for assessing climate change impact on water resources in Thailand. *Modeling Earth Systems and Environment*, *4*(2), 825–839. Available from https://doi.org/10.1007/s40808-018-0428-y, Springer Science and Business Media Deutschland GmbH, Thailand., springer.com/journal/40808.

Deb, P., Kiem, A. S., & Willgoose, G. (2019a). A linked surface water-groundwater modelling approach to more realistically simulate rainfall-runoff non-stationarity in semi-arid regions. *Journal of Hydrology*, *575*, 273–291. Available from https://doi.org/10.1016/j.jhydrol.2019.050.039, Elsevier B.V., Australia. http://www.elsevier.com/inca/publications/store/5/0/3/3/4/3.

Deb, P., Kiem, A. S., & Willgoose, G. (2019b). Mechanisms influencing non-stationarity in rainfall-runoff relationships in southeast Australia. *Journal of Hydrology*, *571*, 749–764. Available from https://doi.org/10.1016/j.jhydrol.2019.020.025, Elsevier B.V., Australia. http://www.elsevier.com/inca/publications/store/5/0/3/3/4/3.

Dolganova, I., Mikosch, N., Berger, M., Núñez, M., Müller-Frank, A., & Finkbeiner, M. (2019). The water footprint of European agricultural imports: hotspots in the context of water scarcity. *Resources*, *8*(3). Available from https://doi.org/10.3390/resources8030141.

D'Odorico, P., Carr, J., Laio, F., & Ridolfi, L. (2012). Spatial organization and drivers of the virtual water trade: a community-structure analysis. *Environmental Research Letters*, *7*(3). Available from https://doi.org/10.1088/1748-9326/7/3/034007.

Ebenstein, A. (2012). The consequences of industrialization: evidence from water pollution and digestive cancers in China. *Review of Economics and Statistics*, *94*(1), 186–201. Available from https://doi.org/10.1162/rest_a_00150.

Ene, S. A., Teodosiu, C., Robu, B., & Volf, I. (2013). Water footprint assessment in the winemaking industry: a case study for a Romanian medium size production plant. *Journal of Cleaner Production*, *43*, 122–135. Available from https://doi.org/10.1016/j.jclepro.2012.110.051.

Ercin, A. E., Aldaya, M. M., & Hoekstra, A. Y. (2012). The water footprint of soy milk and soy burger and equivalent animal products. *Ecological Indicators*, *18*, 392–402. Available from https://doi.org/10.1016/j.ecolind.2011.120.009.

Eriksen, M., Lebreton, L. C. M., Carson, H. S., Thiel, M., Moore, C. J., Borerro, J. C., Galgani, F., Ryan, P. G., & Reisser, J. (2014). Plastic pollution in the world's oceans: more than 5 trillion plastic pieces weighing over 250,000 tons afloat at sea. *PLoS ONE*, *9*(12). Available from https://doi.org/10.1371/journal.pone.0111913, Public Library of Science, United States. http://www.plosone.org/article/fetchObject.action?uri = info%3Adoi%2F10.1371%2Fjournal.pone.0111913&representation = PDF.

Esetlili, M. T., Serbeş, Z. A., Çolak Esetlili, B., Kurucu, Y., & Delibacak, S. (2022). Determination of water footprint for the cotton and maize production in the Küçük menderes basin. *Water (Switzerland), 14*(21). Available from https://doi.org/10.3390/w14213427, MDPI, Turkey. http://www.mdpi.com/journal/water.

Evans, D. K., & Miguel, E. (2007). Orphans and schooling in Africa: a longitudinal analysis. *Demography, 44*(1), 35–57. Available from https://doi.org/10.1353/dem.2007.0002, Duke University Press, United States. https://www.dukeupress.edu/demography.

Fang, K., Heijungs, R., & De Snoo, G. R. (2014). Theoretical exploration for the combination of the ecological, energy, carbon, and water footprints: overview of a footprint family. *Ecological Indicators, 36*, 508–518. Available from https://doi.org/10.1016/j.ecolind.2013.080.017, Elsevier B.V., Netherlands. http://www.elsevier.com/locate/ecolind.

Fang, L., & Wu, F. (2020). Can water rights trading scheme promote regional water conservation in China? Evidence from a time-varying DID analysis. *International Journal of Environmental Research and Public Health, 17*(18), 1–14. Available from https://doi.org/10.3390/ijerph17186679, MDPI AG, China. https://www.mdpi.com/1660-4601/17/18/6679/pdf.

Florea, R. M., Stoica, A. I., Baiulescu, G. E., & Capotă, P. (2005). Water pollution in gold mining industry: a case study in Roşia Montană district, Romania. *Environmental Geology, 48*(8), 1132–1136. Available from https://doi.org/10.1007/s00254-005-0054-7.

Flores, R. A., & Ghisi, E. (2022). Water benchmarking in buildings: a systematic review on methods and benchmarks for water conservation. *Water (Switzerland), 14*(3). Available from https://doi.org/10.3390/w14030473, MDPI, Brazil. https://www.mdpi.com/2073-4441/14/3/473/pdf.

Fonseca, L. M., Domingues, J. P., & Dima, A. M. (2020). Mapping the sustainable development goals relationships. *Sustainability (Switzerland), 12* (8). Available from https://doi.org/10.3390/SU12083359, MDPI, Portugal. https://www.mdpi.com/2071-1050/12/8/3359.

Fox, G. A., Collins, B., Hayakawa, E., Weseloh, D. V., Ludwig, J. P., Kubiak, T. J., & Erdman, T. C. (1991). Reproductive outcomes in colonial fish-eating birds: a biomarker for developmental toxicants in Great Lakes food chains: II. Spatial variation in the occurrence and prevalence of bill defects in young double-crested cormorants in the Great Lakes, 1979–1987. *Journal of Great Lakes Research, 17*(2), 158–167. Available from https://doi.org/10.1016/S0380-1330(91)71353-1.

Galli, A., Wiedmann, T., Ercin, E., Knoblauch, D., Ewing, B., & Giljum, S. (2012). Integrating ecological, carbon and water footprint into a "footprint family" of indicators: definition and role in tracking human pressure on the planet. *Ecological Indicators, 16*, 100–112. Available from https://doi.org/10.1016/j.ecolind.2011.060.017.

Gare, A. (2001). Narratives and the ethics and politics of environmentalism: the transformative power of stories. *ICAAP, undefined Theory and Science, 2*(1), http://theoryandscience.icaap.org.

Geller, E. S., Erickson, J. B., & Buttram, B. A. (1983). Attempts to promote residential water conservation with educational, behavioral and engineering strategies. *Population and Environment, 6*(2), 96–112. Available from https://doi.org/10.1007/BF01362290.

Gerbens-Leenes, W., & Hoekstra, A. Y. (2012). The water footprint of sweeteners and bio-ethanol. *Environment International, 40*(1), 202–211. Available from https://doi.org/10.1016/j.envint.2011.060.006, Elsevier Ltd, Netherlands. http://www.elsevier.com/locate/envint.

Gleeson, T., Wada, Y., Bierkens, M. F. P., & Van Beek, L. P. H. (2012). Water balance of global aquifers revealed by groundwater footprint. *Nature, 488*(7410), 197–200. Available from https://doi.org/10.1038/nature11295.

Gunkel, G., Kosmol, J., Sobral, M., Rohn, H., Montenegro, S., & Aureliano, J. (2007). Sugar cane industry as a source of water pollution—case study on the situation in Ipojuca River, Pernambuco, Brazil. *Water, Air, and Soil Pollution, 180*(1–4), 261–269. Available from https://doi.org/10.1007/s11270-006-9268-x.

Guo, C., Gao, J., Zhou, B., & Yang, J. (2021). Factors of the ecosystem service value in water conservation areas considering the natural environment and human activities: a case study of Funiu Mountain, China. *International Journal of Environmental Research and Public Health, 18*(21). Available from https://doi.org/10.3390/ijerph182111074.

Guo, Q., Han, Y., Yang, Y., Fu, G., & Li, J. (2019). Quantifying the impacts of climate change, coal mining and soil and water conservation on streamflow in a coal mining concentrated watershed on the Loess Plateau, China. *Water (Switzerland), 11*(5). Available from https://doi.org/10.3390/w11051054, MDPI AG, China. https://res.mdpi.com/water/water-11-01054/article_deploy/water-11-01054.pdf?filename = &attachment = 1.

Ho, L. T., & Goethals, P. L. M. (2019). Opportunities and challenges for the sustainability of lakes and reservoirs in relation to the Sustainable Development Goals (SDGs). *Water (Switzerland), 11*(7). Available from https://doi.org/10.3390/w11071462, MDPI AG, Belgium. https://res.mdpi.com/water/water-11-01462/article_deploy/water-11-01462-v2.pdf?filename = &attachment = 1.

Hoekstra, A. Y. (2012). The hidden water resource use behind meat and dairy. *Animal Frontiers, 2*(2), 3–8. Available from https://doi.org/10.2527/af.2012-0038, Oxford University Press, Netherlands. https://academic.oup.com/af.

Hoekstra, A. Y. (2014). Water scarcity challenges to business. *Nature Climate Change, 4*(5), 318–320. Available from https://doi.org/10.1038/nclimate2214, Nature Publishing Group, Netherlands. http://www.nature.com/nclimate/index.html.

Hoekstra, A. Y. (2015). The water footprint of industry. *Assessing and Measuring Environmental Impact and Sustainability*, 221–254. Available from https://doi.org/10.1016/B978-0-12-799968-5.00007-5, Elsevier Inc., Netherlands. http://www.sciencedirect.com/science/book/9780127999685.

Hoekstra, A. Y. (2017). Water footprint assessment: evolvement of a new research field. *Water Resources Management, 31*(10), 3061–3081. Available from https://doi.org/10.1007/s11269-017-1618-5, Springer Netherlands, Netherlands. http://www.wkap.nl/journalhome.htm/.

Hoekstra, A. Y., & Chapagain, A. K. (2007). The water footprints of Morocco and the Netherlands: Global water use as a result of domestic consumption of agricultural commodities. *Ecological Economics, 64*(1), 143–151. Available from https://doi.org/10.1016/j.ecolecon.2007.020.023.

Hoekstra, A. Y., & Mekonnen, M. M. (2012). The water footprint of humanity. *Proceedings of the National Academy of Sciences of the United States of America, 109*(9), 3232–3237. Available from https://doi.org/10.1073/pnas.1109936109Netherlands, http://www.pnas.org/content/109/9/3232.full.pdf + html.

Hoekstra, A. Y., Chapagain, A. K., & van Oel, P. R. (2017). Advancing water footprint assessment research: challenges in monitoring progress towards sustainable development goal 6. *Water (Switzerland)*, *9*(6). Available from https://doi.org/10.3390/w9060438, MDPI AG, Netherlands. http://www.mdpi.com/2073-4441/9/6/438/pdf.

Hossain, L., & Khan, M. S. (2020). Water footprint management for sustainable growth in the bangladesh apparel sector. *Water (Switzerland)*, *12*(10). Available from https://doi.org/10.3390/w12102760, MDPI AG, Bangladesh., http://www.mdpi.com/journal/water.

Hsu, M. J., Selvaraj, K., & Agoramoorthy, G. (2006). Taiwan's industrial heavy metal pollution threatens terrestrial biota. *Environmental Pollution*, *143*(2), 327–334. Available from https://doi.org/10.1016/j.envpol.2005.110.023.

Huang, J., Zhang, H. L., Tong, W. J., & Chen, F. (2012). The impact of local crops consumption on the water resources in Beijing. *Journal of Cleaner Production*, *21*(1), 45–50. Available from https://doi.org/10.1016/j.jclepro.2011.090.014.

Ignjatović, L. Đ., Krstić, V., Radonjanin, V., Jovanović, V., Malešev, M., Ignjatović, D., & Đurđevac, V. (2022). Application of cement paste in mining works, environmental protection, and the sustainable development goals in the mining industry. *Sustainability (Switzerland)*, *14*(13). Available from https://doi.org/10.3390/su14137902, MDPI, Serbia., https://www.mdpi.com/2071-1050/14/13/7902/pdf?version = 1656427074.

Juanga-Labayen, J. P., Labayen, I. V., & Yuan, Q. (2022). A review on textile recycling practices and challenges. *Textiles*, *2*(1), 174–188. Available from https://doi.org/10.3390/textiles2010010.

Keesstra, S., Mol, G., de Leeuw, J., Okx, J., Molenaar, C., de Cleen, M., & Visser, S. (2018). Soil-related sustainable development goals: four concepts to make land degradation neutrality and restoration work. *Land*, *7*(4). Available from https://doi.org/10.3390/land7040133.

Knuth, M. J., Behe, B. K., Huddleston, P. T., Hall, C. R., Fernandez, R. T., & Khachatryan, H. (2020). Water conserving message influences purchasing decision of consumers. *Water (Switzerland)*, *12*(12), 1–21. Available from https://doi.org/10.3390/w12123487, MDPI AG, United States., http://www.mdpi.com/journal/water.

Kongboon, R., & Sampattagul, S. (2012). The water footprint of sugarcane and cassava in northern Thailand. *Procedia—Social and Behavioral Sciences*, *40*, 451–460. Available from https://doi.org/10.1016/j.sbspro.2012.030.215.

Kumar, R., Verma, A., Shome, A., Sinha, R., Sinha, S., Jha, P. K., Kumar, R., Kumar, P., Shubham., Das, S., Sharma, P., & Prasad, P. V. V. (2021). Impacts of plastic pollution on ecosystem services, sustainable development goals, and need to focus on circular economy and policy interventions. *Sustainability (Switzerland)*, *13*(17). Available from https://doi.org/10.3390/su13179963, MDPI, India., https://www.mdpi.com/2071-1050/13/17/9963/pdf.

Landrigan, P. J., Fuller, R., Acosta, N. J. R., Adeyi, O., Arnold, R., Basu, N., Baldé, A. B., Bertollini, R., Bose-O'Reilly, S., Boufford, J. I., Breysse, P. N., Chiles, T., Mahidol, C., Coll-Seck, A. M., Cropper, M. L., Fobil, J., Fuster, V., Greenstone, M., Haines, A., ... Zhong, M. (2018). The Lancet Commission on pollution and health. *The Lancet*, *391*(10119), 462–512. Available from https://doi.org/10.1016/S0140-6736(17)32345-0, Lancet Publishing Group, United States., http://www.journals.elsevier.com/the-lancet/.

Li, Y., Lu, L., Tan, Y., Wang, L., & Shen, M. (2017). Decoupling water consumption and environmental impact on textile industry by using water footprint method: a case study in China. *Water (Switzerland)*, *9*(2). Available from https://doi.org/10.3390/w9020124, MDPI AG, China., http://www.mdpi.com/2073-4441/9/2/124/pdf.

Li, Y., Luo, Y., Wang, Y., Wang, L., & Shen, M. (2017). Decomposing the decoupling of water consumption and economic growth in China's textile industry. *Sustainability*, *9*(3). Available from https://doi.org/10.3390/su9030412.

Lian, Z., Hao, H., Zhao, J., Cao, K., Wang, H., & He, Z. (2022). Evaluation of remote sensing ecological index based on soil and water conservation on the effectiveness of management of abandoned mine landscaping transformation. *International Journal of Environmental Research and Public Health*, *19*(15). Available from https://doi.org/10.3390/ijerph19159750, MDPI, China., http://www.mdpi.com/journal/ijerph.

Licker, R., Ekwurzel, B., Doney, S. C., Cooley, S. R., Lima, I. D., Heede, R., & Frumhoff, P. C. (2019). Attributing ocean acidification to major carbon producers. *Environmental Research Letters*, *14*(12). Available from https://doi.org/10.1088/1748-9326/ab5abc.

Lin, H. H., Lee, S. S., Perng, Y. S., & Yu, S. T. (2018). Investigation about the impact of tourism development on a water conservation area in Taiwan. *Sustainability (Switzerland)*, *10*(7), 20711050. Available from https://doi.org/10.3390/su10072328, MDPI, Taiwan., http://www.mdpi.com/2071-1050/10/7/2328/pdf.

Lu, Y., Song, S., Wang, R., Liu, Z., Meng, J., Sweetman, A. J., Jenkins, A., Ferrier, R. C., Li, H., Luo, W., & Wang, T. (2015). Impacts of soil and water pollution on food safety and health risks in China. *Environment International*, *77*, 5–15. Available from https://doi.org/10.1016/j.envint.2014.120.010, Elsevier Ltd, China. http://www.elsevier.com/locate/envint.

Luján-Ornelas, C., Güereca, L. P., Franco-García, M. L., & Heldeweg, M. (2020). A life cycle thinking approach to analyse sustainability in the textile industry: a literature review. *Sustainability (Switzerland)*, *12*(23), 1–19. Available from https://doi.org/10.3390/su122310193, MDPI, Netherlands. https://www.mdpi.com/2071-1050/12/23/10193/pdf.

Ma, X., Wu, D., & Zhang, S. (2018). Multiple goals dilemma of residential water pricing policy reform: increasing block tariffs or a uniform tariff with rebate? *Sustainability*, *10*(10). Available from https://doi.org/10.3390/su10103526.

Mamun, A. A., Bormon, K. K., Rasu, M. N. S., Talukder, A., Freeman, C., Burch, R., & Chander, H. (2022). An assessment of energy and groundwater consumption of textile dyeing mills in Bangladesh and minimization of environmental impacts via long-term key performance indicators (KPI) baseline. *Textiles*, *2*(4), 511–523. Available from https://doi.org/10.3390/textiles2040029.

McIntyre, L., Williams, J. V. A., Lavorato, D. H., & Patten, S. (2013). Depression and suicide ideation in late adolescence and early adulthood are an outcome of child hunger. *Journal of Affective Disorders*, *150*(1), 123–129. Available from https://doi.org/10.1016/j.jad.2012.110.029.

Mekonnen, M. M., & Hoekstra, A. Y. (2011). The green, blue and grey water footprint of crops and derived crop products. *Hydrology and Earth System Sciences*, *15*(5), 1577–1600. Available from https://doi.org/10.5194/hess-15-1577-2011.

Mekonnen, M. M., & Hoekstra, A. Y. (2012). A global assessment of the water footprint of farm animal products. *Ecosystems, 15*(3), 401–415. Available from https://doi.org/10.1007/s10021-011-9517-8.

Mekonnen, M. M., & Hoekstra, A. Y. (2014). Water conservation through trade: the case of Kenya. *Water International, 39*(4), 451–468. Available from https://doi.org/10.1080/02508060.2014.922014, Routledge, Netherlands. http://www.tandfonline.com/toc/rwin20/current.

Mhlanga, D. (2021). Artificial intelligence in the Industry 4.0, and its impact on poverty, innovation, infrastructure development, and the sustainable development goals: lessons from emerging economies? *Sustainability, 13*(11). Available from https://doi.org/10.3390/su13115788.

Miao, X., Tang, Y., Wong, C. W. Y., & Zang, H. (2015). The latent causal chain of industrial water pollution in China. *Environmental Pollution, 196*, 473–477. Available from https://doi.org/10.1016/j.envpol.2014.110.010, Elsevier Ltd, China. http://www.elsevier.com/inca/publications/store/4/0/5/8/5/6.

Miller, D. B., & O'Callaghan, J. P. (2015). Biomarkers of Parkinson's disease: present and future. *Metabolism: Clinical and Experimental, 64*(3), S40–S46. Available from https://doi.org/10.1016/j.metabol.2014.100.030.

Mingkhwan, R., & Worakhunpiset, S. (2018). Heavy metal contamination near industrial estate areas in Phra Nakhon Si Ayutthaya province, Thailand and human health risk assessment. *International Journal of Environmental Research and Public Health, 15*(9). Available from https://doi.org/10.3390/ijerph15091890.

Moglia, M., Cook, S., & Tapsuwan, S. (2018). Promoting water conservation: where to from here? *Water (Switzerland), 10*(11). Available from https://doi.org/10.3390/w10111510. MDPI AG, Australia. https://www.mdpi.com/2073-4441/10/11/1510/pdf.

Mpofu, F. Y. (2022). Industry 4.0 in financial services: mobile money taxes, revenue mobilisation, financial inclusion, and the realisation of sustainable development goals (SDGs) in Africa. *Sustainability (Switzerland), 14*(14). Available from https://doi.org/10.3390/su14148667. MDPI, South Africa. http://www.mdpi.com/journal/sustainability/.

Murata, K., & Sakamoto, M. (2019). Minamata disease. *Encyclopedia of Environmental Health*, 401–407. Available from https://doi.org/10.1016/B978-0-12-409548-9.02075-3, Elsevier, Japan.

Octavia, D., Suharti, S., Murniati, I. W. S., Dharmawan, H. Y. S. H., Nugroho, B., Supriyanto, D., Rohadi, G. N., Njurumana, I., Yeny, A., Hani, N., Mindawati., Suratman., Adalina, Y., Prameswari, D., Hadi, E. E. W., & Ekawati, S. (2022). Mainstreaming smart agroforestry for social forestry implementation to support sustainable development goals in indonesia: a review. *Sustainability (Switzerland), 14*(15). Available from https://doi.org/10.3390/su14159313. MDPI, Indonesia. http://www.mdpi.com/journal/sustainability/.

Oki, T., & Kanae, S. (2006). Global hydrological cycles and world water resources. *Science, 313*(5790), 1068–1072. Available from https://doi.org/10.1126/science.1128845.

Papadopoulou, C. A., Papadopoulou, M. P., & Laspidou, C. (2022). Implementing water-energy-land-food-climate nexus approach to achieve the sustainable development goals in Greece: indicators and policy recommendations. *Sustainability (Switzerland), 14*(7). Available from https://doi.org/10.3390/su14074100, MDPI, Greece., https://www.mdpi.com/2071-1050/14/7/4100/pdf.

Peng, H., Jia, Y., Tague, C., & Slaughter, P. (2015). An eco-hydrological model-based assessment of the impacts of soil and water conservation management in the Jinghe River Basin, China. *Water (Switzerland), 7*(11), 6301–6320. Available from https://doi.org/10.3390/w7116301. MDPI AG, China. http://www.mdpi.com/2073-4441/7/11/6301/pdf.

le Roux, B., van der Laan, M., Vahrmeijer, T., Bristow, K. L., & Annandale, J. G. (2017). Establishing and testing a catchment water footprint framework to inform sustainable irrigation water use for an aquifer under stress. *Science of the Total Environment, 599–600*, 1119–1129. Available from https://doi.org/10.1016/j.scitotenv.2017.040.170, Elsevier B.V., South Africa. http://www.elsevier.com/locate/scitotenv.

Sheng, P., Dong, Y., & Vochozka, M. (2020). Analysis of cost-effective methods to reduce industrial wastewater emissions in China. *Water (Switzerland), 12*(6). Available from https://doi.org/10.3390/w12061600. MDPI AG, China. https://res.mdpi.com/d_attachment/water/water-12-01600/article_deploy/water-12-01600.pdf.

Shi, J. (2022). Study on the decoupling relationship and rebound effect between agricultural economic growth and water footprint: a case of Yangling agricultural demonstration zone, China. *Water (Switzerland), 14*(6). Available from https://doi.org/10.3390/w14060991, MDPI, China., https://www.mdpi.com/2073-4441/14/6/991/pdf.

Sopper, W. E. (1992). Irrigation with treated sewage effluent. *Soil Science, 153*(3), 258–259. Available from https://doi.org/10.1097/00010694-199203000-00010.

Trubetskaya, A., Horan, W., Conheady, P., Stockil, K., & Moore, S. (2021). A methodology for industrial water footprint assessment using energy–water–carbon nexus. *Processes, 9*(2), 1–24. Available from https://doi.org/10.3390/pr9020393, MDPI AG, Ireland., https://www.mdpi.com/2227-9717/9/2/393/pdf.

Wang, L., Li, Y., & He, W. (2017). The energy footprint of China's textile industry: perspectives from decoupling and decomposition analysis. *Energies, 10*(10). Available from https://doi.org/10.3390/en10101461.

Wang, X., Chen, X., Cheng, Y., Zhou, L., Li, Y., & Yang, Y. (2020). Factorial decomposition of the energy footprint of the Shaoxing Textile Industry. *Energies, 13*(7). Available from https://doi.org/10.3390/en13071683.

Weerasooriya, R. R., Liyanage, L. P. K., Rathnappriya, R. H. K., Bandara, W. B. M. A. C., Perera, T. A. N. T., Gunarathna, M. H. J. P., & Jayasinghe, G. Y. (2021). Industrial water conservation by water footprint and sustainable development goals: a review. *Environment, Development and Sustainability, 23*(9), 12661–12709. Available from https://doi.org/10.1007/s10668-020-01184-0.

Wu, B., Wang, P., Xiao, S., Yu, X., Shu, W., Zhang, H., & Ding, M. (2021). Effects of soil and water conservation measures on soil bacterial community structure in citrus orchards. *The Research of Environmental Sciences, 34*(2), 419–430. Available from https://doi.org/10.13198/j.issn.1001-6929.2021.01.02, http://www.hjkxyj.org.cn/ch/index.aspx.

Zhang, X., Sun, F., Wang, H., & Qu, Y. (2020). Green biased technical change in terms of industrial water resources in China's Yangtze River economic belt. *International Journal of Environmental Research and Public Health, 17*(8). Available from https://doi.org/10.3390/ijerph17082789.

Zhao, J., Wang, Y., Zhang, X., & Liu, Q. (2022). Industrial and agricultural water use efficiency and influencing factors in the process of urbanization in the middle and lower reaches of the Yellow River Basin, China. *Land*, *11*(8), 1248.

Zhou, C., & Etzkowitz, H. (2021). Triple helix twins: a framework for achieving innovation and UN sustainable development goals. *Sustainability*, *13*(12). Available from https://doi.org/10.3390/su13126535.

Zhuo, L., Feng, B., & Wu, P. (2020). Water footprint study review for understanding and resolving water issues in China. *Water (Switzerland)*, *12*(11), 1–14. Available from https://doi.org/10.3390/w12112988, MDPI AG, China., http://www.mdpi.com/journal/water.

Chapter 8

Sustainable water allocation and water footprints

Ananya Roy Chowdhury[1,*] and Achintya Das[2]
[1]*Department of Botany, Chakdaha College, Nadia, West Bengal, India,* [2]*Department of Physics, Mahadevananda Mahavidyalaya, Barrackpore, West Bengal, India*
**Corresponding author. e-mail address: ananya.chakdaha1@gmail.com.*

8.1 Introduction

Water, an indispensable commodity for all organisms, is currently entangled in a worldwide predicament that has extensive consequences for ecosystems, societies, and economies. These days, sustainable practices demand things like a circular economy (Das et al., 2023), and efficient water management (Nova, 2023). As a result of overexploitation, pollution, and climate change, the availability and quality of water are experiencing an unprecedented level of strain. The objective of this chapter is to furnish a thorough comprehension of sustainable water allocation and water footprint. This is accomplished by acquainting the reader with contemporary obstacles, guiding principles, and inventive resolutions that guarantee the sustainable and equitable administration of water resources.

8.1.1 Context of the water crisis

The current worldwide water crisis poses critical and imminent risks to both human and ecological welfare. Water security threats affect approximately 80% of the global population. This predicament is further compounded by the inadequate implementation of preventive measures to safeguard biodiversity in aquatic habitats. In areas where substantial technological investments have been made, affluent countries can mitigate water stressors without confronting their root causes. Conversely, these hazards continue to menace less affluent nations (Vörösmarty et al., 2010). This underscores the critical nature of tackling the worldwide water crisis, which must be addressed not only for the sake of resource management but also to ensure global security and preserve biodiversity.

8.1.2 Climate change and water scarcity

Water scarcity and climate change are intricately and multifacetedly intertwined. Extreme weather patterns, which are influenced by climate change, alter the distribution of precipitation and worsen water scarcity in regions that are already susceptible to such conditions. The global community is deeply concerned about the effects of climate change on water resources. Particular regional issues, such as the depletion of phosphorus reserves, which are crucial in agricultural fertilizers, pose additional risks to both food security and water availability (Cordell et al., 2009). Furthermore, the global incidence of chronic renal disease, which is frequently associated with substandard water quality and environmental contaminants, emphasizes the detrimental health effects of water scarcity and the necessity for sustainable water management (Jha et al., 2013).

8.1.3 Historical context

Reflecting the vital role that water continues to play in the development and survival of societies, water management practices have a history as long as human civilization. This subsection explores the progressive development of these customs throughout various civilizations all over history.

Water Footprints and Sustainable Development. DOI: https://doi.org/10.1016/B978-0-443-23631-0.00008-X

8.1.3.1 At the time of ancient Greece

The origins of urban water management can be historically situated in ancient Greece, specifically in the Minoan civilization of Crete (c.2000–1500 BCE). In Minoan Palaces and other settlements, water supply and effluent management systems underwent remarkable development. Subsequently, these technologies were transmitted to the Greek mainland, where they underwent development throughout the Mycenaean, Archaic, Classical, Hellenistic, and Roman eras.

The study by A. Angelakis and D. Spyridakis, "A brief history of water supply and wastewater management in Ancient Greece," offers a comprehensive examination of the hydraulic systems and associated technologies that were utilized during the time of the ancient Greek civilizations (Angelakis and Spyridakis, 2010).

The advanced water management techniques of the Minoan Civilization are particularly noteworthy. Urban areas were occasionally reliant on water sources located at great distances. In regards to the sanitation of potable water and water treatment technologies, the Minoans made notable advancements. Sklivaniotis's scholarly article, "Water for human consumption throughout the history," provides a methodical account and discourse on the significant developments in water history, commencing with the Minoan period (Sklivaniotis & Angelakis, 2006).

8.1.3.2 From antiquity to the present

The progression of effluent systems from ancient Greece to the present day constitutes an additional pivotal element in the chronicles of water management. Historically, engineering institutions have exhibited a predilection for centralized approaches to wastewater management, frequently disregarding the potential of compact decentralized systems. Decentralized systems enable a more sustainable water cycle and necessitate reduced capital expenditures.

The article by Andreas N. Angelakis, A. Capodaglio, and E. Dialynas, "Wastewater management: from ancient greece to modern times and the future," is a comprehensive examination of the extensive history of wastewater systems, including an analysis of the evolution of natural and physical systems into modern technology. In addition, future trends are deliberated upon, with a particular focus on the sustainability and technological adaptation of decentralized systems (Angelakis et al., 2023).

8.1.3.3 Sociohydrological perspective

"A socio-hydrological perspective on the economics of water resources development and management" by S. Pande et al. demonstrates that a sociohydrological perspective is critical for comprehending the history of water management. Sociohydrology and societies from ancient times are examined, with an emphasis on the coevolution of institutions, migration, technology, and population growth in water-dependent societies. The significance of culture as an emergent characteristic of these dynamics is underscored, with institutions serving as the foundation of culture (Pande et al., 2020).

8.1.4 Chapter overview

Subsequent sections of this chapter will provide a more comprehensive analysis of the notion of water allocation, including its historical implementations, definition, and worldwide allocation of freshwater resources. A discourse will be held regarding the tenets of sustainable water allocation, with a particular focus on the ethical implications of water distribution and the equilibrium between ecological preservation and human requirements. In addition, this chapter will delve into the notion of water footprint, analyze its economic ramifications, and scrutinize the discrepancies that exist in water usage patterns between developed and developing countries. This chapter will emphasize the significance of water footprints in public awareness, corporate responsibility, and policy applications. In addition, it will explore the role that technology plays in promoting sustainable water management. Using case studies and an analysis of obstacles and challenges in water management, this chapter endeavors to furnish a thorough comprehension of the present condition of worldwide water resources as well as the trajectory toward sustainable water practices.

8.2 The concept of water allocation

The allocation of water is a critical factor in determining the utilization and distribution of water resources across various ecosystems and sectors. Various factors must be meticulously considered throughout this complex procedure to guarantee that water is utilized efficiently and sustainably.

8.2.1 Defining water allocation

Water allocation is an essential procedure that encompasses the management and distribution of water resources among diverse consumers and environmental requirements, to guarantee fair and sustainable utilization (Dudgeon et al., 2006). A thorough comprehension of water availability, demand, and the potential ramifications of various allocation scenarios is essential for this procedure. Water management encompasses the establishment of priorities, formulation of decisions, and execution of strategies to ensure the preservation of aquatic ecosystem health while simultaneously addressing societal water demands.

8.2.2 Global freshwater distribution

Freshwater comprises a negligible proportion of the Earth's water, comprising a mere 2.5% of the total. A significant portion of this freshwater is enshrined in glaciers and ice caps (Estes et al., 2011). The availability of accessible freshwater is contingent upon groundwater aquifers, rivers, lakes, and climatic, geological, and geographical conditions. Proximate freshwater resources result from heavy precipitation and extensive river networks, whereas arid regions frequently experience water scarcity. The escalating water requirements resulting from population expansion, urban development, and agricultural progress are exerting strain on these resources, culminating in excessive extraction and deterioration of water quality (Howden et al., 2007).

8.2.3 Historical practices

In the past, water allocation decisions were predominantly influenced by short-term requirements and availability, with limited regard for environmental consequences and long-term sustainability (Gleick, 2003). Water diversion for agricultural and domestic purposes, modification of natural flow regimes, and degradation of aquatic habitats constituted traditional practices. The practice of allocating water rights in a first-come, first-served manner often resulted in water use disputes and an uneven distribution of resources. The increasing acknowledgment of the imperative for sustainable water management has prompted the development of more comprehensive and integrated methodologies that take into account the interests of both humanity and the environment (Pahl-Wostl et al., 2012).

8.2.4 Cultural perspectives

Cultural beliefs and practices play a significant role in shaping water allocation and management practices in different societies.

8.2.4.1 Balinese Subak system, Indonesia

The Subak system, situated in Bali, Indonesia, is an ancient irrigation technique that has been implemented for more than a millennium. The Tri Hita Karana philosophy, which emphasizes the equilibrium between humans, nature, and the gods, serves as its foundation. By utilizing communal water management and allocation, the Subak system ensures that rice cultivation receives an equitable distribution of water. UNESCO has designated this system as a World Heritage Site, underscoring its notable cultural and agricultural importance (Lansing and de Vet, 2012; Stepan, 2017).

8.2.4.2 Zuni Pueblo, USA

The Zuni people residing in Zuni Pueblo, New Mexico, have a profound spiritual affinity toward water, considering it a sacred asset. The allocation and utilization of water are regulated by traditional Zuni customs and laws, which prioritize the preservation and reverence of water sources. The Zuni people have effectively preserved their customary methods of water management in the face of external pressures and challenges related to water scarcity (Cobourn et al., 2014).

8.2.4.3 Rajasthan, India

Traditional water harvesting structures known as "johads" have been utilized to capture and store rainwater for centuries. Local communities administer these structures, and water distribution is determined by customary norms and practices. The resurgence of johads in recent decades has been instrumental in addressing the region's water scarcity and enhancing water security (Hussain et al., 2014).

8.3 Principles of sustainable water allocation

Sustainable water allocation is fundamental to managing water resources in a way that meets the needs of both current and future generations while maintaining the health of ecosystems. It requires a holistic approach, considering social, economic, and environmental factors.

8.3.1 Balancing needs: emphasize the equilibrium between human necessities and ecological balance

Ensuring sustainable water allocation requires a nuanced equilibrium between meeting the demands of human populations and preserving ecological equilibrium. The fundamental idea pertains to the assurance that water resources are employed in a manner that satisfies current demands while safeguarding the capacity of future generations to fulfill their necessities. An important scholarly source in this regard is the research conducted by Pahl-Wostl et al. (2012), which examines the shift toward integrated and adaptive water resources management. The authors emphasize the necessity of striking a balance between the demands of society, the economy, and the environment (Pahl-Wostl et al., 2012).

8.3.2 Ethical dimensions: address issues of equity, justice, and fairness in water allocation

The ethical aspects of water allocation encompass the imperative to guarantee fairness, justice, and equity in the allocation of water resources. It is of utmost importance to acknowledge that water is an inherent human right, and its distribution should be carried out in a way that ensures impartiality toward all communities and groups. In their extensive examination of the ethical implications of water management, Brown and Schmidt (2010) emphasize the importance of ensuring fair and impartial access to and allocation of water resources (Brown and Schmidt, 2007).

8.3.3 Prospects for the future: elucidate the significance of strategic planning that considers the needs and desires of forthcoming generations

A proactive approach is necessary to ensure sustainable water allocation, as it necessitates consideration of the requirements of forthcoming generations. It entails the implementation of policies and strategies designed to guarantee the integrity and availability of water resources in the long run. Gleick (2003) was a trailblazer in delineating the significance of a water strategy that is both sustainable and equitable in the 21st century. He emphasized the necessity of innovative approaches and strategic foresight in the realm of water management (Peter et al., 2003).

8.3.4 Legal frameworks

Legal frameworks and policies are of paramount importance in regulating water allocation on a global scale, guaranteeing the equitable and sustainable distribution of water resources. Frequently, international, national, and local levels are utilized to establish these frameworks to resolve the specific challenges and requirements of various regions.

Publicized in 1959, "The science of 'muddling through'" by Charles E. Lindblom is regarded as a seminal work in the domain of policymaking, encompassing water allocation. Lindblom examines the intricacies of policy formulation and the recursive characteristics of decision-making processes, both of which are profoundly pertinent to water allocation policies (Lindblom, 1959). The paper underscores the significance of gradual modifications and adaptations in policies, following the ever-changing characteristics of water resources and the necessity for flexible legal structures.

The article "Collaborative governance in theory and practice" by Chris Ansell and Alison Gash offers significant contributions to the field of water allocation in the context of collaborative governance, an approach that is gaining increasing recognition as crucial. The study, which was published in 2007, investigates how private and public stakeholders can collaborate to reach decisions based on consensus; this is a crucial element of sustainable water allocation (Ansell and Gash, 2008). The significance of trust establishment, stakeholder commitment, and the formation of a shared understanding is emphasized by the authors.

8.3.5 Conflict resolution

Conflict resolution is an essential element of sustainable water allocation, particularly in the context of transboundary water resources that involve the competing interests of numerous parties. The implementation of efficient conflict resolution tactics is critical to guarantee fair and impartial water distribution and avert the progression of disagreements into more significant hostilities.

While not explicitly concerning water allocation, the framework established by Arthur Kleinman, Leon Eisenberg, and Byron J. Good in their work "Culture, illness, and care" (Kleinman et al., 1978) offers insight into how conflict resolution strategies can be influenced by cultural beliefs and practise. The authors address the relevance of recognizing and respecting diverse cultural perspectives when addressing healthcare issues; this principle can also be applied to the resolution of water allocation disputes. Through comprehension of the cultural milieu in which conflicts over water allocation emerge, involved parties can formulate conflict resolution strategies that are both more efficacious and culturally attuned.

In brief, the tenets of sustainable water allocation are crucial in ensuring the fair and effective utilization of water resources while also considering the requirements of future generations, the environment, and human beings. Critical components of this process include effective conflict resolution strategies that are culturally sensitive, robust legal frameworks, and collaborative governance. The cited literature provides significant contributions in the form of frameworks and insights that aid in the formulation and execution of water management policies and practices by these principles. As a result, a more comprehensive understanding is gained. By doing so, one not only guarantees the satisfaction of immediate necessities but also the preservation of water resources for posterity.

8.4 Water footprint: shedding light on consumption

The concept of water footprint is pivotal in understanding and managing our water usage more sustainably. It provides a comprehensive view of water consumption and pollution, tracing the water used across the entire supply chain of a product or service.

8.4.1 Definition and components: dive deep into the realms of green, blue, and gray water footprints

The water footprint of a product or service is the total volume of freshwater used to produce the product or service, measured over the full supply chain. It is a multidimensional indicator, showing water consumption volumes by source and polluted volumes by type of pollution; it also shows the locations. Hoekstra et al. (2011) provide a comprehensive guide on the water footprint assessment, explaining the three components of the water footprint: green, blue, and gray (Hoekstra et al., 2011). The green water footprint refers to the rainwater consumed, the blue water footprint refers to the surface and groundwater consumed, and the gray water footprint is an indicator of pollution and refers to the volume of freshwater required to assimilate pollutants.

8.4.2 Personal water footprint

An individual's water footprint comprises the entirety of the freshwater utilized in the production of the products and services that an individual consumes. Not only food and drink, but also water used in the residence, garden, and in the production of all consumed products are encompassed within this category. A multitude of digital calculators are accessible to assist users in determining their water footprint. These calculators incorporate variables such as dietary patterns, energy consumption, and consumer behavior (Hoekstra, 2019).

To mitigate their water footprint, individuals can implement a range of strategies. These include consuming less meat, particularly beef, due to the water-intensive nature of livestock husbandry; purchasing products manufactured with less water; and reducing waste, as the production of discarded goods also wastes water. Furthermore, even modest measures such as repairing leaks, implementing water-efficient appliances, and reducing daily water consumption can yield substantial results (Hoekstra, 2019).

8.4.3 Industry-specific footprints

Water resources are significantly influenced by a variety of industries, but energy production, manufacturing, and agriculture are among the most prominent.

8.4.3.1 *Agriculture*

Approximately 70% of freshwater withdrawals are accounted for by agriculture, the greatest consumer of water worldwide. Agricultural products' water footprints are contingent upon the climate, soil, and variety of crops, as well as the cultivation techniques employed (Mekonnen and Hoekstra, 2011).

8.4.3.2 *Manufacturing*

In addition, the manufacturing sector has a significant water footprint, especially in the paper, textile, and chemical industries. As an illustration, 1 ton of marble tiles manufactured demands the use of 3.62 m^3 of water, which has substantial environmental repercussions throughout the product's life cycle (Ahmad et al., 2022).

8.4.3.3 *Energy production*

Energy production is a significant consumer of water, particularly in the context of thermoelectric power plants and biofuel manufacturing. The water footprint associated with energy production is contingent upon the fuel type utilized and the power plant's cooling system. Renewable energy sources, including solar and wind, have an exceptionally small water footprint in comparison to biofuels and fossil fuels (Mielke et al., 2010).

8.4.4 Economic implications: discuss the economic facets of water footprints and how they impact trade and commerce

The economic implications of water footprints are vast, affecting sectors from agriculture to industry, and influencing trade and commerce on a global scale. Aldaya et al. (2010) discusses the water footprint of crop and animal products and its implication for sustainable agrifood systems, highlighting the economic aspects of water usage in agriculture, which is the largest consumer of water worldwide (Aldaya et al., 2010). In addition, the concept of virtual water trade, which refers to the hidden flow of water if food or other commodities are traded from one place to another, has significant economic implications, as discussed by Allan (1998).

8.4.5 Developed versus developing nations: analyze disparities in water footprints and consumption patterns

Developed nations exhibit significant differences in water footprints and consumption patterns when compared to developing nations. Developed nations generally exhibit greater water footprints per capita because of elevated levels of industrialization and increased consumption patterns. On the contrary, developing nations encounter obstacles to water scarcity and contamination, despite potentially having smaller water footprints per capita. Hoekstra and Mekonnen (2012) present an extensive and high-resolution evaluation of the environmental impact of water usage across countries, delineating the disparities in water resources at the international scale (Hoekstra and Mekonnen, 2012). It is imperative to comprehend the water footprint and its constituent parts to illuminate consumption patterns and the ramifications that ensue. Trade and commerce are significantly impacted by the economic dimensions of water footprints, whereas the inequalities between developed and developing countries underscore the necessity for fair and balanced approaches to water usage and management. The works cited offer significant perspectives and empirical evidence regarding these facets of water footprints.

8.5 Applications and implications of water footprints

Understanding and managing water footprints is crucial for achieving sustainable water usage at individual, corporate, and governmental levels. It provides a comprehensive view of water consumption and pollution, helping to identify areas for improvement and promoting responsible water usage.

8.5.1 Policies and management

Globally, there is a growing awareness among governments regarding the significance of water footprints when devising management strategies and policy initiatives. The water footprint data plays a crucial role in providing insight into the overall quantity of potable water consumed during the manufacturing process of products and services. An extensive investigation was undertaken by Hoekstra and Mekonnen (2012) to compute the water footprints of countries; the results of this study were instrumental in informing policy development and resource administration (Hoekstra and Mekonnen, 2012). To identify water-intensive industries, promote water-efficient practices, and implement regulations to ensure sustainable water usage, governments are utilizing this information. The European Water Framework Directive serves as an illustration of a policy endeavor that seeks to elevate the condition of everybody of water within the European Union, in part by promoting the implementation of water footprint assessments across industrial sectors (European Commission, 2000).

8.5.2 Corporate responsibility

The reduction of water footprints is currently a priority for many businesses, which play a vital role in promoting sustainable water usage. Companies increasingly utilize ecolabeling and certifications, such as the endorsement of the Water Footprint Network, to showcase their dedication to sustainable water management. As an illustration, the highly water-intensive beverage industry has made investments in water efficiency measures and provided support for water stewardship initiatives (Jones et al., 2015). These endeavors not only serve to promote water conservation but also bolster the corporate reputation and competitive edge of the organizations.

8.5.3 Public awareness

Raising public awareness about the water footprint of products can lead to behavioral change and more sustainable consumption patterns. The water footprint label, similar to the carbon footprint label, provides consumers with information about the amount of water used in the production of goods. This transparency can encourage consumers to make more informed choices, favouring products with lower water footprints. Educational campaigns and tools like the Water Footprint Calculator are also effective in increasing public understanding of individual water usage and its global impacts (Water Calculator).

8.5.4 Policy implementation

Water footprint data has been increasingly utilized in policy and governance to promote sustainable water usage and management. Here are some examples of successful implementation.

8.5.4.1 Agricultural sector

Governments have implemented policies to optimize water usage in agriculture, which is a major consumer of water. For instance, the European Water Framework Directive aims to achieve a good status for all water bodies, and it encourages the adoption of water-saving irrigation techniques in agriculture (European Commission, 2000).

8.5.4.2 Industrial sector

Industries, particularly those in water-intensive sectors like textiles and beverages, are now required to disclose their water footprint and have been encouraged to adopt water-efficient processes. The Carbon Disclosure Project's Water Program is an example where companies voluntarily disclose their water usage and management strategies (Carbon Disclosure Project, 2020).

8.5.4.3 Urban planning

Cities have started integrating water footprint data in their urban planning to ensure sustainable water supply and demand management. Singapore's Active, Beautiful, Clean Waters (ABC Waters) Programme is an example where urban water management strategies are employed to create a clean and sustainable water supply (Public Utilities Board Singapore, 2020).

8.5.5 Challenges in application

Despite the potential benefits, there are several challenges and limitations in applying water footprint data across different sectors.

8.5.5.1 Data accuracy and availability

The accuracy of water footprint data is crucial for its effective application. However, in many regions, especially in developing countries, there is a lack of reliable and comprehensive water usage data (Hoekstra et al., 2011).

8.5.5.2 Complexity of water footprint assessment

Water footprint assessment involves various parameters and can be complex. The lack of standardized methodologies can lead to inconsistent results and make it difficult for policymakers to make informed decisions (Berger & Finkbeiner, 2010).

8.5.5.3 Economic and social implications

Reducing water footprint often requires changes in production practices or consumer behavior, which can have economic and social implications. For example, shifting to water-efficient crops might impact the livelihoods of farmers accustomed to traditional farming practices (Allan, 2003).

8.5.5.4 Integration with other sustainability goals

Water footprint reduction needs to be balanced with other sustainability goals. For instance, policies aimed at reducing water usage in agriculture should also consider the impact on food security and livelihoods (Foley et al., 2011).

8.6 Technologies driving sustainable water management

In the face of growing water scarcity and the urgent need for sustainable water management, technological innovations are playing an increasingly crucial role. These technologies range from time-tested methods to cutting-edge solutions, all aimed at ensuring the efficient and equitable usage of water resources.

8.6.1 Innovative approaches

Sustainable water management is crucial for ensuring the availability of water resources for future generations. Innovative approaches, both traditional and modern, play a significant role in achieving this goal. Traditional methods such as rainwater harvesting, terracing, and check dams, have been used for centuries in various cultures to manage water sustainably. These methods are cost-effective and eco-friendly, making them viable options for communities worldwide. On the other hand, modern innovations such as smart irrigation systems, which use sensors and automation to optimize water usage, are becoming increasingly popular. These systems help in reducing water waste and ensuring that crops receive the right amount of water at the right time (Qazi et al., 2022; Das et al., 2023).

8.6.2 Predictive analysis

Predictive analysis using advanced algorithms has the potential to revolutionize water management by forecasting water demand and availability, thus helping to prevent water crises. Machine learning models can analyze patterns in water usage, climate data, and other relevant factors to predict future water needs and identify potential risks. For example, using satellite imagery and remote sensing technology can help in monitoring water bodies, predicting droughts, and managing water resources more efficiently. These technologies enable timely decision-making and resource allocation, ensuring that water is used sustainably (Choi and Kim, 2018).

8.6.3 Real-time innovations

Real-time innovations in water management provide instant data and solutions, enabling more efficient and effective water usage. Smart water meters, for instance, provide real-time data on water consumption, helping consumers and utilities monitor and manage water usage more effectively. In addition, real-time monitoring systems can detect leaks and other issues in water distribution networks, reducing water loss and ensuring that water is delivered where it is needed most. These innovations are crucial for sustainable water management, as they provide the data and tools needed to make informed decisions and optimize water usage (Kumar et al., 2022).

8.7 Case studies

8.7.1 Global perspectives

Examining diverse case studies from different parts of the world provides a holistic view of the challenges and innovations in sustainable water management.

8.7.1.1 Case study 1

Australia's Murray–Darling Basin Plan, initiated in 2012, is a remarkable example of water management on a continental scale. The plan aims to ensure that water is shared between all users, including the environment, sustainably. It involves water recovery programs, water trading, and infrastructure investments. The Murray–Darling Basin Authority reports regular progress and adaptations to the plan, showcasing a transparent and adaptive management approach (Murray–Darling Basin Authority).

8.7.1.2 Case study 2

Israel's Water Management Israel is known for its innovative water management practices, particularly in the field of water recycling and desalination. The country has invested heavily in technology to treat and reuse wastewater for agricultural irrigation. In addition, Israel's national water company, Mekorot, has developed advanced water purification and desalination processes to provide clean drinking water. These initiatives have turned Israel into a leader in water management, even in a region with scarce water resources (Israel Ministry of Foreign Affairs. n.d.).

8.7.2 Learning from Failures

Understanding failed initiatives in water management provides valuable lessons for future planning and policymaking.

8.7.2.1 Case study

The Colorado River Compact is an agreement signed in 1922 among seven US states to allocate the waters of the Colorado River. The compact, however, was based on overestimations of the river's flow, leading to allocations that the river could not sustain. The river's flow has further been reduced due to climate change, resulting in water shortages and legal disputes. The case of the Colorado River Compact highlights the importance of accurate data and the consideration of long-term variability in water management planning (Kenney, 2005).

8.8 Challenges and roadblocks

Addressing the global water crisis requires overcoming numerous challenges and roadblocks, ranging from data collection issues to political and social influences.

8.8.1 Data dilemmas

Accurate and comprehensive data collection is crucial for effective water management, but it presents significant challenges. The variability of water availability due to climate change, the lack of standardized measurement techniques, and inadequate infrastructure for data collection in many parts of the world all contribute to data dilemmas. The World Bank emphasizes the importance of investing in hydrometeorological services and data infrastructure to improve water management and reduce uncertainties (World Bank, 2016). Furthermore, the integration of traditional knowledge and community-based monitoring can complement scientific data, providing a more holistic understanding of water resources (Berkes 2009).

8.8.2 Transboundary conflicts

Water does not adhere to political boundaries, leading to complexities in water allocation among countries that share water resources. The Nile River, shared by 11 countries, is a prime example. The construction of the Grand Ethiopian Renaissance Dam has been a source of tension between Ethiopia, Sudan, and Egypt. While the dam promises significant benefits for Ethiopia, downstream countries express concerns over potential reductions in water availability. The situation underscores the need for cooperative transboundary water management and the role of international law in facilitating negotiations and agreements (Cascão 2009).

8.8.3 Sociopolitical influences

Sustainable water allocation is often influenced by a complex interplay of lobby groups, public opinion, regional politics, and economic interests. In California, for instance, agricultural lobby groups exert significant influence over water policies, often at the expense of environmental conservation and equitable water distribution. The situation highlights the challenges of balancing various interests and the importance of transparent governance and public participation in decision-making processes (Hanak and Lund, 2012).

8.8.4 Economic impacts

The economic impacts of water management challenges are particularly profound in developing countries. These impacts manifest in various ways, affecting agriculture, industry, and human health.

8.8.4.1 Agriculture

In many developing countries, agriculture is heavily dependent on water availability. Inefficient water management can lead to water scarcity, affecting crop yield and food security. The economic cost of water scarcity in agriculture can be substantial, leading to loss of livelihood for farmers and increased food prices. A study by Dinar et al. (2015) highlights the economic implications of water scarcity in agriculture and suggests strategies for efficient water usage.

8.8.4.2 Industry

Water is a critical resource for various industries, and its mismanagement can lead to increased production costs and loss of competitiveness. Industries such as textiles, agriculture, and energy are particularly water-intensive, and disruptions in water supply can have significant economic repercussions.

8.8.4.3 Human health

The lack of access to clean water and sanitation facilities can lead to waterborne diseases, which have a direct economic impact through healthcare costs and indirect impacts through loss of productivity. The World Health Organization has reported on the economic benefits of investing in water and sanitation, emphasizing that every $1 invested in water and sanitation provides a $4.30 return in the form of reduced healthcare costs (Hulton et al., 2012).

8.8.5 Social equity

Issues of social equity and justice in water allocation are critical, particularly in regions where water resources are scarce or unevenly distributed.

8.8.5.1 Access to water

Ensuring equitable access to water is a major challenge. Disparities in water access can be observed across different socioeconomic groups, with marginalized communities often having limited access to clean water. Addressing these disparities requires a focus on inclusive water governance and policies that prioritize the needs of vulnerable populations.

8.8.5.2 Participation in decision-making

Equitable water management also involves ensuring that all stakeholders, including marginalized communities, have a voice in decision-making processes related to water allocation. This is essential for fostering social justice and ensuring that water policies are inclusive and equitable.

8.8.5.3 Addressing historical inequities

In many regions, historical inequities have resulted in unequal distribution of water resources. Addressing these inequities requires acknowledging past injustices and implementing policies that aim to redress these imbalances.

8.9 The way forward

Addressing the global water crisis necessitates innovative strategies, collaborative efforts, and a commitment to sustainable practices.

8.9.1 Community initiatives

Grassroots movements and local solutions are pivotal in advancing sustainable water practices. Indigenous communities, especially those in remote areas, often rely on government support for essential services. However, collaboration and partnership in resource planning and management can provide multiple benefits, including robust decisions, relationship-building, and community empowerment. For instance, in Australia, research has shown that while technocratic approaches to community engagement dominate, genuine collaboration is constrained by various institutional, governance, technical, and cultural factors (Jackson et al., 2019).

8.9.2 Funding and incentives

Economic models and financial incentives are essential tools to encourage the adoption of sustainable water practices. In tourist-heavy islands like Santorini, the local economy heavily relies on water. The shift from traditional rainwater harvesting cisterns to modern methods like desalinization and well withdrawals has been significant. However, research suggests that rehabilitating these cisterns and integrating them into the island's centralized water systems could foster more sustainable models of tourism development (Enriquez et al., 2017).

8.9.3 Global collaboration

International cooperation plays a vital role in addressing global water challenges. The mining industry, especially in countries like Chile, faces challenges in maintaining sustainable production due to water and energy constraints. Environmental Management Initiatives have been adopted by several mining companies, with strategies ranging from implementing Environmental Management Systems to water recycling and the use of Nonconventional Renewable Energy. Such initiatives highlight the importance of global collaboration in addressing water and energy challenges in the mining sector (Leiva González & Onederra, 2022).

8.9.4 Innovative solutions

The journey toward sustainable water management is paved with innovative solutions and approaches being developed globally. Constructed wetlands technology stands out as a green, multipurpose option for water management and wastewater treatment, offering numerous environmental and economic advantages. These systems serve various functions, from water treatment plants and habitat creation sites to recreational facilities and landscape engineering areas, highlighting the synergies between green technology and urban areas. This integration promotes circularity in urban contexts and applies innovative wetland designs for landscape infrastructure and water treatment solutions, contributing to the mitigation of urban landscape degradation (Stefanakis, 2019).

The Internet of Things (IoT) also plays a crucial role in water management, offering innovative approaches for environmental quality monitoring and sustainable resource management. IoT applications in wastewater and stormwater management, water quality assessment, and treatment are becoming increasingly prevalent, providing new knowledge and innovative solutions for sustainable water resource management (Salam, 2020).

Furthermore, the Smart City concept introduces self-powered, low-cost water sensors and Apache Spark data aggregation for high-frequency water flow monitoring, automated leak detection, and shutdown, all without the need for utility electrical power (Domoney et al., 2016).

8.9.5 Role of education

Education and awareness are pivotal in promoting sustainable water practices. Farmers, especially in Poland, play a central role in ensuring a sustainable agricultural system, with their knowledge and implementation of sustainable practices varying across different types of farms (Gebska et al., 2020).

Masters-level students in the United Kingdom, representing the future workforce and decision-makers, have shown a significant underestimation of their water usage and a preference for adopting water-saving technologies over changing user behavior (Hunt & Shahab, 2021). This highlights the need for enhanced educational efforts to improve awareness and understanding of personal water consumption and the benefits of sustainable water usage practices.

The role of teachers in a sustainable university is also crucial, requiring a shift in competencies toward more holistic, ecological conceptualizations and a future-oriented, sustainable university mission (Markauskaite et al., 2023).

Lastly, households play a significant role in natural resource conservation and environmental protection. Promoting sustainable household consumption practices related to food, energy, and water through education and awareness can contribute to cleaner environments and the achievement of the United Nation's Sustainable Development Goals (Shahbaz et al., 2022).

8.10 Conclusions

As we reflect on the journey of understanding and managing our water resources, it becomes evident that the path forward is laden with challenges yet filled with opportunities.

8.11 Future glimpse

Predicting the state of global water resources in the next 50 years involves considering various factors such as climate change, population growth, agricultural practices, and technological advancements. According to the World Water Development Report by the UNESCO, water scarcity is expected to intensify in many regions due to increased demand and the impacts of climate change. The report emphasizes the need for integrated water resources management and the adoption of sustainable practices to ensure water security for future generations (UNESCO, 2020).

8.12 Final thoughts

Water is the lifeblood of our planet, and its sustainable management is crucial for the survival and prosperity of all living beings. As eloquently stated by Kofi Annan, former Secretary-General of the United Nations, "Access to safe water is a fundamental human need and, therefore, a basic human right. Contaminated water jeopardizes both the physical and social health of all people. It is an affront to human dignity." The centrality of water to human existence cannot be overstated, and it is our collective responsibility to safeguard this precious resource for present and future generations.

8.13 Call to action

Water is the essence of life, and its sustainable management is paramount for the survival and prosperity of future generations. As we stand at the crossroads of increasing water scarcity and growing demand, the responsibility falls upon each one of us to act. It is not just the duty of policymakers, industries, or communities but also individuals. Every drop saved contributes to a more sustainable future. We must recognize the interconnectedness of our actions and the ripple effects they create in the global water ecosystem. By adopting sustainable water practices in our daily lives, supporting policies that prioritize equitable water distribution, and advocating for innovative solutions, we can collectively ensure that water remains a source of life and not conflict. Let this be our clarion call: to champion the cause of water sustainability and to pass on a legacy of abundance, not scarcity, to the generations that follow.

8.14 Future research

The field of sustainable water management is vast and ever-evolving. As we navigate the challenges of the present, it is essential to invest in research that anticipates the needs of the future. Some potential areas for future exploration include:

- **Technological innovations:** With the advent of technologies like artificial intelligence, machine learning, and the Internet of Things, there is immense potential to revolutionize water management systems. Research into how these technologies can be harnessed for predictive analysis, leak detection, and efficient distribution can pave the way for smarter water management.
- **Behavioral studies:** Understanding human behavior and its impact on water consumption can provide insights into designing effective awareness campaigns and interventions. Studies that delve into the psychology of water usage can offer valuable perspectives.

- **Climate change and water:** As the effects of climate change become more pronounced, research into its impact on water resources, including changing rainfall patterns, melting glaciers, and rising sea levels, will be crucial.
- **Alternative water sources:** With freshwater resources depleting, exploring alternative sources like desalination, rainwater harvesting, and wastewater treatment can provide supplementary sources of water.
- **Ecosystem restoration:** Research into restoring damaged ecosystems, such as wetlands and forests, can offer natural solutions to water purification and conservation.

In conclusion, the journey toward sustainable water management is a collective endeavor, requiring the collaboration of researchers, policymakers, industries, and individuals. By prioritizing research and innovation, we can equip ourselves with the knowledge and tools to navigate the challenges ahead and ensure a water-secure future for all (Gleick, 2003).

References

Ahmad, T., Hussain, M., Iqbal, M., Ali, A., Manzoor, W., Bibi, H., Ali, S., Rehman, F., Rashedi, A., Amin, M., Tabassum, A., Raza, G., & Shams, D. F. (2022). Environmental, energy, and water footprints of marble tile production chain in a life cycle perspective. *Sustainability, 14*(14), 8325. Available from https://doi.org/10.3390/su14148325, 2071-1050.

Aldaya, M. M., Allan, J. A., & Hoekstra, A. Y. (2010). Strategic importance of green water in international crop trade. *Ecological Economics, 69*(4), 887–894. Available from https://doi.org/10.1016/j.ecolecon.2009.11.001, 09218009.

Allan, J. A. (1998). Virtual water: a strategic resource global solutions to regional deficits. *Ground Water, 36*(4), 545–546. Available from https://doi.org/10.1111/j.1745-6584.1998.tb02825.x, 0017467X. Blackwell Publishing Ltd, United Kingdom. Available from: http://onlinelibrary.wiley.com/journal/10.1111/(ISSN)1745-6584.

Allan, J. A. (2003). Virtual water—the water, food, and trade nexus: Useful concept or misleading metaphor? *Water International, 28*.

Angelakis, A. N., & Spyridakis, D. S. (2010). A brief history of water supply and wastewater management in ancient Greece. *Water Supply, 10*(4), 618–628. Available from https://doi.org/10.2166/ws.2010.105, 1606-9749.

Angelakis, A. N., Capodaglio, A. G., & Dialynas, E. G. (2023). Wastewater management: From ancient Greece to modern times and future. *Water (Switzerland), 15*(1). Available from https://doi.org/10.3390/w15010043, MDPI, Greece. 20734441. Available from: http://www.mdpi.com/journal/water.

Ansell, C., & Gash, A. (2008). Collaborative governance in theory and practice. *Journal of Public Administration Research and Theory, 18*(4), 543–571. Available from https://doi.org/10.1093/jopart/mum032, 1477-9803.

Berger, M., & Finkbeiner, M. (2010). Water footprinting: how to address water use in life cycle assessment? *Sustainability, 2*(4), 919–944. Available from https://doi.org/10.3390/su2040919. 20711050. MDPI, Germany. Available from: http://www.mdpi.com/2071-1050/2/4/919/pdf.

Berkes, F. (2009). Community conserved areas: policy issues in historic and contemporary context. *Conservation Letters, 2*(1), 20–25. Available from https://doi.org/10.1111/j.1755-263x.2008.00040.x, 1755-263X.

Brown, P. G., & Schmidt, J. J. (2007). *Water ethics: The case for water recycling and reuse. Water ethics: Marcelino Botin water forum* (pp. 153–168). CRC Press.

Carbon Disclosure Project. (2020).

Cascão, A. E. (2009). Changing power relations in the Nile river basin: Unilateralism vs. cooperation? *Water Alternatives Association, United Kingdom, 2*(2), 245–268, 19650175. Available from. Available from http://www.water-alternatives.org/index.php?option = com_docman& task = doc_download&gid = 52.

Choi, J., & Kim, J. (2018). Analysis of water consumption data from smart water meter using machine learning and deep learning algorithms. *Journal of the Institute of Electronics and Information Engineers, 55*(7), 31–39. Available from https://doi.org/10.5573/ieie.2018.55.7.31, 2287-5026.

Cobourn, Kelly M., Edward, R., Landa, Gail E., Wagner., et al. (2014). *Of silt and ancient voices: Water and the Zuni land and people*. Buffalo: National Center for Case Study Teaching in Science.

Cordell, D., Drangert, J.-O., & White, S. (2009). The story of phosphorus: global food security and food for thought. *Global Environmental Change, 19*(2), 292–305. Available from https://doi.org/10.1016/j.gloenvcha.2008.10.009, 09593780.

Das, A., & Chowdhury, A.R. (2023). Energy decarbonization via material-based circular economy, pp. 263–295. doi: 10.1007/978-3-031-42220-1_15. Springer Science and Business Media LLC.

Domoney, W.F., Ramli, N., Alarefi, S., & Walker, S.D. (2016). Smart city solutions to water management using self-powered, low-cost, water sensors and apache spark data aggregation. In *Proceedings of the 2015 IEEE International Renewable and Sustainable Energy Conference, IRSEC 2015*, Institute of Electrical and Electronics Engineers Inc., United Kingdom, April 18, 2016. doi: 10.1109/IRSEC.2015.7455036. 9781467378949.

Dudgeon, D., Arthington, A. H., Gessner, M. O., Kawabata, Z. I., Knowler, D. J., Lévêque, C., Naiman, R. J., Prieur-Richard, A. H., Soto, D., Stiassny, M. L. J., & Sullivan, C. A. (2006). Freshwater biodiversity: importance, threats, status and conservation challenges. *Biological Reviews of the Cambridge Philosophical Society, 81*(2), 163–182. Available from https://doi.org/10.1017/S1464793105006950, 1469185X.

Enriquez, J., Tipping, D., Lee, J.-J., Vijay, A., Kenny, L., Chen, S., Mainas, N., Holst-Warhaft, G., & Steenhuis, T. (2017). Sustainable water management in the tourism economy: linking the Mediterranean's traditional rainwater cisterns to modern needs. *Water, 9*(11), 868. Available from https://doi.org/10.3390/w9110868, 2073-4441.

Estes, J. A., Terborgh, J., Brashares, J. S., Power, M. E., Berger, J., Bond, W. J., Carpenter, S. R., Essington, T. E., Holt, R. D., Jackson, J. B. C., Marquis, R. J., Oksanen, L., Oksanen, T., Paine, R. T., Pikitch, E. K., Ripple, W. J., Sandin, S. A., Scheffer, M., Schoener, T. W., ... Wardle, D. A. (2011). Trophic downgrading of planet Earth. *Science*, *333*(6040), 301–306. Available from https://doi.org/10.1126/science.1205106, American Association for the Advancement of Science, United States. 10959203. Available from, http://www.sciencemag.org/content/333/6040/301.full.pdf.

Foley, J. A., Ramankutty, N., Brauman, K. A., Cassidy, E. S., Gerber, J. S., Johnston, M., Mueller, N. D., O'Connell, C., Ray, D. K., West, P. C., Balzer, C., Bennett, E. M., Carpenter, S. R., Hill, J., Monfreda, C., Polasky, S., Rockström, J., Sheehan, J., Siebert, S., ... Zaks, D. P. M. (2011). Solutions for a cultivated planet. *Nature*, *478*(7369), 337–342. Available from https://doi.org/10.1038/nature10452, 14764687.

Gebska, M., Grontkowska, A., Swiderek, W., & Golebiewska, B. (2020). Farmer awareness and implementation of sustainable agriculture practices in different types of farms in Poland. *Sustainability*, *12*(19), 8022. Available from https://doi.org/10.3390/su12198022, 2071-1050.

Gleick, P. H. (2003). Water use. *Annual Review of Environment and Resources*, *28*(1), 275–314. Available from https://doi.org/10.1146/annurev.energy.28.040202.122849, 1543-5938.

Gleick, P. H. (2003). Global freshwater resources: soft-path solutions for the 21st century. *Science (New York, N.Y.)*, *302*(5650), 1524–1528. Available from https://doi.org/10.1126/science.1089967, 0036-8075.

Hanak, E., & Lund, J. R. (2012). Adapting California's water management to climate change. *Climatic Change*, *111*(1), 17–44. Available from https://doi.org/10.1007/s10584-011-0241-3, 0165-0009.

Hoekstra, A. Y. (2019). *The water footprint of modern consumer society* (pp. 1–294). Netherlands: Taylor and Francis. Available from: https://www.routledge.com/The-Water-Footprint-of-Modern-Consumer-Society/Hoekstra/p/book/9781138354784. doi: 10.4324/9780429424557.

Hoekstra, A.Y., Chapagain, A.K., Aldaya, M.M., & Mekonnen, M.M., et al. (2011). The water footprint assessment manual: setting the global standard. Earthscan.

Hoekstra, A. Y., & Mekonnen, M. M. (2012). The water footprint of humanity. *Proceedings of the National Academy of Sciences*, *109*(9), 3232–3237. Available from https://doi.org/10.1073/pnas.1109936109, 0027-8424.

Howden, S. M., Soussana, J. F., Tubiello, F. N., Chhetri, N., Dunlop, M., & Meinke, H. (2007). Adapting agriculture to climate change. *Proceedings of the National Academy of Sciences of the United States of America*, *10916490*(50), 104. Available from https://doi.org/10.1073/pnas.0701890104Australia, 19691–19696. Available from, http://www.pnas.org/cgi/reprint/104/50/19691.

Hulton, G. (2012). Global costs and benefits of drinking-water supply and sanitation interventions to reach the MDG target and universal coverage.

Hunt, D. V. L., & Shahab, Z. (2021). Sustainable water use practices: understanding and awareness of masters level students. *Sustainability*, *13*(19), 10499. Available from https://doi.org/10.3390/su131910499.

Hussain, J., Husain, I., & Arif, M. (2014). Water resources management: traditional technology and communities as part of the solution. *Proceedings of the International Association of Hydrological Sciences*, *364*, 236–242. Available from https://doi.org/10.5194/piahs-364-236-2014, 2199-899X.

Jackson, M., Stewart, R., Fielding, K., Cochrane, J., & Beal, C. (2019). Collaborating for sustainable water and energy management: assessment and categorisation of indigenous involvement in remote Australian communities. *Sustainability*, *11*(2), 427. Available from https://doi.org/10.3390/su11020427, 2071-1050.

Jha, V., Garcia-Garcia, G., Iseki, K., Li, Z., Naicker, S., Plattner, B., Saran, R., Yee-Moon Wang, A., & Yang, C.-W. (2013). Chronic kidney disease: global dimension and perspectives. *The Lancet.*, *382*(9888), 260–272. Available from https://doi.org/10.1016/s0140-6736(13)60687-x, 01406736.

Jones, P., Hillier, D., & Comfort, D. (2015). Water stewardship and corporate sustainability: a case study of reputation management in the food and drinks industry. *Journal of Public Affairs*, *15*(1), 116–126. Available from https://doi.org/10.1002/pa.1534.

Israel Ministry of Foreign Affairs. (n.d.). Water Sector in Israel. Retrieved from Israel Ministry of Energy and infrastructure, https://www.gov.il/en/pages/water_main, accesss 9th Nov 2023

Kenney, D.S. (2005). Prior appropriation and water rights reform in the American West. In *Water rights reform: Lessons for institutional design* (pp. 103–131). International Food Policy Research Institute.

Kleinman, A., Eisenberg, L., & Good, B. (1978). Culture, illness, and care. Clinical lessons from anthropologic and cross-cultural research. *Annals of Internal Medicine*, *88*(2), 251–258. Available from https://doi.org/10.7326/0003-4819-88-2-251.

Kumar, B.S., Soumiya, S., Ramalingam, S., Yogeswari, S., & Balamurugan, S. (2022). Water management and control systems for smart city using IoT and artificial intelligence. In *Proceedings of the International Conference on Edge Computing and Applications, ICECAA 2022*, Institute of Electrical and Electronics Engineers Inc. India, January 2022, pp. 653–657. doi: 10.1109/ICECAA55415.2022.9936166. 9781665482325. http://ieeexplore.ieee.org/xpl/mostRecentIssue.jsp?punumber = 9935819.

Lansing, J. S., & de Vet, T. A. (2012). The functional role of Balinese water temples: a response to critics. *Human Ecology*, *40*(3), 453–467. Available from https://doi.org/10.1007/s10745-012-9469-4.

Leiva González, J., & Onederra, I. (2022). Environmental management strategies in the copper mining industry in Chile to address water and energy challenges—review. *Mining*, *2*(2), 197–232. Available from https://doi.org/10.3390/mining2020012, 2673-6489.

Lindblom, C. E. (1959). The science of 'muddling through'. *Public Administration Review*, *19*, 79–88. Available from https://doi.org/10.2307/973677.

Markauskaite, L., Carvalho, L., & Fawns, T. (2023). The role of teachers in a sustainable university: from digital competencies to postdigital capabilities. *Educational Technology Research and Development*, *71*(1), 181–198. Available from https://doi.org/10.1007/s11423-023-10199-z, Springer, Australia. Available from, https://www.springer.com/journal/11423.

Mekonnen, M. M., & Hoekstra, A. Y. (2011). The green, blue and grey water footprint of crops and derived crop products. *Hydrology and Earth System Sciences*, *15*(5), 1577–1600. Available from https://doi.org/10.5194/hess-15-1577-2011, 16077938.

Mielke, E., Diaz Anadon, L., & Narayanamurti, V. (2010). Water consumption of energy resource extraction, processing, and conversion. Belfer Center for Science and International Affairs.

Nova, K. (2023). AI-enabled water management systems: an analysis of system components and interdependencies for water conservation. *Eigenpub Review of Science and Technology*, *7*, 105–124.

Pahl-Wostl, C., Lebel, L., Knieper, C., & Nikitina, E. (2012). From applying panaceas to mastering complexity: toward adaptive water governance in river basins. *Environmental Science and Policy*, *23*, 24–34. Available from https://doi.org/10.1016/j.envsci.2012.07.014, 18736416.

S. Pande, M. Roobavannan, J. Kandasamy, M. Sivapalan, D. Hombing, H. Lyu, L. Rietveld, et al. (2020). A socio-hydrological perspective on the economics of water resources development and management. doi: 10.1093/acrefore/9780199389414.013.657.

Public Utilities Board Singapore. (2020). ABC Waters Programme, ABC Waters Programme.

Qazi, S., Khawaja, B. A., & Farooq, Q. U. (2022). IoT-equipped and AI-enabled next generation smart agriculture: a critical review, current challenges and future trends. *IEEE Access*, *10*, 21219–21235. Available from https://doi.org/10.1109/ACCESS.2022.3152544, Institute of Electrical and Electronics Engineers Inc., Pakistan. Available from, http://ieeexplore.ieee.org/xpl/RecentIssue.jsp?punumber = 6287639.

Salam, A. (2020). *Internet of Things in water management and treatment. Internet of Things* (pp. 273–298). United States: Springer. Available from: http://www.springer.com/series/11636. doi: 10.1007/978-3-030-35291-2_9.

Shahbaz, P., ul Haq, S., Abbas, A., Samie, A., Boz, I., Bagadeem, S., Yu, Z., & Li, Z. (2022). Food, energy, and water nexus at household level: do sustainable household consumption practices promote cleaner environment? *International Journal of Environmental Research and Public Health*, *19*(19). Available from 10.3390/ijerph191912945, MDPI, Turkey. 16604601. Available from, http://www.mdpi.com/journal/ijerph.

Sklivaniotis, M., & Angelakis, A.N. (2006).Water for human consumption through the history.

Stefanakis, A. (2019). The role of constructed wetlands as green infrastructure for sustainable urban water management. *Sustainability*, *11*(24), 6981. Available from https://doi.org/10.3390/su11246981, 2071-1050.

Stepan, L. (2017). *Balinese wet rice agriculture in transition: water knowledge between a sentient ecology and the pursuit of development. Water, Knowledge and the Environment in Asia: Epistemologies, Practices and Locales* (pp. 192–209). Germany: Taylor and Francis. Available from: http://www.tandfebooks.com/doi/book/10.4324/9781315543161. doi: 10.4324/9781315543161.

Vörösmarty, C. J., McIntyre, P. B., Gessner, M. O., Dudgeon, D., Prusevich, A., Green, P., Glidden, S., Bunn, S. E., Sullivan, C. A., Liermann, C. R., & Davies, P. M. (2010). Global threats to human water security and river biodiversity. *Nature*, *467*(7315), 555–561. Available from https://doi.org/10.1038/nature09440, Nature Publishing Group, United States. Available from, http://www.nature.com/nature/index.html.

World Bank. (2016). High and dry: climate change, water, and the economy.

Chapter 9

Equitable water uses and environmental sustainability

Tabinda Amtul Bari[1,*], Javed Rimsha[1], Mahmood Adeel[2] and Yasar Abdullah[1]

[1]*Sustainable Development Study Centre, Government College University, Lahore, Pakistan,* [2]*Department of Environmental Sciences, GC Women University Sialkot, Pakistan*

**Corresponding author. e-mail address: amtulbaritabinda@gcu.edu.pk.*

9.1 Introduction

Water is a crucial resource for socioeconomic development and the preservation of ecosystems in recent decades. "Water resources which are managed properly are seen as an essential component of development that lowers inequality and poverty" (Wong et al., 2020). Water is a vital component in every aspect of life. Water management is fundamental to human economic activity and is predicted to become progressively critical as demand increases and the quantity and quality of the resources affected by climate change (Wan Rosely & Voulvoulis, 2023). According to the Fourth Climate Assessment report, "The quality and quantity of water available for use by people and ecosystems across the country are being affected by climate change, increasing risks and costs to agriculture, energy production, industry, recreation, and the environment" (Guo, 2023). Globally, the most critical issue faced by civilization is the lack of access to clean, reliable, and pure water. Around the world, 4.5 million additional people lack access to appropriate sanitation facilities, and over 2 billion people lack access to clean water for personal consumption. However, the absolute quantity of fresh water poses a threat to biodiversity, ecosystem function and services, human health, infrastructure, economic growth, traditional heritage, and inequitable distribution of water resources. The regions with the least resource adaptation costs are due to the interaction of spatialized economic disparities, population expansion, and unsustainable land use (Peters-Lidard et al., 2021). Government regulations and economic variables, which allocated resources to the privileged at the cost of the disadvantaged, have been determined by geographic and demographic patterns of water distribution (Keeler et al., 2020).

As the population becomes more dependent on water availability and managing water resources, still earliest times of limited water usage. The exploratory stage has been used to describe this phase. A phase of expansion usually follows, during which there is a significant rise in infrastructure, total conjunctive water supply, and regulatory incentives for resource use (Kumar et al., 2023). However, if the overall water supply is depleted, water demand could potentially grow and gradually slow down as the consumption of water becomes reduced. A reduction in total water extraction toward sustainable levels may be necessary due to supply constraints, population growth, climatic events, changes in public opinion, and availability of new knowledge, which is referred to as contraction stages. Water scarcity still appears to be a global issue in some major cities like Cape Town, Sao Paulo, Mexico City, Phoenix, Harare, and Chennai who have experienced severe water shortages (Loch et al., 2020).

The availability of water distinguishes Earth from other planets in the universe. However, the availability of freshwater in the domestic region as well as for economic and environmental needs is inadequate. In some areas, there is a shortage of clean water for sanitation and drinking (Galang et al., 2023; Nirmale et al., 2023). Hence limitation of clean water for sole purposes also effects directly and indirectly on human health. One of the main reasons for the inadequate availability of clean water in this region is poor water management (Fig. 9.1).

Other circumstances such as globalization and urbanization not only reduce the availability of water but also cause climate change. We are altering and compromising the quality as well as the quantity of our freshwater and freshwater resources. With the passage of time, technological, demographical, and economic trends have changed the environment in such a way we cannot imagine. Water is a renewable resource but it is not infinite (Seigerman et al., 2023).

Water Footprints and Sustainable Development. DOI: https://doi.org/10.1016/B978-0-443-23631-0.00009-1

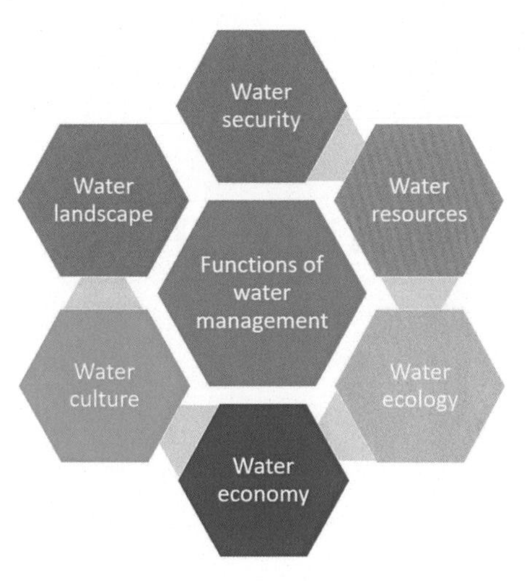

FIGURE 9.1 Functions of urban water management.

9.1.1 Global water distribution

Recently, the percentage of drinking water at a global scale has been reduced due to overpopulation. This resulted in water stress globally, especially in areas where water availability was less than usual even before the water stress. Hence, the availability of water including restricted areas is unsustainable. Day by day water supply and water demand have been changing. As the population is increasing the demand for consumption of water is increasing twice than usual. It has been estimated that in 2050, there will be an increment of more than 2.5 billion people and the world has to feed and facilitate this population with energy and water, but the result would be unsatisfied due to inadequate availability of water (Aivazidous et al., 2021).

9.1.2 Importance of equitable water use and environmental sustainability

Equity may also be promoted as a tool for economic growth and development (Jahanshahi et al., 2023) or employed "to cloak self-interest." Therefore, the consequences of water access by consumers with a range of power and water demands are greatly impacted by either implicit or explicit principles of equality in water distribution.

Conflicts between water consumers have been growing rapidly, which highlights the absence of a rational, equitable, and transparent system for allocating water. This is particularly evident in areas where human appropriation of water is at unsustainable levels and demand comes from users with a diverse range of requirements and degrees of authority (Bozorgzadeh & Mousavi, 2023). Technical and allocative interpretations are the two basic approaches to the concept of water resource efficiency. Comparing the amount of water provided to final consumers with the amount that is lost in the distribution system enables an analysis of efficiency in water distribution systems (WDSs). The Pareto principle states that the optimal allocation corresponds to the situation in which no individual can be made more prosperous without making other people less fortunate that achieved when the scarce resource provides the most quantity of monetary value in utility terms (Perera et al., 2023). According to Pareto, efficiency disregards the unequal distribution of wealth and the declining marginal utilities of wealth and assumes that all commodities, products, or resources have the same monetary worth for everyone. Therefore, equity, equality, and requirements are at the core of cultural efforts for social justice. Both distributive and procedural equity offer effective perspective benefits in allocating limited resources (Richards et al., 2023). Issues of justice, in particular, equity, are addressed by distributive justice. It has been shown that societies emphasize distributive justice concerns over the criteria including efficiency for key resources for water. Equality refers to the distribution that leads to equality among group members while equity refers to the proportionality either directly or indirectly between inputs and outcomes that support the unequal distribution of resources based on merit with limited access (Wilfong et al., 2023).

9.1.3 Major challenges in water management

Some of the major challenges faced while managing water are described below.

9.1.3.1 *Fresh water stress*

The term water stress refers to a condition in which the requirement for freshwater has increased with the accessible source in a specific region or within a particular time. In other words, it can also be described as an imbalance between the demand and availability of water. Its occurrence leads to environmental, social, and economic penalties (Li et al., 2020; Fig. 9.2).

Water stress can be categorized into the following three levels.

9.1.3.1.1 Low water stress

This term is described as the situation in which the water resources exceed within a particular region and excess amount of water will be available in future. This area has a favorable margin of sustainability and availability of water.

9.1.3.1.2 Moderate water stress

It is the condition in areas where the requirement for water reaches the limit for the accessible supply of water. In these areas, the water fulfills the current needs, but in future there is little room to fulfill water needs due to increase in population and change in climate.

9.1.3.1.3 High water stress

This condition is the most crucial among all categories of water stress where the demand for water exceeds the available reserve. In such a condition, there is shortage of water availability and supply, leading to water scarcity, drought, and water-related issues to meet the requirements for urbanization, population, and agriculture. The increase in competition for water resources is the dominant factor in this situation (Wilfong et al., 2023).

Factors that influence the water stress are over population, climate change, agricultural practices water pollution, and water infrastructure. Nowadays, people are more concerned with the production of the energy and food while ignoring such other factors like global warming, climate change, overpopulation, and water scarcity. A survey conducted by the Global Risk Perception among the experts of World Economic forum reported that over the next 10 years, the highest societal impact will be from water problems (Ferrans et al., 2022).

9.1.3.2 *Water scarcity*

Water scarcity is one of the global issues because of the increasing demand for water due to overpopulation, industry, and agriculture, leading to drought, climate change, and water stress. As the population and anthropogenic activities are increasing everyday, water is also getting polluted directly and indirectly by these factors. Climate change has negative implications on water patterns of water and the infrastructure of water bodies, increases the intensity of floods and droughts, and affects the availability of the water. In most of the areas, the structures used to distribute, store, and supply water are very old and outdated, resulting in imbalanced water distribution and leakage, and hence, poor management of water. Sticking to great and pure water quality is the continuous challenge at present because of the limited quality of fresh water. Water polluted by heavy metals, pathogens, and other organic pollutants needs advanced and modern techniques for proper removal.

Impact due to ecosystem changed its pattern due to many natural and anthropogenic disasters. Alteration in the flow of water and quality harms the living organisms and results in the poor management of water. Overextraction of groundwater, less access to clean water, intersectoral competition, data gaps, and institutional governance are also the major challenges to water management (Cosgrove and Loucks, 2015).

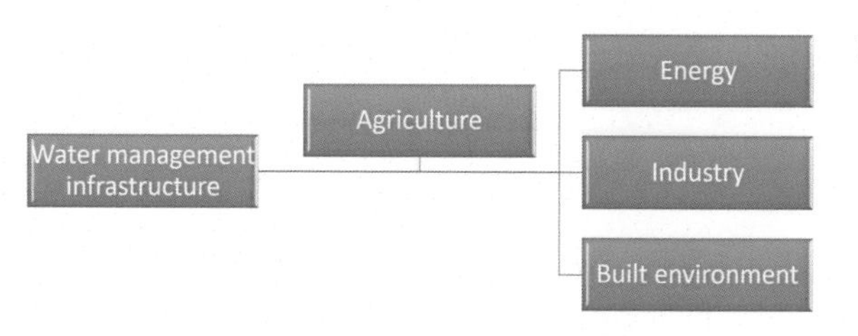

FIGURE 9.2 Water intense management infrastructures.

9.1.4 Impact of climate change on water scarcity

Climate change is a major challenge faced globally, driven by the buildup of gases such as greenhouse gases in the Earth's atmosphere. Its vast changing effects touch nearly every phenomenon of our lives directly and indirectly. Climate change overthrobbed the world's phenomenon, lasting the alteration in the Earth's pattern permanently. The gases CO_2 and CH_4 are trapped by the heat, which leads to an increase in the temperature, thus resulting in global warming. The increase of temperature results in the melting of glaciers, changing the Earth's pattern of water flow, direction and availability of water, level shifts, and floods. Climate change has deep impact on the water scarcity, pressurizing the existing issues globally. Global warming changes the temperature and patterns of precipitation, increases the rate of evaporation, as well as alters the distribution and availability of water, thus leading to severe water scarcity (Abbass et al., 2022).

9.1.4.1 Alteration in the pattern of precipitation

The change in climate leads to the shifts in distribution, intensity, and schedule of precipitation. Some of the areas affected by climate change suffer heavy rainfall and flooding. This disrupts the change in the natural renewal system of the water, leading to scarcity of water (Xiang et al., 2021).

9.1.4.2 Increased droughts

Severe droughts are caused due to continuous climate changes. Less precipitation in certain regions results in intense drought. The droughts make water-deficit areas thus leading to water scarcity. There is less availability of water for sole purposes like drinking and agricultural practices (Aivazidou et al., 2021).

9.1.4.3 Snowpack and glacial retreat alteration

The presence of snowpack in the mountainous regions plays very important role. They serve as a reservoir of water when they melt during summer and winter. But due to climate change, the glaciers and snowpack are not existing as much as they were present in past (Fischer et al., 2021). Warmer temperature causes the snowpacks and glaciers to melt earlier, which disrupts the natural phenomenon of the water availability, thus increasing the risk of water scarcity. This leads to shrinkage of freshwater resources.

9.1.4.4 Rise in sea level

Due to climate change, the sea level is also rising day by day, which leads to the intrusion of saltwater into the coastal aquifers, thus contaminating the freshwater resources. This change leads to the limited resources for drinking water and agriculture (Liu et al., 2019; Loch et al., 2020).

9.1.4.5 Change in water availability

Due to climate change, the water availability patterns are also changing. Some areas are experiencing excess amount of water and some areas are facing less amount of available water resources. The water-deficit areas are facing water scarcity.

9.1.4.6 Increased water demands

Higher temperatures due to climate change and increasing water demands for drinking and other purposes lead to water scarcity.

9.1.4.7 Displacement and migration

Water scarcity cause by climate change can contribute to forced displacement and migration of biodiversity from areas facing water scarcity to the areas where water is available.

9.1.4.8 Energy production

Many forms of energy are produced by water, such as hydropower and thermal power plants coolers. The water scarcity caused by climate change disrupts the production of energy and results in higher cost of energy (Buck et al., 2020).

Abbass et al. (2022) demonstrated in their study about the effect of agriculture and other factors on climate change. They further described the effect of climate change on human lives. Over the previous 10 years, natural catastrophes have claimed an average of 60,000 lives worldwide each year. In other words, around 0.1% of the worldwide fatalities, according to the report. Fig. 1 shows the yearly differences in the quantity and proportion of mortalities from natural catastrophes in recent decades. Less than 10,000 fatalities have been reported, which is as little as 0.01% of the total fatalities. But shock occurrences can have a disastrous effect. This study showed that climate change not only effect ecosystem funtionality, but also effects directly and indirectly on the mental and physical health of humans, along with these anthropogenic activities affecting the sustainability of the environment and the human health. Another problem is that it may result in the low quality and high prices of food, and less yield of crops. The different climatic circumstances like droughts, flash floods, melting of glaciers, and precipitation cause threat to the world's forest. It is the time we should make policies and strategies to lower the anthropogenic activity to save the Earth and climate. Abbass et al. (2022) described the short- and long-term mitigation approaches with the long- and short-term impact strategies (Fig. 9.3).

9.1.5 Water pollution and degradation of ecosystem

Water pollution refers to the contamination or any changes in the chemical, physical, and biological water that can harm the environment, human health, and other living organisms depending on it. Water pollution occurs in many natural reserves such as rivers, sea, lakes, groundwater, tap water, and oceans. Water pollution is caused by the contaminants released into such reserves directly or indirectly, resulting in unfavorable effects on human health and aquatic ecosystem (Quiñones et al., 2019). The ecosystems of the Earth are the complex and dynamic arrases of life that contain and support all living organisms on the planet. These ecosystems facilitate living beings with clean air, food, fresh water, and moderate climate to live. Increase in the anthropogenic activities sited unprecedented burden on these intricate ecosystems. Ecosystem degradation has been declared as the most critical global apprehension. The term ecosystem degradation is the corrosion of the ecosystem via reduction of natural resources of air, water, and land, and destruction of habitat, wildlife, and pollution. In other words, any change or disruption in the ecosystem that is undesirable is called ecosystem degradation (Fischer et al., 2021).

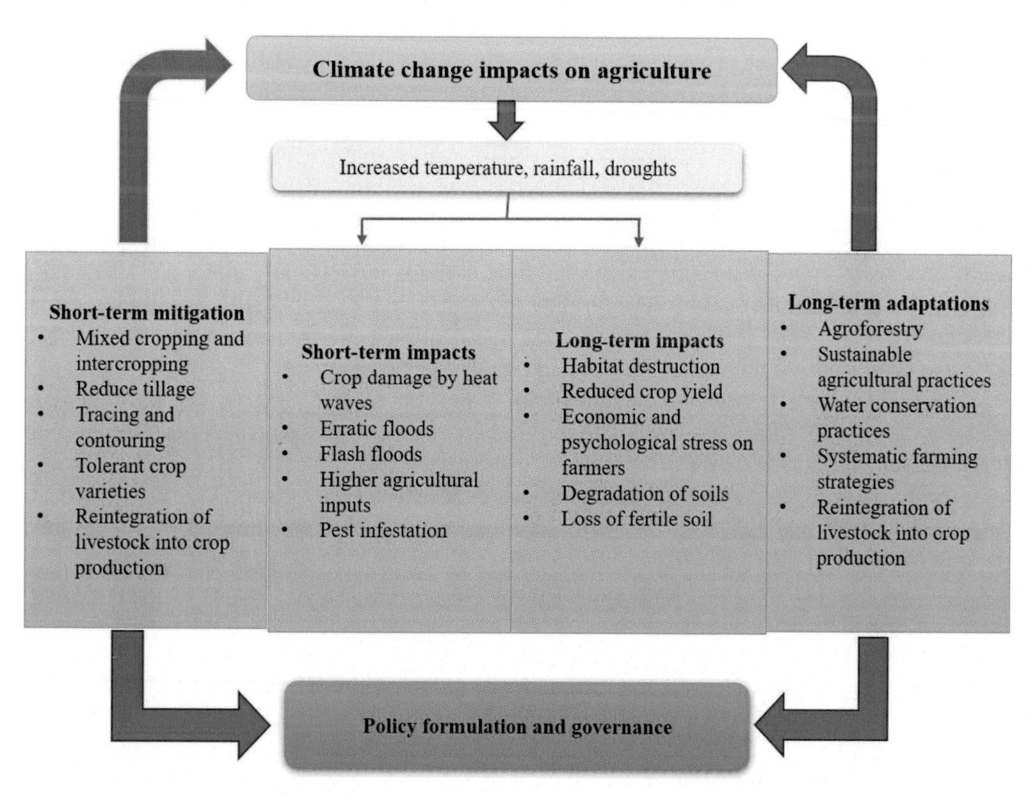

FIGURE 9.3 Approaches and strategies to mitigate short-term and long-term climate change effects on agriculture.

9.1.5.1 *Type of pollutants*

Heavy metals, such as lead, mercury, and cadmium, released from the industries, pesticide, pharmaceuticals, and plastic manufaturing units, are known as chemical pollutants. These chemical pollutants reduce the quality of water and ultimately detoriates human health and aquatic ecosystem. While microorganisms like bacteria and viruses that cause water pollution are called biological pollutants.

Microorganisms like bacteria and viruses that cause water pollution are called biological pollutants. High nutrient load is caused by eutrophication process that can occur when nutrients, notably nitrogen and phosphorus, are present in excess. As a result, aquatic ecosystems suffer from algae blooms, decreased oxygen levels, and other problems. Soil erosion and building operations can bring silt into water bodies, increasing turbidity, decreasing light penetration, and harming aquatic ecosystems. Thermal pollution is caused when hot water from industrial activities or power plants is released into natural water bodies, changing the temperature conditions significantly, and thus impacting aquatic life (Galang et al., 2023).

Destruction of habitat, pollution, invasive species, climate change, overpopulation, and overexploitation of resources are the factors that cause the degradation of the ecosystems of the Earth. Two of these factors are described below.

Habitat destruction is the most prominent factor of ecosystem degradation that occurs when a natural piece of land is destructed by anthropogenic activities such as urbanization, industrialization, and agriculture. The construction of roads cause the fragmentation of the natural biodiversity habituating in that area. Hence, these activities disrupt and degrade the ecosystem.

Pollution of water air, and land is the factor that collides with the existing natural ecosystem. This not only harms the environment but also effects directly and indirectly on living organisms.

In 2018, a case study was conducted by Pakistani and Korean students in Korea in which they investigated the quality of the Ravi River in Punjab. Iqbal et al. (2018) reported that Punjab is the most populous city of Pakistan, comparing the water parameters of the Ravi with two other districts of Pakistan. The Ravi is polluted due to the industrial and residential waste water discharge. The polluted water enters the Ravi by specific channels and links.

In addition, a numerical model was also included (Water Quality Simulation Program) to find the strategies to improve the water quality of the Ravi in Lahore. This model simulator discovered that if the head water flow and link canals will increase to 50% and with 75% improvement in pollution control of the drains by treatment facilities, it is possible to treat and improve the water quality of the Ravi to an acceptable limit and according to water quality standards. The presence of uncontrollable parameters in these water bodies results in the destruction of natural ecosystems, causing water pollution. The situation of the Ravi nowadays is threatening the sustainability of the ecosystem of the river. Iqbal et al. (2018) suggested that waste water treatment plants in countries like Korea, China and Malysia which efficiently treat pollutants, must need to mimic these treatment technologies in less developed nations.

9.2 Principles of equitable water use

The International Law Association did not implement the ground-breaking Helsinki Rules until 1966. These rules were the first to be codified at the international level and relate to the use and conservation of common water resources. It took the UN General Assembly until 1997 to establish the United Nations Watercourses Convention (UNWC), the first worldwide customary agreement on the use of shared international water resources. The area of international law governing the use and conservation of transboundary aquifers is less developed, despite the enormous significance of groundwater resources for supplying human needs around the world. The International Water Law, recently developed according to three fundamental principles: Articles 5 and 6 of the UNWC principle of equitable and appropriate use, which are usually recognized as the field's most important guideline (ILC, 1994). Article 7 of the UNWC provides obligation to prevent major transboundary harm. Furthermore, the obligation to collaborate in the management of shared waterways described in Article 8 of the UNWC.

The route that catchment water resource development often has is a topic of dispute between consumers and riparian nations. Farmers in the catchment's downstream have frequently complained about interbasin transfers occurring upstream. In addition, there has been ambiguity regarding the impact of environmental flow releases (McManaus et al., 2021). The law of the nonnavigational uses of international watercourses, the International Law Commission struggled for more than 10 years with the interaction between the principles of equitable use and no harm. The primary significant rule of International Water Law is the principle of equitable water utilization, which calls for the reasonable and equitable sharing of the beneficial uses of an international watercourse water. The total damage caused by the use of such water is of important consideration when determining whether a particular use is reasonable and equitable. The

Watercourses Convention is interpreted in a way that is consistent with conventional international water law (López et al., 2021).

The practice of states, as demonstrated by the disputes that have arisen regarding this matter, appears to now acknowledge that each state concerned has a right to have a river system taken into account as a whole and to have its own interests balanced against those of the other states, and that no one state may claim to use the waters in a manner that causes material injury to the interests of another, or to oppose their use by another state unless this causes material injury. An international watercourse must be used, developed, and protected by the watercourse States in a fair and reasonable manner. According to the current Convention, such participation encompasses both the right to use the watercourse and the obligation to work with others to maintain and develop Article 5(2) of the UNWC. When significant harm though inflicted on another watercourse State, Article 7(2) of the UNWC states that the States whose use of the watercourse causes the harm shall, in the absence of agreement to such use, take all appropriate measures, having due regard for the provisions of Articles 5 and 6, in consultation with the affected State, to eliminate or mitigate such harm and take appropriate compensation (Tanzi, 2020). Instead, compensation linked to the need for consultation and involved the States fairly under the State liability framework.

9.3 Sustainable water management

To improve water usage efficiency and making sustainable use of water resources are key components of reducing ecological and environmental stress and enhancing human well-being. Two essential steps to encourage sustainable development are effectively controlling water demand and increasing water usage efficiency (Iqbal et al., 2019).

9.3.1 Conservation of water resources

The United States has an abundance of natural resources, including groundwater, streams, rivers, lakes, and reservoirs. It provides freshwater for drinking, agriculture, and industry, as well as being a source of scenic beauty (Loch et al., 2020). The primary element governing the composition and operation of both managed and unmanaged ecosystems is water. The hydrologic cycle includes important processes such precipitation, irrigation, interception, infiltration, surface runoff, subsurface flow, soil water balance, evapotranspiration, drainage, and deep leaching to groundwater. The hydraulic conductivity of the soil and the hydraulic gradient control the transport, or percolation, of water through soil (Peters-Lidard et al., 2021).

Water conservation needs to be prioritized by all people, communities, and nations. Finding strategies to encourage rainfall to percolate into the soil as opposed to letting it run off into streams and rivers is a crucial strategy. In order to conserve water before it reaches streams, rivers, and lakes, for instance, it is possible to capture and reduce water runoff by 10%–20% by increasing the usage of trees and bushes (Kumar et al., 2023).

The maintenance of agriculture, livestock, and forest production necessitates the conservation of all available water supplies, including rainfall. Surveillance of soil water content and advance water application requirements to a particular crop are two practical ways that prevent water loss for agricultural production, using organic mulches to avoid water loss and to enhance water percolation by reducing runoff and evaporated water, avoiding the elimination of biomass from the land, and increasing the use of trees and shrubs to slow drainage. Humans should avoid clearcutting in forest regions and instead practice sustainable forest management. Urban areas, whose runoff rates are thought to be 72% higher than in places with forest cover, can also gain from trees (BASIN, 2003). In addition, water can be collected in reservoirs and manmade ponds because runoff from roofs, driveways, roadways, and parking are huge influx of water recharge (Wong et al., 2020).

9.3.2 Protection and restoration of ecosystem

Numerous services, including the provision of clean water, agricultural activities, and climate regulation, are provided by ecosystems. Maintaining these services, which are critical for human welfare, requires protecting and repairing ecosystems.

9.3.2.1 Ecosystem-based adaptation

Restoration and preservation of an ecosystem based on built-in responses like mangroves, coral reefs, and dunes against climate change is a part of ecosystem-based adaptation techniques.

9.3.2.2 Restoration of ecosystems

To ensure ecosystems restoration, ecosystem health is preserved by choosing native species, restoring habitat structure, and tracking progress.

9.3.2.3 Economic value

Ecosystems provides many services both in monetary and nonmonetary terms. For instance, woods provide both timber and nontimber products while wetlands that cleanse water, cutting down on treatment expenses and providing natural water filtration process. Understanding the worth of ecosystems economically can help to justify conservation (Seigerman et al., 2023).

9.3.3 Efficient water technologies

More than "50% of the world population will experience severe water scarcity by 2050, predicts the United Nation World Water Development Report. Water resources must be handled more securely in order to prevent water stress. Water conservation is aided by efficient water technologies, which are installed to maximize its usage and reduce waste. It is crucial for addressing the water shortage, encouraging sustainability, and reducing the negative effects on the environment. With the ability to identify any anomalies, such as nonrevenue water (NRW) losses and water contamination in the water distribution system (WDS), smart water systems (SWS) use sensor, information, and communication technology (ICT) to provide real-time monitoring of data such as pressure, water flow, water quality, moisture, etc. It improves the efficiency of the use of water and energy in every sector" (Richards et al., 2023).

Here are some examples of efficient water technologies.

Systems for collecting and storing rainwater are used for nonpotable purposes like flushing toilets, washing cars, and irrigation of outside areas. The need for freshwater resources is lessened.

9.3.3.1 Graywater recycling

Wastewater from sinks, showers, and laundry is referred to as graywater. In order to utilize graywater for irrigation or toilet flushing after cleaning and filtration, graywater recycling systems reduce pressure on existing municipal water sources.

9.3.3.2 Smart irrigation systems

These systems optimize irrigation schedules using weather information and sensors for soil moisture, ensuring that plants receive the correct amount of water and prevent water waste (Fig. 9.4).

9.3.3.3 Water-efficient appliances

Through better design and technology, energy-efficient washing machines and dishwashers not only save power but also use less water.

9.3.3.4 Desalination technologies

Reverse osmosis is one desalination technology that can convert saltwater to freshwater, opening up previously unusable water sources for human consumption. Desalination, however, can be expensive and energy-intensive. In addition to helping to mitigate water scarcity, conserve ecosystems, and lower water-related expenses for people, businesses, and governments, efficient water technologies are essential to sustainable water management (Galang et al., 2023).

9.3.4 Integrated water resources management

IWRM is a process that encourages the coordinated development and management of water, land, and related resources in order to increase economic and social welfare in an equitable manner without compromise the sustainability of essential ecosystems. An IWRM approach focuses on four pillars: establishing the institutional framework through which to put the policies, strategies, and legislation into practice; establishing the management instruments required by these institutions to perform their functions; and developing the financial tools required to implement the developed instruments (Bozorgzadeh & Mousavi, 2023; Wilfong et al., 2023).

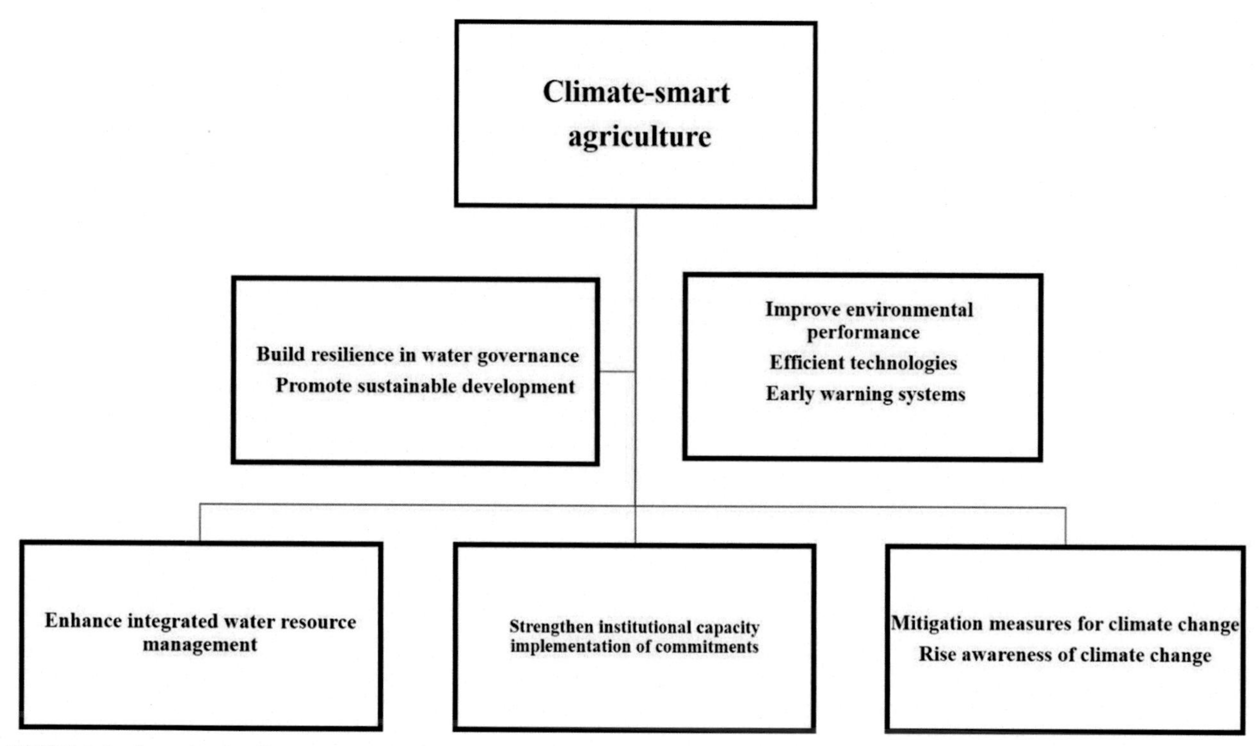

FIGURE 9.4 Strategies for climate-smart agriculture.

Key principles and components of Integrated Water Resource Management (IWRM) include the following.

IWRM takes a holistic approach, taking into account both surface and groundwater resources as well as their relationships. The full hydrological cycle considered including precipitation, runoff, infiltration, and evaporation.

9.3.4.1 *Engagement of stakeholders*

IWRM promotes the active participation of all relevant stakeholders, including public institutions, communities, businesses, and nonprofit organizations.

9.3.4.2 *Protection of ecosystems*

IWRM acknowledges the significance of preserving healthy aquatic ecosystems. In order to maintain water quality and support biodiversity, it aims to protect and restore natural watercourses and habitats. IWRM depends on accurate and current data and information regarding water resources. Systems for monitoring, gathering data, and disseminating information are essential for decision-making.

9.3.4.3 *Resolution of conflicts*

IWRM offers procedures for settling disputes between various water users and stakeholders. Finding solutions that are agreeable to both parties frequently involves mediation and negotiation processes. IWRM is not a one-size-fits-all strategy; rather, it is customized for each region or watershed. The United Nations and other organizations have pushed it, and it is regarded as a crucial foundation for sustainable water management, especially in the impact of climate change (Sanchez-Plaza et al., 2021).

9.4 Water governance and policy

To establish a functional water governance structure, address the water issues, particularly floods and droughts, and have a strong coordination system. Furthermore, because of corruption, continuous violence, and political instability, it is difficult to prioritize, manage, and provide prospective investment opportunities (Mourad, 2020). Water management and development influenced by a group of political, social, economic, and administrative institutions is known as "water governance." Multiple scales, including local, national, regional, and global levels, are involved in water governance. In this regard, there are several obstacles and a limited capability for managing existing or future investment possibilities. At the national, regional, and international levels, water may encourage collaboration among multiple organizations and agencies. At the federal, interministerial and local levels, duties and responsibilities must be defined. In order to collaborate among funding communities on initiatives and projects, a framework must be established. The water governance cycle consists of four parts as described in Fig. 9.5.

9.5 Ecosystem based approach for equitable water

Freshwater ecosystem sustainability is being endangered on global scale. According to McManus et al. (2021), managing freshwater ecosystems, particularly water quality, using "command-and-control" strategy is no longer feasible. Environmental water quality an integrated approach which links an aquatic supply qualities and its response structure, function, and processes of ecosystem. Physical and chemical factors analyzed to determine the quality of the water using the traditional physicochemical method and to estimate the impacts on living organisms (Cosgrove & Loucks, 2015). A "new water paradigm" is being created as a result, which emphasizes on expanded stakeholder involvement, integration of sectors, issues, and increased understanding of the economic, ecological, and cultural significance of water (Jahanshahi et al., 2023). In the case of water management, interest in an ecological or ecosystem-centered approach to natural resource management has been strongly emphasized for many years. Recently, there has been increased attention accorded to the discovery and evaluation of ecosystem services, as well as to the possibility of using the value of these services as a foundation for better management of natural and human-related systems, although the adoption of an ecosystem services approach still faces many obstacles, for instance, conflict with fundamental principles of international water law, such as equitable and reasonable usage (Tangworachai et al., 2023).

Tourism has unacceptably high consequences on the hydroecology and water supply in many popular tourist sites. Due to the relative power imbalances among the many stakeholders, the problem of water inequity is more acute in developing nations, and management become more challenging as catchment regions are impacted by deforestation (Bresney et al., 2023). There are two types of relationships between tourism and water: consumptive and nonconsumptive. Although the lines between consumptive and nonconsumptive might occasionally disintegrate the available literature on potable water resources for consumption and its management. While acknowledging the significance of virtual

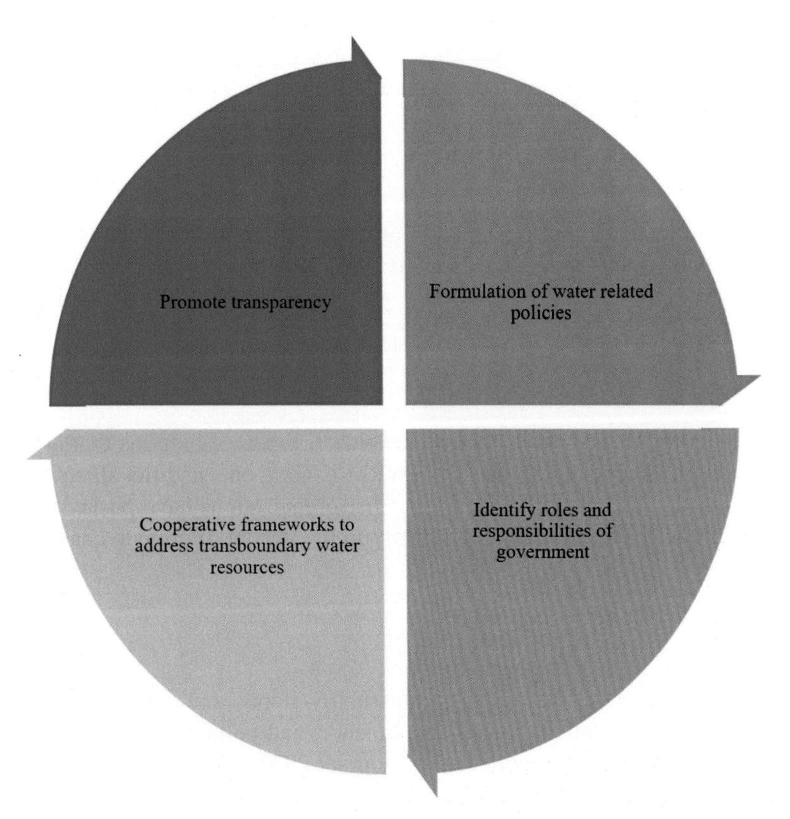

FIGURE 9.5 Framework for water governance and policy formation.

water consumption that is embodied in every process, such as during the building process of a hotel, in the use of fuel, or in the manufacture of food. Although water problem has been recognized, many government stakeholders are still unaware of its scope (Sadoff et al., 2020).

Groundwater extraction by farmers to complement irrigation demand has been increased. As a consequence, groundwater has reduced to a critical level, which poses negative externality and has increased resource costs. To promote equitable groundwater distribution and more effective agricultural water use in response to growing water crisis, informal groundwater markets have steadily developed (Razzaq et al., 2019). One of the major issue faced by developing agricultural nations is the lack of irrigation water. Farmers progressively turned groundwater to complement irrigation needs as a result of the inadequate and irregular surface water supply (Iqbal et al., 2018). Many South Asian nations, where agriculture now uses the majority of groundwater, are the most vulnerable to water shortage. However, over time, groundwater shortages have largely replaced the socioecological benefits of the groundwater irrigation success (Rambonilaza et al., 2023). The majority of studies on well-established water markets either concentrate on how these markets function or solely took economic advantages into account. In addition, some research on the groundwater markets used of the outcomes of simulations to examine the economic viability of these markets (Mourad, 2020). Water markets thought to increase water productivity by distributing water to the most productive users who would provide the highest marginal returns. For those who are involved in the groundwater markets, this influence will manifest. The informal groundwater markets provide additional advantage to resource-poor farmers who are unable to invest in their wells because of financial restrictions and improve water efficiency (Razzaq et al., 2019).

9.6 Economic instruments for equitable water distribution

Globally, water allocation policies have been modified by incorporating economic tools. Among European nations, the effectiveness of economic instrumentation used in water allocation has been changed as related to prior design goals. To manage water resources economically and beneficially, cautious and responsible action is required. Economic instruments gained popularity in European nations with varying degrees of water scarcity, allocation schemes, and regulations. These instruments reveal general and site-specific difficulties for successful implementation the water policy goals that must be resolved (Abbass et al., 2022). While expanding water resource infrastructure to increase the supply, river basin closure is a process that ultimately causes the restricted water resources capacity to sustain advanced flows at reasonable costs to be exhausted (Quiñones et al., 2019). The vital sectors for water demand are agriculture, which is the greatest consumer of water worldwide, and the industry with the highest concentration of marginal water resource uses (Food and Agriculture Organization, 2013). Economic tools serve as a means of achieving a goal. Any benefit that arises from economic tools is accomplished by social benefits from the success of water policy goals (Keeler et al., 2020). Economic instruments are created in combination with other policies; these cannot be developed independently. To ensure success and prevent externalities, complementing actions must be strengthened and promoted. For instance, the Australian reform was frequently used as an illustration of global best practices (OECD, 2015). In order to prevent unwanted socioeconomic and environmental effects, economic instruments and water policy as a whole vigilantly developed. That implies owners of water rights are only permitted to transfer water, saved as a result of a decrease in the consumption (Escriva-Bou et al., 2017). Understanding the costs and benefits for users, the environment, and other stakeholders is necessary for the effective use of economic tools. Economic tools must take into account legislation, customs, and political authority dynamics that define the expense of transactions and the possible range actions for institutions. In order to determine the overall cost of economic instruments, aggregation of the institutional transaction costs, third-party impacts, and net costs to consumers is essential (Ferrans et al., 2022).

Water and irrigated agriculture are essential for Pakistan economic development. According to Abbass et al. (2022), 75% population of Pakistan is dependent on agriculture, which generates 60% of the nation foreign exchange profits, employs 44% of the labor force, and contributes about 20% of the GDP of the nation. Effective management of both surface water and groundwater required to meet demands. Numerous potential to enhance water management. However, provinces experienced constant instability as a result of the increasing supply-demand disparities. More violence and devastation can result from intra-state or inter-provincial water disputes than from international conflicts. International water disputes typically receive significantly more attention than intra-state water issues (Mourad, 2020). Conflict on the social and political levels results from the majority of water being wasted and spread inequitably in developing nations. For instance, inefficient use of water occurs in irrigation, industry, and provincial water distribution because unequal allocation makes some social groups feel constrained and unsatisfied (Xiang et al., 2021). When compared to conventional irrigation methods, lowest agricultural productivity was recorded while yield increased from 20% to 100% along with water savings that ranged from 40% to 70%. This strategy significantly improves water usage efficiency. Other water-saving technologies, such as zero tillage, precise field leveling, and bed-and-furrow planting, can also help to increase water productivity (Tariq et al., 2020).

9.7 Conclusions

Equitable water use and sustainability are interconnected principles aimed at ensuring fair access to water resources while safeguarding the environment and future generations. Equitable water use entails distributing water fairly among various stakeholders, prioritizing basic human needs, and considering social and economic factors. Sustainability, on the other hand, focuses on managing water resources in a way that preserves their availability and quality for current and future generations. Immediate actions are necessary to prevent environmental catastrophes in the near future because of precedented environmental concerns like climate change and mass extinction. In order to support construction of a functional water governance system, encompassing water legislation, vision, policy, strategy, and staff capability in IWRM, these systems must be put in place. Our current issues with water scarcity and drought should be considered as a governance issue that is being and will continue to get worse due to climate change. Proper policies and strategies should be implemented to lower the anthropogenic activity to save the Earth and climate. It is the need of hour to understand and raise awareness by addressing the water stress by implementing and structuring the policies and strategies to manage the available water while promoting the sustainability and importance of water resource. In conclusion, sustainablity in ecosystem water usage can be achieved in future through responsible resource management, efficient water allocation, and addressing social and environmental concerns.

References

Abbass, K., Qasim, M. Z., Song, H., Murshed, M., Mahmood, H., & Younis, I. (2022). A review of the global climate change impacts, adaptation, and sustainable mitigation measures. *Environmental Science and Pollution Research, 29*(28), 42539–42559. Available from https://doi.org/10.1007/s11356-022-19718-6, https://link.springer.com/journal/11356.

Aivazidou, E., Banias, G., Lampridi, M., Vasileiadis, G., Anagnostis, A., Papageorgiou, E., & Bochtis, D. (2021). Smart technologies for sustainable water management: an urban analysis. *Sustainability, 13*(24), 13940. Available from https://doi.org/10.3390/su132413940.

Bozorgzadeh, E., & Mousavi, S. J. (2023). Water-constrained green development framework based on economically-allocable water resources. *Scientific Reports, 13*(1). Available from https://doi.org/10.1038/s41598-023-31550-7, https://www.nature.com/srep/.

Bresney, S., Chalmers, D., Coleoni, C., Dyke, A., Escobar, M., Farnan, R., & Santos, T. (2023). Guidelines for equitable participation in water decision-making. https://doi.org/10.51414/sei2023.030.

Buck, H. J., Martin, L. J., Geden, O., Kareiva, P., Koslov, L., Krantz, W., Kravitz, B., Noël, J., Parson, E. A., Preston, C. J., Sanchez, D. L., Scarlett, L., & Talati, S. (2020). Evaluating the efficacy and equity of environmental stopgap measures. *Nature Sustainability, 3*(7), 499–504. Available from https://doi.org/10.1038/s41893-020-0497-6, http://www.nature.com/natsustain/.

Cosgrove, W. J., & Loucks, D. P. (2015). Water management: current and future challenges and research directions. *Water Resources Research, 51* (6), 4823–4839. Available from https://doi.org/10.1002/2014WR016869, http://onlinelibrary.wiley.com/journal/10.1002/(ISSN)1944-7973.

Escriva-Bou, A., Pulido-Velazquez, M., & Pulido-Velazquez, D. (2017). Economic value of climate change adaptation strategies for water management in Spain's Jucar Basin. *Journal of Water Resources Planning and Management, 143*(5). Available from https://doi.org/10.1061/(ASCE)WR.1943-5452.0000735, http://ojps.aip.org/wro/.

Ferrans, P., Torres, M. N., Temprano, J., & Rodríguez Sánchez, J. P. (2022). Sustainable Urban Drainage System (SUDS) modeling supporting decision-making: A systematic quantitative review. *Science of the Total Environment, 806*. Available from https://doi.org/10.1016/j.scitotenv.2021.150447, http://www.elsevier.com/locate/scitotenv.

Fischer, J., Riechers, M., Loos, J., Martin-Lopez, B., & Temperton, V. M. (2021). Making the UN Decade on ecosystem restoration a social-ecological endeavour. *Trends in Ecology and Evolution, 36*(1), 20–28. Available from https://doi.org/10.1016/j.tree.2020.08.018, https://www.sciencedirect.com/journal/trends-in-ecology-and-evolution.

Food and Agriculture Organization. (2013). *The State of Food and Agriculture*. Rome, Italy: FAO.

Galang, E. I. N. E., Rosas, A. J., & Samonte, P. (2023). *Local knowledge on water use and water-related ecosystem services in lowland, midland, and upland villages in Mindanao, Philippines. Indigenous and Local Water Knowledge, Values and Practices* (pp. 199–218). Canada: Springer Nature. Available from https://link.springer.com/book/10.1007/978-981-19-9406-7, http://doi.org/10.1007/978-981-19-9406-7_12.

Guo, Q. (2023). Strategies for a resilient, sustainable, and equitable Mississippi River basin. *River, 2*(3), 336–349. Available from https://doi.org/10.1002/rvr2.60, 2750-4867.

ILC. (1994). United Nations Yearbook of the international law commission 1(2).

Iqbal, M. M., Shoaib, M., Agwanda, P., & Lee, J. L. (2018). Modeling approach for water-quality management to control pollution concentration: A case study of Ravi River, Punjab, Pakistan. *Water, 10*(8). Available from https://doi.org/10.3390/w10081068, http://www.mdpi.com/2073-4441/10/8/1068/pdf.

Jahanshahi, S., Kerachian, R., & Emamjomehzadeh, O. (2023). A leader-follower framework for sustainable water pricing and allocation. *Water Resources Management, 37*(3), 1257–1274. Available from https://doi.org/10.1007/s11269-023-03428-w, https://www.springer.com/journal/11269.

Keeler, B. L., Derickson, K. D., Waters, H., & Walker, R. (2020). Advancing water equity demands new approaches to sustainability science. *One Earth, 2*(3), 211–213. Available from https://doi.org/10.1016/j.oneear.2020.03.003, http://www.cell.com/one-earth.

Kumar, A., Button, C., Gupta, S., & Amezaga, J. (2023). Water sensitive planning for the cities in the global south. *Water, 15*(2), 235. Available from https://doi.org/10.3390/w15020235, 2073-4441.

Li, M., Xu, Y., Fu, Q., Singh, V. P., Liu, D., & Li, T. (2020). Efficient irrigation water allocation and its impact on agricultural sustainability and water scarcity under uncertainty. *Journal of Hydrology, 586*. Available from https://doi.org/10.1016/j.jhydrol.2020.124888, http://www.elsevier.com/inca/publications/store/5/0/3/3/4/3.

Liu, G., Wang, W., & Li, K. W. (2019). Water footprint allocation under equity and efficiency considerations: a case study of the yangtze river economic belt in China. *International Journal of Environmental Research and Public Health, 16*(5). Available from https://doi.org/10.3390/ijerph16050743, https://www.mdpi.com/1660-4601/16/5/743/pdf.

Loch, A., Adamson, D., & Dumbrell, N. P. (2020). The fifth stage in water management: policy lessons for water governance. *Water Resources Research, 56*(5). Available from https://doi.org/10.1029/2019WR026714, http://agupubs.onlinelibrary.wiley.com/hub/journal/10.1002/(ISSN)1944-7973/.

López, S. T., De Los Angeles Barrionuevo, M., & Rodríguez-Labajos, B. (2021). A new operational approach for understanding water-related interactions to achieve water sustainability in growing cities. *Environment, Development and Sustainability, 25*(1), 122–137. Available from https://doi.org/10.1007/s10668-021-02045-0.

McManus, P., Pandey, C., Shrestha, K., Ojha, H., & Shrestha, S. (2021). Climate change and equitable urban water management: critical urban water zones (CUWZs) in Nepal and beyond. *Local Environment, 26*(4), 431–447. Available from https://doi.org/10.1080/13549839.2021.1892045, http://www.tandf.co.uk/journals/titles/13549839.asp.

Mourad, K. A. (2020). A water compact for sustainable water management. *Sustainability, 12*(18), 7339. Available from https://doi.org/10.3390/su12187339, 2071-1050.

Nirmale, G., Shinde, D., & Shinde, S. (2023). Bhandare, evaluation and planning of an equitable water supply for the rural water distribution network at Undri village in Kolhapur district. Maharashtra. *International Journal of Science and Research, 12*(5). Available from https://doi.org/10.21275/SR23506105358.

OECD. (2015). Water resources allocation: Sharing risks and opportunities.

Perera, E. D., Moglia, M., & Glackin, S. (2023). Beyond "community-washing": effective and sustained community collaboration in urban waterways management. *Sustainability (Switzerland), 15*(5). Available from https://doi.org/10.3390/su15054619, http://www.mdpi.com/journal/sustainability/.

Peters-Lidard, C. D., Rose, K. C., Kiang, J. E., Strobel, M. L., Anderson, M. L., Byrd, A. R., Kolian, M. J., Brekke, L. D., & Arndt, D. S. (2021). Indicators of climate change impacts on the water cycle and water management. *Climatic Change, 165*(1-2). Available from https://doi.org/10.1007/s10584-021-03057-5, http://www.wkap.nl/journalhome.htm/0165-0009.

Quiñones, R.A., Fuentes, M., Montes, R.M., Soto, D., & León-Muñoz, J. (2019). Environmental issues in Chilean salmon farming: a review. *Reviews in Aquaculture*. 375–402. Wiley-Blackwell Chile. http://onlinelibrary.wiley.com/journal/10.1111/(ISSN)1753-513111, https://doi.org/10.1111/raq.12337175351312.

Rambonilaza, T., Rulleau, B., & Assouan, E. (2023). On sharing the costs of public drinking water infrastructure renewal among users with different preferences. *Utilities Policy, 85*101661. Available from https://doi.org/10.1016/j.jup.2023.101661, 09571787.

Razzaq, A., Qing, P., Naseer, M. Au. R., Abid, M., Anwar, M., & Javed, I. (2019). Can the informal groundwater markets improve water use efficiency and equity? Evidence from a semi-arid region of Pakistan. *Science of the Total Environment, 666*, 849–857. Available from https://doi.org/10.1016/j.scitotenv.2019.02.266, http://www.elsevier.com/locate/scitotenv.

Richards, C. E., Tzachor, A., Avin, S., & Fenner, R. (2023). Rewards, risks and responsible deployment of artificial intelligence in water systems. *Nature Water, 1*(5), 422–432. Available from https://doi.org/10.1038/s44221-023-00069-6.

Sadoff, C. W., Borgomeo, E., & Uhlenbrook, S. (2020). Rethinking water for SDG 6. *Nature Sustainability, 3*(5), 346–347. Available from https://doi.org/10.1038/s41893-020-0530-9, http://www.nature.com/natsustain/.

Sanchez-Plaza, A., Broekman, A., Retana, J., Bruggeman, A., Giannakis, E., Jebari, S., Krivograd-Klemenčič, A., Libbrecht, S., Magjar, M., Robert, N., & Verkerk, P. J. (2021). Participatory evaluation of water management options for climate change adaptation in river basins. *Environments, 8*(9), 93. Available from https://doi.org/10.3390/environments8090093, 2076-3298.

Seigerman, C. K., McKay, S. K., Basilio, R., Biesel, S. A., Hallemeier, J., Mansur, A. V., Piercy, C., Rowan, S., Ubiali, B., Yeates, E., & Nelson, D. R. (2023). Operationalizing equity for integrated water resources management. *Journal of the American Water Resources Association, 59*(2), 281–298. Available from https://doi.org/10.1111/1752-1688.13086, http://onlinelibrary.wiley.com/journal/10.1111/(ISSN)1752-1688.

Tangworachai, S., Wong, W. K., & Lo, F. Y. (2023). Determinants of water consumption in Thailand: sustainable development of water resources. *Emerald Publishing, Taiwan Studies in Economics and Finance*. Available from https://doi.org/10.1108/SEF-06-2022-0310, http://www.emeraldinsight.com/info/journals/sef/sef.jsp.

Tanzi, A. M. (2020). The inter-relationship between no harm, equitable and reasonable utilisation and cooperation under international water law. *International Environmental Agreements: Politics, Law and Economics, 20*(4), 619–629. Available from https://doi.org/10.1007/s10784-020-09502-7, http://ww.kluweronline.com/issn/1567-9764.

Tariq, M. A. U. R., van de Giesen, N., Janjua, S., Shahid, M. L. U. R., & Farooq, R. (2020). An engineering perspective of water sharing issues in Pakistan. *Water (Switzerland)*, *12*(2). Available from https://doi.org/10.3390/w12020477, https://res.mdpi.com/d-attachment/water/water-12-00477/article-deploy/water-12-00477-v2.pdf.

Wan Rosely, W. I. H., & Voulvoulis, N. (2023). Systems thinking for the sustainability transformation of urban water systems. *Critical Reviews in Environmental Science and Technology*, *53*(11), 1127–1147. Available from https://doi.org/10.1080/10643389.2022.2131338, http://www.tandf.co.uk/journals/titles/10643389.asp.

Wilfong, M., Paolisso, M., Patra, D., Pavao-Zuckerman, M., & Leisnham, P. T. (2023). Shifting paradigms in stormwater management—hydrosocial relations and stormwater hydrocitizenship. *Journal of Environmental Policy and Planning*, *25*(4), 429–442. Available from https://doi.org/10.1080/1523908X.2023.2169262, http://www.tandf.co.uk/journals/titles/1523908x.html.

Wong, T. H. F., Rogers, B. C., & Brown, R. R. (2020). Transforming cities through water-sensitive principles and practices. *One Earth*, *3*(4), 436–447. Available from https://doi.org/10.1016/j.oneear.2020.09.012, http://www.cell.com/one-earth.

Xiang, X., Li, Q., Khan, S., & Khalaf, O. I. (2021). Urban water resource management for sustainable environment planning using artificial intelligence techniques. *Environmental Impact Assessment Review*, *86*, 106515. Available from https://doi.org/10.1016/j.eiar.2020.106515, 01959255.

Chapter 10

Water footprints in power generation: challenges and strategies for a sustainable environment

Fayaz A. Malla[1,*], Sonia Grover[2], Afzal Hussain[2], Mohamed F. Alajmi[2], Waseem A. Wani[3], Afaan A. Malla[4], Nazir A. Sofi[5] and Foozia Majeed[6]

[1]*Department of Environmental Science, Govt. Degree College Tral, Jammu and Kashmir, India,* [2]*Department of Pharmacognosy, College of Pharmacy, King Saud University, Riyadh, Saudi Arabia,* [3]*Department of Chemistry, Govt. Degree College Tral, Kashmir, Jammu and Kashmir, India,* [4]*Department of Environmental Science, Government Degree College Baramulla, Indira Gandhi National Open University, Baramulla, Jammu and Kashmir, India,* [5]*Department of Agriculture Research Information System, Sher-e-Kashmir University of Agricultural Sciences and Technology, Shalimar Campus, Srinagar, Jammu and Kashmir, India,* [6]*Conservation Biology Lab Sos in Zoology Jiwaji University Gwalior, Gwalior, Madhya Pradesh, India*

**Corresponding author. e-mail address: nami.fayaz@gmail.com.*

10.1 Introduction

The concept of a water footprint is an important framework for analyzing the total amount of water utilized directly and indirectly in the creation of goods and services. This can be done by looking at a product's "water footprint." It offers an all-encompassing perspective on the effect that human activities have had, both in terms of consumption and pollution, on the world's water supplies. Blue, green, and gray are the three different sorts of water footprints that can be left behind.

10.1.1 Blue water footprint

The volume of freshwater taken in during the process is represented by the blue water footprint. This includes both surface water and groundwater. It encompasses water that has been extracted from bodies of water such as rivers, lakes, and aquifers for uses such as irrigation, manufacturing, and household use (Senthil Kumar & Janet Joshiba, 2019). The blue water footprint is a crucial indication for determining the severity of water shortage and the level of rivalry for water among various industries. It is a reflection of the physical usage of water resources.

10.1.2 Green water footprint

The amount of precipitation that is stored in the soil and used for agricultural purposes is factored into the "green water footprint." This is especially important in farming, as rain is essential for the survival of rain-fed crops like wheat, maize, and rice. To optimize agricultural practices for sustainable water use and crop yields, knowledge of the green water footprint is vital (Olivera Rodriguez et al., 2021).

10.1.3 Gray water footprint

The volume of water needed to dilute contaminants from an industrial process to acceptable levels is estimated by the greywater footprint (Dong *et al.*, 2021). It is a significant indicator for gauging the ecological sustainability of various activities, particularly in industries and wastewater management, as it quantifies the impact of pollution on the environment. The pollution of both surface and groundwater is reflected in what is known as the "greywater footprint."

Water Footprints and Sustainable Development. DOI: https://doi.org/10.1016/B978-0-443-23631-0.00010-8

Differentiating between these three water footprints provides stakeholders with a more complete picture of how water is used and impacted by human activities, allowing for more responsible management and distribution of water resources. The energy industry relies heavily on water because of its long history of use in power generation. Hydroelectric dams, thermal power plants, and nuclear reactors are all examples of the numerous systems that contribute to its ability to generate electricity (Abdelkareem *et al.*, 2018). The importance of water all over the world is discussed, along with the numerous techniques and real-world applications that have emerged as a result.

10.1.4 Hydropower generation

Hydroelectricity provides for about 16% of global electricity production and is the greatest renewable energy source worldwide (Strielkowski *et al.*, 2021). Electricity is produced by hydropower facilities by harnessing the kinetic energy of water in motion. Electricity is produced when water is released from behind a dam and passes through a turbine. One of the most well-known and eco-friendly ways to produce electricity is through hydropower. It generates power by utilizing the water's potential and kinetic energy. The Hoover Dam and the Three Gorges Reservoir in China are two of the most well-known examples of large dams and reservoirs in the world. The electricity generated by these dams is an important part of the regional energy mix. They can also be used to store water for irrigation, flood prevention, and even leisure activities. The Three Gorges Dam contains the world's most extensive hydroelectric facility. It can produce 22,500 MW of electricity (Moragoda *et al.*, 2023). The Yangtze River's kinetic energy is converted into power at the dam. Hoover Structure, on the Colorado River, is a massive structure that generates electricity. It can produce 2080 MW of electricity. The dam harnesses the potential energy of Lake Mead's falling water to produce power (Karambelkar, 2018).

10.1.5 Thermal power plants

Many thermal power plants rely on water for their cooling systems, particularly those that run on fossil fuels or nuclear reactors. The water is used to absorb the heat produced by the power generation process, serving as a coolant. This hot water is then discharged into waterways like rivers and lakes, where it can harm aquatic life. Nuclear power stations like Arizona's Palo Verde Nuclear Generating Station consume massive amounts of water for cooling systems (Epiney *et al.*, 2019).

10.1.6 Geothermal power generation

Geothermal energy uses steam or hot water to tap into the Earth's internal heat. It has a little carbon footprint and a very long life span. The abundance of geothermal resources has led to the widespread use of geothermal electricity in countries like Iceland and New Zealand. Wairakei Power Station in New Zealand and the Blue Lagoon Geothermal Power Plant in Iceland are two examples of power plants that successfully use hot water to generate electricity (Nyaga, 2019).

10.1.7 Ocean and tidal energy

Ocean and tidal energy, in which the flow of water is converted into electricity, further demonstrates water's importance. The Sihwa Lake Tidal Power Station in South Korea is an example of a tidal power plant that generates electricity from the ebb and flow of the tides (Park & Lee, 2021). One more novel method that makes use of ocean water temperature differences is ocean thermal energy conversion. These strategies hold great promise as a source of sustainable power.

10.1.8 Solar desalination

Solar desalination, the process of turning salty ocean water into drinkable fresh water using only the power of the sun, can not happen without water. This not only helps with water scarcity but also helps with power generation in a roundabout way by reducing the need for as much energy as conventional desalination methods. Desalination plants like Jordan's MEDRC and Saudi Arabia's Al Khafji solar desalination plants are two prominent examples (Kharraz *et al.*, 2017).

10.1.9 Biomass power plants

To turn biomass into usable electricity, many biomass power plants require large amounts of water. These facilities may require water for the conveyance of feedstock or the operation of cooling systems. One such biomass power plant that makes effective use of water is the Stockholm Exergi Värtaverket facility in Sweden (Dittrich & Lillieroth, 2019).

10.1.10 Nuclear power

Nuclear fission in power plants produces heat, which is used to boil water into steam, which is then used to turn a turbine and produce electricity (Zohuri, 2019). In a nuclear power plant, water is utilized to keep the reactor at a safe temperature. In 2011 a tsunami hit Japan and caused extensive damage to the Fukushima Daiichi Nuclear Power Plant. The plant's reactors were cooled by water circulation. The cooling systems at the plant were destroyed by the tsunami, leading to reactor overheating.

There is no denying that the growing worries about water scarcity are a worldwide dilemma with far-reaching effects. Water scarcity impacts many different sectors, including agriculture, urbanization, climate change, and international relations, as these instances show. Effective water management, conservation, and the introduction of sustainable agricultural and urban design practices are only some of the measures that must be taken to lessen the impact of water scarcity. For the sake of present and future generations, this issue must be resolved so that everyone has fair access to water.

The rising global demand for energy, the rising worries about water scarcity, and the urgency to shift to more sustainable energy sources all highlight the importance of water footprint evaluation in power generation. Understanding the intricate relationship between water supply and electricity production is impossible without water footprint evaluations. The World Resources Institute found that in 2021, thermoelectric power generation consumed 41% of all water withdrawn in the United States, demonstrating the industry's high water demand (Jin *et al.*, 2022). The Middle East is an example of a water-scarce region where environmental degradation and heightened geopolitical tensions have resulted from competition between the energy industry and agriculture for scarce water resources (Ward & Ruckstuhl, 2017). Furthermore, the implementation of dry cooling systems at the Ivanpah Solar Power Facility in California is an example of how water footprint assessments motivate development in the energy sector (Nouri, *et al.*, 2019). To ensure a sustainable future for power production while protecting water resources for larger societal demands, these evaluations provide a solid foundation for policymakers and industry stakeholders to maximize water usage efficiency and minimize environmental impacts.

Multiple goals can be attained by writing a chapter on "Water Footprints in Power Generation" for a book. The primary goal of this chapter is to present a high-level overview of the connection between water use and electricity production, illuminating the many energy-related processes and technologies that rely on water supplies. Second, it hopes to educate people on the notion of "water footprints" and its importance in gauging the sustainability of energy production, as well as the environmental and social repercussions of water usage in power generation. This chapter also makes an effort to use international examples and case studies to demonstrate the many consequences and difficulties related to water use in power generation. The goal of this chapter is to add to the body of knowledge on this important topic so that policymakers, researchers, and industry stakeholders may make more effective judgments and adopt more environment-friendly procedures in the arena of electricity generation.

10.2 Water use in traditional power generation

10.2.1 Fossil fuel power plants

To function properly, power plants rely heavily on water, especially those that use fossil fuels like coal and natural gas. It plays an essential role in cooling, producing steam, controlling pollutants, and other power-plant operations. The importance of water and the various ways it is used in coal and natural gas power plants are discussed here, along with some global case studies.

10.2.1.1 Water requirements for coal power plants

10.2.1.1.1 Cooling systems

Water is typically used for cooling purposes in coal power plants. The most typical techniques are once-through systems and cooling towers (CTs). Water from a local source is extracted, cooled, and then discharged back into the system, where its temperature may have increased (Kehinde, 2019). The Drax Power Station in the United Kingdom, for instance, uses once-through cooling and draws a lot of water from the River Ouse.

10.2.1.1.2 Steam generation

Coal-fired power plants rely heavily on steam generation, which requires an abundance of water. Its main purpose is to generate steam under high pressure, which then turns the turbines. Power plants like Taiwan's Taichung Power Plant, one of the world's largest coal-fired facilities, use vast volumes of water in this process (Pioro & Duffey, 2019).

10.2.1.1.3 Emissions control

Flue gas desulfurization (FGD) and selective catalytic reduction (SCR) are two examples of emissions control techniques used by power plants that involve the use of water to reduce pollution. A lot of water is needed for these systems. The W.A. Parish Power Plant in Texas uses FGD to reduce sulfur dioxide emissions; this process requires large amounts of water (Kaur, 2020).

10.2.1.2 Water requirements for natural gas power plants

10.2.1.2.1 Cooling systems

Natural gas power facilities, like coal power plants, require water for cooling the gas turbines to keep them running efficiently. Cooling at Kuwait's Sabiya Power Plant is provided by seawater, with once-through pipes discharging the heated water back into the Arabian Gulf (Prata & Simões-Moreira, 2019).

10.2.1.2.2 Combined cycle plants

Water is used to produce steam in combined cycle natural gas power plants, which use both gas and steam turbines. One such combined cycle facility that uses water for both cooling and steam generation is the Marcellus Power facility in Pennsylvania, United States (Mallapragada *et al.*, 2018).

10.2.1.2.3 Emissions control

Emissions control using water is common in natural gas power plants as well, especially when doing combustion adjustments or implementing emission reduction technology like SCR systems (Mac Kinnon et al., 2018). The California power plant in El Segundo uses water for both cooling and reducing emissions.

Coal and natural gas power plants have large water needs, which is a cause for alarm in arid and semiarid locations. Water usage in power generation is an issue that needs to be addressed because of the potential negative effects on the environment, including thermal pollution from once-through cooling systems, pollution, and damage to local ecosystems. As the need for electricity grows around the world, the energy industry must adopt more eco-friendly methods that reduce water use and soften their environmental impact. The supplied examples highlight the necessity for efficient water management and ecologically responsible solutions in power generation by illustrating the different water needs of coal and natural gas power plants.

Water use is reduced at natural gas power plants compared to coal facilities. Natural gas plants consume 584% less water than conventional plants, per data from the US Energy Information Administration (EIA) (Dufour, 2018). The water extraction intensity of natural gas combined-cycle production in 2020 was 2793 gal $(\mathrm{MW\ h})^{-1}$, while that of coal was 21,406.

The average amount of water used by a coal power station to generate 1 MW h of electricity is 19,185 gal. A maximum of 3.5 m^3 of water per MWh in water use was mandated for plants in India that were operational before January 1, 2017. The 3 m^3 $(\mathrm{MW\ h})^{-1}$ standard was enforced for plants built after January 1, 2017. Compared to coal, the technology used in natural gas plants is more efficient (Chaturvedi et al., 2018).

10.2.2 Nuclear power plants

By using the heat created by nuclear fission to generate electricity, nuclear power plants play a crucial role in the world's overall energy production. Water is essential to the operation of these facilities in several ways, including cooling, steam generation, radiation shielding, and safety features. In this note, we will look at the unique water needs of nuclear power plants, as well as their relevance and potential environmental impacts.

10.2.2.1 Cooling systems

To dissipate the massive heat produced during the nuclear fission process, nuclear power facilities use a variety of cooling technologies. Once-through chillers and CTs make up the bulk of the cooling equipment used today. Once-through cooling involves removing water from its source, passing it through the reactor to absorb heat, and then releasing it back into its original location. On the other hand, CTs dissipate heat by evaporation into the air. Diablo Canyon Nuclear Power Plant in California is only one example of a nuclear power plant that uses CTs to conserve water (Nelson & Ramana, 2023).

10.2.2.2 Steam generation

A nuclear power plant cannot generate steam without access to water. Power is generated by using the high-pressure steam created by the reactor's heat to turn turbines. Since water is constantly being recycled between the reactor and the turbines, it is crucial to have a reliable source of clean, well-regulated water. There is a high water cost associated with this procedure.

10.2.2.3 Radiation shielding and safety

In nuclear power facilities, water is also used as a radiation barrier. It serves as a shield around the reactor core to prevent leakage of radioactive material. Emergency cooling systems utilized in the case of a reactor breakdown or overheating rely heavily on water.

There are large and potentially dangerous water needs for nuclear power facilities. Thermal pollution and an increase in water temperature are two ways that once-through cooling systems can damage aquatic habitats. In addition, strict safety procedures and monitoring are required due to the possibility of radioactive contamination of cooling water. Some nuclear power facilities have implemented measures to reduce water consumption in response to these worries. Some buildings, for instance, have replaced inefficient once-through cooling with modern CTs. Pressurized water reactors and boiling water reactors are two examples of how reactor technology has improved water recycling and conservation (Morales Pedraza, 2017).

The water needs of nuclear power plants are very high. One nuclear reactor uses anything from 1514 to 2725 L of water per megawatt-hour (MWh), according to the Nuclear Energy Institute (Vaseashta, 2021). This amounts to millions of cubic feet of water each year. Water use of nuclear power plants is 20%–83% more than that of coal-fired plants. Depending on the size of the reactor, water usage can range from 35 to 65 million liters per day, or 13–24 billion liters annually. There are 104 reactors in operation in the United States, and 60% of them employ once-through cooling while 40% use closed-cycle cooling (Pan *et al.*, 2018). To date, all planned new plants have adopted the usage of closed-cycle cooling. The concrete pool in a nuclear power plant is normally roughly 40 ft deep. In most situations, there is more than 25 ft of water above the spent fuel, which means that it is entirely submerged (Wolfson & Dalnoki-Veress, 2021).

10.3 Renewable energy and water footprints

10.3.1 Hydroelectric power

Energy from moving water is converted into electricity through hydroelectric power generation. This water-based renewable energy source is both eco-friendly and sustainable.

10.3.1.1 Water flow and head

The water's volume and flow rate are crucial in hydroelectric power generation. To operate turbines and produce energy, there needs to be a sufficient volume of water available.

Power generation is also affected by the "head," or the vertical decrease in water level from the dam to the turbine. More of your body's gravitational potential energy can be converted into electricity if you raise your head higher.

10.3.1.2 Reservoirs and flow control

Reservoirs, which are made by damming rivers, are common components of hydroelectric power facilities. These water storage reservoirs enable more manageable power generation.

Power plants can create electricity on demand by controlling the flow of water from the reservoir, which helps to satisfy varying energy demands. The largest concrete construction in the United States, the Grand Coulee Dam, is an example of how reservoirs can be used to generate energy (Jaramillo, 2022).

Ecosystems in rivers, habitats, and the natural flow of water may all be affected by hydroelectric power facilities. One of the greatest hydroelectric projects in the world, the Three Gorges Dam in China has been linked to environmental issues due to its effects on aquatic habitats and on populations that were forced to relocate.

10.3.1.3 Hydrological administration

Optimizing hydroelectric power generation relies heavily on effective water management. Water for power generation must be carefully planned to meet environmental and downstream water needs without interruption. An example of efficient water management in hydroelectric power generation is the Hoover Dam on the Colorado River in the United States, which serves multiple purposes (including generating electricity, supplying water, and preventing flooding) (Karambelkar, 2018). Not all hydroelectric plants require large storage reservoirs; these are called "run-of-river" projects. Instead, they go with the river's flow to cause as little ecological impact as possible. One such facility is the Shoshone Falls Hydroelectric Plant in Idaho, United States.

The average amount of water used to produce 1 kWh of electricity from hydropower is calculated to be 63.8 L. This comprises both multipurpose and dedicated hydroelectric dams, as well as run-of-river (ROR) facilities.

The water consumption for each type of facility is:

- ROR facilities: Negligible
- Dedicated reservoirs: 10.2 gal $(kW\ h)^{-1}$
- Multipurpose reservoirs: 22.7 gal $(kW\ h)^{-1}$

For both thermoelectric and hydroelectric power generation, the national weighted average of water use is 2.0 gal (7.6 L) of evaporated water per kWh of energy (Massetti, et al., 2017).

10.3.2 Solar power

Solar panels can be used to conserve water in the long run. They can run entirely on dry fuel, and they need neither water nor air to stay cool.

About 20 gal of water per megawatt-hour (gal $(MW\ h)^{-1}$) are needed for cleaning solar gathering and reflection surfaces in solar power technology. About 2.5 L of water each cleaning cycle is needed for each panel (Boeing, 2018). Plants' operational and maintenance water needs are scale- and location-specific. The projected amount of water needed for operation and maintenance in India ranges from 7000 to 20,000 L MW^{-1} per wash (Lahmouri *et al.*, 2019).

The effects of solar power plants on nearby water supplies are similar to and frequently smaller than, those of conventional fossil fuel power plants. Different energy processes and technologies have vastly different requirements for both the quantity and quality of water. Solar power does not release any harmful gases like carbon dioxide or other "greenhouse" gases when it is used. The harmful effects on the environment from extracting fossil fuels like mining or drilling are avoided. Drilling for fossil fuels causes soil degradation, erosion, and water pollution.

10.3.3 Wind power

Electricity generated by wind turbines does not require the use of water. However, water is needed for the production of wind turbines and the upkeep of existing machinery. Water consumption per kilowatt hour (L $(kW\ h)^{-1}$) for wind turbines is negligible to quite low at 0.64 L. Water consumption ranges from 0.2% to 245% for hydropower, 0% to 0.11% for solar PV, and 2.5% to 6.8% for geothermal power (Jia *et al.*, 2021). Wind turbines do not contribute to air or water pollution because they do not produce any pollutants. They can stay cool without any water. Wind turbines have the potential to decrease the use of fossil fuels in electricity generation, hence decreasing overall pollution levels and carbon dioxide output.

10.4 Advanced power generation technologies

10.4.1 Carbon capture and storage

When combined with carbon capture and storage (CCS), advanced power generation technologies offer the potential to reduce water use in the energy industry. Water is used more efficiently in cooling and other processes because of technology like supercritical and ultrasupercritical steam cycles in coal-fired power plants and combined cycle gas turbine systems in natural gas power plants (Tramošljika *et al.*, 2021). Water utilized in the capture process can also be captured and reused thanks to CCS technology. To effectively capture carbon emissions while decreasing overall water use, the Kemper County Energy Facility in Mississippi, United States, uses both cutting-edge power generating and CCS (Nomoto, 2017). The energy sector may aid in sustainable water management and lessen the ecological impact of power generation by optimizing water consumption and reusing water resources inside these cutting-edge technologies.

10.4.2 Advanced nuclear technologies

The nuclear power industry might drastically improve its water use efficiency with the use of advanced nuclear technologies, including next-generation reactors and novel cooling systems. Large amounts of water are needed for cooling purposes in conventional nuclear reactors, which can put a burden on local water supplies and ecosystems. High-temperature gas-cooled reactors and molten salt reactors are two examples of modern nuclear technology that run at greater temperatures, using less water for cooling (Ho *et al.*, 2019). For instance, the sodium-based coolant used in Terra Power's Natrium reactor, which was developed in collaboration with Bill Gates and Warren Buffet, significantly improves safety while significantly reducing water use. These emerging nuclear technologies hold great promise for a more environment-friendly and water-wise nuclear energy future.

10.5 Managing water footprints in power generation

10.5.1 Water-energy nexus

When it comes to the long-term viability and robustness of contemporary societies, the Water-energy nexus is an essential component of power generation. Water is used in many steps of power generation, while energy is needed to extract, transport, treat, and distribute water, highlighting the complex connection between these two resources. One of the oldest and most used renewable energy sources, hydropower, uses the kinetic energy of moving water to create electricity. In this situation, water is the resource from which energy is generated. Thermoelectric power plants, which rely mostly on fossil fuels, require significant amounts of water for cooling. Water is also needed in large quantities for other purposes at nuclear power plants, such as cooling reactors. These instances highlight the importance of water in energy production and the need for energy in water-related procedures.

When we think about the environmental repercussions and the issues brought by climate change, the water-energy nexus becomes even clearer. Power plants, especially thermoelectric and nuclear facilities, can be hampered by water scarcity and temperature changes due to climate change, resulting in decreased power generation and increased operational costs. Alterations in precipitation patterns and diminished snowmelt in mountainous locations, which affect water availability for electricity generation, provide difficulties for hydropower. Furthermore, thermal discharge or contamination from fossil fuel combustion, two power generation operations that pollute water bodies, can affect aquatic ecosystems and compromise water quality for human consumption.

Policymakers, engineers, and academics need to take a systemic approach to the water-energy nexus, one that maximizes water efficiency and reduces energy use in the water sector while simultaneously encouraging energy savings and cleaner technology in the power sector. To reduce negative environmental effects and guarantee the long-term viability of energy production in a world with scarcer supplies of water, an interdisciplinary approach is essential.

10.5.2 Water footprint assessment

To assess and lessen the negative effects on the environment from the use of water needed to generate electricity, a water footprint assessment (WFA) is an essential analytical tool in the power generation sector. This technique calculates the total amount of freshwater needed to produce 1 kWh of electricity, including all steps in the process from water extraction to water transportation to water treatment to release (Hoekstra, 2017). Coal-fired power plants are an essential part of many energy grids, but they have a terrible reputation for the amount of water they waste. Coal-based electricity production has negative implications on local aquatic ecosystems and water quality due to the high water use of coal mining and the high water use of cooling processes. This results in significant freshwater withdrawals and thermal discharge into nearby water bodies. However, renewable energy sources like wind and solar have far smaller water footprints because they do not require water to produce electricity. These two instances underline the need of addressing the environmental effects of water use in power generation and the potential of WFA to steer policymakers and industry stakeholders towards sustainable energy options (Kelly *et al.*, 2019). To better understand the water-energy nexus, to enable evidence-based decision-making, and to steer the power generation sector towards more environmentally responsible practices in a world grappling with escalating water scarcity and climate change impacts, the WFA methodology must be further developed and applied.

10.5.3 Sustainable power generation

For thermal power plants, efficient water use is necessary to abide by government regulations. Based on the notification issued by the Ministry of Environment Forest and Climate Change, all the existing thermal power plants are expected to reduce specific water consumption by up to 3.5 m^3 $(MW\,h)^{-1}$, while the new ones established after 2017 are expected to have specific water consumption of 2.5 m^3 $(MW\,h)^{-1}$ (Sugandha, 2021).

There are direct and indirect ways of potentially reducing specific water consumption in thermal power plants. Thermal power plants need significant volumes of water for cooling purposes ash handling and demineralized water. Besides, this water is also used for drinking, coal handling, fire-fighting purposes, and sundry others.

A significant volume of water is used in CT to dissipate the heat of the hot water received from the condensers. Cooling accounts for the greatest use of water in electrical generation operations, but the amount of water used depends upon the type of cooling system used.

Once-through is a cheaper option in economic terms but heavier in water use terms. It has an environmental impact also and thus there has been a shift from this process to a closed-cycle system. Closed-cycle cooling systems use significantly less water than once-through systems, making them a preferred option.

A significant amount of water is used in the power plants for handling the ash generated from the combustion of coal. Water is used to carry the ash from the plant to the far-off ash dykes in the form of slurry. Water is often lost in significant quantities in this process

Some of the key approaches to reduce water use in the two processes (cooling and ash handling) are presented next.

10.5.3.1 Dry cooling system

A portion of the circulating water evaporates as hot water is cooled in a traditional wet CT through direct mixing with ambient air and make-up water is needed to compensate for this loss. Dry cooling systems do not need replenishment water because they reject heat to the atmosphere via sensible cooling.

Direct dry cooling systems and indirect dry cooling systems are the two main types of dry cooling systems. The direct dry cooling system employs mechanical draught fans or a natural draught hyperbolic tower to cool the exhaust steam from the LP turbine with air from the surrounding environment. Indirect dry cooling systems use a combination of a surface or jet condenser to cool the turbine's exhaust steam and finned tube bundles to chill the hot water with the use of mechanical draught fans or a naturally drafted hyperbolic tower.

Other than a direct cooling system which has feasibility conditions, water consumption in the CT can be decreased by increasing the cycle of concentration through chemical treatment and improvements in the operation and and maintenance of the CTs.

10.5.3.2 Dry bottom ash handling

Historically, most slurry made up of the plant's fly ash and bottom ash was dumped into the ash pond. Ash generation in thermal plants has been linked to environmental issues, prompting several steps to be implemented, including the dry disposal of fly ash and the reduction of water needed for wet ash disposal. However, the dry bottom ash handling has the potential to decrease the water consumption in this process to nill. It does not involve the use of any water. In ash handling, one method is to decrease water consumption by increasing the recirculation of water from ash dykes, or, switching to dry handling of ash which requires no water

10.5.3.3 Reusing/recycling water in the plant from the various process

Other than the abovementioned approaches, the wastewater generated can be reused and recycled for various activities within the plant like for horticulture or the ash handling process either directly or through regeneration (Bohra et al., 2022).

10.6 Future trends and challenges

10.6.1 Emerging technologies

Future power-generating technologies are an innovative and game-changing part of the global energy landscape that is driven by the need for long-term sustainability. The importance of modern materials and processes in shaping the future of energy production must be emphasized in this academic setting. Next-generation photovoltaics, in particular perovskite solar cells, are a prime example of these developments; thanks to their low production costs, adaptability, and increased energy efficiency, they have the potential to completely transform solar energy conversion. In addition, there is hope that intermittency problems with renewable energy sources can be resolved by the creation of improved energy storage technologies, such as solid-state batteries and flow batteries, which would provide a consistent and reliable energy supply. The development of small modular reactors in the nuclear industry also represents a step forward in terms of safer and more efficient nuclear energy generation; these reactors can be easily scaled up to meet the world's expanding energy demands without posing as great a threat to the environment. Innovative geothermal energy technologies, such as enhanced geothermal systems and binary cycle power plants, hold the potential to develop these untapped geothermal resources and turn them into a reliable, low-carbon electricity supply. The potential for cleaner, more efficient, and sustainable power production systems is demonstrated by these cutting-edge technologies. Such systems are necessary to meet the problems of climate change and the ever-increasing global energy demand. Successful integration of these emerging technologies into the future energy landscape is dependent on extensive research and development in these areas, as well as strong policy and investment support, ushering in a new era of environmentally responsible and resilient power generation.

10.6.2 Water scarcity and climate change

With its rapidly decreasing freshwater resources as per the estimates, India has about 16% of the global population but it has only 4% of the total water resource. Approximately 85% of the water is used by the agriculture sector followed by industries at 9% and the domestic sector at 6%. The power sector among all industries is the major user of fresh water at 87.8%. Water and electricity generation are part of the intricate nexus of water and energy. Both resources are heavily interdependent. On the one hand, electricity is needed to access water, while on the other hand, water is needed for electricity generation.

However, with the changing climate, stress on water resources is increasing and given that the National Water Policy accords priority to drinking water before any other use, the functioning of power plants would be impacted. There have been some examples in the past where power plants were shut down due to water scarcity. One such example is of 1130 MW Parli power station in Maharashtra state which was shut down in July 2015 due to lack of water (Luo *et al.*, 2018).

A study by WRI for India reported that 40% of the country's thermal power plants are located in areas facing high water stress (Fig. 10.1), a problem since these plants use water for cooling (Zhang *et al.*, 2021).

The projected climate change impacts suggest that temperatures would rise and the rainfall pattern would be variable. One of the significant impacts of climate change would be an increase in the frequency of extreme events such as droughts, floods, etc. In 2016 India reported the highest number of deaths due to extreme weather (2119 fatalities) and suffered losses of more than $ 21 billion in property damage.

A total of 20 river basins of India are water-stressed due to increasing water demand from various sectors, including the industrial sector which comprises the energy sector (Thokchom, 2020). With growing demand and changing climate, water in these basins would decline at a rapid pace which would pose a serious threat to the social and economic development of any nation. With ongoing climate change, the competition between the different water users and users will increase and thus there is a need for all the users to become resilient and adaptive.

As per the International Energy Agency projections, an average global temperature increase of 2°C by 2050 compared to preindustrial levels will change demand for cooling and heating enormously (Méjean *et al.*, 2019). This would increase the demand for cooling significantly. The largest change in cooling demand as a result of climate change would be in China, followed by the United States, the Middle East, and India. Fig. 10.2 presents an indicative summary of the potential impacts of climate hazards on the electricity supply chain. It is important to note that the magnitude of impact will vary by location and by event (WBCSD, 2014).

Every scientific report is proving that human-induced climate change is growing stronger. There are efforts taken to mitigate rising temperatures but equally important is to learn how to adapt to the consequences of global warming.

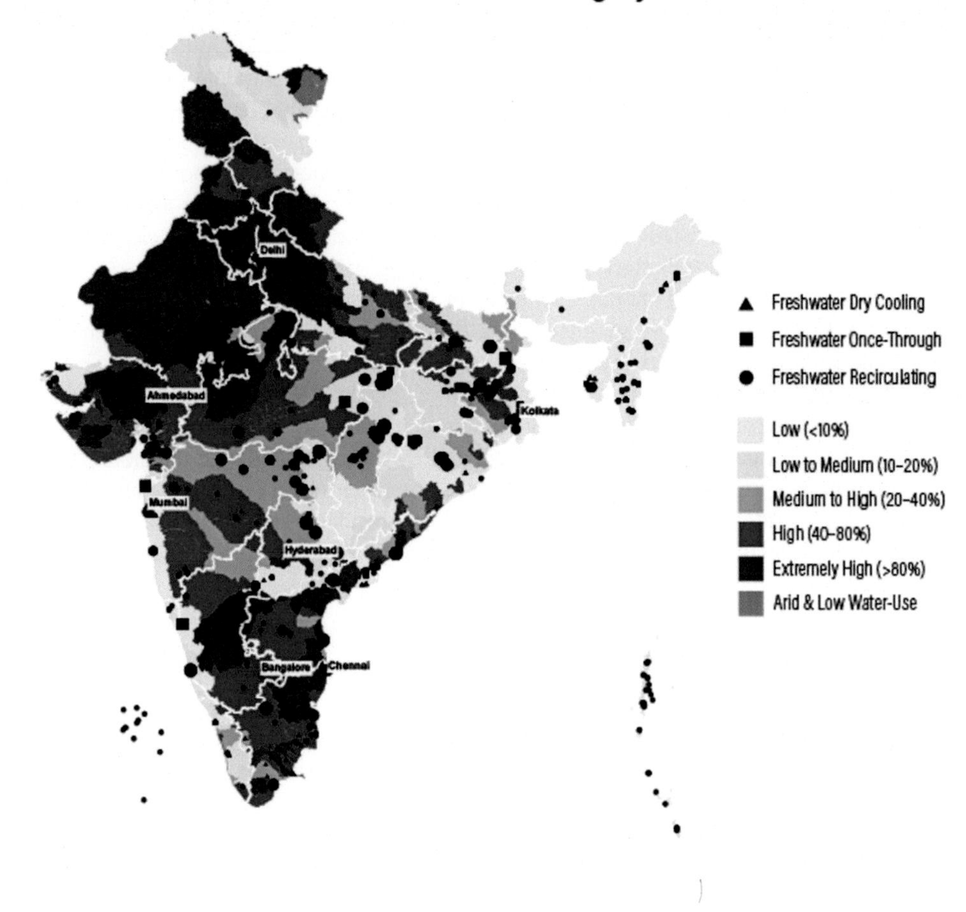

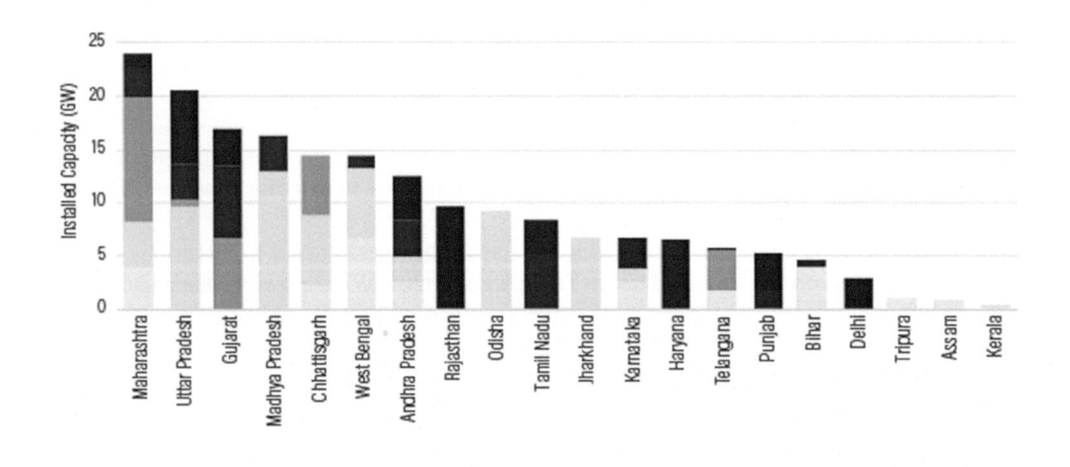

FIGURE 10.1 Thermal power plants and water stress areas. *World Resources Institute.*

	Generation				T&D		Customers
	Thermal	Hydro	Wind/PV	Biomass	Lines	Stations	
Air temperature	●	●		●			●
Water temperature	●			●			
Water availability	●	●	●	●			
Wind speed			●		●		
Sea level	●	●	●	●	●	●	
Floods	●	●	●	●	●	●	
Heat waves	●		●			●	●
Drought	●	●		●			
Storms					●	●	
	Impacts						

FIGURE 10.2 Potential impacts of climate hazards on the electricity supply chain. *Adapted from "Building a Resilient Power Sector, WBCSD, 2014."*

TABLE 10.1 Adaptation strategies for thermal power generation.

Climate change effect	Likely Impacts	Responses
Higher temperatures	Effect efficiency	Design production facilities less sensitive to air and water temperature. Build new plants in locations with lower temperatures. Decentralize generation. Avoid refueling nuclear plant and plant maintenance during summer
	Output may be affected by regulatory limits on the temperature of discharged cooling water	Build a "helper" cooling tower, if space is available. Redesign the cooling system to increase the pumping capacity and achieve a higher cooling flow rate.
Inadequate water supplies for cooling	Affect generating efficiency	Increase water efficiency, using recirculation, dry (air-cooled) or hybrid cooling – offsetting increased consumption by improving overall generation efficiency. Use lower-quality water resources
Raised sea levels and flooding	Damage to coastal infrastructure	Coastal sites remain preferable because seawater temperature is more stable, but any development design will need to raise the level of structures, including flood defences, improved drainage and protection for fuel storage
Other extreme weather events	Damage to infrastructure	Apply higher structural standards, anticipating gradual sea level rise, more storm events, and associated tidal surges.

Source: Adapted from "Building a Resilient Power Sector, WBCSD, 2014."

Thus, for the power sector, it is important to understand the risks and then prepare a response as a resilience plan. Risks are generally in the form of extreme events like high winds, heat waves, heavy downpours, etc. Heatwave is one of the major risks to infrastructure. It also impacts water temperature and water availability and on the demand side, it increases cooling demand as customers respond to increasing temperatures.

One of the responsive or adaptation strategies (Table 10.1) for such risks includes crisis planning with a focus on daily operations. It is advanced preparation of emergency plans, including lessons learned from previous crises.

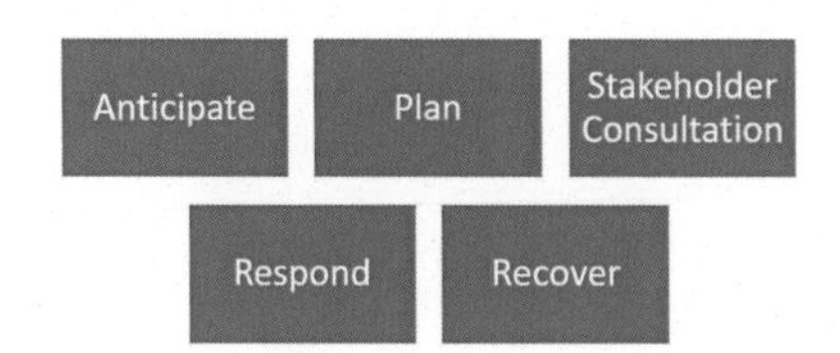

FIGURE 10.3 Response to extreme events. *Adapted from "Building a Resilient Power Sector, WBCSD, 2014."*

In the long run, changes in temperature and precipitation would impact the value chain of power production. It needs comprehensive long-term planning to identify the requirements due to potential impacts and scenarios of climate change. Management has to make decisions regarding the lifetime of assets, whether retrofitting or refurbishment is needed, etc. Such planning would help the power plants to become resilient and better adapt to changing climate.

In general, the process for responding to extreme events is as presented in Fig. 10.3.

10.6.3 Policy and regulation

The government plays a crucial role in the sustainable management and operation of power production. However, there is a need for all the key stakeholders to work on building resilient power-generating systems that are water efficient or taking a step forward can become water neutral power-generating units. Stakeholders like the industry need to build their expertise in analyzing climate information to better understand risks, especially downscaling global climate models to a more local level and also in undertaking risk assessment for developing adaptation plans. Industry could take up initiatives of water conservation in the watersheds they are based in and can offset their water footprints by undertaking various water conservation measures outside their premises and set an example of being water-neutral or water-positive power generating utility. The earlier section discussed ways to efficiently use water in electricity production. But, the power plants can go outside their boundary into the watershed to offset their water footprints and can take up interventions such as rainwater harvesting, groundwater recharge, pond rejuvenation, or promoting water-efficient practices in the watershed such as microirrigation for agricultural users. Through such interventions, power plants can augment the freshwater availability in their watershed and reduce the competition between the users. This would also help in developing cordial social relationships with the competing users. While policymakers should back a climate-resilient corporate model that encourages utilities to spend in adaptation by providing financial incentives. The government should encourage water-neutral or water-positive power-generating units.

10.7 Conclusions

A water footprint is a framework for analyzing the total amount of water used in the creation of goods and services, consisting of three types: blue, green, and gray. It is essential for responsible management and distribution of water resources. Water scarcity is a global issue affecting sectors like agriculture, urbanization, climate change, and international relations. To address this issue, effective water management, conservation, and sustainable agricultural and urban design practices are essential. Coal and natural gas power plants rely heavily on water for cooling, steam generation, emissions control, and other operations. Nuclear power plants use heat from nuclear fission to generate electricity, requiring significant water resources for cooling, steam generation, radiation shielding, and safety features. Hydroelectric power generation relies on effective water management to meet environmental and downstream water needs without interruption. The water-energy nexus is crucial for the long-term viability of contemporary societies, as water is used in various steps of power generation, including hydropower and nuclear power plants. Climate change can exacerbate this issue, causing water scarcity and temperature changes, affecting power generation and operational costs. A systemic approach is needed to maximize water efficiency and reduce energy use in the water sector while encouraging cleaner technology. Emerging technologies like next-generation photovoltaics, solid-state batteries, and flow batteries are transforming the global energy landscape, offering cleaner, more efficient, and sustainable power production systems. However, India faces water scarcity and climate change, with 32% of coal-fired power plants globally exposed to water scarcity for at least five months per year (Abdelkareem et al., 2018; Boeing, 2018; Chaturvedi, 2018; Dong et al., 2021; Hoekstra, 2017; Jaramillo, 2022; Jia et al., 2021; Jin et al., 2022; Karambelkar, 2018; Kaur, 2020; Kehinde, 2019; Kelly et al., 2019; Lahmouri et al., 2019; Luo et al., 2018; Mac Kinnon et al., 2018; Mallapragada et al., 2018; Méjean et al., 2019; Moragoda et al., 2023; Nelson & Ramana, 2023; Nouri et al., 2019; Olivera Rodriguez et al., 2021; Pan et al., 2018; Park & Lee, 2021; Pioro & Duffey, 2019; Prata & Simões-Moreira, 2019; Senthil Kumar & Janet Joshiba, 2019; Strielkowski et al., 2021; Sugandha, 2021; Thokchom, 2020; Tramošljika et al., 2021;

Vaseashta, 2021; Ward & Ruckstuhl, 2017; Wolfson & Dalnoki-Veress, 2021; Zhang et al., 2021; Zohuri, 2019; Bohra, et al., 2021; Dittrich & Lillieroth, 2019; Dufour, 2018; Epiney, et al., 2019; Ho, et al., 2019; Kharraz, et al., 2017; Massetti, 2017; Morales Pedraza, 2017; Nomoto, 2017; Nyaga, 2019).

References

Abdelkareem, M. A., El Haj Assad, M., Sayed, E. T., & Soudan, B. (2018). Recent progress in the use of renewable energy sources to power water desalination plants. *Desalination, 435*, 97–113. Available from https://doi.org/10.1016/j.desal.2017.11.018, https://www.journals.elsevier.com/desalination, 00119164.

WBCSD. (2014). Building a Resilient Power Sector. Available from https://docs.wbcsd.org/2014/03/Building_A_Resilient_Power_Sector.pdf. (2014).

Bohra, V., Ahamad, K.U., Kela, A., Vaghela, G., Sharma, A., & Deka, B.J. (2021). Energy and resources recovery from wastewater treatment systems. *Clean Energy and Resource Recovery*, 17–36. Available from: https://www.sciencedirect.com/book/9780323901789, https://doi.org/10.1016/B978-0-323-90178-9.00007-X.

Boeing, A. L. (2018). The impact of waters of low quality on soiling removal from photovoltaic panels [Diss]. *University of Nevada – Las Vegas.*

Chaturvedi, V., et al. (2018). *Energy-water nexus and efficient water-cooling technologies for thermal power plants in.* In India, (pp. 19–22). Council on Energy, Environment and Water.

Dong, H., Zhang, L., Geng, Y., Li, P., & Yu, C. (2021). New insights from grey water footprint assessment: An industrial park level. *Journal of Cleaner Production, 285*, 124915. Available from https://doi.org/10.1016/j.jclepro.2020.124915, 09596526.

Dittrich, L., & Lillieroth, S. (2019). *The role of bioenergy for achieving a fossil fuel free Stockholm by 2040.*

Dufour, F. (2018). The Costs and Implications of Our Demand for Energy: A Comparative and comprehensive Analysis of the available energy resources *The costs and implications of our Demand for energy: A comparative and comprehensive analysis of the available energy resources, 2018.*

Epiney, A.S., Rabiti, C., Talbot, P., & Richards, J. (2019). Economic assessment of nuclear hybrid energy systems: Nuclear-renewable-water integration in Arizona. *Transactions of the American Nuclear Society.* INL/CON-19-52394-Rev000.

Ho, M., Obbard, E., Burr, P.A., & Yeoh G. (2019). A review on the development of nuclear power reactors. *Energy Procedia, 160*, 459–466.

Hoekstra, A. Y. (2017). Water footprint assessment: Evolvement of a new research field. *Water Resources Management, 31*(10), 3061–3081. Available from https://doi.org/10.1007/s11269-017-1618-5, http://www.wkap.nl/journalhome.htm/0920-4741, 15731650.

Jaramillo, G. S. (2022). In a different light: Performative 'power' generation of the historic light show at Grand Coulee Dam. *Comparative American Studies An International Journal, 19*(4), 321–340. Available from https://doi.org/10.1080/14775700.2022.2154109, 1477–5700.

Jia, X., Klemeš, J. J., & Tan, R. R. (2021). Overview of water use in renewable electricity generation. *Italian Association of Chemical Engineering – AIDIC, Czech Republic Chemical Engineering Transactions, 89*, 403–408. Available from https://doi.org/10.3303/CET2189068, https://www.aidic.it/cet/21/89/068.pdf, 22839216.

Jin, Y., Scherer, L., Sutanudjaja, E. H., Tukker, A., & Behrens, P. (2022). Climate change and CCS increase the water vulnerability of China's thermoelectric power fleet. *Energy, 245*, 123339. Available from https://doi.org/10.1016/j.energy.2022.123339, 03605442.

Karambelkar, S. (2018). Hydropower operations in the Colorado River Basin: Institutional analysis of opportunities and constraints. Hydropower Foundation.

Kaur, H. (2020). Impacts of flue gas desulfurization gypsum application on water quality and crop production in Southern Illinois. Southern Illinois University at Carbondale.

Kehinde, A. G. (2019). *Feasibility of once-through cooling for 50MWe solar thermal power plant on/near the lower Orange River.* Stellenbosch University.

Kelly, C., Onat, N. C., & Tatari, O. (2019). Water and carbon footprint reduction potential of renewable energy in the United States: A policy analysis using system dynamics. *Journal of Cleaner Production, 228*, 910–926. Available from https://doi.org/10.1016/j.jclepro.2019.04.268, https://www.journals.elsevier.com/journal-of-cleaner-production, 09596526.

Kharraz,J.A., RichardsB.S., & Schäfer, A.I. (2017). *Autonomous solar-powered desalination systems for remote communities desalination sustainability: A technical, socioeconomic, and environmental approach.* Elsevier Inc., Germany Elsevier Inc., Germany, 75–125, Available from: http://www.sciencedirect.com/science/book/9780128097915, https://doi.org/10.1016/B978-0-12-809791-5.00003-1.

Mac Kinnon, M. A., Brouwer, J., & Samuelsen, S. (2018). The role of natural gas and its infrastructure in mitigating greenhouse gas emissions, improving regional air quality, and renewable resource integration. *Progress in Energy and Combustion Science, 64*, 62–92. Available from https://doi.org/10.1016/j.pecs.2017.10.002, 03601285.

Lahmouri, M., Drewes, J. E., & Gondhalekar, D. (2019). Analysis of greenhouse gas emissions in centralized and decentralized water reclamation with resource recovery strategies in Leh town, Ladakh, India, and potential for their reduction in context of the water–energy–food nexus. *Water, 11*(5), 906. Available from https://doi.org/10.3390/w11050906, 2073–4441.

Luo, T., Krishnan, D., & Sen, S. (2018). *Parched power: Water demands, risks, and opportunities for India's power sector.* Water Resources Institute.

Mallapragada, D. S., Reyes-Bastida, E., Roberto, F., McElroy, E. M., Veskovic, D., & Laurenzi, I. J. (2018). Life cycle greenhouse gas emissions and freshwater consumption of liquefied Marcellus shale gas used for international power generation. *Journal of Cleaner Production, 205*, 672–680. Available from https://doi.org/10.1016/j.jclepro.2018.09.111, https://www.journals.elsevier.com/journal-of-cleaner-production, 09596526.

Moragoda, N., Cohen, S., Gardner, J., Muñoz, D., Narayanan, A., Moftakhari, H., & Pavelsky, T. M. (2023). Modeling and analysis of sediment trapping efficiency of large dams using remote sensing. *Water Resources Research, 59*(6). Available from https://doi.org/10.1029/2022wr033296, 0043–1397.

Massetti, E. et al. (2017). Environmental quality and the US power sector: Air quality, water quality, land use and environmental justice (pp. 1–169). Oak Ridge National Laboratory 772.

Morales Pedraza, J. (2017). Advanced nuclear technologies and its future possibilities. Springer Science and Business Media LLC, 35–122. Available from https://doi.org/10.1007/978-3-319-52216-6_2.

Méjean, A., Guivarch, C., Lefèvre, J., & Hamdi-Cherif, M. (2019). The transition in energy demand sectors to limit global warming to 1.5°C. *Energy Efficiency*, *12*(2), 441–462. Available from https://doi.org/10.1007/s12053-018-9682-0, http://www.springer.com/environment/journal/12053, 15706478.

Nelson, S., & Ramana, M. V. (2023). Managing decline: Devaluation and just transition at Diablo Canyon nuclear power plant. *Environment and Planning A*, *55*(8), 1951–1969. Available from https://doi.org/10.1177/0308518X231167865, https://journals.sagepub.com/home/EPN, 14723409.

NomotoH. (2017). Advanced ultra-supercritical pressure (A-USC) steam turbines and their combination with carbon capture and storage systems (CCS). In: Advances in steam turbines for modern power plants. Elsevier Inc., Japan Elsevier Inc., Japan, 501–519, Available from: http://www.sciencedirect.com/science/book/9780081003145, https://doi.org/10.1016/B978-0-08-100314-5.00021-X.

Nouri, N., Balali, F., Nasiri, A., Seifoddini, H., & Otieno, W. (2019). Water withdrawal and consumption reduction for electrical energy generation systems. *Applied Energy*, *248*, 196–206. Available from https://doi.org/10.1016/j.apenergy.2019.04.023, https://www.journals.elsevier.com/applied-energy, 03062619.

Nyaga, V.N. (2019). *Feasibility study including preliminary design of a power project for the utilization of a low temperature production well in Grafarbakki, Iceland* [Diss.].

Olivera Rodriguez, P., Holzman, M. E., Degano, M. F., Faramiñán, A. M. G., Rivas, R. E., & Bayala, M. I. (2021). Spatial variability of the green water footprint using a medium-resolution remote sensing technique: The case of soybean production in the Southeast Argentine Pampas. *Science of the Total Environment*, *763*, 142963. Available from https://doi.org/10.1016/j.scitotenv.2020.142963, 00489697.

Pan, S. Y., Snyder, S. W., Packman, A. I., Lin, Y. J., & Chiang, P. C. (2018). Cooling water use in thermoelectric power generation and its associated challenges for addressing water-energy nexus. *Water-Energy Nexus*, *1*(1), 26–41. Available from https://doi.org/10.1016/j.wen.2018.04.002, https://www.sciencedirect.com/journal/water-energy-nexus, 25889125.

Park, E. S., & Lee, T. S. (2021). The rebirth and eco-friendly energy production of an artificial lake: A case study on the tidal power in South Korea. *Energy Reports*, *7*, 4681–4696. Available from https://doi.org/10.1016/j.egyr.2021.07.006, http://www.journals.elsevier.com/energy-reports/, 23524847.

Pioro, I., & Duffey, R. (2019). *Current status of electricity generation in the world and future of nuclear power industry.' Managing global warming* (pp. 67–114). Academic Press.

Prata, J. E., & Simões-Moreira, J. R. (2019). Water recovery potential from flue gases from natural gas and coal-fired thermal power plants: A Brazilian case study. *Energy*, *186*. Available from https://doi.org/10.1016/j.energy.2019.07.110, https://www.journals.elsevier.com/energy, 03605442.

Senthil Kumar, P., & Janet Joshiba, G. (2019). *Water Footprint of Agricultural Products. Environmental footprints and eco-design of products and processes* (pp. 1–19). India: Springer. Available from http://springer.com/series/13340.

Strielkowski, W., Civín, L., Tarkhanova, E., Tvaronavičienė, M., & Petrenko, Y. (2021). Renewable energy in the sustainable development of electrical power sector: A REVIEW. *Energies*, *14*(24), 8240. Available from https://doi.org/10.3390/en14248240, 1996-1073.

Sugandha, A. (2021). *Water in efficient power: Implementing Water Norms and Zero Discharge in India's Coal-Power Fleet*. Centre for Science and Environment.

Thokchom, B. (2020). *Water-related problem with special reference to global climate change in India. Water Conservation and Wastewater Treatment in BRICS Nations: Technologies, Challenges, Strategies and Policies* (pp. 37–60). India: Elsevier. Available from https://www.sciencedirect.com/book/9780128183397.

Tramošljika, B., Blecich, P., Bonefačić, I., & Glažar, V. (2021). Advanced ultra-supercritical coal-fired power plant with post-combustion carbon capture: Analysis of electricity penalty and CO_2 emission reduction. *Sustainability*, *13*(2), 801. Available from https://doi.org/10.3390/su13020801.

Vaseashta, A. (2021). *Introduction to water safety, security and sustainability. Advanced sciences and technologies for security applications* (pp. 3–22). United States: Springer. Available from https://link.springer.com/bookseries/5540.

Ward, C., & Ruckstuhl, S. (2017). *Water scarcity, climate change and conflict in the Middle East: Securing livelihoods, building peace. Bloomsbury Publishing*.

Wolfson, R., & Dalnoki-Veress, F. (2021). *Nuclear choices for the twenty-first century: A citizen's guide*. The MIT Press. Available from http://doi.org/10.7551/mitpress/11993.001.0001.

Zhang, C., Yang, J., Urpelainen, J., Chitkara, P., Zhang, J., & Wang, J. (2021). Thermoelectric power generation and water stress in India: A spatial and temporal analysis. *Environmental Science and Technology*, *55*(8), 4314–4323. Available from https://doi.org/10.1021/acs.est.0c08724, http://pubs.acs.org/journal/esthag, 15205851.

Zohuri, B. (2019). *Heat pipe applications in fission driven nuclear power plants heat pipe applications in fission driven nuclear power plants* (pp. 1–362). United States: Springer International Publishing. Available from https://doi.org/10.15344/2456-351X/2019/166.

Chapter 11

Water footprints and thermal power generation

Reshma Shinde[1], Anand B. Rao[1,2,*] and Shastri Yogendra[1,3]

[1]Interdisciplinary Program in Climate Studies, Indian Institute of Technology Bombay, Mumbai, India, [2]Centre for Technology Alternatives for Rural Areas, Indian Institute of Technology Bombay, Mumbai, India, [3]Department of Chemical Engineering, Indian Institute of Technology Bombay, Mumbai, India

Corresponding author. e-mail address: a.b.rao@iitb.ac.in

11.1 Introduction

Water footprint studies, particularly for power generation, have been receiving greater attention over the last two decades. While the agricultural sector dominates the discussion on water footprint, water consumption/footprint of the industrial sector is equally important but not explored much (Zhu et al., 2020). Water particularly plays a crucial role in power generation. Both thermal and hydropower technologies require large quantities of water. The average consumptive water footprint of electricity and heat generation is reported to be 378 billion m^3 per year globally (Mekonnen et al., 2015).

More than 81% of the world's electricity comes from water-dependent thermal power plants (Mekonnen et al., 2015). The average operational water footprint of thermal power generation ranges between 61 and 1410 $m^3\ TJ_e^{-1}$ per year globally (Mekonnen et al., 2015). Water is consumed in several phases of power generation. It is used in the extraction, processing, and transportation of coal; the most commonly used fuel for power generation. Water is also used for coal washing and dust suppression. In a nuclear plant, water is required during uranium ore mining and is converted to uranium fluoride (Mekonnen et al., 2015). Though water is required in several phases of thermal power generation, 80%–85% requirement is for cooling water (Behrens et al., 2017). The operation stage of fossil fuel, nuclear, geothermal, and hydropower plants accounts for 70%–94% of water footprints from a life cycle-based water footprints perspective (Wang et al., 2020). This cooling water is typically acquired from nearby water resources such as lakes, rivers, and oceans. There is a strong nexus between water-thermal power generation.

Although there are several studies available on the life cycle water footprints of thermal power generation, this chapter encompasses the direct water footprints involved in electricity generation on a plant-level basis. A key focus of this chapter is to show the interaction of the water-thermal power generation nexus. In this chapter, it will be understood where the world stands with respect to its thermal capacity and total water withdrawals. This chapter will also discuss the factors that play a crucial role in establishing water needs for thermal power generation such as the type of cooling technology, the age of the plant, and the plant configuration. The addition of environmental pollution control technologies such as flue gas desulfurization (FGD) and carbon capture and storage (CCS) further alters the water needs for a thermal power plant. Furthermore, the phenomenon of climate change is likely to affect the overall water availability and its quality due to increased evaporation losses and increased water temperature. This in turn may result in future blackouts and brownouts leading to potential financial losses. Therefore this chapter emphasizes the need for systematic planning of future thermal power expansion regionally and considers the need for suitable adaptation measures. The remaining chapter is arranged as follows. Section 11.2 provides some background information regarding various metrics to quantify water utilization. Section 11.3, will look at thermal power generation and water withdrawals on a global scale as well as for India. Section 11.4 describes factors affecting water usage in thermal power generation. The impacts of climate change and water availability are discussed in Section 11.5. The chapter concludes with Section 11.6 describing a summary and some key conclusions.

Water Footprints and Sustainable Development. DOI: https://doi.org/10.1016/B978-0-443-23631-0.00011-X

11.2 Water utilization metric

It is first necessary to understand the terms regarding water usage such as water withdrawal, water consumption, and water footprints. "Water withdrawal" is the quantity of water that is abstracted from the nearest water body such as a river, or lake, and does not immediately return to the same watershed. The term "water consumption" essentially conveys the amount of water lost in evaporative cooling in the case of thermal power generation. Both water withdrawal and consumption depend on the type of cooling technology used in the plant and hence it determines further sensitivity of the plant to the changes in water availability. "Water footprint" indicates the direct and indirect water used to produce a product or a service. Thus it can be over a full supply chain. Moreover, water footprint also accounts for potential water required to handle the pollution generated. There are different types of water footprints such as green, blue, and gray. Green water footprints account for the consumption of direct rainwater. Blue water footprints represent the usage of surface and groundwater whereas water required to manage the pollution is indicated by gray water footprints. Since, for thermal power generation, the power plants mainly utilize surface water, and then groundwater, reclaimed municipal water, and treated potable water, it can be categorized as "consumptive blue water footprints." There are many studies available in the literature which particularly consider the life cycle water footprints of various thermal power plants (Ali & Kumar, 2016; Jin et al., 2019; Wang et al., 2020; Zhu et al., 2020). By defining system boundaries, these studies evaluate the water footprints of power generation in a particular region for upstream and downstream processes. The extraction, processing, and transport of fuels in electricity generation could add as much as 10% to the life cycle water consumption of coal-fired power plants and 20% to that of nuclear plants with wet cooling systems (Meldrum et al., 2013). Therefore these water footprint numbers vary from region to region, the extent of system boundary involved, and the approach used such as cradle-to-gate, cradle-to-grave.

11.3 Thermal power and water withdrawal

Global electricity generation is dominated by thermal capacity. The global water withdrawal for thermal power generation was 500,000 million m^3 (MCM) in the year 2015; out of which the freshwater use was 290,000 MCM (Lohrmann et al., 2019). The United States accounted for 44% of its freshwater withdrawal for thermal power generation in 2005 (Talati et al., 2016). China, the United States, and India are the top three countries that withdraw more water compared to the rest of the world. Moreover, China and the United States had the highest withdrawals with 31.5% and 35.7% of total water used for power generation in the year 2015 (Lohrmann et al., 2019). However, water withdrawal in the United States was reduced by 18% in 2015 with respect to the year 2010 owing to plant closures, fuel switching from coal to gas, and the usage of water-efficient cooling technologies (Peer & Sanders, 2018). The European Union (the EU) withdrew roughly 55% of its water for electricity generation (Eurostat, 2014). Lohrmann et al. (2019) did geographic information system analysis that showed 55.5% of the thermal capacity of the world is situated within 5 km from the main rivers, and lakes, and hence are assumed to be freshwater cooled.

A total of 36 countries around the world already experience high to extremely high-water stress (Reig et al., 2013). According to the World Resources Institute (WRI), water stress is determined by taking the ratio of water demand to the water available. Sometimes, physical water stress occurs when poor quality of water restricts its use. Higher water stress indicates a high level of competition in water use. Around 47% of the world's existing thermal power capacity is situated in the highly water-stressed areas (WRI, 2018). Between the years 2000 and 2015, there have been 43 incidents in the United States where water-related power plant issues were identified (McCall et al., 2016). One of the studies in the European Union showed that the number of regions facing power shortages due to water stress is expected to rise from 47 to 54 basins between 2014 and 2030 (Behrens et al., 2017). Moreover, developing nations in Asia such as China and India may run out of water for power production due to climate change and overexploitation of water resources (Hanumante et al., 2022; Wang et al., 2019). The situation becomes worse when new thermal capacity is added in the areas where water scarcity is already conspicuous.

India ranks 13th for overall water stress and has more than three times the population of the other 17 extremely highly water-stressed countries combined (Hofste et al., 2019). According to WRI, the most water-stressed regions in India are the states of Haryana, Rajasthan, Uttar Pradesh, Gujarat, Punjab, Uttarakhand, Madhya Pradesh, and Jammu and Kashmir. Uttar Pradesh, and Punjab receive sufficient annual rainfall and are usually water abundant. However, unmanaged and poor exploitation of water resources for agricultural practices make these states' water scarce (Verma et al., 2021). 41 GW of India's installed thermal capacity is in drought-affected areas with about 37 GW located in extreme drought areas (Schlissel & Woods, 2019). Insufficient cooling water availability has already created problems such as shutdowns and running plants at lower capacities in India. For instance, the Sipat Power Plant of Chhattisgarh had to be shut down because of the unavailability of cooling water in 2008 (Srinivasan et al., 2018). In the region of Rajasthan, the stored water loss due to evaporation was found to be 30% in the Barsingsar Thermal Power Station (Chandra et al., 2015). So, evaporation losses are also critical in thermal stations in arid regions. Furthermore, there was loss of 14 TWh of potential thermal generation due to water shortages in India in 2016 (Luo & Krishnan, 2018). This indicates that thermal power generation is affected if sufficient cooling water is not available.

Coal power plants are responsible for 70% water usage of industrial water use (Ramanathan et al., 2020). To curb the rising demands for water in the power sector, the Ministry of Environment Forest and Climate Change in India has stipulated old coal plants use 3.5 m^3 $(MW\,h)^{-1}$ of water and new plants installed after 2017; would use 2.5–3 m^3 $(MW\,h)^{-1}$. Still, it was found only half of the plants abide by the stipulations and these numbers are self-reported (IEA, 2021).

There are some studies in the literature that evaluated water consumption of power generation in India (IEA, 2020; Mitra et al., 2014). Here, the water footprint numbers for different power sources were estimated using the data available in the literature for the current and the next two decades. The electricity generation by coal in India is estimated to be 1135, 1343, and 1334 TWh in the stated policy scenario in the year 2019, 2030, and 2040, respectively (IEA, 2020). The average consumptive water footprint for coal power generation is assumed to be 3.5 L $(kW\,h)^{-1}$ (Mekonnen et al., 2015; Zhu et al., 2020). Similar exercise was done for other sources of power generation. The water footprint for other power sources was obtained from different resources in the literature (NREL, 2012; Desai & Bandyopadhyay, 2017; Mekonnen et al., 2015).

Table 11.1 shows the source-specific total water footprint for power generation in India for years 2019, 2030, and 2040. It can be seen that the coal power capacity has more blue water footprints compared to other sources currently and it is expected to rise in the next decade. This is because although decommissioning of coal plants has been accelerating in developed economies such as the United States and the EU, the same is not true for developing nations like India and China. The water footprint is expected to slightly reduce in 2040 due to more renewable additions. It is important to note that the water footprint numbers for bioenergy generation consider both blue and green water footprints and they almost get doubled in the year 2040 compared to the present scenario.

The water needs are going to rise in the future for thermal as well as for all the other power generation technologies. However, increasing water stress in many countries is likely to make this demand difficult to meet. Therefore it is necessary to identify the factors that influence the water withdrawals in the thermal power plants and use the corrective measures to reduce the water requirements.

11.4 Factors affecting water usage in thermal power plants

In a thermal power plant, water is used for various purposes. The primary use is boiler feed and cooling water needs. The other usages include dust suppression, ash pond, service water, horticulture, potable water, and water getting used in traditional environmental control systems. Thermal power generation utilizes coal, gas, biomass, oil, or nuclear power as fuel sources. The fuel source gives the amount of heat required $(kJ\,h^{-1})$ to generate a unit of electricity, namely, often termed "heat rate" (HR, kJ $(kW\,h)^{-1}$). The waste heat generated in the process of electricity generation is removed by using cooling water. The quantity of water getting used in a thermal power plant depends upon the type of cooling technology used in the plant, regional climate factors, seasonal changes, plant configuration, environmental control technologies used as well as the age of the plant. The subsequent points will discuss how these technologies affect water footprints.

TABLE 11.1 Estimated water footprints based on the values reported in the literature (billion m^3/year).

Source/year	Coal	NG	Nuclear	Hydro	Wind	Solar	Bio	Total
2019	3.97	0.07	0.097	4.88	0.0033	0.003	50.56	59.58
2030	4.70	0.105	0.263	6.3	0.0098	0.024	97.51	108.91
2040	4.66	0.153	0.537	8.55	0.026	0.078	145.67	159.67

11.4.1 Type of cooling technology

Primarily, there are three main cooling technologies—once-through (open-loop), recirculating (closed-loop), and dry cooling. Hybrid cooling essentially uses both dry and recirculating cooling. The choice of these technologies depends upon regional water availability, weather conditions, and affordability. As of 2016, 80% of cooling in thermal plants took place by freshwater recirculating technology and 7% by freshwater once-through technology in India (Luo & Krishnan, 2018). Freshwater dry cooling constituted only 2% of the cooling whereas rest of the cooling needs utilized seawater. Almost 90% of India's thermal power generation depends on freshwater cooling requirements.

11.4.1.1 Once-through cooling (open-loop coolings)

Once-through cooling uses large quantities of water from nearby water resources, passes it through the heat exchanger to cool the steam, and dumps the same water in the same source but at an elevated temperature. Hence, water withdrawal is higher in this technology whereas consumption or evaporation losses are fairly negligible (Dodder, 2014).

While this type of cooling is considered energy efficient and requires low infrastructure, it causes thermal pollution endangering aquatic life. Therefore environmental agencies enforce an upper bound on the temperature of the water to be discharged in the source. Thermal power plants are penalized if they fail to meet the temperature regulations of the discharge water. A typical water withdrawal range of once-through cooling thermal power plants lies between 75,000 and 1,89,000 liters per Megawatt hour (L $(\mathrm{MW\,h})^{-1}$) (Dorfman & Haren, 2014). Once-through plants use around 40–80 times higher water than recirculating plants (Srinivasan et al., 2018). Though the United States had more once-through plants than the other cooling types, they are now getting shifted to closed-loop cooling and dry cooling. In India, freshwater-cooled once-through plants are banned since 1999 (IEA, 2015). Only seawater once-through plants are operational in the coastal region of the country. Once-through cooling technology is vulnerable to an increased intake of water temperature due to climate change since water is used at ambient temperatures. Moreover, water scarcity may make these plants more vulnerable.

11.4.1.2 Recirculating cooling (closed-loop cooling)

Recirculating cooling technology uses water in a repetitive cycle. Therefore water withdrawn is comparatively less compared to once-through cooling. However, water consumption is higher. The cooling tower is an important component of closed-loop cooling. This cooling tower could be a natural draft or a forced draft. The main job of a cooling tower is to reject heat into the atmosphere. Hot water coming out of the condenser after cooling the turbine exhaust steam is passed through the packings in a cooling tower. This packing is made of wood or *polyvinyl chloride* (PVC) which gives more surface area for effective (Fig. 11.1). It should be noted that water cannot be cooled below the local wet bulb temperature (WBT). The losses that occurred during cooling such as blowdown, drift, and evaporative losses are made up using makeup water. The Cycle of Concentration is one of the important parameters while dealing with make-up water requirements. It represents the ratio of total dissolved solids in the cooling tower water to the dissolved solids in the makeup water. It is representative of the frequency of the quantity of freshwater getting added in the cooling cycle. The water withdrawal numbers for recirculating cooling technology range between 2000 and 5000 l/MWh whereas the consumption numbers lie between 1500 and 2200 L $(\mathrm{MW\,h})^{-1}$ (Macknick et al., 2012). However, it is observed in India that the older plants withdraw as much as 7000–9000 L $(\mathrm{MW\,h})^{-1}$ (Central Electricity Authority, 2012). For closed-loop cooling also, the water withdrawal numbers depend on the design of the cooling tower, the age of the plant, and the climatology of the local region.

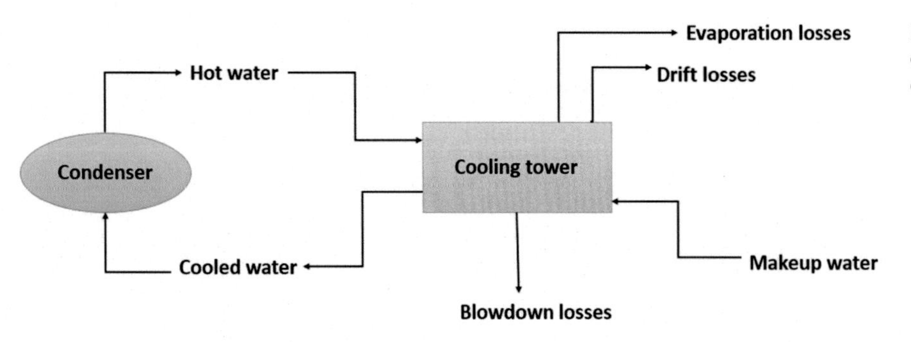

FIGURE 11.1 Schematic representation of recirculating cooling (closed-loop cooling).

The efficiency of a thermal power plant reduces by 2%–5% when recirculating technology (closed-loop) is used (Zhou et al., 2018). It is because of the auxiliary power consumption of water pumps and ventilators. Moreover, they are 40% more expensive than once-through systems (Luo & Krishnan, 2018). During cooling in a cooling tower, 2%–5% of water is evaporated due to evaporation (Zhang et al., 2021). However, a cooling tower saves 25%–30% of natural water (Zhang et al., 2021). It avoids the thermal pollution of natural water bodies.

11.4.1.3 Dry cooling technology

Dry cooling does not require water for cooling as the name suggests and turbine exhaust steam is cooled in a heat exchanger that is externally cooled by a cold airflow. The key to maximizing the effect of dry cooling is the airflow and the surface area. There are two types of dry cooling systems namely, direct and indirect cooling. Dry cooling systems use little to no water, but they are three to four times more expensive than wet cooling technologies (European Space Agency, 2002). Dry cooling also reduces power efficiency considerably, namely, around 7%–8% (Chaturvedi et al., 2018).

In a direct dry cooling tower, turbine exhaust steam is condensed straight through an air-cooled condenser (ACC). The turbine exhaust steam passes directly in the ACC as shown in Fig. 11.2. This ACC has the tubes where the steam is passed and fans are provided to increase the airflow on the tubes' surface.

In indirect dry cooling, steam exhaust from the turbine is condensed in a surface heat exchanger where heat transfer occurs from turbine steam to the cold water pumped from the air-cooled exchanger located inside the tower. The heat gained by the cooling water in the condenser is pumped back to the heat exchanger inside the tower. Airflow is used on the heat exchanger surface to cool the hot water inside the exchanger tubes. The cooled water is then pumped back into the condenser. Since turbine exhaust is condensed by circulating water than direct ambient air, it is called indirect dry cooling.

Hybrid cooling is the combination of dry and evaporative cooling to provide heat rejection and reduce water and energy consumption. It reduces water consumption by 80% by utilizing wet cooling when it is hot and dry cooling when it is not (Zhai et al., 2011). Hybrid cooling implements elements of both dry and wet cooling concurrently or separately. It should be noted though that both dry and hybrid cooling are characterized by high capital costs and reduced net power efficiency. Hence, recirculating and once-through cooling methods are always preferred compared to dry and hybrid.

11.4.2 Regional climate factors and seasonal changes

Regional climate factors such as ambient temperature, relative humidity, and WBT do affect the performance of a cooling tower and hence the subsequent water needs. It is an important aspect while designing a cooling tower. WBT is a controlling factor considering the minimum cold-water temperature to which water can be cooled by the evaporative method. Air at higher WBTs has the capability of picking up more heat. As far as ambient dry bulb temperature is considered, as per the increase in the temperature, the cooling water needs also rise because of more evaporation losses. Therefore more water is required for cooling in the summer than in the winter months. The increased water temperatures in summer also give rise to more water requirements per unit of electricity generation in once-through cooling systems too. Increments in the air and water temperatures can result into reductions in the efficiency of power plants subsequently (Colman, 2013). It should also be noted that cooling water availability is also season-specific and there is a high disparity in spatial and temporal water resources (Chowdhury et al., 2021). Therefore evaluation of seasonal

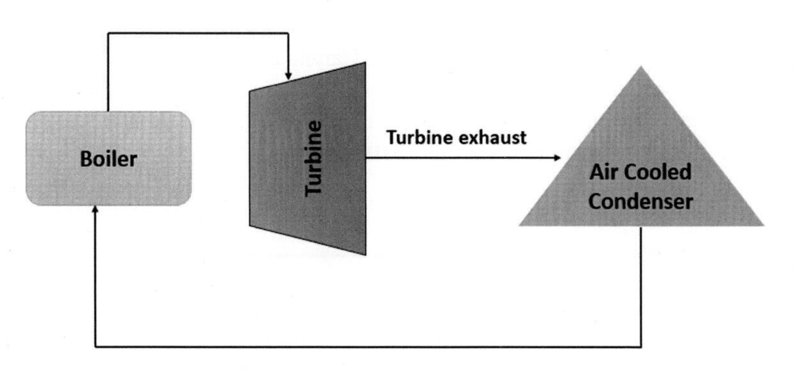

FIGURE 11.2 Schematic representation of direct dry cooling system.

water footprints is also necessary (Chini & Delorit, 2021). Not only it will give an estimate of available water in advance but planning for shortages will also be accounted for.

11.4.3 Type of boiler technology

Power plant designers always tend to look at how to produce electricity at a lower cost of generation. This cost increases with the increase in the fuel cost. Supercritical or ultrasupercritical boilers require less fuel than the traditional subcritical drum boilers which make them more efficient and economical. Boilers produce high-pressure high-temperature steam which eventually passes through turbines to generate electricity. The energy in the fuel cannot be completely transferred to the steam due to incomplete combustion and exhaust heat, resulting in energy loss. The efficiency of around 90%–94% can be attained using modern boilers (Zhang, 2020). The type of boiler used in a thermal power plant has a considerable impact on cooling water needs. Moving from subcritical to ultrasupercritical boiler, the water withdrawals are reduced significantly, and the efficiency of the plant is improved too. The ultra-supercritical boilers with wet cooling towers in India need comparatively less water in the range of 1000–2000 L $(MW\ h)^{-1}$ (Chaturvedi et al., 2018) The subcritical power plants, which are more common in India, withdraw more water in the range of 4000–9000 L $(MW\ h)^{-1}$ (Luo & Krishnan, 2018). The water withdrawals for ultrasupercritical power plants reduce by almost 700–1000 L $(MW\ h)^{-1}$ and that for supercritical reduce by 200–500 L $(MW\ h)^{-1}$ compared to a subcritical coal plant (Macknick et al., 2012). However, it should be noted that older plants withdraw more water compared to new plants. Hence, they are more vulnerable to underperformance caused by water unavailability or availability of water at higher temperatures. Pan et al. (2018) have collated more data about water withdrawal and water consumption for various boiler technologies and cooling systems.

11.4.4 Ancillary environmental control technologies

Environmental control technologies are used in thermal power plants to reduce air, water, and land pollution. These technologies can further exacerbate water withdrawals. The ash handling system also requires significant water. Some of the common environmental control technologies used are FGD, selective catalytic reduction, and carbon capture and sequestration (CCS) technology.

Coal plants in India are the largest emitters of SO*x* and NO*x*. Based on the feasibility studies, coal power plants in India are required to install an FGD system to reduce SO*x* emissions. Various elements of the FGD system require water such as the wash system of the mist eliminator, the absorber system, the limestone grinding and slurry preparation system, and the gypsum dewatering system. For the cooling requirements of FGD, demineralized water is required. The FGD system for a typical 500 MW coal plant is expected to utilize 200–300 L $(MW\ h)^{-1}$ of additional freshwater (Cordoba P., 2015; Gupta, 2020). The wastewater from FGD is expected to be treated to achieve zero liquid discharge. FGD systems can also use process water, such as from cooling tower blowdown for processes like grinding limestone. Wet FGD systems need more water withdrawals compared to semidry scrubbers or dry scrubbers.

CCS is used to reduce the CO_2 emissions in the atmosphere and maintain global warming below 1.5°C. Several studies in the literature have evaluated water footprints for CCS systems (Rao & Kumar, 2014; Zhai et al., 2011; Davis et al., 2018; Sharma & Mahapatra, 2018; Srinivasan et al., 2018). The anthropogenic water footprints are anticipated to be doubled with the deployment of CCS technology. Depending upon the types of CCS, they can have water footprints ranging between 0.74 and 575 m^3 of water per ton of CO_2 removed (Rosa et al., 2021). Bioenergy with CCS has the highest water footprints per ton of CO_2 removed among all CCS technologies (Jin et al., 2019). This is because of more water requirements associated with transpiration. Postcombustion CO_2 capture systems used in coal-fired power plants use considerable amounts of water. The amine-based postcombustion CO_2 capture technology is a paramount commercial CO_2 removal system. The amine-based CO_2 capture process will double the water withdrawn in the wet cooling system (Carter, 2010; Zhai et al., 2011). Almost 90% CO_2 capture and storage systems in the coal-fired power plant would increase water use by 20%–50% (Talati et al., 2016). Postcombustion CCS technologies are preferred and could be economically viable because they can be installed in the existing plant infrastructure and need not require decommissioning (Davis et al., 2018).

11.5 Impacts of climate change

Assessments of the Intergovernmental Panel on Climate Change and other studies have shown that the energy sector not only contributes to climate change but is also vulnerable to the impacts of climate change (Cronin et al., 2018; Pörtner, 2022). India already has one of the highest water footprints in the world, and changing climate conditions only aggravate the whole situation (Asoka et al., 2017; Preethi et al., 2019; Schneider et al., 2014). As climate change manifests itself, the corresponding increment in ambient temperatures would also give rise to the increase in cooling water stream temperatures. As per Morrill et al. (2005) for every 1°C rise in the atmospheric temperature, the river stream temperature rises by 0.7°C–0.8°C. This will reduce the cooling capacity of water, thereby resulting in increased total withdrawal. Simultaneously, demands for space cooling are increasing and it may give rise to higher peak loads too. This would result in more water requirements, especially in the summer seasons. Higher temperatures not only affect the freshwater-dependent power plants but those which use seawater for cooling also get affected too. There are instances in Sweden where the nuclear power plants had to reduce the capacity output due to an unanticipated increase in seawater temperatures (Patel, 2018). The rise in the ambient temperature, the increased water temperatures, and water unavailability are the major climate change impacts the thermal sector is vulnerable to (Förster & Lilliestam, 2010; Spanger-Siegfried, 2012; Wang et al., 2019). Shallow bodies of water are more affected by higher air temperatures, and this can lead to a rise in water temperature. In general, a rise in ambient cooling water temperature of 1°C could cause a drop in power output of 0.15%–0.5% (Cook et al., 2015). Globally, there are several examples of blackouts and brownouts that happened because of extreme heat/cold events (Añel et al., 2017). The effect is not only limited to the temperatures of cooling water and its availability. But It also has "infrastructural impacts" on the electricity supply such as transformers and transmission and distribution lines. These impacts directly affect economic activities and hence are the reasons for fiscal losses.

Reclaimed water could be one of the solutions to overcome the problems of water shortages for thermal power generation. It constitutes treated water from effluent streams, rainwater, process water from industries. These options are comparatively cheaper than seawater desalination. In the recent years, the potential of using treated brackish groundwater for cooling is proposed (Kahsar, 2020; Wu et al., 2023). However, the potential is significant for countries like the United States because of its underutilization. Another approach to deal with the increased water requirements is to opt for alternate cooling technologies such as dry and hybrid cooling as discussed earlier. The renewable power generation technologies like solar and wind have practically the lowest operational water footprints. The adaptation measures like these would be important to bring about regional security in satiating corresponding power demands.

Water availability will be a crucial factor in uninterrupted power generation. Loew et al. (2020) suggest that in the future, the engineering configurations of the plants should be such that they should be able to operate at different peak temperatures. The authors conclude that the cooling system designs are responsible for the vulnerability of power plants to rising temperatures. Cronin et al. (2018) imply that the thermal capacity of India would tend to rise till mid-century even if climate change persists in the country. The resilience of thermoelectric power plants to climate change and increased demand would be crucial in the coming decades. Therefore alternative measures and suitable state-of-the-art adaptation strategies must be implemented.

11.6 Summary and conclusions

There is a strong nexus between water and thermal power generation. There are various metrics which define the water utilization for a product or a service such as green water footprint, blue water footprint. The water getting used in a thermal power generation can be categorized as "consumptive blue water footprint." Cooling water needs in a thermal power generation amounts to 80%–85% of the total water needs. A considerable thermal power generation was adversely affected because of cooling water shortages globally as well as in India leading to significant fiscal losses. These cooling water requirements by thermal power plants depend on several factors such as the age of the plant, type of cooling technology, type of boiler technology, local and seasonal climate. Once-through cooling withdraws more water compared to any other cooling technology and is responsible for the thermal pollution of water bodies. Though recirculating, dry, and hybrid cooling technologies have lower operational water footprints, they have higher capital costs as well as they reduce the net power output. Transitioning from the subcritical to the supercritical boiler technology, the corresponding water withdrawals get drastically reduced. As more environmental control technologies are added, the operational water footprints increase. The vulnerability of the thermal power plants to climate change–driven water stress can be reduced by undertaking suitable adaptation measures such as shifting to recirculating/dry cooling, and the use of renewable sources of energy such as solar and wind. Even though the world is shifting to more renewable energy sources, due to their intermittent nature, there still is a need to depend upon conventional electricity generation sources such as coal and natural gas. With population growth and economic developments, the electricity demand would only rise in the future, and hence the sectoral water demands are also going to rise. Subannual studies are essential at a national as well as a regional level, because the demands for electricity vary as per the seasonal changes.

References

Ali, B., & Kumar, A. (2016). Development of life cycle water footprints for gas-fired power generation technologies. *Energy Conversion and Management*, *110*, 386–396. Available from https://doi.org/10.1016/j.enconman.2015.12.048.

Asoka, A., Gleeson, T., Wada, Y., & Mishra, V. (2017). Relative contribution of monsoon precipitation and pumping to changes in groundwater storage in India, Nature Publishing Group*India Nature Geoscience*, *10*(2), 109–117. Available from https://doi.org/10.1038/ngeo2869, http://www.nature.com/ngeo/index.html.

Añel, J., Fernández-González, M., Labandeira, X., López-Otero, X., & De la Torre, L. (2017). Impact of cold waves and heat waves on the energy production sector. *Atmosphere*, *8*(11), 209. Available from https://doi.org/10.3390/atmos8110209.

Behrens, P., van Vliet, M. T. H., Nanninga, T., Walsh, B., & Rodrigues, J. F. D. (2017). Climate change and the vulnerability of electricity generation to water stress in the European Union. *Nature Energy*, *2*(8). Available from https://doi.org/10.1038/NENERGY.2017.114, http://www.nature.com/nenergy/.

Carter, N. T. (2010). *Energy's water demand: trends, vulnerabilities, and management*. CRS Report for Congress. Congressional Research Service.

Central Electricity Authority. (2012). CEA Annual report 2012, Ministry of Power, New Delhi, India. Available from https://cea.nic.in/wp-content/uploads/2020/03/annual_report-2013.pdf

Chandra, B., Bhati, P., Kumar, S., Sangeetha, A., Siddhartha, S., Ramanathan, S., & Rudra, A. (2015). *Heat on power: Green rating of coal-based power plants*. Centre for Science and Environment.

Chaturvedi., Sugam, R., Koti, N., & Neog, K. (2018). Energy-water nexus and efficient water-cooling technologies for thermal power plants in India: An analysis within an integrated assessment modelling framework. *Environment And Water*, 19–22.

Chini, C. M., & Delorit, J. D. (2021). Opportunities for robustness of water footprints in electricity generation, John Wiley and Sons Inc, United States. *Earth's Future*, *9*(7), 23284277. Available from https://doi.org/10.1029/2021EF002096, http://onlinelibrary.wiley.com/journal/10.1002/(ISSN)2328-4277.

Chowdhury, A. F. M. K., Dang, T. D., Nguyen, H. T. T., Koh, R., & Galelli, S. (2021). The Greater Mekong's climate-water-energy nexus: How ENSO-triggered regional droughts affect power supply and CO_2 emissions, John Wiley and Sons Inc, Singapore. *Earth's Future*, *9*(3), 23284277. Available from https://doi.org/10.1029/2020EF001814, http://onlinelibrary.wiley.com/journal/10.1002/(ISSN)2328-4277.

Colman, J. (2013). *The effect of ambient air and water temperature on power plant efficiency*. Report. https://dukespace.lib.duke.edu/dspace/bitstream/handle/10161/6895/.

Cook, M. A., King, C. W., Davidson, F. T., & Webber, M. E. (2015). Assessing the impacts of droughts and heat waves at thermoelectric power plants in the United States using integrated regression, thermodynamic, and climate models, Elsevier Ltd, United States. *Energy Reports*, *1*, 193–203. Available from https://doi.org/10.1016/j.egyr.2015.10.002, http://www.journals.elsevier.com/energy-reports/.

Cronin, J., Anandarajah, G., & Dessens, O. (2018). Climate change impacts on the energy system: A review of trends and gaps, Springer Netherlands, United Kingdom. *Climatic Change*, *151*(2), 79–93. Available from https://doi.org/10.1007/s10584-018-2265-4, http://www.wkap.nl/journalhome.htm/0165-0009.

Davis, S. J., Lewis, N. S., Shaner, M., Aggarwal, S., Arent, D., Azevedo, I. L., Benson, S. M., Bradley, T., Brouwer, J., Chiang, Y. M., Clack, C. T. M., Cohen, A., Doig, S., Edmonds, J., Fennell, P., Field, C. B., Hannegan, B., Hodge, B. M., Hoffert, M. I., ... Caldeira, K. (2018). Net-zero emissions energy systems, American Association for the Advancement of Science, United States. *Science (New York, N.Y.)*, *360*(6396), 10959203. Available from https://doi.org/10.1126/science.aas9793, http://science.sciencemag.org/content/360/6396/eaas9793/tab-pdf.

Desai, N. B., & Bandyopadhyay, S. (2017). *Sustainability in power generation systems* Elsevier, India*Encyclopedia of sustainable technologies* (pp. 157–163). India: Elsevier. Available from http://www.sciencedirect.com/science/book/9780128047927, https://doi.org/10.1016/B978-0-12-409548-9.10045-4.

Dodder, R. S. (2014). A review of water use in the U.S. electric power sector: Insights from systems-level perspectives, Elsevier Ltd, United States*Current Opinion in Chemical Engineering*, *5*, 7–14. Available from https://doi.org/10.1016/j.coche.2014.03.004, http://www.elsevier.com/wps/find/journaldescription.cws_home/725837/description#description.

Dorfman, M., & Haren, A. (2014). *Testing the waters*. Natural Resources Defense Council.

European Space Agency. (2002). Galileo Joint Undertaking., & European Commission. Directorate-General for Energy and Transport. Unit E.4 Satellite Navigation System (Galileo), I. Transport. (2002). *Galileo: the European programme for global navigation services*. ESA Publications Division. Available from https://www.esa.int/Newsroom/Press_Releases/Galileo_break_in_the_negotiations.

Eurostat. (2014). *Energy, transport and environment indicators*. Pocketbook 2013. Available from http://epp.eurostat.ec.europa.eu/cache/ITY_OFFPUB/KS-DK-13-001/EN/KS-DK-13-001-EN.PDF. Accessed 16.08.21.

Förster, H., & Lilliestam, J. (2010). Modeling thermoelectric power generation in view of climate change. *Regional Environmental Change*, *10*(4), 327–338. Available from https://doi.org/10.1007/s10113-009-0104-x.

Hanumante, N., Shastri, Y., Nisal, A., Diwekar, U., Cabezas, H., & Añel, J. A. (2022). Integrated model for food-energy-water (FEW) nexus to study global sustainability: The water compartments and water stress analysis. *PLoS One*, *17*(5), e0266554. Available from https://doi.org/10.1371/journal.pone.0266554.

Hofste, R., Reig, P., & Leah, S. (2019). 17 countries, home to one-quarter of the world's population, face extremely high water stress., World Resource Institute, webpage. Available from https://thecityfix.com/blog/17-countries-home-one-quarter-worlds-population-face-extremely-high-water-stress-rutger-willem-hofste-paul-reig-leah-schleifer/.

IEA. (2015). International Energy Agency report. World Energy Outlook 2015, IEA, Paris. License: Creative Commons BY 4.0. Available from https://www.iea.org/reports/world-energy-outlook-2015.

IEA. (2020). International Energy Agency report. World Energy Outlook 2020. IEA, Paris. License: Creative Commons BY 4.0. Available from https://www.iea.org/reports/world-energy-outlook-2020.

IEA. (2021). International Energy Agency report. World Energy Outlook 2021, IEA, Paris. Available from https://www.iea.org/reports/world-energy-outlook-2021.

Jin, Y., Behrens, P., Tukker, A., & Scherer, L. (2019). Water use of electricity technologies: A global meta-analysis. *Renewable and Sustainable Energy Reviews*, *115*, 109391. Available from https://doi.org/10.1016/j.rser.2019.109391.

Kahsar, R. (2020). The potential for brackish water use in thermoelectric power generation in the American southwest. *Energy Policy*, *137*, 111170. Available from https://doi.org/10.1016/j.enpol.2019.111170.

Loew, A., Jaramillo, P., Zhai, H., Ali, R., Nijssen, B., Cheng, Y., & Klima, K. (2020). Fossil fuel–fired power plant operations under a changing climate, Springer Science and Business Media B.V., United States. *Climatic Change*, *163*(1), 619–632. Available from https://doi.org/10.1007/s10584-020-02834-y, http://www.wkap.nl/journalhome.htm/0165-0009.

Lohrmann, A., Farfan, J., Caldera, U., Lohrmann, C., & Breyer, C. (2019). Global scenarios for significant water use reduction in thermal power plants based on cooling water demand estimation using satellite imagery, Nature Research, Finland. *Nature Energy*, *4*(12), 1040–1048. Available from https://doi.org/10.1038/s41560-019-0501-4, http://www.nature.com/nenergy/.

Luo,T., & Krishnan,D. (2018). *Parched power: Water demands, risks and opportunities for India's power sector*.

Macknick, J., Newmark, R., Heath, G., & Hallett, K. C. (2012). Operational water consumption and withdrawal factors for electricity generating technologies: A review of existing literature. *Environmental Research Letters*, *7*(4), 045802. Available from https://doi.org/10.1088/1748-9326/7/4/045802.

McCall, J., Jordan, M., & Hillman, D. (2016). *Water-related power plant curtailments: An overview of incidents and contributing factors*.

Mekonnen, M. M., Gerbens-Leenes, P. W., & Hoekstra, A. Y. (2015). The consumptive water footprint of electricity and heat: A global assessment, Royal Society of Chemistry, Netherlands*Environmental Science: Water Research and Technology*, *1*(3), 285–297. Available from https://doi.org/10.1039/c5ew00026b, http://pubs.rsc.org/en/journals/journalissues/ew#!recentarticles&adv.

Meldrum, J., Nettles-Anderson, S., Heath, G., & Macknick, J. (2013). Life cycle water use for electricity generation: a review and harmonization of literature estimates. *Environmental Research Letters*, *8*(1), 015031. Available from https://doi.org/10.1088/1748-9326/8/1/015031.

Mitra, B., Bhattacharya, A., & Zhou, X. (2014). *A critical review of long-term water energy nexus in India nexus 2014: Water, food, climate and energy conference*. A conference proceeding article. https://www.iges.or.jp/en/pub/critical-review-long-term-water-energy-nexus/en.

Morrill, J. C., Bales, R. C., & Conklin, M. H. (2005). Estimating stream temperature from air temperature: Implications for future water quality. *Journal of Environmental Engineering*, *131*(1), 139–146. Available from https://doi.org/10.1061/(ASCE)0733-9372(2005)131:1(139).

Pan, S. Y., Snyder, S. W., Packman, A. I., Lin, Y. J., & Chiang, P. C. (2018). Cooling water use in thermoelectric power generation and its associated challenges for addressing water-energy nexus, KeAi Communications Co., Taiwan*Water-Energy Nexus*, *1*(1), 26–41. Available from https://doi.org/10.1016/j.wen.2018.04.002, https://www.sciencedirect.com/journal/water-energy-nexus.

Patel, S. (2018). Intense summer heatwaves rattle world's power plants. *POWER Magazine*, *162*(10). Available from https://www.powermag.com/intense-summer-heatwaves-rattle-worlds-power-plants/.

Peer, R. A. M., & Sanders, K. T. (2018). The water consequences of a transitioning US power sector, Elsevier Ltd, United States*Applied Energy*, *210*, 613–622. Available from https://doi.org/10.1016/j.apenergy.2017.08.021, http://www.elsevier.com/inca/publications/store/4/0/5/8/9/1/index.htt.

Preethi, B., Ramya, R., Patwardhan, S. K., Mujumdar, M., & Kripalani, R. H. (2019). Variability of Indian summer monsoon droughts in CMIP5 climate models, Springer Verlag*India Climate Dynamics*, *53*(3–4), 1937–1962. Available from https://doi.org/10.1007/s00382-019-04752-x, http://link.springer.de/link/service/journals/00382/index.htm.

Pörtner, H. (2022). Climate change 2022: Impacts, adaptation and vulnerability. In: *Working group II contribution to the sixth assessment report of the Intergovernmental Panel on Climate Change*.

Ramanathan, S., Arora, S., & Trivedi,V. (2020). Coal-based power norms: Where do we stand today. A CSE Report. Centre for Science and Environment.

Rao, A.B., & Kumar, P. (2014). Cost implications of carbon capture and storage for the coal power plants in India. *Energy Procedia Elsevier Ltd India*, *54*, 431–438. https://doi.org/10.1016/j.egypro.2014.07.285.

Reig, P., Maddocks, A., & Gassert, F. (2013). *World's 36 most water-stressed countries*. World Resource Institute. Web article. https://www.wri.org/insights/worlds-36-most-water-stressed-countries.

Rosa, L., Sanchez, D. L., Realmonte, G., Baldocchi, D., & D'Odorico, P. (2021). The water footprint of carbon capture and storage technologies, Elsevier Ltd, United States*Renewable and Sustainable Energy Reviews*, *138*. Available from https://doi.org/10.1016/j.rser.2020.110511, https://www.journals.elsevier.com/renewable-and-sustainable-energy-reviews.

Schlissel, D., & Woods, B. (2019). *Risks growing for india's coal sector*. Researcher, Applied Economics Clinic. Web article. http://asar.co.in/.

Schneider, T., Bischoff, T., & Haug, G. H. (2014). Migrations and dynamics of the intertropical convergence zone. *Nature*, *513*(7516), 45–53. Available from https://doi.org/10.1038/nature13636.

Sharma, N., & Mahapatra, S. S. (2018). A preliminary analysis of increase in water use with carbon capture and storage for Indian coal-fired power plants, Elsevier B.V., United States*Environmental Technology and Innovation*, *9*, 51–62. Available from https://doi.org/10.1016/j.eti.2017.10.002, http://www.journals.elsevier.com/environmental-technology-and-innovation/.

Spanger-Siegfried, E. (2012). *If You Can't Take the Heat: How Summer 2012 Strained U.S. Power Plants*. Blog article. The Equation. Web article. Available from https://ucsusa.org/erika-spanger-siegfried/if-you-cant-take-the-heat-how-summer-2012-strained-u-s-power-plants/.

Srinivasan, S., Kholod, N., Chaturvedi, V., Ghosh, P. P., Mathur, R., Clarke, L., Evans, M., Hejazi, M., Kanudia, A., Koti, P. N., Liu, B., Parikh, K. S., Ali, M. S., & Sharma, K. (2018). Water for electricity in India: A multi-model study of future challenges and linkages to climate change mitigation, Elsevier Ltd, India. *Applied Energy*, *210*, 673–684. Available from https://doi.org/10.1016/j.apenergy.2017.04.079, http://www.elsevier.com/inca/publications/store/4/0/5/8/9/1/index.htt.

Talati, S., Zhai, H., Kyle, G. P., Morgan, M. G., Patel, P., & Liu, L. (2016). Consumptive water use from electricity generation in the southwest under alternative climate, technology, and policy futures, American Chemical Society, United States. *Environmental Science and Technology*, *50*(22), 12095–12104. Available from https://doi.org/10.1021/acs.est.6b01389, http://pubs.acs.org/journal/esthag.

Verma, A., Chhokra, A., Geete, A., Mandelkar, S., & Agarwal Malhotra, M. (2021). *Global Governance Initiative. India's water woes: Looming crisis in Punjab and Uttar Pradesh.* Web article. https://www.Globalgovernanceinitiative.Org/Post/India-s-Water-Woes-Looming-Crisis-in-Punjab-and-Uttar-Pradesh.

Wang, Y., Byers, E., Parkinson, S., Wanders, N., Wada, Y., Mao, J., & Bielicki, J. M. (2019). Vulnerability of existing and planned coal-fired power plants in developing Asia to changes in climate and water resources, Royal Society of Chemistry, United States. *Energy and Environmental Science*, *12*(10), 3164–3181. Available from https://doi.org/10.1039/c9ee02058f, http://pubs.rsc.org/en/journals/journal/ee.

Wang, L., Fan, Y. V., Varbanov, P. S., Alwi, S. R. W., & Klemeš, J. J. (2020). Water footprints and virtual water flows embodied in the power supply chain, MDPI AG, Czech Republic. *Water (Switzerland)*, *12*(11), 1–21. Available from https://doi.org/10.3390/w12113006, http://www.mdpi.com/journal/water.

WRI. (2018). WRI Global Power Plant Database, Global Coal Plant Tracker. Available from https://www.wri.org/data/47-worlds-thermal-power-capacity-highly-water-stressed-areas.

Wu, Z., Zhai, H., Grol, E. J., Able, C. M., & Siefert, N. S. (2023). Treatment of brackish water for fossil power plant cooling. *Nature Water*, *1*(5), 471–483. Available from https://doi.org/10.1038/s44221-023-00072-x.

Zhai, H., Rubin, E. S., & Versteeg, P. L. (2011). Water use at pulverized coal power plants with postcombustion carbon capture and storage. *Environmental Science and Technology*, *45*(6), 2479–2485. Available from https://doi.org/10.1021/es1034443.

Zhang, C., Yang, J., Urpelainen, J., Chitkara, P., Zhang, J., & Wang, J. (2021). Thermoelectric power generation and water stress in India: A spatial and temporal analysis, American Chemical Society, China. *Environmental Science and Technology*, *55*(8), 4314–4323. Available from https://doi.org/10.1021/acs.est.0c08724, http://pubs.acs.org/journal/esthag.

Zhang, T. (2020). Methods of improving the efficiency of thermal power plants. *Journal of Physics: Conference Series*, *1449*(1), 012001. Available from https://doi.org/10.1088/1742-6596/1449/1/012001.

Zhou, Q., Hanasaki, N., & Fujimori, S. (2018). Economic consequences of cooling water insufficiency in the thermal power sector under climate change scenarios. *Japan Energies*, *11*(10). Available from https://doi.org/10.3390/en11102686, https://www.mdpi.com/1996-1073/11/10/2686/pdf.

Zhu, Y., Jiang, S., Zhao, Y., Li, H., He, G., & Li, L. (2020). Life-cycle-based water footprint assessment of coal-fired power generation in China. *Journal of Cleaner Production*, *254*, 120098. Available from https://doi.org/10.1016/j.jclepro.2020.120098.

Further reading

Cordoba, P. (2015). Status of flue gas desulphurisation (FGD) systems from coal-fired power plants: Overview of the physic-chemical control processes of wet limestone FGDs. *Fuel*, *144*.

IEA. (2012). World Energy Outlook 2012, IEA, Paris, License: Creative Commons BY 4.0. Available from https://www.iea.org/reports/world-energy-outlook-2012.

Jović, M., Laković, M., & Banjac, M. (2018). Improving the energy efficiency of a 110 MW thermal power plant by low-cost modification of the cooling system. *Energy & Environment*, *29*(2), 245–259. Available from https://doi.org/10.1177/0958305x17747428.

Linnerud, K., Mideksa, T. K., & Eskeland, G. S. (2011). The impact of climate change on nuclear power supply, International Association for Energy Economics, Norway*Energy Journal*, *32*(1), 149–168. Available from https://doi.org/10.5547/ISSN0195-6574-EJ-Vol32-No1-6, http://www.iaee.org/en/publications/journal.aspx.

Nagarkatti, A., & Kolar, A. K. (2021). Advanced coal technologies for sustainable power sector in India. *Electricity Journal*, *34*(6). Available from https://doi.org/10.1016/j.tej.2021.106970, https://www.sciencedirect.com/journal/the-electricity-journal.

Yadav, N. K., & Arora, S. (2021). *Water-inefficient power: Implementing water norms and zero discharge in India's coal power fleet* (pp. 12–20). Centre for Science and Environment.

Chapter 12

Potentials and limitations of water footprints for gauging environmental sustainability

Cayetano Navarrete-Molina[1,*], María de los Ángeles Sariñana-Navarrete[1], Cesar Alberto Meza-Herrera[2], José Luis Rodríguez-Álvarez[3] and Raúl Alejandro Cuevas-Jacquez[3]

[1]Department of Chemical area Environmental Technology, Technological University of Rodeo, Rodeo, Durango, Mexico, [2]Regional Universitary Unit on Arid Lands, Chapingo Autonomous University, Bermejillo, Durango, Mexico, [3]Campus Region of the Llanos, National Technological of Mexico, Guadalupe Victoria, Durango, Mexico

**Corresponding author. e-mail address: z42namoc@uco.es.*

12.1 Introduction

The accelerated transformation that the Earth has undergone in recent decades due to intense human activity has not only been positive, but in many cases negative: negatively affecting vital life cycles by decreasing the availability of resources in a faster and more decisive way compared to any other period in history (Morandi, 2022). This decline increases the uncertainty of achieving environmental sustainability (ES), considering that the growing demand for natural resources (NRs) is key not only for sustaining the planet's population but also for sustaining economic development (Sarkodie, 2021). In this sense, ES is indispensable to ensure the availability of NRs, not only for this generation but also for the well-being of future generations (Bishop, 2021; Dube et al., 2021). However, ES involves several components, which are closely interrelated and often complement each other to approximate the balance between environmental conservation and human development (Alsaad & Abdul-Fariji, 2021; Galli et al., 2020; Hanifah et al., 2020; Jain & Mohapatra, 2023).

The development of ES goes hand in hand with society and the proper management of the resources that make human progress and evolution possible, such as conservation and sustainable use (SU) of NRs, planning and development of new infrastructure, social and economic development, government policies on environmental management, and the promotion of environmental education (Yue et al., 2020); however, each of these components is closely related to the use, management, and conservation of water. In many countries, the search for new sources of water supply has intensified in the face of water scarcity (WS), considering that water is key for domestic and industrial and agricultural activities (IAA) (Di Baldassarre et al., 2018), and that scarcity leads to overexploitation. Given this and with the support of current technological resources, it is necessary to improve the integrated management of water resources (WRs) by implementing actions that promote the efficient use and increase the resilience of surface and groundwater basin resources, as well as forests and wetlands (Scanlon et al., 2023), which are key to the capture, regeneration, supply, and maintenance of water quality (Levizzani & Cattani, 2019).

Considering that integrated WR management is the key to maintaining the balance in water basins, in the early 2000s, Hoekstra (2003) introduced the term "water footprints" (WFs) as an indicator of a nation's overall WR use in terms of food (Mekonnen & Gerbens-Leenes, 2020) and consumption patterns (Hoekstra, 2017). The WF encompasses the use of the WR itself, without emphasizing where and when it was used, and what type of water it was. Thus, the concepts of green water footprint (GWF), indicative of rainwater consumption; blue water footprint (BWF), indicative of surface or groundwater consumption; and grey water footprint (GrWF), indicative for water consumption (WC) needed to assimilate a pollution load, were included (Hoekstra, 2017). The WFs are of particular interest, considering

Water Footprints and Sustainable Development. DOI: https://doi.org/10.1016/B978-0-443-23631-0.00012-1

that one of the goals of the United Nations 2030 Agenda is to "Ensure availability and sustainable management of water." In this sense, the WFs contribute to identifying trade-offs and synergies between strategies, not only to achieve water security (WSe), but also energy and food security (Berger et al., 2021). Therefore, this chapter explores the main potentialities and limitations of the WFs as indicators of ES.

12.2 Development of water footprints as an indicator of environmental sustainability

The development of human activities in recent years has generated environmental impacts (EIs) in the process, consuming resources at a faster rate than nature itself can regenerate (Lovarelli et al., 2016). One of the most overexploited resources, whose scarcity leads to environmental, social, and economic problems, is freshwater, estimating an approximate demand of more than 60 trillion cubic meters by 2050 (Shahzad et al., 2017; Wang et al., 2021), causing the creation of methodologies, whose purpose is to analyze and quantify water use (WU) (Lovarelli et al., 2016; Vollmer & Harrison, 2021). In this sense, the calculation of WFs arises as part of the procedures focused on the study of water and the consumption pattern of a person, community, or country (Gerbens-Leenes et al., 2021; Hoekstra, 2017). Its proper calculation contributes to the study and questions the main challenges facing ES, considering that ES focuses on limiting human activity to the regenerative capacity of ecosystems in conjunction with the quality of human life (Olawumi & Chan, 2018), and the link between both concepts arises from the need to address the perspectives around the assessment of the sustainability of WR use (Liu et al., 2020).

Fig. 12.1 shows the three components into which the WF has been divided, where the use of WRs in all aspects is conceptualized: GWF, BWF, and GrWF (De Girolamo et al., 2019; Hoekstra, 2017). In technical terms, the WF is expressed as the volume of water used to produce a product (m^3/t) or as the water volume of an area, individual, or

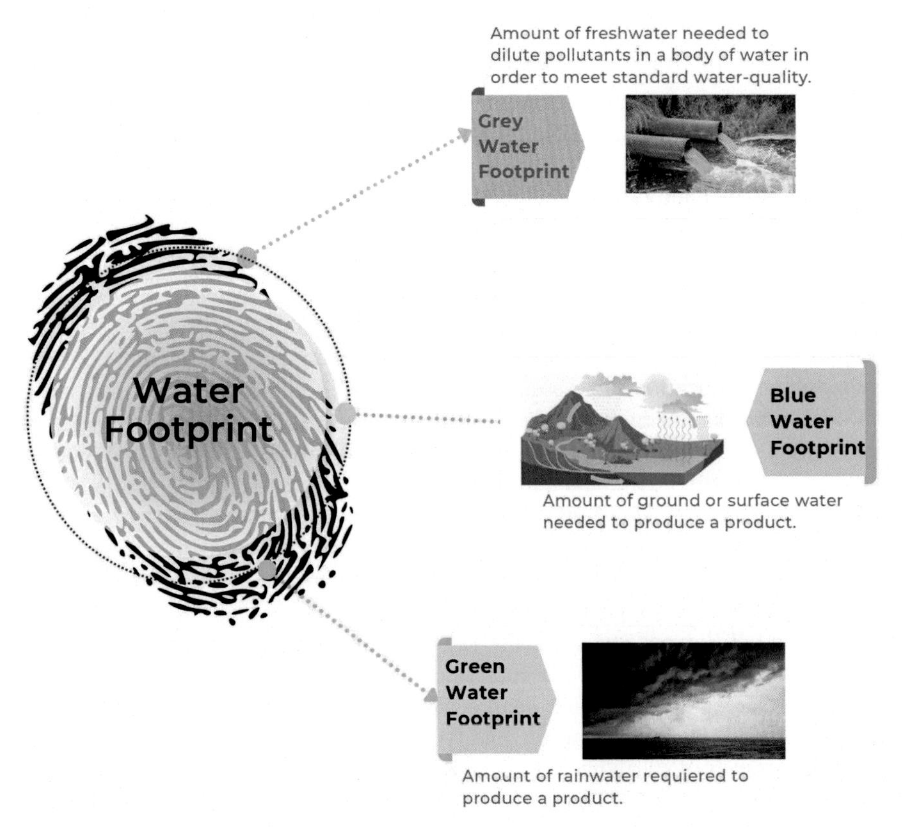

FIGURE 12.1 Components of the water footprint.

community (m^3/year) (Lovarelli et al., 2016). The study and continuous measurement of the WFs are necessary to estimate the quality and safeguard the quantity of WRs available for future generations (Pellicer-Martínez & Martínez-Paz, 2016). To this end, indicators such as the Water Stress Index, Water Exploration Index, Exploitable Water Resources, and Water Footprint Network, among others, have been developed and used to assess the WFs, which also include indicators of extreme situations, such as WS or drought (D'Ambrosio et al., 2020).

12.2.1 Water footprints calculation

Initially, the calculation of WF focused on water used directly in the production processes of goods and services (Chenoweth et al., 2014). However, over time, the limitations of using this methodology directly, excluding concepts such as "virtual water" and a nation's stock of natural water capital, were recognized (Wu et al., 2022a). The introduction of these new terms, in conjunction with the original definition of WF calculation, has enabled the development of two- and three-dimensional scale model studies that help to differentiate sustainable from unsustainable WRs (Wu et al., 2022b). Such studies of WFs have been conducted from various perspectives and using specific indicators. Some case studies and the most relevant results are presented below (Table 12.1).

12.2.2 Standardization and evolution of water footprints

The main objective of the calculation of the WF is to focus on WRs management. However, it excludes the impact of WU by anthropogenic activities, leading to the emergence of terms such as eutrophication, acidification, and ecotoxicity (Ansorge, 2020). These concepts contribute to assessing the impact of the WFs, and were initially embodied in the ISO 14046 international standard, whose focus is to "*specify principles, requirements, guidelines for water footprint calculation, assessment and reporting*" (Deepa et al., 2021). The Water Footprint International Standard (ISO-WF) maintains the original definition of the WF. However, it proposes a broader overview of the impacts generated on the water in terms of quantity and quality (Diaz, 2021). The methodological process for the calculation of the WFs through the ISO 14046 international standard encompasses more specific concepts to improve the understanding of the impact of the use of WRs, as well as to optimize their use, using assessment techniques that can be used at an international level (Fig. 12.2) (Forin et al., 2020; Mansouri et al., 2022; Pierrat et al., 2023).

The following is a brief commentary on each of the main phases covered in the ISO 14046 2014 standard.

12.2.2.1 Goal and scope

The definition of the objective and scope is a function of the information that is expected to be obtained and sets the limits of the analytical processes in the first instance. It will also have an impact on the accuracy of the results and their interpretation (International Organization for Standardization, 2014).

12.2.2.2 Inventory analysis

According to the international standard, data collection should include water abstracted and released to the environment without human transformation. It should also consider the input and output of any substances, forms of use, location of abstractions, and flow releases, following the established objective and scope (International Organization for Standardization, 2014).

12.2.2.3 Impact assessment

At this stage, the EIs of the inventory analysis are assessed. However, although the international standard does not provide specific guidance on which elements should be chosen, it does establish certain requirements such as EI category indicators, characterization models, geographical dimension, and those that reflect the scarcity and degradation footprint (International Organization for Standardization, 2014).

12.2.2.4 Results analysis

The interpretation of the results identifies the potentially significant EIs, in addition to identifying the limitations exhibited during the study, leading to conclusions and recommendations, which should be consistent with the results of the study (International Organization for Standardization, 2014).

In this context, the ISO 14046:2014 international standard has been used as a worldwide reference for the WFs calculation, Table 12.2 shows some research carried out in this regard, as well as its main findings.

TABLE 12.1 Study of the components of the water footprint with different evaluation models.

Country/area	Model of the water footprint	Component(s) of water footprint calculated	Highlight (reference)
China/surface water basin	Water Quality Index (WQI)	Grey Water Footprint (GrWF)	Improvement of the traditional WQI model by breaking down an additional component, the dilution required to obtain the desired water quality (Lahlou et al., 2023).
Italy/Agriculture	Water Footprint Network	All components	The Blue Water Footprint (BWF) is not sustainable to meet irrigation requirements, nor to dilute the pollutant load of surface water (D'Ambrosio et al., 2020).
China/agricultural, industrial, and domestic	3D-WF Model	BWF and GrWF	The largest BWF is for the agricultural sector. The model covers more calculation indices, therefore greater certainty of water footprint (WF) and sustainable development (Wu et al., 2022a).
Global/Agriculture		BWF	More than 50% of the BWF of crops is unsustainable. Countries in the Middle East and Central Asia have the largest unsustainable BWF. Wheat and cotton production have the highest BWF (Mekonnen & Hoekstra, 2020).
China/Surface Water Basin	Logarithmic mean divisia index model	GrWF	The economy played a fundamental role in the comprehensive development of environmental protection. The industrial sector generates the largest load of GrWF to the Yangtze River basin (Fu et al., 2022).
Australia/Agriculture	Water Footprint Assessment	All components	Unlike vegetables, the production of fruit species was considered as a possible threat of scarcity of water resources, generating the highest WF (Hossain et al., 2021).
Germany and Netherlands/Surface Water Basin	Water Footprint Assessment	GrWF	Contamination by pharmaceutical sources has the potential to become a future global contamination threat. Reducing the consumption of foods of animal origin would favor the reduction of contamination by pharmaceuticals (Wöhler et al., 2020).
Global/Agriculture	Water Footprint Assessment	BWF and GrWF	The WF generated by avocado production in Mexico is three times larger than other producing countries, representing an income of $2.8 billion. This WF is presented as a source of water scarcity. Sustainable development initiatives are needed (Sommaruga & Eldridge, 2021).
Europe/Industrial	Water Footprint Assessment	All components	During the COVID-19 lockdown, the WF of power plants decreased by 16%, and the import of virtual water increased. Electricity was generated using combined technologies that consume less water (Roidt et al., 2020).
United States of America/Industrial	Water Footprint Assessment	General Water Footprint	The exploration of new sources of oil and gas has generated an increase in the volume of water used, and the volume of wastewater generated, exceeding 700% in a period of 6 years (Kondash et al., 2018).
England/Industrial	Water Footprint Assessment	All components	The production of a brick generates an expenditure of 2.02 L of water. The reuse of water used for brick production reduces the consumption of clean water by 15.6%, presenting itself as a solid and effective strategy for sustainable consumption in this industrial sector (Skouteris et al., 2018).

The Water Footprint Assessment (WFA), using specific indicators, or through the ISO 14046 protocol, generates relevant indicators of WU and the EIs that arise. In a single study, the use of concepts such as eutrophication, acidification, ecotoxicity, WS footprint, and virtual water, favours understanding and interpretation, in addition to providing clarity regarding preventive measures, resource management, and good practices aimed at ES.

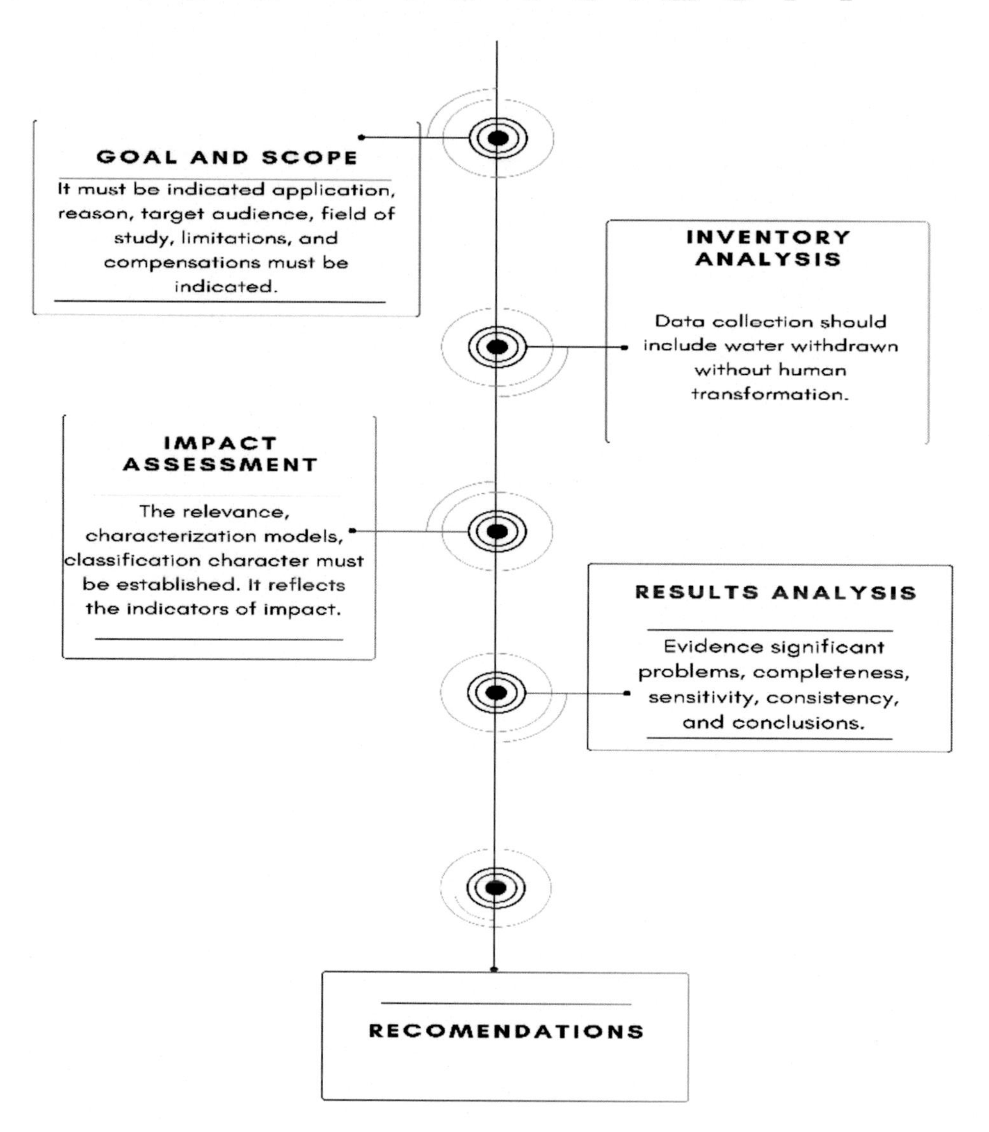

FIGURE 12.2 Phase's main of water footprint calculation using the standard ISO 14046:2014.

12.3 Potentials of water footprints for gauging environmental sustainability

The unsustainability of human activities and the EI that has been generated has led to the creation of tools to measure the impacts caused. Derived from the above, the WFs arose from the need to assess the use of WRs, allowing a holistic measurement throughout the cycle of a product or service, within domestic activities or IAA (Hoekstra, 2017). However, if a real measurement of these impacts is desired, it is necessary to consider other indicators of ES, to realize a true approximation in the calculation of the impacts of human activities. In this context, methodologies have been developed to quantify other indicators, such as the carbon footprint, ecological footprint, energy footprint, land footprint, materials footprint, resource footprint, nitrogen footprint, phosphorus footprint, chemical footprint, biodiversity footprint, and emission footprint (Laurent & Owsianiak, 2017). The calculation of the WFs is specifically studied as an indicator of ES, where the most frequently used methods are analyzed, without suffering any type of modification, and that researchers can make minor modifications to the method, aligning them with their research interests, as well as the available resources.

TABLE 12.2 Main results of some research carried out using the standard ISO 14046:2014.

Country/Area	Highlight (Reference)
China/Textile industry	The ecotoxicity found was mainly due to the excessive load of Zn^{2+} and CS_2, rather than conventional contaminants. The water footprint of scarcity in China would be greater if the viscous pulp manufacturing process is carried out in an area with limited water resources. The disposal of pollutants is freely discharged into clean and polluted bodies of water, without distinction (Zhu et al., 2020).
Brazil/Agriculture	The activities that contribute most to the carbon and water footprint were fertilization and irrigation in green coconut cultivation. The use of vegetable oils as insecticides reduced the ecotoxic impact, improving the profile of the water footprint. Adopting good production practices such as companion crops can reduce environmental impact (Sampaio et al., 2021).
Malaysia/Agriculture	The production of 1 kg of rice requires 1800–2600 liters of water. 42% of the water used during the cycle is used during the vegetative stage of the crop. Global demand for rice by 2030 is expected to increase to 533 million tons (Han et al., 2021).
United States of America/Agriculture	Water consumption in the peanut growth and development stage is different between the counties considered in the assessment. Incorporating crop-specific characterization factors will improve local assessments for proper water resources management (Deepa et al., 2022).
China/Mining	Gold production generates large emissions of chromium and arsenic derived from energy consumption. The pollution of water basins is mainly due to emissions of chromium and copper. The environmental impact categories considered would be reduced by more than 10% if hydropower and biodiesel sources are used (Chen et al., 2020).
Iran/Steel industry	The extraction of water from the Zayandeh-Rud river basin for the maintenance of steel plants is not sustainable, estimating a shortage rate of 145%. Illegal water extraction has caused a drop in groundwater level of 5 m on average, leading to salinization of watersheds intended for farmland (Nezamoleslami & Hosseinian, 2020).

To appreciate the importance of assessing the use of WRs, it is necessary to mention that freshwater represents only 2.5% of all available water on earth, and the increase in population has generated a greater demand for drinking water and food sources. Given this scenario, ensuring the availability of WRs for the population is one of the main limiting factors for sustainable development (SD) (Ansorge et al., 2022). Therefore, it is necessary to evaluate the WFs by the method proposed in 2003 by Hoekstra et al., reported as WFA or through the guidelines of the ISO 14046 international standard, for the generation of comprehensive assessments that show the sustainability index of WU and the environmental footprint generated by the use of WRs, by providing real-time information on WC associated with everyday goods, services, and IAA. Such evidence is the starting point for establishing environmental management policies and sustainable WR management, as well as identifying areas of vulnerability, contributing to the generation of detailed maps on climatic conditions, agricultural practices, and trade statistics, which strengthen WFAs and contribute to reducing those areas and practices where WFs are unsustainable (Hogeboom, 2020).

Such assessments are important, considering that more than four billion people in the world experience a lack of WRs, which is evident in developing regions, where climate variability makes them prone to suffer the consequences of WS (Keune & Miralles, 2019). In these areas, the WFA and ISO-WF allow the detection of areas where the use of WRs is unsustainable, that is, located outside ecological limits (Hogeboom et al., 2020). In this sense, it is estimated that more than 90% of the world's WFs are related to agricultural activities, where approximately 40% of the use of water supplies is unsustainable, risking the availability of WRs, leading to situations such as food insecurity (Qasemipour & Abbasi, 2019). To reduce water unsustainability, Yan et al. (2023), highlighted the importance of forest cover from the point of view of WFs, and proposed a special emphasis on the relationship between regional biodiversity conservation and SD of WRs, stating that resource management would be improved by planning with the availability of watersheds (Yan et al., 2023).

12.4 Limitations of water footprints for gauging environmental sustainability

Today, the calculation of WFs has become an essential tool for understanding and addressing the challenges related to water availability and management in today's world. However, the comparability and transparency of methodologies used to calculate the WF present significant challenges (Rodríguez et al., 2023). In this regard, current methodologies for calculating WFs vary in their approach and complexity, often making it difficult to directly compare different studies and analyses. This is essential for a proper understanding of WFs, and for its calculation, it is key to consider both quantitative and qualitative aspects of the water used (Chuang et al., 2023). In this regard, one of the key challenges in comparing WF calculations is the lack of transparency in data collection and processing; similarly, data availability, and quality can vary significantly between different studies and regions (Guan et al., 2022). In addition, a lack of standardization in methodological calculation procedures is evident, as well as almost nonexistent public access to the underlying databases, which raises important doubts about the accuracy and reliability of the results, thus creating important challenges that need to be addressed by the different existing WFs calculation methodologies (Tompa et al., 2022).

Water as an NR is necessary for social and economic development, as it is involved in production processes. Therefore, the study of WFA and ISO-WF are presented as excellent methodologies to estimate the consequences of WU, as well as tools to develop strategies for the proper management of WRs (Chen et al., 2021). However, such assessments are not without obstacles during their methodological development. In this context, the estimation of the WF, either overall, or any of its components (GWF, BWF, or GrWF) tends to suffer from biases during the monitoring of water inflows and outflows. This is most common in hydrological basins, considering that rainfall regimes are uncertain during the seasons. In addition, these basins are partially monitored, and estimates are generated that need to be strengthened to assess the impacts that may be created (Taffarello et al., 2020). Another interference, which is identified during the WFA, is the difference between the assessment methods: while the WFA Manual (Hoekstra, 2003) allows to assessment of any of its components separately, the ISO-WF frames all components appropriately delimited in a general description of the assessment procedure, generating contributions whose results cannot be comparable (Rodríguez et al., 2023). Also, the complexity of the procedure may restrict the method, as large amounts of information are required and access to real-time WC data is sometimes limited (Chuang et al., 2023).

However, some key points that limit the WFA in different areas, as well as limitations related to the methodology used for its calculation, have been identified (Table 12.3).

To address these challenges and improve the comparability and transparency of the procedures and methodologies used to calculate WFs, it is essential to take concrete steps. These measures should include promoting the standardization of calculation methodologies and data collection at the international level. This would help to ensure that the results of different studies are comparable and reliable. In this sense, another recommended measure would be to promote transparency by providing public access to the input databases and algorithms used in WF calculations. This would promote an independent review of the results and increase the level of confidence in the accuracy of the results. In addition, it is important to encourage collaboration between governments, companies and international organizations to share data and experience in measuring and managing WFs. If progress is to be made in understanding and mitigating the water-related impacts of anthropogenic activities on the planet, a collaborative approach and open access to the information used in the calculations made to quantify WFs need to be encouraged.

TABLE 12.3 Limiting scenarios in the utilization of the standard ISO 14046:2014.

Country/Area	Water footprint methodology	Limitations (Reference)
Hungry/Food production	Water Footprint Network Database	(a) Lack of homogeneity in research methodologies; (b) Low availability of information (Tompa et al., 2022).
Turkey/Agriculture	Water Footprint Assessment	(a) Spatio-temporal irrigation data were not available; (b) The calculation of the grey water footprint was considered as a constant; (c) No prescribed data were used (Muratoglu et al., 2022).
Mexico/Food production	Water Footprint Assessment	(a) Virtual water flows were not considered; (b) Other environmental indicators can reinforce the study of WF and; (c) The relationship of diets with the environment in the area studied (Lares-Michel et al., 2022).
China/Agriculture	Water Footprint Evaluation Handbook	Limited access to real-time data and measurements, creates a gap between research data and reality (Guan et al., 2022).
Global/ International trade	Water Stress Index	Currently, water scarcity indices do not consider water quality, so water with a high pollutant load is excluded when estimating scarcity indices (Wu et al., 2022b).

12.5 Water footprints calculation thresholds and approach

The calculation of the WFs was designed to quantify the magnitude of pressure on the finite resource water, without necessarily comparing the results with absolute limits or thresholds derived from the ecological limits of the planet or environmental sustainability objectives. For its calculation to be useful in decision-making contexts, it is important to link the obtained value of WFs of the analyzed system and the maximum allowable value of environmental pressure that the system can withstand (WFs threshold), above which the system is environmentally unsustainable in absolute terms (Laurent & Owsianiak, 2017).

Therefore, the purpose of the calculation of the WFs is to provide timely and meaningful information on WU and management, to be assertive in making decisions on the SU of WRs, and the protect the environment. In this scenario, Ruddell (2018), proposes the introduction of the threshold-based Volumetric WF indicator, which sets a limit for the BWF, establishing a threshold called free water footprint, where the use of resources is not associated with unfavorable EIs. Likewise, the use of indicators that favor the development of ecosystems raises awareness of environmental limits, that is, the maximum load of resource exploitation that can regenerate naturally, as suggested by Scherer and Pfister (2016), through the Environmental Water Stress Footprints, which makes it possible to estimate the limits, not only of WS but also of excess water that can damage the ecosystem (Fig. 12.3). The establishment of these limits (thresholds) goes hand in hand with the objectives set out in the different studies that evaluate the WFs. Therefore, the WFA and the ISO-WF establish clear and consistent procedures when setting targets for the study area (Hoekstra et al., 2011).

The calculation and valuation of WFs in different study areas represent a fundamental concept in the sustainable management of WRs and in the evaluation of the EI of human activity. In the same way, it contributes to identifying those risk areas, where the use of WRs is unsustainable, and generates negative EIs. However, it is necessary to incorporate other ES indicators, which show an integral vision of present and future EIs, allowing the generation of coordinated actions in the sustainable management of NRs (Pryor et al., 2022). In their research, Pryor et al. (2022), propose an association of indicators for estimating WS, from the data collected from Lesotho, Africa; they integrate regional climate models, hydrological models of hydrographic basins, and the analysis of WFs. The results of the study provide an estimate of the blue water availability of any hydrological basin under study upto 2100, besides presenting scenarios with moderate and severe rates of WS and contributing to determining, as closely as possible, WF thresholds and thus helping society's efforts against climate change, which continues to represent a major challenge. Such WF thresholds

FIGURE 12.3 Indicators that define the environmental water stress footprints.

must be based on the ecological limits of the planet. However, only in the case of the BWF, there is a direct coincidence between its calculation value and the estimated value of the ecological limit of the planet (Paterson et al., 2015; Venetoulis & Talberth, 2008). Both have the same units and are based on almost the same underlying data and modeling approaches. For example, at the global level, a threshold for BWF of 4000 km^3/year has been proposed (Laurent & Owsianiak, 2017).

12.6 Perspectives of water footprints in the assessment of environmental sustainability

Ensuring WSe and achieving the SD goals by 2030 requires a change in water management (WM) and a better understanding of water issues, which promotes a more sustainable, efficient, fair, and equitable use of water (Meza-Herrera et al., 2022; Ornelas-Villarreal et al., 2022a,b). In this sense, the calculation of WF stands as a new perspective to address the great challenges in terms of WSe, considering the perspective of preserving ecosystems, in a context of peace and political stability (Pellicer-Martínez & Martínez-Paz, 2016). This perspective is of great relevance, considering that the calculation of the WF proposes to explore the management of WRs and other relationships of WU, with sustainable consumption and virtual water transfers of goods and services that occur beyond the borders of the countries, showing a real situation of the country's water (Sun et al., 2019). Therefore, the determination of WF can provide a new perspective by complementing other indicators, thereby guiding public policies and assisting in decision-making, not only governmental but also business, social and personal, to promote more sustainable, efficient and fair WM (Han et al., 2021).

This perspective has gained relevance in the scientific community in recent years, considering that the WFs calculation methodologies and their results have provided important guidelines for water policies, which should be promoted in different countries, highlighting that international trade can help to achieve more efficient use of water on a global scale, for example, exporting products with a high water requirement from a country with high water productivity (lower volume of water per ton of product) to a country with lower productivity (Liu et al., 2020). However, it is necessary to assess whether these products contribute to the food self-sufficiency of the poorest strata of the exporting country. Therefore, the calculation of WFs should be considered as an indicator that has the potential to guide public policies that promote the protection and SU of WRs, which should be carried out from the perspective of territorial planning and development, and the impacts and benefits of WU should be contextualized (Hua et al., 2020). Although the calculation and analysis of WFs must be complemented by other indicators if the major water challenges are to be addressed, it provides a nontraditional approach, which allows the problem to be viewed from a more holistic perspective and has the potential to generate a response aimed at sustainability (Navarrete-Molina et al., 2019a, 2019b, 2020; Rios-Flores et al., 2014).

12.7 Conclusions

Global water problems show a close relationship between the overexploitation and contamination of water with the increase in population and globalization. In addition, a disconnection is evident between exporters and importers of "virtual water" and how the calculation of the WFs reveals patterns of WU, generating social and environmental costs. Therefore, it is important to recognize and determine the potentials and limitations of WFs for gauging ES, considering that calculating the WFs is important if you want to evaluate sustainability, as well as guide public policies. In addition, it is necessary to promote a holistic perspective to raise awareness about the conscious use of water, modify consumption patterns and anticipate the repercussions of political and commercial decisions.

In this context, it is evident to recognize the usefulness of calculating the WFs as a communication and measurement tool for ES. However, it is necessary to improve and standardize calculation methodologies, considering aspects such as the equitable allocation of WRs and the inclusion of factors, such as the water assimilation capacity of water bodies. In addition, the calculation of the WFs must be complemented with other indicators to obtain a complete vision of ES. Due to the above, it is necessary to propose improvements in the calculation of the WFs, promoting collaboration between researchers, academics, companies, and governments toward the development of research aimed at understanding the real situation regarding WS and pollution at the local, regional, and global levels. For this reason, it is important to identify existing knowledge gaps, as well as coordinate research areas and facilitate the exchange of knowledge to comprehensively address water-related challenges.

Global water problems show a close relationship between the overexploitation and contamination of water with the increase in population and globalization. In addition, a disconnection is evident between exporters and importers of "virtual water" and how the calculation of the WFs reveals patterns of WU, generating social and environmental costs. Therefore, it is important to recognize and determine the potentials and limitations of WFs for gauging ES, considering that calculating the WFs is important if you want to evaluate sustainability, as well as guide public policies. In addition, it is necessary to promote a holistic perspective to raise awareness about the conscious use of water, modifying consumption patterns, and anticipating the repercussions of political and commercial decisions.

References

Alsaad, M. A. S., & Abdul-Fariji, J. (2021). The impact of the university's environmental sustainability strategy on the continuous improvement of the overall university performance: an exploratory study on a sample of Basra University teachers. *Journal of Social Science For Policy Implications*, *9*(2). Available from https://doi.org/10.15640/jsspi.v9n2a2.

Ansorge, L. (2020). Water footprint: two different methodologies. *Tecnura*, *24*(66), 119–121. Available from https://doi.org/10.14483/22487638.15903.

Ansorge, L., Stejskalová, L., Vološinová, D., & Dlabal, J. (2022). Limitation of water footprint sustainability assessment: a review. *European Journal of Sustainable Development*, *11*(2), 1. Available from https://doi.org/10.14207/ejsd.2022.v11n2p1.

Berger, M., Campos, J., Carolli, M., Dantas, I., Forin, S., Kosatica, E., Kramer, A., Mikosch, N., Nouri, H., Schlattmann, A., Schmidt, F., Schomberg, A., & Semmling, E. (2021). Advancing the water footprint into an instrument to support achieving the SDGs—recommendations from the "Water as a Global Resources" Research Initiative (GRoW). *Water Resources Management*, *35*(4), 1291–1298. Available from https://doi.org/10.1007/s11269-021-02784-9, http://www.wkap.nl/journalhome.htm/0920-4741.

Bishop, C. P. (2021). Sustainability lessons from appropriate technology. *Current Opinion in Environmental Sustainability*, *49*, 50–56. Available from https://doi.org/10.1016/j.cosust.2021.02.011, http://www.elsevier.com/wps/find/journaldescription.cws_home/718675/description#description.

Chen, F., He, W., Tian, Z., & Wang, L. (2020). Impacts of silk garment production on water resources and environment. *Environmental Science and Engineering*, 311–319. Available from https://doi.org/10.1007/978-3-030-45263-6_28, http://www.springer.com/series/7487, Springer China.

Chen, W., Hong, J., Wang, C., Sun, L., Zhang, T., Zhai, Y., & Zhang, Q. (2021). Water footprint assessment of gold refining: case study based on life cycle assessment. *Ecological Indicators*, *122*, 107319. Available from https://doi.org/10.1016/j.ecolind.2020.107319.

Chenoweth, J., Hadjikakou, M., & Zoumides, C. (2014). Quantifying the human impact on water resources: a critical review of the water footprint concept. *Hydrology and Earth System Sciences*, *18*(6), 2325–2342. Available from https://doi.org/10.5194/hess-18-2325-2014, http://www.hydrol-earth-syst-sci.net/volumes_and_issues.html.

Chuang, W. K., Lin, Z. E., Lin, T. C., Lo, S. L., Chang, C. L., & Chiueh, P. T. (2023). Spatial allocation of LID practices with a water footprint approach. *Science of the Total Environment*, *859*. Available from https://doi.org/10.1016/j.scitotenv.2022.160201, http://www.elsevier.com/locate/scitotenv.

D'Ambrosio, E., Gentile, F., & De Girolamo, A. M. (2020). Assessing the sustainability in water use at the basin scale through water footprint indicators. *Journal of Cleaner Production*, *244*, 118847. Available from https://doi.org/10.1016/j.jclepro.2019.118847.

Deepa, R., Anandhi, A., Bailey, N. O., Grace, J. M., Betiku, O. C., & Muchovej, J. J. (2022). Potential environmental impacts of peanut using water footprint assessment: a case study in Georgia. *Agronomy*, *12*(4). Available from https://doi.org/10.3390/agronomy12040930, https://www.mdpi.com/2073-4395/12/4/930/pdf.

Deepa, R., Anandhi, A., & Alhashim, R. (2021). Volumetric and impact-oriented water footprint of agricultural crops: a review. *Ecological Indicators*, *130*, 108093. Available from https://doi.org/10.1016/j.ecolind.2021.108093.

De Girolamo, A. M., Miscioscia, P., Politi, T., & Barca, E. (2019). Improving grey water footprint assessment: accounting for uncertainty. *Ecological Indicators*, *102*, 822–833. Available from https://doi.org/10.1016/j.ecolind.2019.03.040, http://www.elsevier.com/locate/ecolind.

Diaz, L. (2021). *Water footprint: a sustainability tool for industries*. IWA Publishing, pp. 333–342. Available from https://doi.org/10.2166/9781789060676_0333.

Di Baldassarre, G., Wanders, N., AghaKouchak, A., Kuil, L., Rangecroft, S., Veldkamp, T. I. E., Garcia, M., van Oel, P. R., Breinl, K., & Van Loon, A. F. (2018). Water shortages worsened by reservoir effects. *Nature Sustainability*, *1*(11), 617–622. Available from https://doi.org/10.1038/s41893-018-0159-0, http://www.nature.com/natsustain/.

Dube, O. P., Brondizio, E. S., & Solecki, W. (2021). Technology innovations and environmental sustainability in the Anthropocene. *Current Opinion in Environmental Sustainability*, *49*, A1–A2. Available from https://doi.org/10.1016/j.cosust.2021.08.001.

Forin, S., Mikosch, N., Berger, M., & Finkbeiner, M. (2020). Organizational water footprint: a methodological guidance. *The International Journal of Life Cycle Assessment*, *25*(2), 403–422. Available from https://doi.org/10.1007/s11367-019-01670-2.

Fu, T., Xu, C., Yang, L., Hou, S., & Xia, Q. (2022). Measurement and driving factors of grey water footprint efficiency in Yangtze River Basin. *Science of the Total Environment*, *802*, 149587. Available from https://doi.org/10.1016/j.scitotenv.2021.149587.

Galli, A., Iha, K., Moreno Pires, S., Mancini, M. S., Alves, A., Zokai, G., Lin, D., Murthy, A., & Wackernagel, M. (2020). Assessing the ecological footprint and biocapacity of Portuguese cities: critical results for environmental awareness and local management. *Cities*, *96*. Available from https://doi.org/10.1016/j.cities.2019.102442, http://www.elsevier.com/inca/publications/store/3/0/3/9/6/.

Gerbens-Leenes, W., Berger, M., & Allan, J. A. (2021). Water footprint and life cycle assessment: the complementary strengths of analyzing global freshwater appropriation and resulting local impacts. *Water (Switzerland)*, *13*(6). Available from https://doi.org/10.3390/w13060803.

Guan, D., Wu, L., Cheng, L., Zhang, Y., & Zhou, L. (2022). How to measure the ecological compensation threshold in the upper Yangtze River basin, China? An approach for coupling InVEST and grey water footprint. *Frontiers in Earth Science, 10.* Available from https://doi.org/10.3389/feart.2022.988291, https://www.frontiersin.org/journals/earth-science.

Han, X., Zhao, Y., Gao, X., Jiang, S., Lin, L., & An, T. (2021). Virtual water output intensifies the water scarcity in Northwest China: current situation, problem analysis and countermeasures. *Science of the Total Environment, 765*, 144276. Available from https://doi.org/10.1016/j.scitotenv.2020.144276.

Hanifah, M., Mohmadisa, H., Yazid, S., Nasir, N., Samsudin, S., & Balkhis, N. S. (2020). Determination of physical geographical components in the construction of environmental sustainability awareness index of the Malaysian society. *Asia-Pacific Social Science Review, 20*(3), 142–152. Available from http://apssr.com/wp-content/uploads/2020/09/RB-2.pdf.

Hoekstra, A. Y. (2003). Virtual water trade: proceedings of the international expert meeting on virtual water trade, value of water research report series No. UNESCO-IHE. 12.

Hoekstra, A. Y. (2017). Water footprint assessment: evolvement of a new research field. *Water Resources Management, 31*(10), 3061–3081. Available from https://doi.org/10.1007/s11269-017-1618-5, http://www.wkap.nl/journalhome.htm/0920-4741.

Hoekstra, A. Y., Chapagain, A. K., Aldaya, M. M., Mekonnen, M. M. (2011).The water footprint assessment manual. Setting the global standard. Water Footprint Network. Daugherty Water for Food Global Institute: Faculty Publications. vol. 77.

Hogeboom, R., de Bruin, D., Schyns, J. F., Krol, M., & Hoekstra, A. Y. (2020). Capping human water footprints in the world's river basins. *Earth's Future, 8*(2). Available from https://doi.org/10.1029/2019EF001363, http://onlinelibrary.wiley.com/journal/10.1002/(ISSN)2328-4277.

Hogeboom, R. J. (2020). The water footprint concept and water's grand environmental challenges. *One Earth, 2*(3), 218–222. Available from https://doi.org/10.1016/j.oneear.2020.02.010, http://www.cell.com/one-earth.

Hossain, I., Imteaz, M. A., & Khastagir, A. (2021). Water footprint: applying the water footprint assessment method to Australian agriculture. *Journal of the Science of Food and Agriculture, 101*(10), 4090–4098. Available from https://doi.org/10.1002/jsfa.11044, http://onlinelibrary.wiley.com/journal/10.1002/(ISSN)1097-0010.

Hua, E., Wang, X., Engel, B. A., Sun, S., & Wang, Y. (2020). The competitive relationship between food and energy production for water in China. *Journal of Cleaner Production, 247*, 119103. Available from https://doi.org/10.1016/j.jclepro.2019.119103.

International Organization for Standardization. (2014). Environmental management—water footprint—principles, requirements and guidelines.

Jain, N., & Mohapatra, G. (2023). A comparative assessment of Composite Environmental Sustainability Index for emerging economies: a multidimensional approach. *Management of Environmental Quality: An International Journal, 34*(5), 1314–1331. Available from https://doi.org/10.1108/MEQ-12-2022-0330, http://www.emeraldinsight.com/info/journals/meq/meq.jsp.

Keune, J., & Miralles, D. G. (2019). A precipitation recycling network to assess freshwater vulnerability: challenging the watershed convention. *Water Resources Research, 55*(11), 9947–9961. Available from https://doi.org/10.1029/2019WR025310, http://agupubs.onlinelibrary.wiley.com/hub/journal/10.1002/(ISSN)1944-7973/.

Kondash, A. J., Lauer, N. E., & Vengosh, A. (2018). The intensification of the water footprint of hydraulic fracturing. *Science Advances, 4*(8). Available from https://doi.org/10.1126/sciadv.aar5982, http://advances.sciencemag.org/content/4/8/eaar5982.

Lahlou, F. Z., Mackey, H. R., & Al-Ansari, T. (2023). Towards the development of an improved mass balance and water quality index based grey water footprint model. *Environmental and Sustainability Indicators, 18.* Available from https://doi.org/10.1016/j.indic.2023.100236, http://www.journals.elsevier.com/environmental-and-sustainability-indicators.

Lares-Michel, M., Housni, F. E., Aguilera Cervantes, V. G., Reyes-Castillo, Z., Michel Nava, R. M., Llanes Cañedo, C., & López Larios, Md. J. (2022). The water footprint and nutritional implications of diet change in Mexico: a principal component analysis. *European Journal of Nutrition, 61*(6), 3201–3226. Available from https://doi.org/10.1007/s00394-022-02878-z, https://www.springer.com/journal/394.

Laurent, A., & Owsianiak, M. (2017). Potentials and limitations of footprints for gauging environmental sustainability. *Current Opinion in Environmental Sustainability, 25*, 20–27. Available from https://doi.org/10.1016/j.cosust.2017.04.003, http://www.elsevier.com/wps/find/journal-description.cws_home/718675/description#description.

Levizzani, V., & Cattani, E. (2019). Satellite remote sensing of precipitation and the terrestrial water cycle in a changing climate. *Remote Sensing, 11*(19), 2301. Available from https://doi.org/10.3390/rs11192301.

Liu, J., Zhao, D., Mao, G., Cui, W., Chen, H., & Yang, H. (2020). Environmental sustainability of water footprint in Mainland China. *Geography and Sustainability, 1*(1), 8–17. Available from https://doi.org/10.1016/j.geosus.2020.02.002, http://www.journals.elsevier.com/geography-and-sustainability.

Lovarelli, D., Bacenetti, J., & Fiala, M. (2016). Water footprint of crop productions: a review. *Science of the Total Environment, 548–549*, 236–251. Available from https://doi.org/10.1016/j.scitotenv.2016.01.022, http://www.elsevier.com/locate/scitotenv.

Mansouri, H.-E.M. H.E., Belaitouche, F., Hamiche, N.B., Arbaoui, S.A. Abdelghani Aieb, A. Aouchiche, T. Atmaniou, M. Khenteche S. Madani K. (2022). Water footprint assessment of the Tichi-Haf dam waters (Soummam Valley, Bejaia, Algeria) according to ISO 14044 and ISO 14046 under the 6 and 12 UN-SDGs. In *Advances in science, technology and innovation*, pp. 287–297. Springer Nature Algeria. Available from https://doi.org/10.1007/978-3-030-76081-6_35, https://www.springer.com/series/15883.

Mekonnen, M. M., & Gerbens-Leenes, W. (2020). The water footprint of global food production. *Water, 12*(10), 2696. Available from https://doi.org/10.3390/w12102696.

Mekonnen, M. M., & Hoekstra, A. Y. (2020). Sustainability of the blue water footprint of crops. *Advances in Water Resources, 143.* Available from https://doi.org/10.1016/j.advwatres.2020.103679, http://www.elsevier.com/inca/publications/store/4/2/2/9/1/3/index.htt.

Meza-Herrera, C. A., Navarrete-Molina, C., Luna-García, L. A., Pérez-Marín, C., Altamirano-Cárdenas, J. R., Macías-Cruz, U., de la Peña, C. García, & Abad-Zavaleta, J. (2022). Small ruminants and sustainability in Latin America & the Caribbean: regionalization, main production systems, and a combined productive, socio-economic & ecological footprint quantification. *Small Ruminant Research, 211*, 106676. Available from https://doi.org/10.1016/j.smallrumres.2022.106676.

Morandi, A. (2022). *Sustainable agricultural development to achieve SDGs: the role of livestock and the contribution of gis in policy-making process. Drones and Geographical Information Technologies in Agroecology and Organic Farming: Contributions to Technological Sovereignty* (pp. 45–71). Italy: CRC Press. Available from https://www.taylorfrancis.com/books/edit/10.1201/9780429052842, 10.1201/9780429052842-4.

Muratoglu, A., Iraz, E., & Ercin, E. (2022). Water resources management of large hydrological basins in semi-arid regions: spatial and temporal variability of water footprint of the Upper Euphrates River basin. *Science of the Total Environment, 846*, 157396. Available from https://doi.org/10.1016/j.scitotenv.2022.157396.

Navarrete-Molina, C., Meza-Herrera, C. A., Herrera-Machuca, M. A., Lopez-Villalobos, N., Lopez-Santos, A., & Veliz-Deras, F. G. (2019a). To beef or not to beef: Unveiling the economic environmental impact generated by the intensive beef cattle industry in an arid region. *Journal of Cleaner Production, 231*, 1027–1035. Available from https://doi.org/10.1016/j.jclepro.2019.05.267.

Navarrete-Molina, C., Meza-Herrera, C. A., Ramirez-Flores, J. J., Herrera-Machuca, M. A., Lopez-Villalobos, N., Lopez-Santiago, M. A., & Veliz-Deras, F. G. (2019b). Economic evaluation of the environmental impact of a dairy cattle intensive production cluster under arid lands conditions. *Animal, 13*(10), 2379–2387. Available from https://doi.org/10.1017/S175173111900048X.

Navarrete-Molina, C., Meza-Herrera, C. A., Herrera-Machuca, M. A., Macias-Cruz, U., & Veliz-Deras, F. G. (2020). Not all ruminants were created equal: environmental and socio-economic sustainability of goats under a marginal-extensive production system. *Journal of Cleaner Production, 255*, 120237. Available from https://doi.org/10.1016/j.jclepro.2020.120237.

Nezamoleslami, R., & Hosseinian, S. M. (2020). An improved water footprint model of steel production concerning virtual water of personnel: the case of Iran. *Journal of Environmental Management, 260*, 110065. Available from https://doi.org/10.1016/j.jenvman.2020.110065.

Olawumi, T. O., & Chan, D. W. M. (2018). A scientometric review of global research on sustainability and sustainable development. *Journal of Cleaner Production, 183*, 231–250. Available from https://doi.org/10.1016/j.jclepro.2018.02.162.

Ornelas-Villarreal, E. C., Navarrete-Molina, C., Meza-Herrera, C. A., Herrera-Machuca, M. A., Altamirano-Cardenas, J. R., Macias-Cruz, U., García-de la Peña, C., & Veliz-Deras, F. G. (2022a). Goat production and sustainability in Latin America & the Caribbean: a combined productive, socio-economic & ecological footprint approach. *Small Ruminant Research, 211*. Available from https://doi.org/10.1016/j.smallrumres.2022.106677, http://www.elsevier.com/inca/publications/store/5/0/3/3/1/7/index.htt.

Ornelas-Villarreal, E. C., Navarrete-Molina, C., Meza-Herrera, C. A., Herrera-Machuca, M. A., Altamirano-Cardenas, J. R., Macias-Cruz, U., la Peña, C. Gd, & Veliz-Deras, F. G. (2022b). Sheep production and sustainability in Latin America & the Caribbean: a combined productive, socio-economic & ecological footprint approach. *Small Ruminant Research, 211*. Available from https://doi.org/10.1016/j.smallrumres.2022.106675, http://www.elsevier.com/inca/publications/store/5/0/3/3/1/7/index.htt.

Paterson, W., Rushforth, R., Ruddell, B. L., Konar, M., Ahams, I. C., Gironas, J., Mijic, A., & Mejia, A. (2015). Water footprint of cities: A review and suggestions for future research. *Sustainability, 7*(7), 8461–8490. Available from https://doi.org/10.3390/su7078461.

Pellicer-Martínez, F., & Martínez-Paz, J. M. (2016). The water footprint as an indicator of environmental sustainability in water use at the river basin level. *Science of the Total Environment, 571*, 561–574. Available from https://doi.org/10.1016/j.scitotenv.2016.07.022, http://www.elsevier.com/locate/scitotenv.

Pierrat, É., Laurent, A., Dorber, M., Rygaard, M., Verones, F., & Hauschild, M. (2023). Advancing water footprint assessments: combining the impacts of water pollution and scarcity. *Science of the Total Environment, 870*, 161910. Available from https://doi.org/10.1016/j.scitotenv.2023.161910.

Pryor, J. W., Zhang, Q., & Arias, M. E. (2022). Integrating climate change, hydrology, and water footprint to measure water scarcity in Lesotho, Africa. *Journal of Water Resources Planning and Management, 148*(1). Available from https://doi.org/10.1061/(ASCE)WR.1943-5452.0001502, https://ascelibrary.org/journal/jwrmd5.

Qasemipour, Ehsan, & Abbasi, Ali (2019). Virtual water flow and water footprint assessment of an arid region: a case study of South Khorasan Province, Iran. *Water, 11*(9), 1755. Available from https://doi.org/10.3390/w11091755.

Rios-Flores, J.L., Rios-Arredondo, B.E., Cantu-Brito, J.E., Rios-Arredondo, H.E., Armendariz-Erives, S., Chavez-Rivero, J.A., Navarrete-Molina, C., Castro-Franco, R. (2014). Analisis de la eficiencia fisica, economica y social del agua en esparrago (*Asparagus officinalis L.*) y uva (*Vitis vinifera*) de mesa del DR-037 Altar. 50.

Rodríguez, J. E., Razo, I., & Lázaro, I. (2023). Water footprint for mining process: a proposed method to improve water management in mining operations. *Cleaner and Responsible Consumption, 8*, 100094. Available from https://doi.org/10.1016/j.clrc.2022.100094.

Roidt, M., Chini, C. M., Stillwell, A. S., & Cominola, A. (2020). Unlocking the impacts of COVID-19 lockdowns: changes in thermal electricity generation water footprint and virtual water trade in Europe. *Environmental Science and Technology Letters, 7*(9), 683–689. Available from https://doi.org/10.1021/acs.estlett.0c00381, http://pubs.acs.org/page/estlcu/about.html.

Ruddell, B. (2018). Threshold based footprints (for water). *Water, 10*(8), 1029. Available from https://doi.org/10.3390/w10081029.

Sampaio, A. P. C., Silva, A. K. P., de Amorim, J. R. A., Santiago, A. D., de Miranda, F. R., Barros, V. S., Sales, M. C. L., & de Figueirêdo, M. C. B. (2021). Reducing the carbon and water footprints of Brazilian green coconut. *International Journal of Life Cycle Assessment, 26*(4), 707–723. Available from https://doi.org/10.1007/s11367-021-01871-8, http://www.springerlink.com/content/0948-3349.

Sarkodie, S. A. (2021). Environmental performance, biocapacity, carbon & ecological footprint of nations: drivers, trends and mitigation options. *Science of the Total Environment, 751*, 141912. Available from https://doi.org/10.1016/j.scitotenv.2020.141912.

Scanlon, B. R., Fakhreddine, S., Rateb, A., de Graaf, I., Famiglietti, J., Gleeson, T., Grafton, R. Q., Jobbagy, E., Kebede, S., Kolusu, S. R., Konikow, L. F., Long, D., Mekonnen, M., Schmied, H. M., Mukherjee, A., MacDonald, A., Reedy, R. C., Shamsudduha, M., Simmons, C. T., ... Zheng, C. (2023). Global water resources and the role of groundwater in a resilient water future. *Nature Reviews Earth and Environment*, *4*(2), 87–101. Available from https://doi.org/10.1038/s43017-022-00378-6, https://www.nature.com/natrevearthenviron/.

Scherer, L., & Pfister, S. (2016). Global water footprint assessment of hydropower. Elsevier Ltd, Switzerland. *Renewable Energy*, *99*, 711–720. Available from https://doi.org/10.1016/j.renene.2016.07.021, http://www.journals.elsevier.com/renewable-and-sustainable-energy-reviews/.

Shahzad, M. W., Burhan, M., Ang, L., & Ng, K. C. (2017). Energy–water–environment nexus underpinning future desalination sustainability. *Desalination*, *413*, 52–64. Available from https://doi.org/10.1016/j.desal.2017.03.009.

Skouteris, G., Ouki, S., Foo, D., Saroj, D., Altini, M., Melidis, P., Cowley, B., Ells, G., Palmer, S., & O'Dell, S. (2018). Water footprint and water pinch analysis techniques for sustainable water management in the brick-manufacturing industry. *Journal of Cleaner Production*, *172*, 786–794. Available from https://doi.org/10.1016/j.jclepro.2017.10.213.

Sommaruga, R., & Eldridge, H. M. (2021). Avocado production: water footprint and socio-economic implications. *EuroChoices*, *20*(2), 48–53. Available from https://doi.org/10.1111/1746-692X.12289, http://onlinelibrary.wiley.com/journal/10.1111/(ISSN)1746-692X.

Sun, S. K., Yin, Y. L., Wu, P. T., Wang, Y. B., Luan, X. B., & Li, C. (2019). Geographical evolution of agricultural production in China and its effects on water stress, economy, and the environment: the virtual water perspective. *Water Resources Research*, *55*(5), 4014–4029. Available from https://doi.org/10.1029/2018WR023379, http://agupubs.onlinelibrary.wiley.com/hub/journal/10.1002/(ISSN)1944-7973/.

Taffarello, D., Bittar, M. S., Sass, K. S., Calijuri, M. C., Cunha, D. G. F., & Mendiondo, E. M. (2020). Ecosystem service valuation method through grey water footprint in partially-monitored subtropical watersheds. *Science of the Total Environment*, *738*, 139408. Available from https://doi.org/10.1016/j.scitotenv.2020.139408.

Tompa, O., Kiss, A., Maillot, M., Sarkadi Nagy, E., Temesi, Á., & Lakner, Z. (2022). Sustainable diet optimization targeting dietary water footprint reduction—a country-specific study. *Sustainability*, *14*(4), 2309. Available from https://doi.org/10.3390/su14042309.

Venetoulis, J., & Talberth, J. (2008). Refining the ecological footprint. *Environment. Development and Sustainability*, *10*, 441–469. Available from https://doi.org/10.1007/s10668-006-9074-z.

Vollmer, D., & Harrison, I. J. (2021). $H_2O \neq CO_2$: framing and responding to the global water crisis. *Environmental Research Letters*, *16*(1), 011005. Available from https://doi.org/10.1088/1748-9326/abd6aa.

Wang, D., Hubacek, K., Shan, Y., Gerbens-Leenes, W., & Liu, J. (2021). A review of water stress and water footprint accounting. *Water*, *13*(2), 201. Available from https://doi.org/10.3390/w13020201.

Wöhler, L., Niebaum, G., Krol, M., & Hoekstra, A. Y. (2020). The grey water footprint of human and veterinary pharmaceuticals. *Water Research X*, *7*, 100044. Available from https://doi.org/10.1016/j.wroa.2020.100044.

Wu, H., Jin, R., Liu, A., Jiang, S., & Chai, L. (2022a). Savings and losses of scarce virtual water in the international trade of wheat, maize, and rice. *International Journal of Environmental Research and Public Health*, *19*(7), 4119. Available from https://doi.org/10.3390/ijerph19074119.

Wu, N., Yin, J., Engel, B. A., Hua, E., Li, X., Zhang, F., & Wang, Y. (2022b). Assessing the sustainability of freshwater consumption based on developing 3D water footprint: a case of China. *Journal of Cleaner Production*, *364*, 132577. Available from https://doi.org/10.1016/j.jclepro.2022.132577.

Yan, Y., Wang, R., Chen, S., Zhang, Y., & Sun, Q. (2023). Three-dimensional agricultural water scarcity assessment based on water footprint: a study from a humid agricultural area in China. *Science of the Total Environment*, *857*, 159407. Available from https://doi.org/10.1016/j.scitotenv.2022.159407.

Yue, S., Munir, I. U., Hyder, S., Nassani, A. A., Qazi Abro, M. M., & Zaman, K. (2020). Sustainable food production, forest biodiversity and mineral pricing: interconnected global issues. *Resources Policy*, *65*. Available from https://doi.org/10.1016/j.resourpol.2020.101583, http://www.elsevier.com/inca/publications/store/3/0/4/6/7/.

Zhu, J., Yang, Y., Li, Y., Xu, Pinghua, & Wang, Laili (2020). Water footprint calculation and assessment of viscose textile. *Industria Textila*, *71*(01), 33–40. Available from https://doi.org/10.35530/it.071.01.1642.

Chapter 13

Strategies for the reduction of water footprints

Fayaz A. Malla[1,*], Mir Tamana[2], Farhana Rahman[2], Afaan A. Malla[3], Suhaib A. Bandh[4], Nazir A. Sofi[5], Mukhtar Ahmed[6] and Showkat Rashid[7]

[1]Department of Environmental Science, Govt. Degree College Tral, Jammu and Kashmir, India, [2]Department of Environmental Science, Higher Education Department Govt. of Jammu and Kashmir, Government Degree College Tral, India, [3]Department of Environmental Science, Government Degree College Baramulla, Indira Gandhi National Open University, Baramulla, Jammu and Kashmir, India, [4]Department of Higher Education, Sri Pratap College, Government of Jammu and Kashmir, Srinagar, India, [5]Department of Agriculture Research Information System, Sher-e-Kashmir University of Agricultural Sciences and Technology, Shalimar Campus, Srinagar, Jammu and Kashmir, India, [6]Meteorological Centre, India Meteorological Department Rambagh Srinagar, India, [7]Department of Economics, Higher Education Department Govt. of Jammu and Kashmir, Government Degree College Tral, India

**Corresponding author. e-mail address: nami.fayaz@gmail.com.*

13.1 Introduction

The world's water supply is under strain due to the needs of a rapidly expanding human population. Nearly half of the world's population will reportedly live in "water-stressed" areas by 2025 when the need for water greatly exceeds the supply (McNabb, 2019). For sustainable water management, it is crucial to cut back on production's overall water consumption, or "water footprint." Reducing water use would have major positive effects on the economy, society, and environment (Butler et al., 2017). If individuals used less water, it would reduce the strain on ecosystems and freshwater settings. Because of the energy required for water pumping and purification, conserving water can also help mitigate global warming. If businesses reduce their water usage, they can save money on their water bills and lessen the effects of a water shortage on their operations (Ahmad et al., 2020).

Water is essential for many aspects of modern life, including agriculture, industry, and human consumption. However, due to the rising water demand brought on by population growth, urbanization, economic development, and the effects of climate change, there is growing worry about water scarcity and its impact on ecosystems and human well-being (Du Plessis & du Plessis, 2019). Analyzing and regulating human water consumption is essential for overcoming this challenge. The water footprint is the total amount of water used to produce goods and services for human consumption in an individual, area, or country. Included are not just municipal and industrial water consumption, but also agricultural, animal, and industrial water consumption. The concept of a water footprint, which was first proposed by Arjen Hoekstra in 2002, has now achieved widespread acceptance as an effective tool for water management and sustainability (Hoekstra, 2017a).

Green, blue, and gray are the subcategories that make up a water footprint. The term "green water" is used to refer to rain that is absorbed by the soil and then used by plants for photosynthesis. Bluewater is both surface and groundwater used for irrigation, domestic use, and industrial activities (Hogeboom et al., 2018). Gray water is the amount of clean water needed to dilute toxins to a point where they pose no threat to human or environmental health. The quantity of water used by a given product, undertaking, or country can be calculated. The water used to produce one cup of coffee includes not just the water needed for brewing, but also the water used in growing, processing, and transporting the coffee beans (Hicks, 2018). The water footprint of a city is calculated by adding up all of its water use, both domestically and through imports and exports. A water footprint is a useful tool for ensuring water is used sustainably. As was previously said, water scarcity is becoming a worldwide issue; the water footprint can help identify areas where there is a disproportionate amount of water stress and guide allocation decisions. Some of the environmental effects of human water use include the depletion of freshwater resources, pollution of water bodies, and loss of biodiversity (Albert et al., 2021). Using the water

Water Footprints and Sustainable Development. DOI: https://doi.org/10.1016/B978-0-443-23631-0.00013-3

footprint, we can target the industries and goods that consume the most water and contribute the most to water pollution. The water footprint of food production is crucial in making sure there is always enough food for everyone (Hoekstra, 2017b). By analyzing the water footprint of various crops and farming systems, we can improve water use efficiency and reduce risks related to food production. A water footprint is a useful tool for managing water-related risks in supply chains and decreasing an organization's overall water usage. Water consumption, distance from water sources, and water use efficiency are just a few of the factors that might affect the water footprint of a given product or activity (Pfister et al., 2017).

Up to 70% of all freshwaters withdrawn annually is used in agriculture, making it the largest consumer of water in the world (Chen et al., 2018). The water usage associated with producing a crop depends on the crop itself, as well as farming methods and irrigation technology. Rice, cotton, and sugarcane are examples of high-water footprint crops. Traditional methods of irrigation, such as flood irrigation, can have a sizable water footprint in comparison to more modern methods like drip irrigation (Singh, 2022). Nearly 20% of all freshwater withdrawals are used by the industrial sector. The amount of water used in manufacturing is affected by the industries involved, the manufacturing processes employed, and the geographic location of the factories. Industries such as textile manufacturing, paper manufacturing, and chemical production tend to have large water footprints. Companies based in dry regions may also have a bigger water footprint as a result of intense competition for limited water resources. About 10% of all freshwaters extracted is used for residential purposes. The water we drink, cook with, and clean with all fall under this category. The amount of water used at home might vary depending on factors like availability, water efficiency, and human behavior. By putting in low-flow toilets and showerheads, homeowners can reduce their demand for the municipal water system (Ebrahimi et al., 2022).

The climate and geography of a place have an impact on the availability and quality of its water resources. Regions with abundant precipitation and adequate water consume less water overall compared to locations with weak rainfall and restricted water supply. The water footprint of a product manufactured in the Middle East, where potable water is in short supply, will be greater than that of an identical product manufactured in Canada, where potable water is not in short supply (Ewaid et al., 2020; Mian et al., 2021). International trade may also have an impact on the water footprints of individual products. Water used in production in the country of export is factored into the total water footprint. Consequently, there may be a greater global water footprint from commodities imported from areas with limited water resources compared to those produced locally (Hoekstra, 2017a). In conclusion, the water footprint of a product or an activity depends on several factors. Water consumption efficiency, geographical location, kind of industry, and agricultural practices all play a role. Long-term water sustainability is more likely if we have a deeper understanding of these factors.

13.2 Water footprint mitigation techniques for households and farms

13.2.1 Simple lifestyle changes

Growing concern over our water footprints can be attributed to both the increasing shortage of water and the pressing requirement for more responsible water management (Fig. 13.1). Modifications to one's way of life, even the smallest ones, can have a big effect on water conservation.

The purpose of this chapter is to examine, from a scientific point of view, various methods for reducing one's water footprint through the implementation of very uncomplicated lifestyle adjustments. Individuals can contribute to water conservation efforts and encourage sustainable water use if they follow these practices and adopt them.

13.2.1.1 Water-efficient appliances and fixtures

1. *Low-flow showerheads and faucets*: When it comes to day-to-day activities like showering and washing your hands, installing low-flow showerheads and faucets can considerably cut down on the amount of water that is used (Rahman, 2019). These fixtures reduce the amount of water flow while preserving both their functionality and the comfort of the user.
2. *Water-efficient washing machines and dishwashers*: Choosing home appliances that have been rated well for their water efficiency can result in significant water-savings. Washing machines and dishwashers that have earned the Energy Star certification are designed to use less water without sacrificing their overall efficiency (Dybeck Carlsson et al., 2020).
3. *Dual-flush toilets*: Dual-flush Toilets provide users with two flushing options, allowing them to choose between a flush that uses less water for liquid waste and a flush that uses more water for solid waste (Kadijk, 2021). The application of this technology helps reduce the amount of water that is wasted in the bathroom.

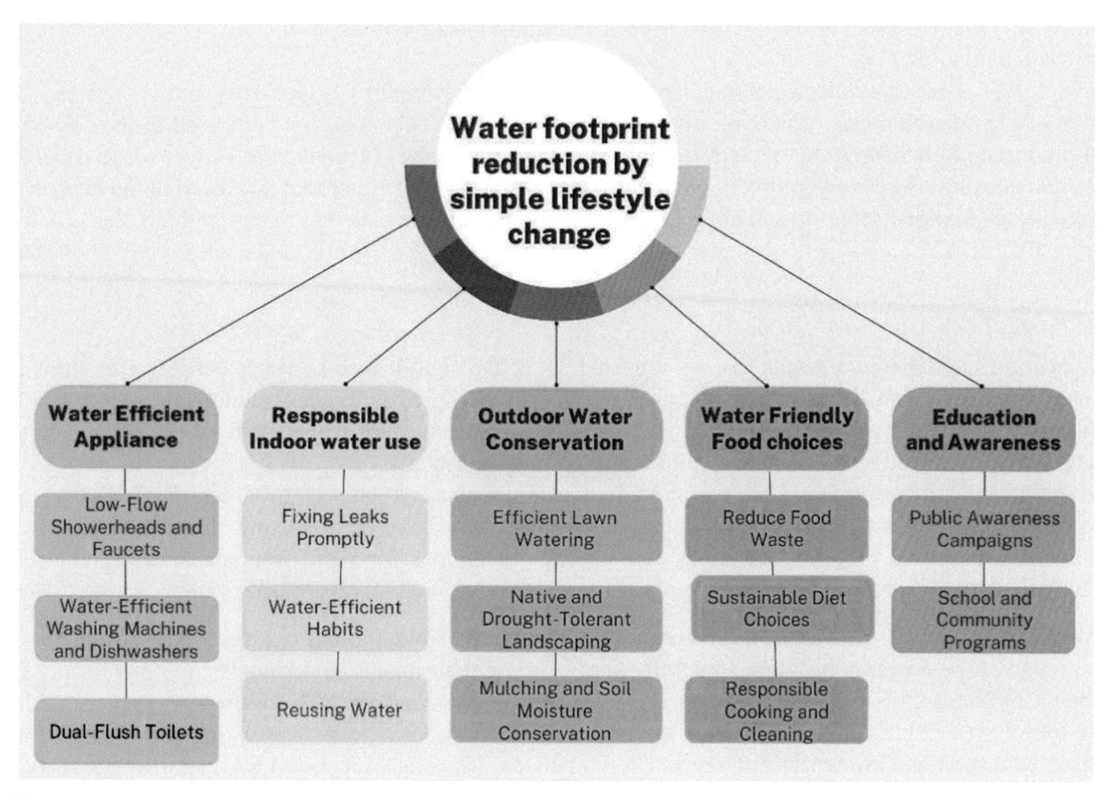

FIGURE 13.1 Water footprint reduction by simple lifestyle changes.

13.2.1.2 Responsible indoor water use

1. *Fixing leaks promptly*: Water loss can be avoided by promptly repairing leaks in plumbing fixtures such as faucets, toilets, and pipes (Seyoum et al., 2017). Fixing even very minor water leaks as soon as possible can result in significant cost savings as well as a reduction in the amount of water lost over time.
2. *Water-efficient habits*: The adoption of water-conscious behaviors, such as turning off the faucet while brushing one's teeth, washing dishes in a basin rather than under-running water, and only running washing machines and dishwashers with full loads, can assist in the conservation of water during day-to-day activities (McCarroll & Hamann, 2020).
3. *Reusing water*: The conservation of water can be helped by the practice of reusing water for reasons other than drinking it. For instance, the water that was used to rinse fruits and vegetables can be collected and repurposed for watering plants, as can the time that was spent waiting for hot water to reach the proper temperature (Fu et al., 2020).

13.2.1.3 Outdoor water conservation

1. *Efficient lawn watering*: It is best to water lawns in the early morning or late evening, when temperatures are lower, to minimize the amount of water that is lost to evaporation. In addition, you may help save water by regulating your sprinklers to reduce the amount of water that is wasted as overspray and by only watering when it is required (El-Beltagi et al., 2022).
2. *Native and drought-tolerant landscaping*: The use of less water in outdoor spaces can be accomplished by planting drought-resistant and native plant species. These plants have adjusted to the climate of the area and, once established, have a lower watering requirement, which contributes to water conservation (Rusyn, 2021).
3. *Mulching and soil moisture conservation*: The use of mulch around plants and in garden beds helps to retain soil moisture, which in turn reduces the frequency with which plants need to be watered. In addition to this, mulch helps prevent the growth of weeds and keeps the temperature of the soil more consistent (El-Beltagi et al., 2022).

13.2.1.4 Water-friendly food choices

1. *Reduce food waste*: In both the manufacturing of food and its subsequent preparation, a substantial amount of water is consumed. Individuals can indirectly contribute to the conservation of water resources by cutting down on the

amount of food that is wasted through practices such as careful meal planning, astute buying, and the storage of left-overs (Scanlon et al., 2017).

2. *Sustainable diet choices*: Producing particular foodstuffs typically requires a greater quantity of water. The consumption of plant-based meals, which, on average, have a reduced water footprint compared to the consumption of animal products, might contribute to the overall effort to conserve water (Mekonnen & Gerbens-Leenes, 2020).
3. *Responsible cooking and cleaning*: It is possible to reduce the amount of water that is wasted in the kitchen by using less water when cooking (e.g., by steaming instead of boiling) and by cleaning using methods that use less water (Foden et al., 2019).

13.2.1.5 Education and awareness

1. *Public awareness campaigns*: People can be educated about the significance of water conservation through public awareness campaigns and educational programs, which can also give individuals actionable advice for lowering their water footprints in their day-to-day lives (McCarroll & Hamann, 2020). These efforts can encourage people to adjust their behaviors and help cultivate a culture that is water-conscious.
2. *School and community programs*: Children and adults can be educated about environmentally friendly water management methods through the inclusion of water conservation subjects in the academic curriculum and participation in community initiatives. Future generations can become champions for more responsible water use if they are instilled with an early appreciation of the importance of water conservation (Onyenankeya & Salawu, 2018).

Individuals can contribute to water conservation initiatives in a meaningful and practical manner by reducing their water footprints via the adoption of a few straightforward lifestyle adjustments. Individuals can dramatically cut down on the amount of water they use by installing water-saving appliances and fixtures, cultivating responsible behaviors around the use of water both inside and outside the home, selecting foods that are less taxing on water resources, and taking part in education and awareness programs. These methods, which are supported by scientific evidence, not only help to save water but also encourage the sustainable management of water resources, which will lead to a more secure future. We can have a positive effect on water resources and guarantee that future generations will have access to them if we work together and each makes a personal commitment to living in a way that is water-conscious.

13.2.2 Landscaping and gardening

The adoption of water footprint mitigation measures in landscaping and gardening is becoming increasingly important as the issue of water scarcity becomes a significant problem (Fig. 13.2). This chapter of the book will investigate methods that are supported by scientific research to cut down on the amount of water used for these activities. Individuals and communities alike may contribute to water conservation and environmentally responsible water management by putting these strategies into practice.

13.2.2.1 Plant selection and design

1. *Native and drought-tolerant plants*: Choosing native species and plants that have already adapted to the conditions of the local climate can considerably cut down on the amount of water needed (Hulme, 2005). Native plants have undergone natural selection to be able to thrive in the local environment and, once established, often require less water. Plants that can survive in arid environments, including succulents and Mediterranean species, have inherent adaptations that allow them to conserve water (Pompelli et al., 2019). Individuals and communities alike may contribute to water conservation and environmentally responsible water management by putting these strategies into practice.
2. *Xeriscaping*: A method of landscaping known as xeriscaping places an emphasis on water-conserving architectural design concepts (Çetin et al., 2018). It entails selecting plants that require less water, combining effective methods of irrigation, and applying mulch to limit the amount of evaporation that occurs and to save soil moisture. Xeriscaping is an approach to landscape design that can drastically cut water usage while preserving the visual appeal of the outdoor space.
3. *Grouping plants by water needs*: Irrigation can be performed more effectively in landscapes that have been designed to cluster plants that have comparable water requirements together (Huo et al., 2017). By generating hydrazones, which are clusters of plants with comparable water requirements, it is possible to target and apply water more effectively, hence minimizing the amount of water that is wasted.

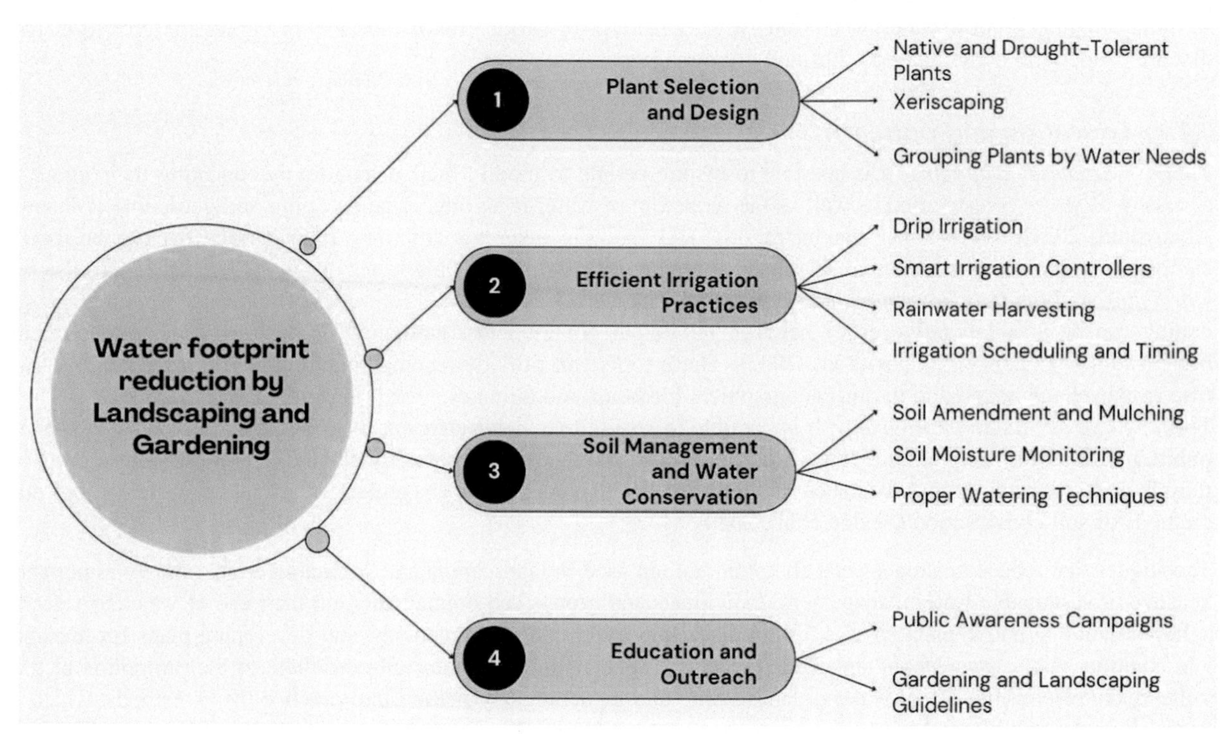

FIGURE 13.2 Water footprint reduction by landscaping and gardening.

13.2.2.2 Efficient irrigation practices

1. *Drip irrigation*: With drip irrigation, water is delivered directly to the root zone, hence lowering the amount of water that is lost to evaporation and wasted (Wang et al., 2020). This technique allows for accurate and effective watering, which helps to ensure that plants receive the appropriate amount of moisture while also reducing the amount of water that is wasted through waste.
2. *Smart irrigation controllers*: The use of intelligent irrigation controllers that modify watering schedules depending on data from weather sensors, soil moisture sensors, and the water requirements of individual plants is one way to increase the productivity of irrigation systems (Goap et al., 2018). These controllers can stop excessive watering, take into consideration the amount of precipitation, and adjust to shifting environmental conditions.
3. *Rainwater harvesting*: It is possible to greatly lessen one's dependency on available freshwater resources by capturing rainwater from rooftops and storing it in cisterns or tanks for later use in irrigation (Abdulla et al., 2021). It is possible to incorporate rainwater collection devices into landscapes to supplement irrigation requirements during dry months.
4. *Irrigation scheduling and timing*: It is best to water your plants during the cooler portions of the day, such as first thing in the morning or the evening, to minimize the amount of water that is lost to evaporation. In addition, modifying the timing of irrigation in accordance with the requirements of the plants, the climate, and the amount of moisture present in the soil can help reduce the waste of water and encourage the development of plants that are in better health (Yousif & Abdalgader, 2022).

13.2.2.3 Soil management and water conservation

1. *Soil amendment and mulching*: When organic matter is worked into the soil, the water-holding capacity of the soil improves, and the soil also infiltrates and retains moisture more effectively (Herawati, 2021). The practice of covering the surface of the soil with organic materials such as wood chips or compost serves to reduce the amount of water lost to evaporation, inhibits the growth of weeds, and keeps soil moisture levels stable.
2. *Soil moisture monitoring*: It is possible to avoid overwatering plants and ensure they receive appropriate irrigation by using soil moisture sensors to monitor the levels of moisture in the soil (Osanaiye et al., 2022). These sensors deliver data on the moisture content of the soil in real time, which enables accurate regulation of irrigation.
3. *Proper watering techniques*: Utilizing methods like deep watering, which stimulates root growth by saturating the soil at a deeper level, is one way to assist plants in gaining access to water more efficiently (Uddin et al., 2018).

Avoiding regular shallow watering not only lowers the risk of surface runoff but also promotes the growth of roots that are deeper and better able to withstand dry conditions.

13.2.2.4 Education and outreach

1. *Public awareness campaigns*: It is possible to inspire people to modify their behaviors by educating them about the necessity of water conservation as well as the reduction of water footprints in landscaping and gardening (Okutan & Akkoyunlu, 2021). Workshops, demonstrations, and teaching materials are some of the tools that can be used in campaigns to promote environmentally sustainable practices and responsible water usage.
2. *Gardening and landscaping guidelines*: Developing rules and best practices for water-efficient landscaping and gardening can be a collaborative effort between municipal governments, gardening associations, and environmental organizations (Lopez-Villalobos et al., 2022). These tools can provide recommendations that are backed by scientific evidence and adapted to various geographical locations and climates.
3. *Training and certification programs*: It is possible to ensure that water-efficient measures are adopted by the general public by providing training and certification programs for landscape professionals and gardeners. These programs may include instruction on issues such as the selection of plants, the management of irrigation systems, and other methods of soil conservation (Abdou et al., 2020).

Techniques that reduce an area's water footprint and are used in landscaping and gardening are extremely important to the practice of sustainable water management. Individuals and groups can dramatically cut their use of water by selecting drought-resistant and native plant species, instituting water-saving irrigation methods, and developing plans for managing soil. In addition, public awareness campaigns, guidelines, and training programs all contribute to the promotion of practices that make responsible use of water in landscaping and gardening. By putting into practice these methods, which are supported by scientific research, we can take steps toward a future that is more water-conscious. This will allow us to guarantee the long-term viability of our landscapes and gardens while also preserving this valuable resource.

13.3 Methods for reducing water footprints in business and industry

13.3.1 Reducing water footprints in business

To alleviate global water scarcity and advance sustainable practices, businesses and industries need to reduce their water footprints. Companies can reduce their water usage by optimizing their processes, implementing water-efficient technologies, and conducting water footprint analyses (Berger et al., 2021). Promoting responsible water management practices across the value chain requires thinking about water in product design, working together, and being good water stewards. Water risk assessments, sustainable supply chain management, and responsible sourcing are some other factors that can lessen environmental impacts caused by water use. Developing a culture of water conservation requires both education and staff participation. Case studies based on real-world events can provide instructive illustrations for others in related fields.

The lack of available water is a serious problem around the world that affects many different types of enterprises. Companies are taking steps to lessen their impact on the water supply as they become more aware of the need to conserve this precious resource (Smol et al., 2020). To help businesses better manage their water resources and reduce their water footprint, this chapter will discuss the scientific methodologies that have proven effective in doing so. Businesses that want to reduce their water footprint can learn a lot from this research since it examines new methods, processes, and technologies that are both efficient and environmentally friendly (Hoekstra, 2019). Businesses must take preventative action to lessen their water usage because of the negative effects of water shortage on ecosystems and human health. An overview of water conservation's significance in the context of corporate sustainability is provided below.

13.3.1.1 Water footprint assessment

For organizations to effectively minimize their water footprints, they must first understand the patterns of water usage they currently engage in and identify areas in which they can improve. Tools and procedures for assessing water footprints provide useful insights into the water consumption of particular processes, products, or entire supply chains (Acquaye et al., 2017).

13.3.1.2 Water-efficient technologies

Recent developments in technology have provided some interesting potential solutions for lowering the amount of water used in corporate operations. This section describes cutting-edge technologies that can considerably cut down on water consumption. Some examples of these technologies include water-efficient fixtures, intelligent irrigation systems, and recycling and reuse systems (Varma, 2022).

13.3.1.3 Process optimization and water management

Improving water management methods and making processes more efficient are both essential steps in the process of lowering water footprints (Skouteris et al., 2018). This section discusses measures that promote responsible water usages, such as water-efficient industrial processes, leak detection and repair, and complete water management plans.

13.3.1.4 Product design and life cycle assessment

When it comes to calculating a company's total water footprint, the product life cycle is one of the most important factors to consider. Companies can find areas where water consumption can be decreased by integrating water considerations into product design and applying life cycle assessment procedures (Bai et al., 2018). These methodologies allow businesses to identify areas such as selecting less water-intensive materials or improving packaging, where water consumption can be reduced.

13.3.1.5 Water stewardship and collaboration

Engaging in collaborative problem-solving with a variety of parties, including stakeholders, local communities, and other enterprises, is an essential part of good water management. Partnerships for water conservation, initiatives for collective action, and participation in communication with local authorities are all examples of possible forms of collaborative work that can be done to develop sustainable water management practices (Teague et al., 2021).

13.3.1.6 Supply chain management

Strategies for reducing a company's water footprint should not be limited to the company's internal activities but should also include the supplier chain (Hoekstra, Chapagain, et al., 2019). This section examines ways to encourage water stewardship throughout the entirety of the value chain, including supplier engagement, responsible sourcing practices, and water risk assessment.

13.3.1.7 Education and employee engagement

Educating workers on the importance of water sustainability and offering training on efficient water management practices are crucial to establishing a culture of water conservation within an organization. The importance of employee involvement and awareness programs in influencing behavior change and encouraging water-aware decision-making is highlighted in this section (Amahmid et al., 2019).

Business sustainability relies heavily on minimizing water usage. In this extensive examination, we looked at how different companies use science to reduce water use and enhance water management. Businesses may make a significant impact on global water conservation efforts by adopting water-efficient technologies, streamlining operations, collaborating, and taking a comprehensive approach to water stewardship (Delgado, 2021).

Sustainable water management strategies in the business sector require more study and constant innovation due to the complexity of water-related concerns. Businesses can make great advances toward decreasing their water footprints and contributing to a more water-secure future by implementing the strategies described in this assessment (Compagnucci & Spigarelli, 2018).

13.3.2 Reducing water footprints in Industries

Because water shortage is a worldwide problem, businesses must take immediate steps to lessen their consumption of the resource. The purpose of this study is to present a synopsis of the scientific approaches taken by businesses to reduce water use and enhance water management. This research provides useful information for businesses that want to reduce their water footprint by investigating new technologies, process optimization methods, environmentally friendly methods, and cooperative efforts. Industries contribute significantly to water pollution since they create pollutants that are hazardous to both humans and ecosystems (Goel, 2006). Waste from many factories is flushed into rivers, lakes, and the ocean using potable water (Anju et al., 2010). Streams are transformed into open sewers when industrial facilities dump oil, dangerous compounds, and other harmful liquids known as effluents into them (Ahmed et al., 2021). The water supply is largely affected by the agricultural sector. Pimentel estimates that 70% of the world's freshwater is put to use in agriculture (Pimentel et al., 2004). Industrial polluters include oil refineries, paper mills, textile and sugar mills, and chemical plants (Kanu & Achi, 2011). Arsenic and lead are two examples of toxic metals that can impair both plant and animal life. There is a considerable demand for water treatment in the brewing and carbonated beverage water industries, as well as the dairy industry. Many industries are negatively impacted by water scarcity. This section gives a primer on water conservation's significance in promoting industrial sustainability and on the necessity for efficient techniques to lessen water use.

13.3.2.1 Water footprint assessment

Industries may cut their water footprints by learning how they currently use water and then finding ways to cut back even more (Marston et al., 2018). Insights on the water consumption of individual industrial processes, goods, or value chains can be gleaned with the help of water footprint assessment tools and procedures.

13.3.2.2 Water-efficient technologies

Industrial water footprints can be decreased with the use of cutting-edge technologies. Precision water management and optimized water use in industrial activities are discussed, along with cutting-edge technologies including water-efficient equipment, sensor-based monitoring systems, and real-time data processing (Saha, 2022).

13.3.2.3 Process optimization and water management

Industrial water use can be reduced through process optimization and wise water management. Water footprint reduction options are discussed in this section (Tayyab et al., 2020). These include water recycling and reuse, water-efficient cooling systems, and modern wastewater treatment technologies.

13.3.2.4 Sustainable water sourcing

Businesses can lessen their impact on water resources by switching to more environmentally friendly methods of obtaining water (Liu & Jensen, 2018). Water-efficient irrigation systems, rainwater collecting, and other alternative water sources (such as treated wastewater) are discussed here for use in the agricultural and horticultural sectors.

13.3.2.5 Industrial water stewardship

Industries practicing good water stewardship acknowledge and address the effects of their operations on local water supplies and actively seek out partnerships with other interested parties and local communities. These good practices include water risk assessments, community partnerships, and water footprint benchmarking, all of which aim to encourage more responsible water usage and safeguard water supplies for the future (Wilson et al., 2018).

13.3.2.6 Energy–water nexus

The energy–water nexus highlights the connection between these two vital sectors. Energy production frequently requires large quantities of water, thus industries can lessen their impact on water resources by switching to more water-efficient energy production methods and alternative energy (Trubetskaya et al., 2021). In this piece, we look at how energy and water management might work together to lessen their respective negative effects on the environment.

13.3.2.7 Policy and regulatory frameworks

To encourage businesses to lessen their impact on water resources, enabling legislative and regulatory frameworks is essential. To encourage businesses to adopt sustainable water management practices, this section covers the value of water pricing systems, water efficiency requirements, and regulatory compliance promotion (Brack et al., 2017).

13.3.2.8 Employee education and training

Educating and training workers is essential to establishing a culture of water-saving inside businesses. Here we see the need of getting the word out about water shortage, teach people how to manage water effectively, and encourage accountability in the workplace.

To achieve sustainable water management and address global water scarcity, industrial water footprints must be drastically reduced (Fig. 13.3). Industries may considerably lessen their water footprints and aid in global water conservation efforts by embracing water-efficient technologies, streamlining processes, adopting sustainable sourcing

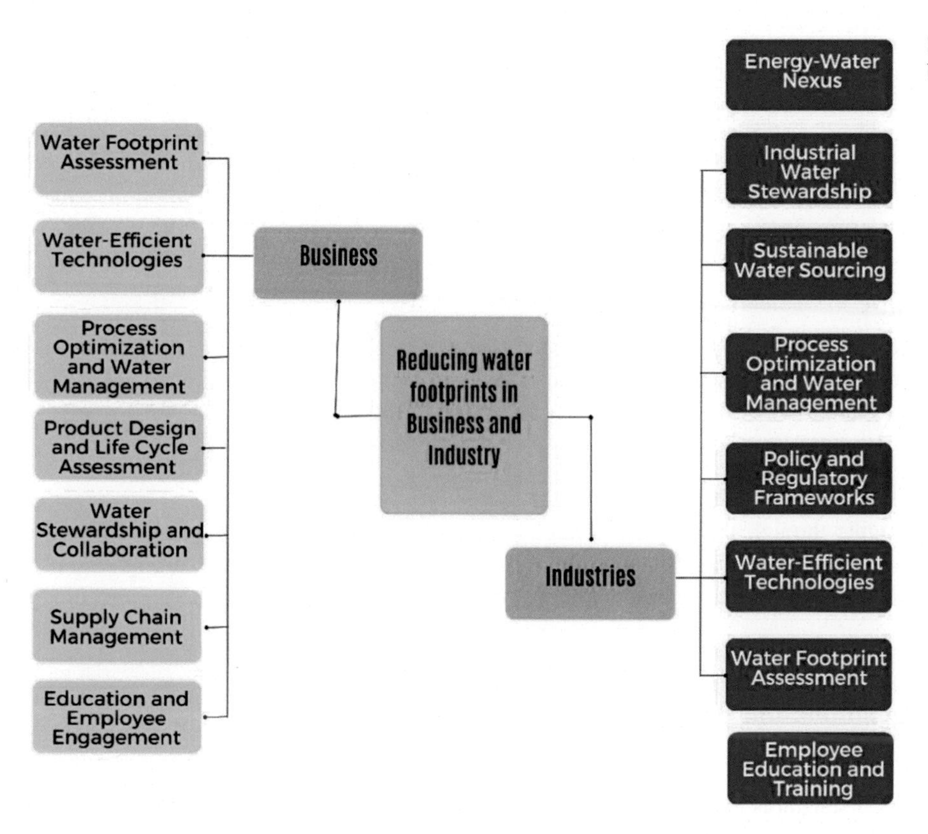

FIGURE 13.3 Water footprint reduction by business and industries.

methods, engaging in collaborative initiatives, and thinking about the energy–water nexus (Nezamoleslami & Hosseinian, 2020). More study and constant invention are needed to create and adopt cutting-edge techniques for decreasing industrial water usage. Industries may play a crucial role in ensuring future generations have access to clean water by implementing the solutions outlined in this review study.

13.4 Water conservation policies

Because of water shortage and the necessity for responsible water management, new water conservation policies have been implemented. Responsible water usage, effective water management, and conservation techniques are all encouraged by these laws, which are grounded in scientific research. From the domestic to the industrial, this chapter examines water conservation policies and practices that can efficiently reduce water footprints at each level. Sustainable water management and solving the problems of water shortage are both possible if these policies are put into place. According to a report published by UNESCO on behalf of UN-Water, 2 billion people worldwide (or 26% of the population) lack access to safe drinking water, and 3.6 billion (or 46% of the population) lack access to adequately managed sanitation (Boretti & Rosa, 2019). In the fight against water scarcity, education is the single most effective tool we have at our disposal. A student's typical water footprint (WF)is 3223 ± 830 L each day (Venckute et al., 2017). More than 10 times as much water is used indirectly as is used directly. The primary direct usage of water footprint among young people is for food (84.4%), with clothing-related water footprint (6.4%) coming in second (Venckute et al., 2017). Meat consumption accounts for the greatest share of a person's total water footprint (39.6%), among the dietary groups considered (Bahn et al., 2019). Direct water use is considerable (264 L per day) for both sexes mainly because of prolonged showers (Venckute et al., 2017). This demonstrates that young people have a larger water footprint than older generations; college and university-level awareness programs are therefore useful; young people need to be educated about the importance of water management to reduce their impact on the environment. Personal water conservation measures, such as just flushing when necessary, can also help minimize water footprint. Effective strategies for lowering water waste include teaching kids early on, holding regular awareness campaigns with eye-catching banners and posters, and emphasizing the value of water in marketing campaigns. Consequently, education is crucial in the fight against water footprint.

13.4.1 Water pricing and economic incentives

13.4.1.1 Increasing block rates

By charging higher rates for higher levels of water usage, increasing block rates for water consumption can encourage water conservation. Increasing block rates are implemented by charging higher rates for higher levels of water usage. This price structure offers both consumers and businesses a financial incentive to cut their water consumption, which helps to protect the environment (Sahin et al., 2017).

13.4.1.2 Water efficiency rebates

Individuals and organizations are encouraged to make investments in water-saving technologies by being offered monetary incentives, such as rebates or subsidies, for the installation of water-efficient appliances, fixtures, and irrigation systems. These types of financial incentives are known as water efficiency incentives (Addo et al., 2018).

13.4.1.3 Water trading and market-based instruments

It is possible to buy and sell water rights through the implementation of water trading schemes, which provides an incentive for water users to more effectively evaluate and manage the amount of water they consume. Through the use of various economic mechanisms, market-based tools, such as cap-and-trade schemes for water usage, can encourage water conservation (Yao et al., 2017).

13.4.2 Water conservation regulations and standards

13.4.2.1 Water-efficient building codes

When water-efficient building norms and standards are enforced, new constructions and renovations may be required to make use of water-saving fixtures, appliances, and landscaping techniques. These regulations ensure that water conservation measures are incorporated into the planning and development of metropolitan areas (Amaral et al., 2020).

13.4.2.2 Water use restrictions

The management of water demand can be helped along by the implementation of water consumption limits during times of drought or water scarcity. These restrictions may include limitations on watering outdoor areas, washing cars, and other nonessential uses of water to promote responsible water consumption (Gude, 2017).

13.4.2.3 Landscape watering guidelines

The development of recommendations for best practices in outdoor watering, including the specification of watering hours and frequency, encourages the use of effective irrigation methods. These recommendations can be adapted to local temperatures and water supplies, which will ensure that landscaping projects make the most efficient use of water (Pratt et al., 2019).

13.4.3 Water management and infrastructure

13.4.3.1 Water loss reduction

Enhancing water efficiency and conservation can be accomplished by the implementation of measures to reduce water losses in distribution systems. These methods can include correcting leaks, strengthening infrastructure, and adopting sophisticated metering systems (Mohapatra & Rath, 2019).

13.4.3.2 Water reuse and recycling

The use of treated wastewater for nonpotable uses, such as irrigation, industrial activities, and toilet flushing, can help to lessen the demand for freshwater resources (Takeuchi & Tanaka, 2020). This can be accomplished by encouraging the use of treated wastewater for these nonpotable reasons. The development of laws and infrastructure for the reuse of water paves the way for more effective water management and conservation.

13.4.3.3 Stormwater management

It is possible to decrease dependency on freshwater sources for irrigation and other nonpotable applications by putting into place policies and infrastructure designed to catch and regulate stormwater runoff (Shimabuku et al., 2018). To achieve the goals of maximizing water retention while simultaneously minimizing runoff, many strategies may be implemented, such as rainwater collecting, green infrastructure, and permeable surfaces.

13.4.4 Agricultural water conservation

13.4.4.1 Irrigation efficiency and best management practices

It is possible to reduce the amount of water that is wasted in agricultural settings by increasing irrigation efficiency. This can be accomplished by implementing technology such as drip irrigation, precision irrigation, and soil moisture monitors (Grafton et al., 2018). Water conservation can be improved by instructing farmers on the most effective management strategies, such as the importance of correct scheduling and monitoring.

13.4.4.2 Crop selection and rotational farming

It is possible to lower the water demand and improve the efficiency of its use by establishing crop rotation practices and encouraging farmers to select crops that are suitable to the conditions of the area (Yu et al., 2022). Crop types that are more resistant to drought and use less water can also contribute to sustainable agriculture.

13.4.4.3 Water-use efficiency programs

Farmers may benefit from educational opportunities, financial incentives, and technical assistance if water-use efficiency programs designed specifically for the agricultural sector are developed and put into action (Uhlenbrook et al., 2022). These projects have the goals of enhancing irrigation methods, implementing technology that saves water, and promoting sustainable water management in agricultural settings.

13.4.5 Education and public awareness

13.4.5.1 Water conservation campaigns

Changing people's behaviors can be facilitated by organizing public awareness campaigns on the significance of water conservation and the role that individual actions play (Thaker et al., 2019). These campaigns may make use of a variety of communication avenues to inform the general public about water footprints and offer useful advice for cutting down on water consumption.

13.4.5.2 Water education in schools

It is helpful to establish a culture of water consciousness in future generations by including water conservation instruction in the standard curriculum of schools (Amahmid et al., 2019). Instilling a sense of responsibility for water conservation in children through educating them about water resources, conservation measures, and sustainable water management can be accomplished through teaching.

To lower water footprints and establish sustainable water management, water conservation policies that are founded on scientific methodologies are necessary. Societies can encourage responsible water usage, effective water management, and conservation practices through the introduction of policies such as water pricing mechanisms, legislation, infrastructural improvements, and educational programs (Khalid, 2018). These policies address the issues posed by a lack of available water and contribute to the long-term sustainability of water resources. We can create a future in which water resources are efficiently managed, water footprints are reduced to the smallest possible size, and the requirements of both the current generation and the generations to come are satisfied by combining scientific understanding with policy actions.

13.5 Conclusion

It is essential to use water footprint reduction strategies to lessen the impact of water scarcity and advance the cause of sustainable water management. This chapter explored a variety of scientifically sound techniques to address water footprints at a variety of levels, including individual houses, landscaping and gardening, straightforward lifestyle modifications, and policy interventions. Among the several scientifically sound options that were examined was gardening. The measures that were considered included a wide variety of different activities, such as water-efficient appliances and fixtures, responsible water use behaviors, selecting native and drought-resistant plant species, water-efficient irrigation systems, water conservation legislation, education and awareness campaigns, and so on. These tactics intend to maximize the efficiency with which water is used, reduce the amount of water that is wasted, and stimulate behavioral shifts that contribute to water conservation. Individuals, communities, and politicians all can contribute to sustainable water management and provide a good impact on water resources by implementing the aforementioned measures.

In light of the ever-increasing demand for sustainable water management and the importance of reducing water footprints, it is of the utmost importance to make these reductions. Approaches that are effective in addressing water footprints across a variety of sectors can be derived from scientifically grounded tactics. These techniques provide actionable solutions for maximizing water use efficiency and encouraging water conservation habits. These solutions range from individual activities to policy initiatives.

At the individual level, water use can be greatly reduced by installing water-efficient equipment, adopting responsible water use behaviors both indoors and outdoors, and making deliberate choices regarding landscaping and gardening. Modifications to one's way of life, such as reducing the amount of food that is wasted and adopting dietary choices that are water-friendly, both contribute to water conservation.

Policy interventions, such as mechanisms for the pricing of water, laws, and upgrades to infrastructure, play an important part in the process of developing water management practices. Policies and regulations pertaining to water conservation incentivize effective water usage and give guidance for the environmentally responsible management of water resources. Individuals are given the ability to make educated decisions and adopt sustainable water practices as a result of education and awareness programs, which result in the creation of a culture of water consciousness.

Collectively, these approaches encourage the use of available water resources that is both more effective and more environmentally friendly. Communities can solve the difficulties of water scarcity, reduce their water footprints, and ensure that there will be water available for future generations if they implement ways that are informed by scientific research. Long-term water sustainability can only be achieved via ongoing research and development of new technologies, as well as through the concerted efforts of individuals, communities, and policymakers working together. We can strive toward a future in which water resources are preserved, protected, and used in a manner that is both responsible and sustainable if we put these methods into practice altogether.

References

Abdou, A. H., Hassan, T. H., & El Dief, M. M. (2020). A description of green hotel practices and their role in achieving sustainable development. *Sustainability*, *12*(22), 9624. Available from https://doi.org/10.3390/su12229624.

Abdulla, F., Abdulla, C., & Eslamian, S. (2021). *Concept and technology of rainwater harvesting. Handbook of water harvesting and conservation: Basic concepts and fundamentals* (pp. 1–16). Wiley.

Acquaye, A., Feng, K., Oppon, E., Salhi, S., Ibn-Mohammed, T., Genovese, A., & Hubacek, K. (2017). Measuring the environmental sustainability performance of global supply chains: A multi-regional input–output analysis for carbon, sulphur oxide and water footprints. *Journal of Environmental Management*, *187*, 571–585. Available from https://doi.org/10.1016/j.jenvman.2016.10.059.

Addo, I. B., Thoms, M. C., & Parsons, M. (2018). Barriers and drivers of household water-conservation behavior: A profiling approach. *Water*, *10*(12), 1794. Available from https://doi.org/10.3390/w10121794.

Ahmad, S., Jia, H., Chen, Z., Li, Q., & Xu, C. (2020). Water-energy nexus and energy efficiency: A systematic analysis of urban water systems. *Renewable and Sustainable Energy Reviews*, *134*, 110381. Available from https://doi.org/10.1016/j.rser.2020.110381.

Ahmed, J., Thakur, A., & Goyal, A. (2021). *Industrial wastewater and its toxic effects. Biological treatment of industrial wastewater* (pp. 1–14). Royal Society of Chemistry.

Albert, J. S., Destouni, G., Duke-Sylvester, S. M., Magurran, A. E., Oberdorff, T., Reis, R. E., Winemiller, K. O., & Ripple, W. J. (2021). Scientists' warning to humanity on the freshwater biodiversity crisis. *Ambio*, *50*(1), 85–94. Available from https://doi.org/10.1007/s13280-020-01318-8.

Amahmid, O., El Guamri, Y., Yazidi, M., Razoki, B., Kaid Rassou, K., Rakibi, Y., Knini, G., & El Ouardi, T. (2019). Water education in school curricula: Impact on children knowledge, attitudes and behaviours towards water use. *International Research in Geographical and Environmental Education*, *28*(3), 178–193. Available from https://doi.org/10.1080/10382046.2018.1513446.

Amaral, R. E. C., Brito, J., Buckman, M., Drake, E., Ilatova, E., Rice, P., Sabbagh, C., Voronkin, S., & Abraham, Y. S. (2020). Waste management and operational energy for sustainable buildings: A review. *Sustainability*, *12*(13), 5337. Available from https://doi.org/10.3390/su12135337.

Anju, A., Ravi, S. P., & Bechan, S. (2010). Water pollution with special reference to pesticide contamination in India. *Journal of Water Resource and Protection*, *2*(5), 432–448.

Bahn, R., EL Labban, S. E. L., & Hwalla, N. (2019). Impacts of shifting to healthier food consumption patterns on environmental sustainability in MENA countries. *Sustainability Science*, *14*(4), 1131–1146. Available from https://doi.org/10.1007/s11625-018-0600-3.

Bai, X., Ren, X., Khanna, N. Z., Zhou, N., & Hu, M. (2018). Comprehensive water footprint assessment of the dairy industry chain based on ISO 14046: A case study in China. *Resources, Conservation and Recycling*, *132*, 369–375. Available from https://doi.org/10.1016/j.resconrec.2017.07.021.

Berger, M., Campos, J., Carolli, M., Dantas, I., Forin, S., Kosatica, E., Kramer, A., Mikosch, N., Nouri, H., Schlattmann, A., Schmidt, F., Schomberg, A., & Semmling, E. (2021). Advancing the water footprint into an instrument to support achieving the SDGs – Recommendations from the Water as a Global Resources' research initiative (GRoW). *Water Resources Management*, *35*(4), 1291–1298. Available from https://doi.org/10.1007/s11269-021-02784-9.

Boretti, A., & Rosa, L. (2019). Reassessing the projections of the world water development report. *npj Clean Water*, *2*(1), 15. Available from https://doi.org/10.1038/s41545-019-0039-9.

Brack, W., Dulio, V., Ågerstrand, M., Allan, I., Altenburger, R., Brinkmann, M., Bunke, D., Burgess, R. M., Cousins, I., Escher, B. I., Hernández, F. J., Hewitt, L. M., Hilscherová, K., Hollender, J., Hollert, H., Kase, R., Klauer, B., Lindim, C., Herráez, D. L., & Vrana, B. (2017). Towards the review of the European Union water Framework Directive: Recommendations for more efficient assessment and management of chemical contamination in European surface water resources. *Science of the Total Environment*, *576*, 720–737. Available from https://doi.org/10.1016/j.scitotenv.2016.10.104.

Butler, D., Ward, S., Sweetapple, C., Astaraie-Imani, M., Diao, K., Farmani, R., & Fu, G. (2017). Reliable, resilient and sustainable water management: The Safe and SuRe approach. *Global Challenges*, *1*(1), 63–77. Available from https://doi.org/10.1002/gch2.1010.

Çetin, N., Mansuroğlu, S., & Önaç, A. K. (2018). Xeriscaping feasibility as an urban adaptation method for global warming: A case study from turkey. *Polish Journal of Environmental Studies*, *27*(3), 1009–1018. Available from https://doi.org/10.15244/pjoes/76678.

Chen, B., Han, M. Y., Peng, K., Zhou, S. L., Shao, L., Wu, X. F., Wei, W. D., Liu, S. Y., Li, Z., Li, J. S., & Chen, G. Q. (2018). Global land-water nexus: Agricultural land and freshwater use embodied in worldwide supply chains. *Science of the Total Environment*, *613–614*, 931–943. Available from https://doi.org/10.1016/j.scitotenv.2017.09.138.

Compagnucci, L., & Spigarelli, F. (2018). Fostering cross-sector collaboration to promote innovation in the water sector. *Sustainability*, *10*(11), 4154. Available from https://doi.org/10.3390/su10114154.

Delgado, A., Rodriguez, D. J., Amadei, C. A., & Makino, M. (2021). *Water in circular economy and resilience*. World Bank.

Du Plessis, A. (2019). *Current and future water scarcity and stress. Water as an inescapable risk: Current global water availability, quality and risks with a specific focus on South Africa* (pp. 13–25). Springer.

Dybeck Carlsson, S., Hamulczuk, D., & Raja, S. (2020). *Exploring a new energy-efficient way to heat water: Design of a heat exchanger for laundry machine applications produced using additive manufacturing*. TRA105 Additive Manufacturing, Chalmers University of Technology Gothenburg, 2020-12-07

Ebrahimi, M., Naghali, B., & Aryanfar, M. (2022). Thermoeconomic and environmental evaluation of a combined heat, power, and distilled water system of a small residential building with water demand strategy. *Energy Conversion and Management*, *258*, 115498. Available from https://doi.org/10.1016/j.enconman.2022.115498.

El-Beltagi, H. S., Basit, A., Mohamed, H. I., Ali, I., Ullah, S., Kamel, E. A. R., Shalaby, T. A., Ramadan, K. M. A., Alkhateeb, A. A., & Ghazzawy, H. S. (2022). Mulching as a sustainable water and soil saving practice in agriculture: A review. *Agronomy*, *12*(8), 1881. Available from https://doi.org/10.3390/agronomy12081881.

Ewaid, S. H., Abed, S. A., Abbas, A. J., & Al-Ansari, N. (2020). Estimation the virtual water content and the virtual water transfer for Iraqi wheat. *Journal of Physics: Conference Series*, *1664*(1). Available from https://doi.org/10.1088/17426596/1664/1/012143.

Foden, M., Browne, A. L., Evans, D. M., Sharp, L., & Watson, M. (2019). The water–energy–food nexus at home: New opportunities for policy interventions in household sustainability. *Geographical Journal*, *185*(4), 406–418. Available from https://doi.org/10.1111/geoj.12257.

Fu, H., Manogaran, G., Wu, K., Cao, M., Jiang, S., & Yang, A. (2020). Intelligent decision-making of online shopping behavior based on internet of things. *International Journal of Information Management*, *50*, 515–525. Available from https://doi.org/10.1016/j.ijinfomgt.2019.03.010.

Goap, A., Sharma, D., Shukla, A. K., & Rama Krishna, C. (2018). An IoT based smart irrigation management system using machine learning and open source technologies. *Computers and Electronics in Agriculture*, *155*, 41–49. Available from https://doi.org/10.1016/j.compag.2018.09.040.

Goel, P. K. (2006). *Water pollution: causes, effects and control*. New Age. International.

Grafton, R. Q., Williams, J., Perry, C. J., Molle, F., Ringler, C., Steduto, P., Udall, B., Wheeler, S. A., Wang, Y., Garrick, D., & Allen, R. G. (2018). The paradox of irrigation efficiency. *Science*, *361*(6404), 748–750. Available from https://doi.org/10.1126/science.aat9314.

Gude, V. G. (2017). Desalination and water reuse to address global water scarcity. *Reviews in Environmental Science and Bio/Technology*, *16*(4), 591–609. Available from https://doi.org/10.1007/s11157-017-9449-7.

Herawati, A., Mujiyo., Syamsiyah, J., Baldan, S. K., & Arifin, I. (2021). Application of soil amendments as a strategy for water holding capacity in sandy soils. *IOP Conference Series: Earth and Environmental Science*, *724*(1).

Hicks, A. L. (2018). Environmental implications of consumer convenience: Coffee as a case study. *Journal of Industrial Ecology*, *22*(1), 79–91. Available from https://doi.org/10.1111/jiec.12487.

Hoekstra, A. Y. (2017a). Water footprint assessment: Evolvement of a new research field. *Water Resources Management*, *31*(10), 3061–3081. Available from https://doi.org/10.1007/s11269-017-1618-5.

Hoekstra, A. Y. (2017b). *The water footprint of animal products. The meat crisis* (pp. 21–30). Routledge.

Hoekstra, A. Y. (2019). *The water footprint of modern consumer society*. Routledge.

Hoekstra, A. Y., Chapagain, A. K., & Van Oel, P. R. (2019). Progress in water footprint assessment: Towards collective action in water governance. *Water*, *11*(5), 1070. Available from https://doi.org/10.3390/w11051070.

Hogeboom, R. J., Knook, L., & Hoekstra, A. Y. (2018). The blue water footprint of the world's artificial reservoirs for hydroelectricity, irrigation, residential and industrial water supply, flood protection, fishing and recreation. *Advances in Water Resources*, *113*, 285–294. Available from https://doi.org/10.1016/j.advwatres.2018.01.028.

Hulme, P. E. (2005). Adapting to climate change: Is there scope for ecological management in the face of a global threat? *Journal of Applied Ecology*, *42*(5), 784–794. Available from https://doi.org/10.1111/j.1365-2664.2005.01082.x.

Huo, X., Yu, A. T. W., & Wu, Z. (2017). A comparative analysis of site planning and design among green building rating tools. *Journal of Cleaner Production*, *147*, 352–359. Available from https://doi.org/10.1016/j.jclepro.2017.01.099.

Kadijk, N. P. (2021). *Make the UT community aware of their toilet flushing behaviour* [BS Thesis, University of Twente].

Kanu, I., & Achi, O. K. (2011). Industrial effluents and their impact on water quality of receiving rivers in Nigeria. *Journal of Applied Technology in Environmental Sanitation*, *1*(1), 7586.

Khalid, R. M. (2018). Review of the water supply management and reforms needed to ensure water security in Malaysia. *International Journal of Business and Society*, *19*(Suppl. 3), 472–483.

Liu, L., & Jensen, M. B. (2018). Green infrastructure for sustainable urban water management: Practices of five forerunner cities. *Cities*, *74*, 126–133. Available from https://doi.org/10.1016/j.cities.2017.11.013.

Lopez-Villalobos, A., Bunsha, D., Austin, D., Caddy, L., Douglas, J., Hill, A., Kubeck, K., Lewis, P., Stormes, B., Sugiyama, R., & Moreau, T. (2022). Aligning to the UN sustainable development goals: Assessing contributions of UBC Botanical Garden. *Sustainability*, *14*(10), 6275. Available from https://doi.org/10.3390/su14106275.

Marston, L., Ao, Y., Konar, M., Mekonnen, M. M., & Hoekstra, A. Y. (2018). High-resolution water footprints of production of the United States. *Water Resources Research*, *54*(3), 2288–2316. Available from https://doi.org/10.1002/2017WR021923.

McCarroll, M., & Hamann, H. (2020). What we know about water: A water literacy review. *Water*, *12*(10), 2803. Available from https://doi.org/10.3390/w12102803.

McNabb, D. E. (2019). *Global pathways to water sustainability*. Springer.

Mekonnen, M. M., & Gerbens-Leenes, W. (2020). The water footprint of global food production. *Water*, *12*(10), 2696. Available from https://doi.org/10.3390/w12102696.

Mian, H. R., Hu, G., Hewage, K., Rodriguez, M. J., & Sadiq, R. (2021). Drinking water quality assessment in distribution networks: A water footprint approach. *Science of the Total Environment*, *775*, 145844. Available from https://doi.org/10.1016/j.scitotenv.2021.145844.

Mohapatra, H., & Rath, A. K. (2019). Detection and avoidance of water loss through municipality taps in India by using smart taps and ICT. *IET Wireless Sensor Systems*, *9*(6), 447–457. Available from https://doi.org/10.1049/iet-wss.2019.0081.

Nezamoleslami, R., & Hosseinian, S. M. (2020). An improved water footprint model of steel production concerning virtual water of personnel: The case of Iran. *Journal of Environmental Management*, *260*, 110065. Available from https://doi.org/10.1016/j.jenvman.2020.110065.

Okutan, P., & Akkoyunlu, A. (2021). Identification of water use behavior and calculation of water footprint: A case study. *Applied Water Science*, *11*(7), 127. Available from https://doi.org/10.1007/s13201-021-01459-5.

Onyenankeya, K., & Salawu, A. (2018). Negotiating water conservation communication through indigenous media. *Communitas*, *23*(1), 178–193. Available from https://doi.org/10.18820/24150525/Comm.v23.12.

Osanaiye, O. A., Mannan, T., & Aina, F. (2022). An IoT-based soil moisture monitor. *African Journal of Science, Technology, Innovation and Development*, *14*(7), 1908–1915. Available from https://doi.org/10.1080/20421338.2021.1988413.

Pfister, S., Boulay, A. M., Berger, M., Hadjikakou, M., Motoshita, M., Hess, T., Ridoutt, B., Weinzettel, J., Scherer, L., Döll, P., Manzardo, A., Núñez, M., Verones, F., Humbert, S., Buxmann, K., Harding, K., Benini, L., Oki, T., Finkbeiner, M., & Henderson, A. (2017). Understanding the LCA and ISO water footprint: A response to Hoekstra (2016) "A critique on the water-scarcity weighted water footprint in LCA". *Ecological Indicators*, *72*, 352–359. Available from https://doi.org/10.1016/j.ecolind.2016.07.051.

Pimentel, D., Berger, B., Filiberto, D., Newton, M., Wolfe, B., Karabinakis, E., Clark, S., Poon, E., Abbett, E., & Nandagopal, S. (2004). Water resources: Agricultural and environmental issues. *BioScience*, *54*(10), 909–918. Available from https://doi.org/10.1641/0006-3568(2004)054[0909:WRAAEI]2.0.CO;2.

Pompelli, M. F., Mendes, K. R., Ramos, M. V., Santos, J. N. B., Youssef, D. T. A., Pereira, J. D., Endres, L., Jarma-Orozco, A., Solano-Gomes, R., Jarma-Arroyo, B., Silva, A. L. J., Santos, M. A., & Antunes, W. C. (2019). Mesophyll thickness and sclerophylly among *Calotropis procera* morphotypes reveal water-saved adaptation to environments. *Journal of Arid Land*, *11*(6), 795–810. Available from https://doi.org/10.1007/s40333-019-0016-7.

Pratt, T., Allen, L. N., Rosenberg, D. E., Keller, A. A., & Kopp, K. (2019). Urban agriculture and small farm water use: Case studies and trends from Cache Valley, Utah. *Agricultural Water Management*, *213*, 24–35. Available from https://doi.org/10.1016/j.agwat.2018.09.034.

Rahman, M. M., Ashiqur Rahman, M., Mahmudul Haque, M., & Rahman, A. (2019). *Sustainable water use in construction. Sustainable construction technologies* (pp. 211–235). Butterworth-Heinemann.

Rusyn, I. (2021). Role of microbial community and plant species in performance of plant microbial fuel cells. *Renewable and Sustainable Energy Reviews*, *152*, 111697. Available from https://doi.org/10.1016/j.rser.2021.111697.

Saha, U. S., Prajapati, J. B., Sompurkar, M. S., & Mohapatra, S. S. (2022). *Water footprint of milk production in India: A case of Anand District of Gujarat* [Working Paper 336]. Institute of Rural Management Anand

Sahin, O., Bertone, E., & Beal, C. D. (2017). A systems approach for assessing water conservation potential through demand-based water tariffs. *Journal of Cleaner Production*, *148*, 773–784. Available from https://doi.org/10.1016/j.jclepro.2017.02.051.

Scanlon, B. R., Ruddell, B. L., Reed, P. M., Hook, R. I., Zheng, C., Tidwell, V. C., & Siebert, S. (2017). The food-energy-water nexus: Transforming science for society. *Water Resources Research*, *53*(5), 3550–3556. Available from https://doi.org/10.1002/2017WR020889.

Seyoum, S., Alfonso, L., van Andel, S. J., Koole, W., Groenewegen, A., & van de Giesen, N. (2017). A Shazam-like household water leakage detection method. *Procedia Engineering*, *186*, 452–459. Available from https://doi.org/10.1016/j.proeng.2017.03.253.

Shimabuku, M., Diringer, S., & Cooley, H. (2018). *Stormwater capture in California: Innovative policies and funding opportunities*. Pacific Institute. Available from https://pacinst.org/publication/stormwater-capture-in-california.

Singh, V. K., Rajanna, G. A., Paramesha, V., & Upadhyay, P. K. (2022). *Agricultural water footprint and precision management. Sustainable Agriculture Systems and Technologies* (pp. 251–266). John Wiley & Sons.

Skouteris, G., Ouki, S., Foo, D., Saroj, D., Altini, M., Melidis, P., Cowley, B., Ells, G., Palmer, S., & O'Dell, S. (2018). Water footprint and water pinch analysis techniques for sustainable water management in the brick-manufacturing industry. *Journal of Cleaner Production, 172*, 786–794. Available from https://doi.org/10.1016/j.jclepro.2017.10.213.

Smol, M., Adam, C., & Preisner, M. (2020). Circular economy model framework in the European water and wastewater sector. *Journal of Material Cycles and Waste Management, 22*(3), 682–697. Available from https://doi.org/10.1007/s10163-019-00960-z.

Takeuchi, H., & Tanaka, H. (2020). Water reuse and recycling in Japan—History, current situation, and future perspectives. *Water Cycle, 1*, 1–12. Available from https://doi.org/10.1016/j.watcyc.2020.05.001.

Tayyab, M., Jemai, J., Lim, H., & Sarkar, B. (2020). A sustainable development framework for a cleaner multi-item multi-stage textile production system with a process improvement initiative. *Journal of Cleaner Production, 246*, 119055. Available from https://doi.org/10.1016/j.jclepro.2019.119055.

Teague, A., Sermet, Y., Demir, I., & Muste, M. (2021). A collaborative serious game for water resources planning and hazard mitigation. *International Journal of Disaster Risk Reduction, 53*, 101977. Available from https://doi.org/10.1016/j.ijdrr.2020.101977.

Thaker, J., Howe, P., Leiserowitz, A., & Maibach, E. (2019). Perceived collective efficacy and trust in government influence public engagement with climate change-related water conservation policies. *Environmental Communication, 13*(5), 681–699. Available from https://doi.org/10.1080/17524032.2018.1438302.

Trubetskaya, A., Horan, W., Conheady, P., Stockil, K., & Moore, S. (2021). A methodology for industrial water footprint assessment using energy-water-carbon nexus. *Processes, 9*(2), 393. Available from https://doi.org/10.3390/pr9020393.

Uddin, S., Löw, M., Parvin, S., Fitzgerald, G. J., Tausz-Posch, S., Armstrong, R., O'Leary, G., & Tausz, M. (2018). Elevated [CO_2] mitigates the effect of surface drought by stimulating root growth to access sub-soil water. *PLoS One, 13*(6), e0198928. Available from https://doi.org/10.1371/journal.pone.0198928.

Uhlenbrook, S., Yu, W., Schmitter, P., & Smith, D. M. (2022). Optimising the water we eat—Rethinking policy to enhance productive and sustainable use of water in agri-food systems across scales. *Lancet Planetary Health, 6*(1), e59–e65. Available from https://doi.org/10.1016/S2542-5196(21)00264-3.

Varma, V. G. (2022). Water-efficient technologies for sustainable development. *Current Directions in Water Scarcity Research, 6*, 101–128.

Venckute, M., Silva, M., & Figueiredo, M. (2017). Education as a tool to reduce the water footprint of young people. *Millenium – Journal of Education, Technologies, and Health, 4*(4), 101–111. Available from https://doi.org/10.29352/mill0204.09.00144.

Wang, Y.-P., Zhang, L., Mu, Y., Liu, W., Guo, F., & Chang, T. (2020). Effect of a root-zone injection irrigation method on water productivity and apple production in a semi-arid region in north-western China. *Irrigation and Drainage, 69*(1), 74–85. Available from https://doi.org/10.1002/ird.2379.

Wilson, N. J., Mutter, E., Inkster, J., & Satterfield, T. (2018). Community-based monitoring as the practice of Indigenous governance: A case study of Indigenous-led water quality monitoring in the Yukon River Basin. *Journal of Environmental Management, 210*, 290–298. Available from https://doi.org/10.1016/j.jenvman.2018.01.020.

Yao, L., Zhao, M., & Xu, T. (2017). China's water-saving irrigation management system: Policy, implementation, and challenge. *Sustainability, 9*(12), 2339. Available from https://doi.org/10.3390/su9122339.

Yousif, J. H., & Abdalgader, K. (2022). Experimental and mathematical models for real-time monitoring and auto watering using IoT architecture. *Computers, 11*(1), 7. Available from https://doi.org/10.3390/computers11010007.

Yu, T., Mahe, L., Li, Y., Wei, X., Deng, X., & Zhang, D. (2022). Benefits of crop rotation on climate resilience and its prospects in China. *Agronomy, 12*(2), 436. Available from https://doi.org/10.3390/agronomy12020436.

Chapter 14

Energy-water nexus in sustainable development

Abhishek Gautam[1,*] and Manoj Sood[2]

[1]Department of Mechanical Engineering, Indian Institute of Technology Bombay, Mumbai, India, [2]Department of Hydro & Renewable Energy, Indian Institute of Technology Roorkee, Roorkee, India

**Corresponding author. e-mail address: abhishekgautam@iitb.ac.in.*

14.1 Introduction

The exploitation of natural resources has reached a certain point from which the accomplishment of present requirements without a sustainable approach will create a disastrous situation for future generations (Gautam et al., 2023a). The execution of anthropogenic activities beyond a limit has disrupted the earth's natural processes substantially and their nasty repercussion can be seen in terms of irregular environmental activities (Bhat et al., 2015). In view of the same, 195 countries of the United Nations Framework Convention on Climate Change (UNFCCC) signed an agreement in the year 2015 to limit the rise in mean global temperature to 1.5°C to reduce the effect of climate change. This agreement is recognized by the name "Paris Agreement (2015)" (UNFCCC, 2015). Various countries and groups are also making their efforts in the same line. Water and energy are the major elements of human life, and their sustainable production and utilization can also play a significant role in the global fight against climate change (Dai et al., 2018).

Globally, people rely upon freshwater resources for a number of applications like drinking, domestic purposes, agricultural needs, and industrial requirements. Nearly 85% of the total water footprint is consumed for agricultural needs (Yu et al., 2021). However, the consumption pattern depends on the rural and urban categorization as domestic and industrial needs are predominant in urban areas, whereas fresh water for agricultural needs predominates in rural areas. Moreover, the accomplishment of freshwater requirements is going to be a big challenge irrespective of the application, and it is expected to enhance due to climate change, particularly in subtropical regions. A slight increment in global precipitation of 2%−3%, uneven shifting in rain patterns, and unexpected changes in the rain intensity are the negative outcomes of climate change. This variability in the precipitation pattern may increase in more natural calamities like drought and floods (Fishman, 2016).

On the other side, energy is another basic need for human existence and comfort. It is used for all types of convenience in today's modern lifestyle from lightening the bulbs to the use of appliances and from running vehicles on the road to flying aircraft in the sky (Gautam & Saini, 2022). In a broader aspect, energy is required for transportation, industrialization, agriculture, and the building sector (Gautam & Saini, 2020). In the current scenario, nearly 63.3% of the energy requirement is accomplished through fossil fuels, whereas the contribution of renewable energy sources (RESs) and nuclear power are around 26.3% and 10.4%, respectively. The utility of fossil fuels is required to be reduced gradually and brought down to zero as the production of energy from the same involves emissions of harmful gases, which are not suitable for a sustainable environment (Gautam & Saini, 2021). Despite this fact, the notion of its security has been neglected for many years, and therefore it is necessary to produce, convert, store, and utilize it sustainably.

Based on the aforementioned discussion, it emerges that water and energy are the two indispensable inputs for the sustainable development of the world. Therefore their supply security and economic efficiency should be undergone through major productive reforms. Moreover, food is also a crucial need in a similar line and the study of food-energy-water nexus will give deeper insights into precious analysis. However, the scope of the present chapter is limited to the energy-water nexus (EWN) as they rely more on each other, and understanding this complex nexus is more beneficial for the think tanks and stakeholders. Considering the necessity of this topic, the authors are motivated to provide comprehensive knowledge about the EWN and its role in sustainable development through this chapter.

Water Footprints and Sustainable Development. DOI: https://doi.org/10.1016/B978-0-443-23631-0.00014-5

In the present chapter, the role of EWN in sustainable development and its various aspects are discussed in detail. The chapter includes a discussion on the necessity of water and energy in human life, which is followed by the introduction of the EWN concept and its importance. Thereafter the hydropower technique used to produce energy through water and its various aspects are discussed. Further, the water desalination techniques are presented to show the energy-to-water aspect. This chapter also includes the environmental, economic, technical, social, and political aspects of EWN to share its importance. Lastly, the chapter ends with the major conclusions drawn from the different aspects discussed in the present chapter. The present chapter will be beneficial for academicians, research & development, industrialists, environmentalists, and other stakeholders involved in the field of sustainable development.

14.2 Energy-water nexus

Water and energy are interrelated with each other, and it is known by the term "Energy-Water Nexus." Each step involved in the water cycle, viz., collection, purification, wastewater treatment, supply, and others, requires a significant amount of energy. Similarly, whether it is petroleum production, coal washing, biofuel growth, and distillation, or waste heat rejection from the steam turbine, almost all energy production approaches require water in any form. The interdependency can be easily understood by the example of running some power plants below their actual capacity due to insufficiency of water during a drought. Similarly, fresh water can only be obtained through reverse osmosis (RO) or a thermal desalination plant when the required amount of energy will be provided to the same. However, the accomplishment of this energy through fossil fuels will result in harmful emissions. Therefore the use of clean/green energy for freshwater production can only be considered a sustainable approach.

There are various links between water and energy, which are divided into three different categories to simplify its understanding, viz., production, transportation, and consumption as shown in Fig. 14.1. In the case of water-energy links for production, the water is used for mining, electricity generation, and energy for desalination (electricity for membrane-based techniques and heat for thermal-based techniques), whereas the transportation linkage between them

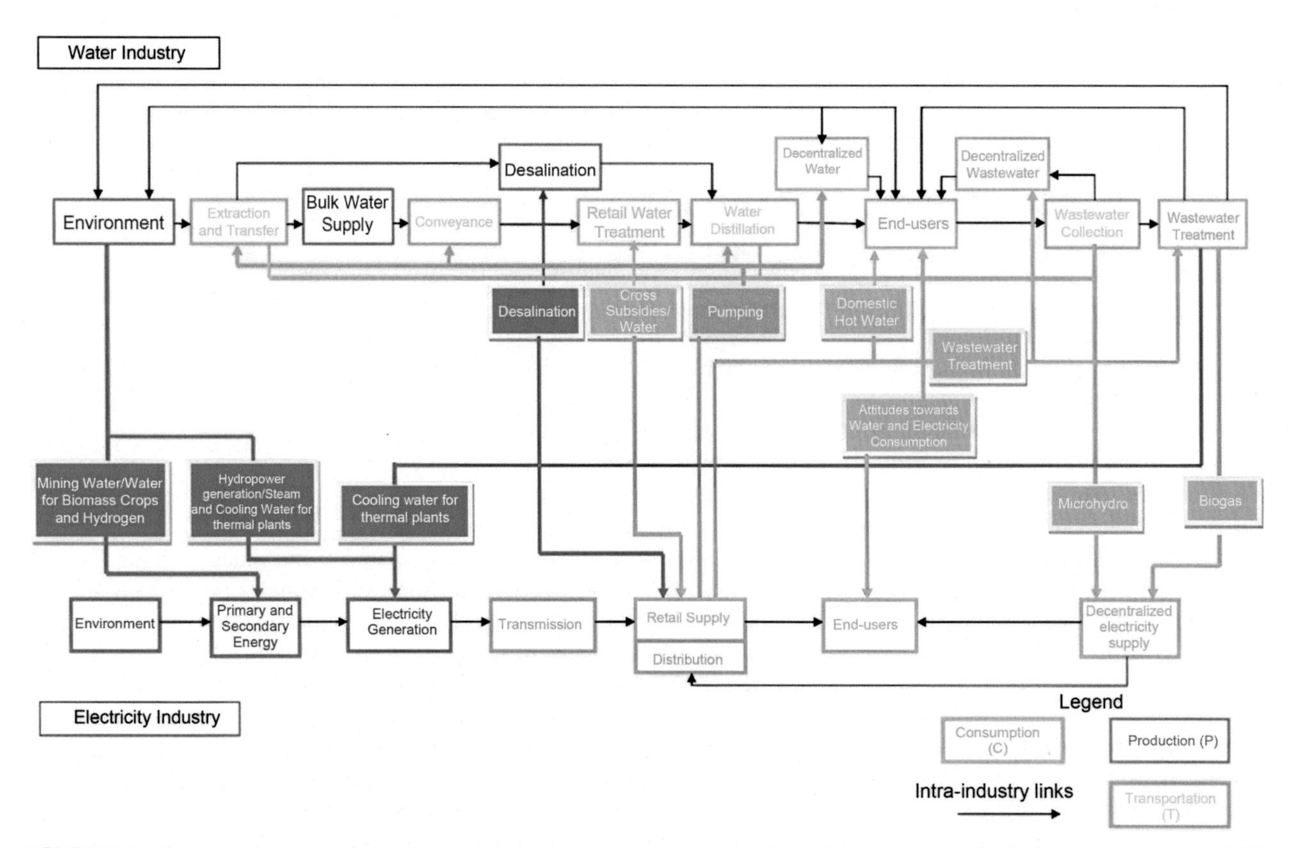

FIGURE 14.1 Energy-water linkage for production, transportation, and consumption. *From Hamiche, A. M., Stambouli, A. B., & Flazi, S. (2016). A review of the water-energy nexus.* Renewable and Sustainable Energy Reviews, 65, *319–331. https://doi.org/10.1016/j.rser.2016.07.020.*

includes energy for groundwater extraction, wastewater collection systems, and surface water distribution. These functions consume nearly 7% of the total energy generated across the world. On the contrary, the amount of water consumed for transporting the energy is less and the potable water consumed by the employees during the energy is a major portion, which is considered negligible in most of the analysis. Moreover, the electricity utilized for domestic water treatment, and potable water purification comes under the category of consumption, which is comparatively less dominant compared to production and transportation linkages. The detailed discussion regarding water-to-energy and energy-to-water linkage from the point of sustainability is given in Sections 14.3 and 14.4, respectively.

14.3 Hydropower techniques for power generation: water to energy

The static approach (conventional methods) and the kinetic approach (upcoming hydrokinetic method) can be used to generate power using water as an energy source. In the former, energy is captured by building a water head, whereas energy is captured by using water currents in the latter method.

14.3.1 Conventional techniques

The section explains all the conventional and upcoming techniques available for hydropower generation. Conventional techniques include run-of-river schemes, canal-based schemes, pumped storage schemes, and dam-based schemes.

14.3.1.1 Run-of-river scheme

Run-of-river hydropower facilities typically lack an impoundment; however, they occasionally have a small amount of storage space. Despite this, they are unable to alter the streamflow regime for the seasons (Egré & Milewski, 2002). As a result, the streamflow pattern of the exploited natural water body directly affects the hydropower generation from these types of plants, making it less dependable and frequently discontinuous. However, compared to dam-based facilities, investment costs are often significantly cheaper, and environmental impacts are typically milder (Magaju et al., 2020). Run-of-river plants are often (but not always) very small in size: installed capacity is frequently between 1 and 50 MW, with some notable outliers (Yildiz & Vrugt, 2019); nevertheless, there are many more appropriate sites, many of which have not yet been utilized (Gernaat et al., 2017; Tefera & Kasiviswanathan, 2022). These reasons help to explain why both private and public investors are now showing increased interest in the tiny run-of-river hydropower facilities that are popping up all over the world, notably in Europe (Couto & Olden, 2018; Quaranta et al., 2022).

There has been much discussion surrounding the design of new run-of-river plants. Some authors have concentrated on the input hydrological data, demonstrating that the water flow distribution at the intake location should be carefully evaluated to avoid designing errors (Barelli et al., 2013; Niadas & Mentzelopoulos, 2008), while various approaches have been suggested for the optimal design of the plant (Sasthav & Oladosu, 2022). A comparison of several design options is typically done, taking into consideration a number of technical, economic, and environmental factors (Santolin et al., 2011), in addition to a few methodologies that rely on an analytical framework (Basso & Botter, 2012). Fig. 14.2 depicts the layout of a typical run-of-river project under consideration.

Run-of-river hydropower projects include the diversion of water from a river without the construction of any kind of water storage, such as a reservoir (Bragalli et al., 2023). These projects use available heads and flowing water to create electricity. A run-of-river plant's production is dependent on the stream's current velocity. Through a power channel, the water that the diversion weir and intake redirect is sent to the forebay. Penstock is used to transport water from the forebay to the powerhouse's turbines, which turn them to produce energy. These projects are divided into three head categories: low (3–20 m), medium (20–60 m), and high head projects (above 60 m). Project designs are distinct for each category. Check out Indian Standard IS: 12800 (Part 3) (Indian Standards, n.d.), for instance. The two main elements of a run-of-river small hydropower plant are electromechanical equipment and civil construction. Due to the enormous volume of water, low-head hydropower facilities require larger electromechanical and civil construction.

Recent case study (Inam Ullah et al., 2023): Geographically, the study region is in Punjab province in the Khyber Pakhtunkhwa province of Pakistan. It covers the area from 72.2597° East longitude to 33.7800° North latitude. The study region experiences a humid, hot summer season, and moderate winter season, with an average annual temperature of 22.2°C and 639 mm of rainfall (Khalid et al., 2017). The location of Ghazi Barotha Hydropower Project, which is built in the Indus River, is situated 10 km in the west of the district of Attock in Punjab state and east of Swabi and Haripur in Khyber Pakhtunkhwa region.

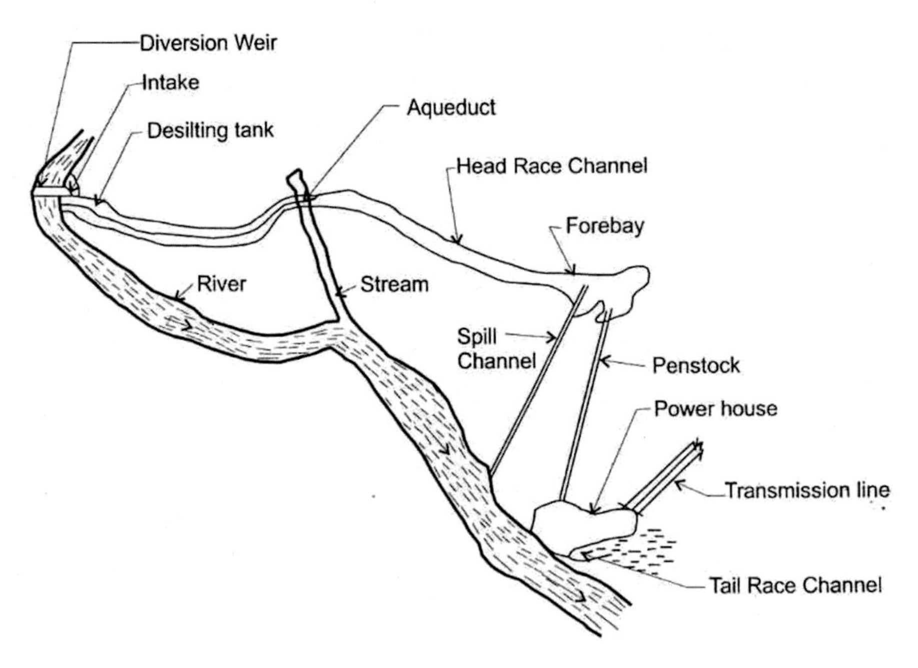

FIGURE 14.2 Layout of a typical run-of-river scheme. *From Singal, S. K., Saini, R. P., & Raghuvanshi, C. S. (2010). Analysis for cost estimation of low head run-of-river small hydropower schemes.* Energy for Sustainable Development, 14 *(2), 117–126. https://doi.org/10.1016/j.esd.2010.04.001.*

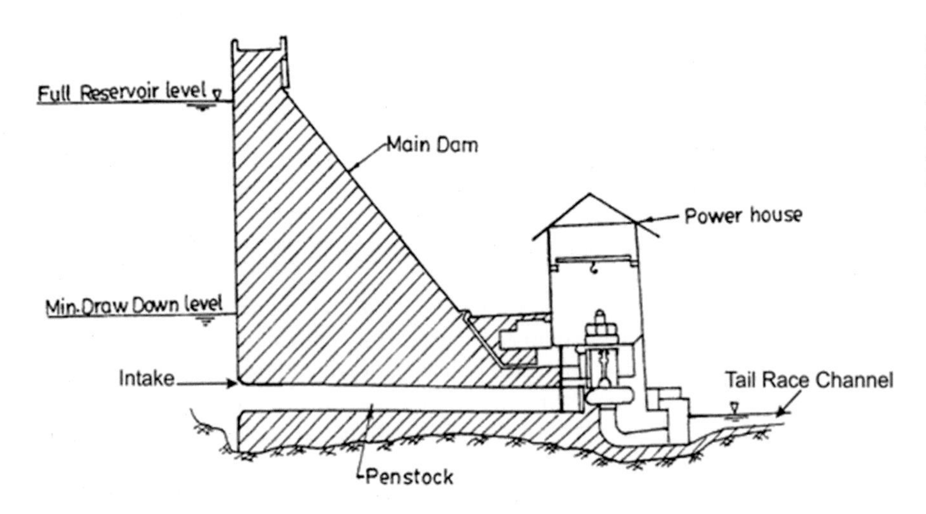

FIGURE 14.3 Layout of a typical dam-toe scheme. *From Singal, S. K., & Saini, R. P. (2008). Cost analysis of low-head dam-toe small hydropower plants based on number of generating units.* Energy for Sustainable Development, 12*(3), 55–60. https://doi.org/10.1016/S0973-0826(08)60439-1.*

14.3.1.2 Dam-based power plant

Dam-based plans involve building a dam across a river to hold water and then using the water's regulated flow to create electricity. In dam-toe plans, the powerhouse is built at the toe of the dam, while the intake system is a component of the main dam as shown in Fig. 14.3. Penstocks built straight through the body of the dam transport water to the turbine. In southern India, these systems are typical. Fig. 14.1 depicts a typical arrangement of the plan; it makes clear what a dam's "toe" refers to. According to the head, small hydropower projects are divided into the following categories: (1) high head, more than 50 m; (2) low head, 3–20 m; and (3) medium head, 20–50 m. These suggested boundaries are only a technique for classifying the locations and are not fixed (Varshney, 1977).

Recent case study: The study area is the Akosombo Hydropower Plant dam site, which has a storage capacity of around 153×10^9 m^3, and offers nearby residents a variety of amenities, including fishing and navigational needs (Alhassan et al., 2023). Additionally, the region receives 4.66 kWh/m^2 of solar energy per day (Fiagbe & Obeng, 2007). The facility supplies the majority of Ghana's electricity or around 38.3%. The station can now generate 1020 MW of electricity. A total of 5137 MW, or around 12.2% greater than in 2018, was the generation capacity in 2020 for hydro, solar, and thermal power (Energy Commission of Ghana, 2020).

14.3.1.3 Pumped storage hydropower plant

Pumped storage hydropower (PSH), one of several energy storage technologies, accounts for 95% of the existing grid-scale energy storage capacity (volume) in the United States and up to 98% of the energy storage capacity globally. PSH has a longer discharge time and a relatively higher power rating. In addition, PSH is less expensive than alternative energy storage options. PSH systems are used widely, particularly in large-scale applications. Despite the benefits, there is little research on PSH system size that is optimal. PSH may be utilized to effectively shift the power supply, balance the grid, and complement other RESs.

PSH enhances freshwater storage in addition to providing long-term energy storage at a reasonably cheap cost. More than 10% of the world's hydropower capacity and around 94% of the world's energy storage capacity are represented by PSH. Over 100 PSH hydro plants exist worldwide, with a 130 GW total generating capacity. Europe now has 44 GW installed pumped hydro storage capacity in just Switzerland and Norway, with a further 7–9 GW planned during the following eight years. PSH's installed capacity is currently 160 GW or so worldwide. With an efficiency of 70%–85%, pumped hydro is the most workable approach on a wide scale.

The conversion of existing conventional hydropower facilities to PSH by either the addition of an upper or lower reservoir has been described as one of the most attractive PSH alternatives in the literature. Therefore, a capital cost-effective method is to supplement existing conventional hydropower (CH) reservoirs with open-loop PSH. The approach to studying energy generation for a complicated configuration of PSH in combination with CH is not included in the scant literature that is currently accessible. In order to establish an equilibrium condition for an overall net positive energy balance, it is important to study PSH and its optimal operation alongside an existing CH by taking into account the power, energy output, and consumption for both installations (Baniya et al., 2023).

14.3.1.4 Canal-based hydropower plant

The canal-based small hydro power (SHP) scheme is planned to generate power by utilizing the fall and flow in the canal. These schemes may be planned as low head (3–20 m), medium head (20–60 m), and high head (above 60 m). These recommended parameters are only a way to group the sites; they are not rigorous. By attaining better conversion efficiencies and eliminating transmission and distribution losses, low-head SHP plants may be situated close to end consumers and produce power reliably and economically. The scheme's components are primarily separated into civil construction and electromechanical machinery.

Construction projects are often site-specific. Diversion and intake, desilting tank, channel, spillway, powerhouse building, and tail race channel are typical SHP plant's civil components in low-head plants as shown in Fig. 14.4. Further, because managing a significant volume of water is required for low-head hydropower plants, the equipment is larger and more expensive. The main divisions of the electromechanical equipment include a turbine, generator, excitation system, transformer, and switchyard.

14.3.2 Upcoming technology in hydro sector

Instream technique is the new upcoming scheme, also known as hydrokinetic technology. This technology works on the flow velocity of water instead of the water head.

14.3.2.1 Instream technology

The motion/current of waves must be used to capture the kinetic energy; therefore building a dam to store water does not cost anything. The kinetic approach has advantages over the static method, such as reduced civil work requirement's, the fact that it does not require a hydraulic head, and the fact that it does not alter the natural pathway to operate, as opposed to the Run-of-River scheme (Khan et al., 2009). Using hydrokinetic devices, hydrokinetic energy is a recently developed renewable technology that makes use of the kinetic energy of flowing water. Fig. 14.5 depicts the hydrokinetic technology's schematic.

The literature has divided resource estimates into three categories: theoretical, technical, and practical. The technical energy is that which can be captured by the existing technology (due to device efficiency, and site covered) and the theoretical energy is the whole yearly hydrokinetic energy available at the location (Behrouzi et al., 2014). The energy in the remaining site will be the practical energy that is accessible since those locations that are not possible for power extraction owing to fish habitats and navigation issues must be disregarded. The arrangement of the number of turbines in an array configuration will be governed by wake recovery distance (Sood & Singal 2019, 2021, 2022).

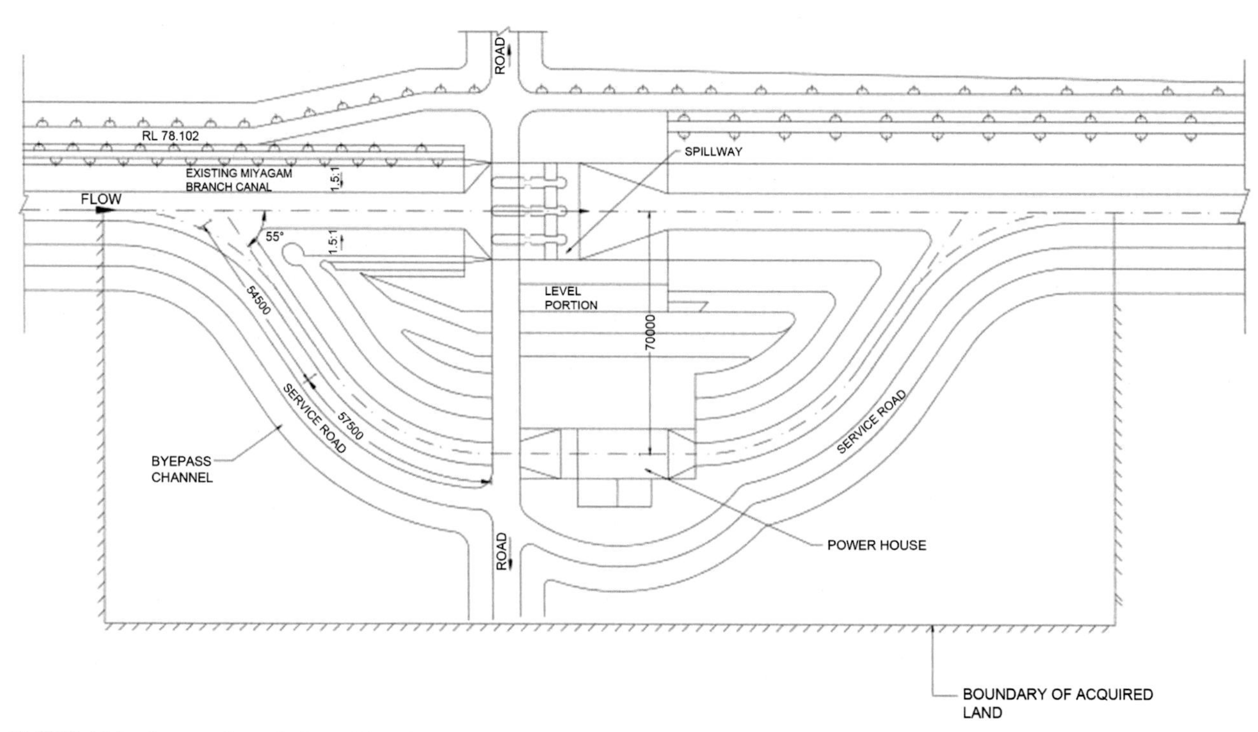

FIGURE 14.4 Layout of a typical canal-based scheme. *From Singal, S. K., & Saini, R. P. (2008). Analytical approach for development of correlations for cost of canal-based SHP schemes.* Renewable Energy, *33(12), 2549–2558. https://doi.org/10.1016/j.renene.2008.02.010.*

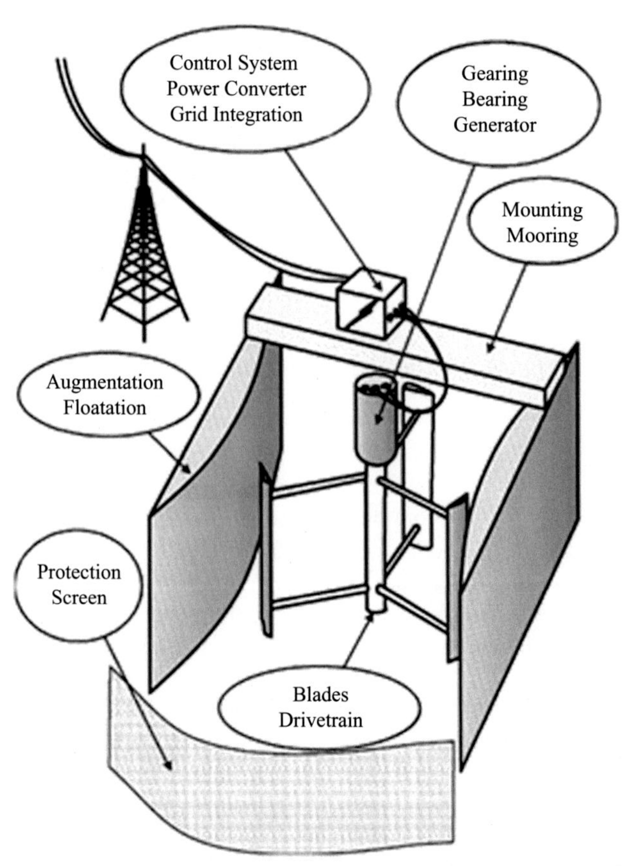

FIGURE 14.5 Hydrokinetic technology system. *From Khan, M. J., Iqbal, M. T., & Quaicoe, J. E. (2008). River current energy conversion systems: Progress, prospects and challenges.* Renewable and Sustainable Energy Reviews, *12(8), 2177–2193. https://doi.org/10.1016/j.rser.2007.04.016.*

14.4 Water desalination techniques: energy to water

The amount of fresh water is decreasing at a very rapid rate due to advancements in industrialization, rising living standards, climate change, as well as groundwater reservoir contamination. The Earth is considered a blue planet as nearly 71% of its surface is occupied by water (Rahmstorf, 2002). However, humanity and other water-dependent species are dealing with water scarcity as 96.5% of its availability on earth is not suitable for drinking due to its high salinity and brackish form (Chauhan et al., 2021). A small fraction of such water can only be used in industries for a few processes such as mining.

United Nations World Water Development Report (2021) reveals that nearly two-third (66.667%) of the global population is confronting water scarcity for at least one month each year (Scientific and Cultural Organization, 2020), whereas the threat is expected to increase for one-fourth of the global population for the whole year by 2040 (VALUING WATER, Scientific and Cultural Organization, 2021). Fig. 14.6 depicts the curves between clean water availability and its forecasted demand (Boretti & Rosa, 2019). Water scarcity has become a threat for future generations even though water is considered a renewed product on earth. The major reason behind it is the overexploitation of water reserves, whereas its natural processes of regeneration are comparatively much slower. Hence, it is necessary to adopt substantial measures now to overcome this issue for the welfare of our upcoming generations.

The distillation and membrane process are the two approaches that can be taken into common practice for converting saline/brackish water into fresh water (Esmaeilion et al., 2021). Desalination of water (saline) is one of the most standard practices to obtain fresh water for utility and a + ve phase-transformation has been seen across the world in the last decade to utilize desalination techniques for handling water scarcity (Ahmed et al., 2021). India, China, Israel, and nations of Arab, Australia, and Africa are developing desalination plants due to the limited availability of fresh water in their region (Esfahani et al., 2016). At the same time, several other countries are also working to develop RESs-based desalination systems. The total capacity of desalination plants worldwide is 97.2 million m^3/day till 2020, and its regional distribution is shown in Fig. 14.7 (Eke et al., 2020).

The desalination methods should go through a set of sequential processes. It includes filtration of water before sending it under a desalination unit to remove the coarse particles, and sand and thereafter, water should undergo the pretreatment by a chemical compound to avoid the formation of scale over the metallic surfaces due to the presence of metallic salts. The posttreatment of water is also necessary to make it drinkable in terms of taste, whereas disinfection is required in the case of low-temperature desalination methods. The flow diagram given in Fig. 14.8 is used to present the general steps to be followed for getting the fresh water through any type of desalination method.

The execution of the desalination process requires energy and various desalination methods are broadly classified into two main categories: thermal energy and nonthermal energy (membrane-based) desalination methods as shown in Fig. 14.9. Irrespective of the operation and type of equipment used, the large-capacity desalination plants are generally based on fossil fuels either to produce vapor pressure or electricity. However, such large-capacity plants of desalination consume a large amount of fossil fuels and result in unviable economic and ecological outcomes. The accomplishment of fresh water through desalination plants based on fossil fuels at a cost of enhancement in environmental-related issues like climate change and global warming is not a smart approach. Hence, efforts are being made by the stakeholders to fulfill the energy requirement of such large desalination plants through RES.

Irrespective of the substantial improvement achieved in the development of various desalination technologies, still energy intensiveness is a major issue. Therefore, stakeholders are looking for cost-effective and sustainable resources to

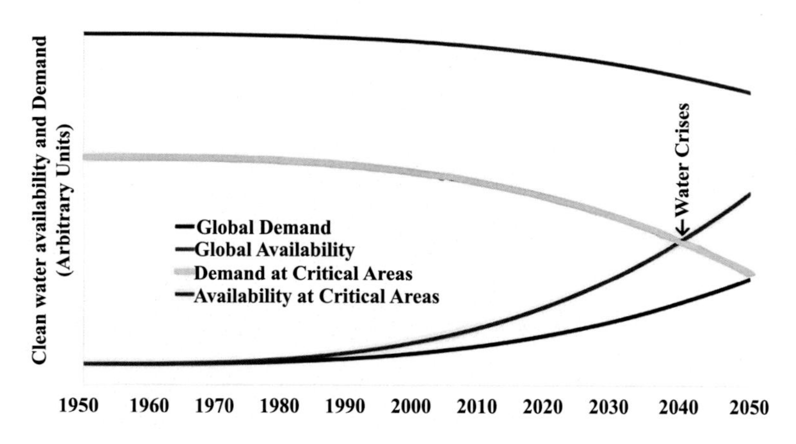

FIGURE 14.6 Demand curve and clean water availability. *From Barelli, L., Liucci, L., Ottaviano, A., & Valigi, D. (2013). Mini-hydro: A design approach in case of torrential rivers.* Energy, 58, *695–706. https://doi.org/10.1016/j.energy.2013.06.038.*

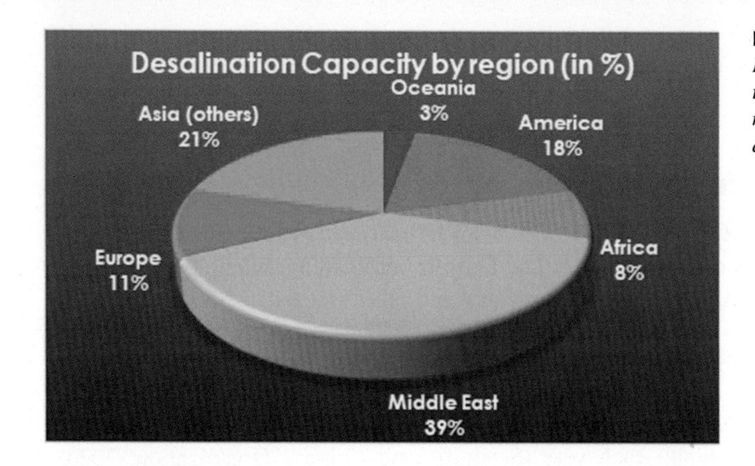

FIGURE 14.7 Regional capacity of desalination plants. *From Eke, J., Yusuf, A., Giwa, A., & Sodiq, A. (2020). The global status of desalination: An assessment of current desalination technologies, plants and capacity.* Desalination, 495, *114633. https://doi.org/10.1016/j.desal.2020.114633.*

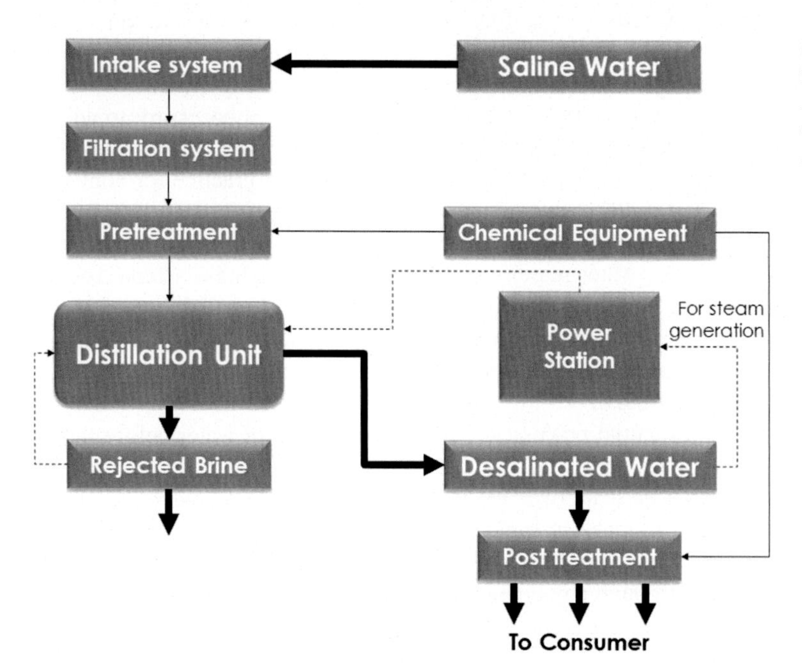

FIGURE 14.8 Flow diagram of the general steps involved in the desalination system. *From Belessiotis, V., Kalogirou, S., & Delyannis, E. (2016).* Desalination methods and technologies—Water and energy *(pp. 1–19). Elsevier BV. https://doi.org/10.1016/b978-0-12-809656-7.00001-5.*

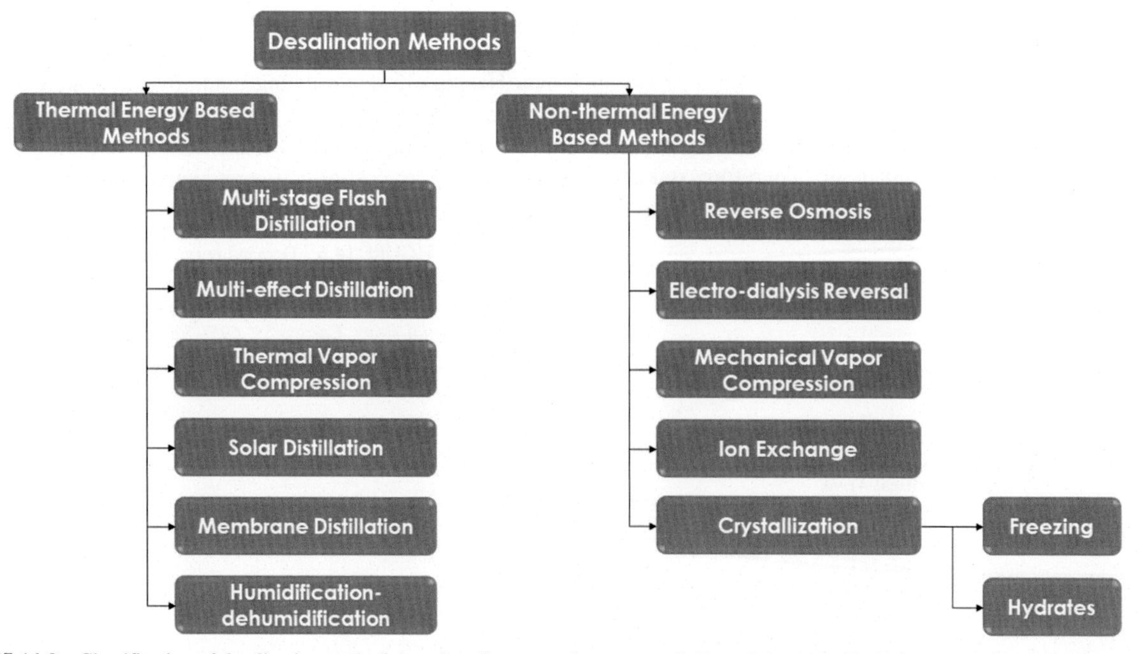

FIGURE 14.9 Classification of desalination methods based on the types of energy used. *From Belessiotis, V., Kalogirou, S., & Delyannis, E. (2016).* Desalination Methods and Technologies—Water and Energy *(pp. 1–19). Elsevier BV. https://doi.org/10.1016/b978-0-12-809656-7.00001-5.*

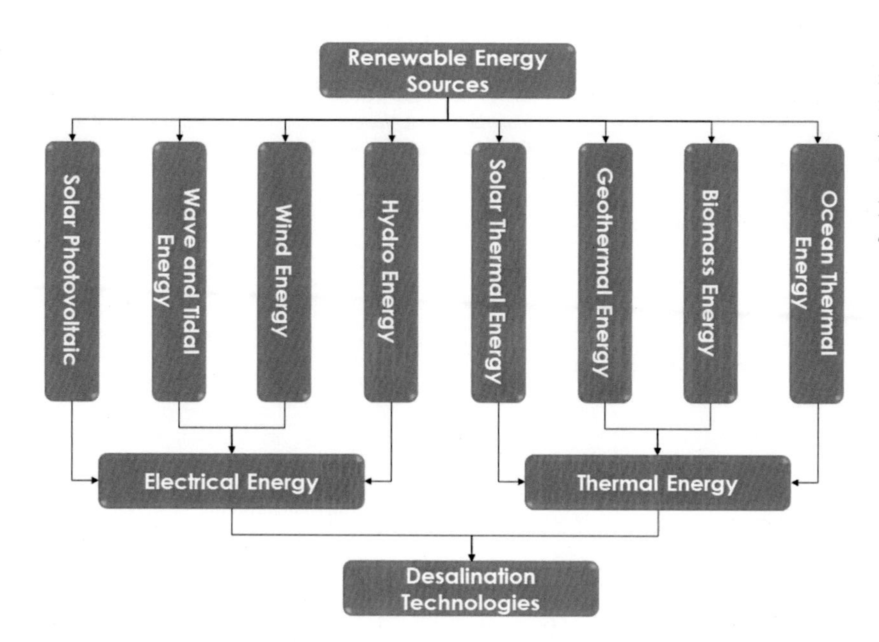

FIGURE 14.10 Various renewable energy resources to accomplish energy demand of desalination technologies. *From Alawad, S. M., Mansour, R. B., Al-Sulaiman, F. A., & Rehman, S. (2023). Renewable energy systems for water desalination applications: A comprehensive review.* Energy Conversion and Management, 286. *https://doi.org/10.1016/j.enconman.2023.117035.*

fulfill the energy requirement of desalination technologies. RESs are found as the most tempting resolution to minimize the carbon footprint produced through the desalination plants at a comparatively lesser cost in the most sustainable manner. Based on the type of technology, the appropriate RES can be integrated to provide energy in the required form, that is, electrical or thermal. The major possible combinations of RESs to accomplish the energy demand of desalination technologies are presented in Fig. 14.10. Despite the technology, the coupling of RES technology with a desalination plant is influenced by other major factors such as geographical location, plant size, feedwater quality, required quality of fresh water, and existing energy infrastructure. However, researchers are still working to identify an efficient energy solution based on RES for desalination plants that can sustain long-duration operations.

The outcomes of integrating various commonly used desalination methods reported in the literature are discussed in the following subsections.

14.4.1 Renewable energy source–based reverse osmosis

RO is a membrane-based technique that involves the application of hydraulic pressure to obtain fresh water through a semipermeable membrane. It is a market-dominant technology in the current scenario due to its comparatively lesser energy requirement and freshwater production cost. However, the energy requirement and hence, the water production cost, is significantly affected by the quality of feedwater. The power consumption in RO systems is majorly influenced by pump and membrane efficiency, the recovery rate of the system, and the energy recovery device. The seawater having total dissolved solids value ranging between 15,000 and 46,000 mg/L is usually associated with a high energy requirement for desalination using RO. For such cases, multistage and multipass RO systems are utilized whose power consumption varies between 2.5 and 4 kWh/m^3. Nearly, 1.7–2.8 $kgCO_2/m^3$ can be saved by replacing fossil fuels from RESs to accomplish the mentioned energy requirement.

Nearly 44% cost of the total cost of water production in the RO plant is constituted by the electrical energy required by the plants, and it is recommended to accomplish the same through, wind, solar, or wave energy (National Research Council, 2004). Experts have recommended wind and solar energy for RO plants due to their comparatively lesser cost, easy accessibility, zero water consumption, and mature technology, whereas bio-energy and hydro are not found appropriate for these plants due to their site dependency (Alawad et al., 2023). However, the intermittent nature of wind and solar energy is the major obstacle to their integration with RO plants. Therefore hybrid systems made up of wind and solar systems along with battery or grid-connected systems are under consideration by the researchers.

14.4.2 Renewable energy source–based electrodialysis desalination system

Electrodialysis (ED) is a water purification method based on the principle of electrochemical separation in which the electric current is used to separate the salt ions from the pure water. As shown in Fig. 14.11, ED contains electrodes

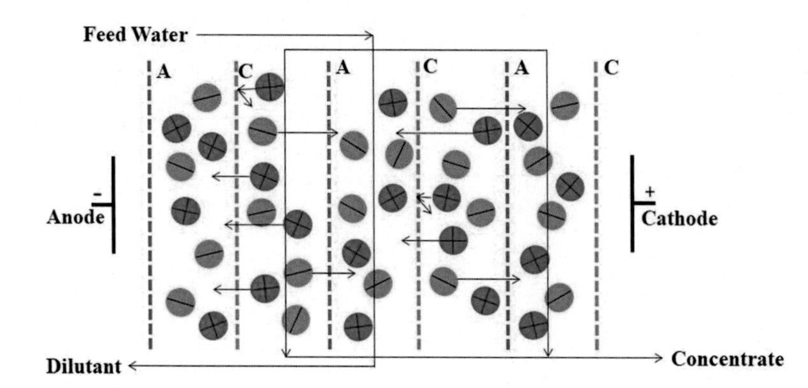

FIGURE 14.11 A typical process of electrodialysis. *From Gautam, A., Dave, T., & Krishnan, S. (2023). Can solar energy help ZLD technologies to reduce their environmental footprint? – A review.* Solar Energy Materials and Solar Cells, 256. *https://doi.org/10.1016/j.solmat.2023.112334.*

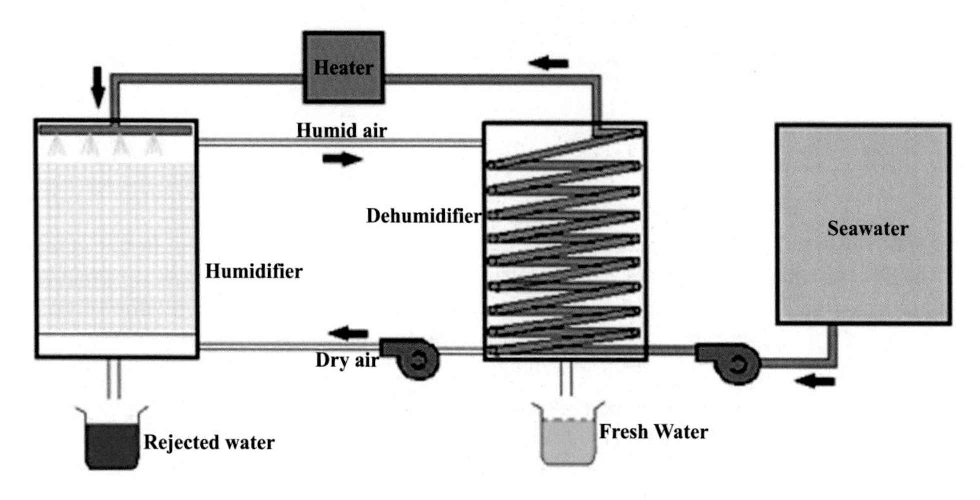

FIGURE 14.12 A typical system of humidification and dehumidification desalination. *From Gautam, A., Dave, T., & Krishnan, S. (2023). Can solar energy help ZLD technologies to reduce their environmental footprint? – A review.* Solar Energy Materials and Solar Cells, 256. *https://doi.org/10.1016/j.solmat.2023.112334.*

connected to the electrical current from an external source and placed into the saltwater container having a selective ion membrane across the channels. As compared to the RO, ED requires lesser pumping power, and therefore its energy consumption is comparatively much lesser. The resistance to fouling deposits over the membrane due to regular switch of polarity is its additional benefit compared to the RO. Moreover, it is found that ED is more energy efficient than RO in the case of low salinity (Patel et al., 2021). However, several associated drawbacks with ED like its ineffective for nonionic solutes, and very high ion-exchange membrane costs have reduced its commercial competitiveness with other technologies (Shi et al., 2020).

The ED technologies require electrical energy for ED electrodes (in DC) and to drive pumps (in AC or DC). The performance of the ED system is not affected by the load variation, and therefore solar photovoltaic (SPV) arrays are found suitable option to provide electricity to the ED electrodes (Shi et al., 2020). In view of the same, various SPV-based ED plants are studied and their reliability-related outcomes are reported in the literature (Lundstrom, 1979; Ortiz et al., 2008).

14.4.3 Renewable energy–based humidification and dehumidification desalination

The humidification and dehumidification (HDH)-based desalination system is based on the natural water cycle that involves the evaporation and condensation processes to obtain fresh water from seawater. A typical HDH desalination system involves a humidifier, a dehumidifier, and a heating source as shown in Fig. 14.12. The humidifier includes a sprinkler, packing material, and a container provided at the bottom to collect the brine, whereas the dehumidifier is a typical shell and tube heat exchanger in which low-temperature feedwater is allowed to pass through the coil and the shell is occupied by the moist air. The fresh water obtained through condensation is collected using a container/tank placed at the bottom of the dehumidifier. Moreover, the different pumps, blowers, and valves were also integrated into the circuit to circulate water, propel air, and control the flow, respectively.

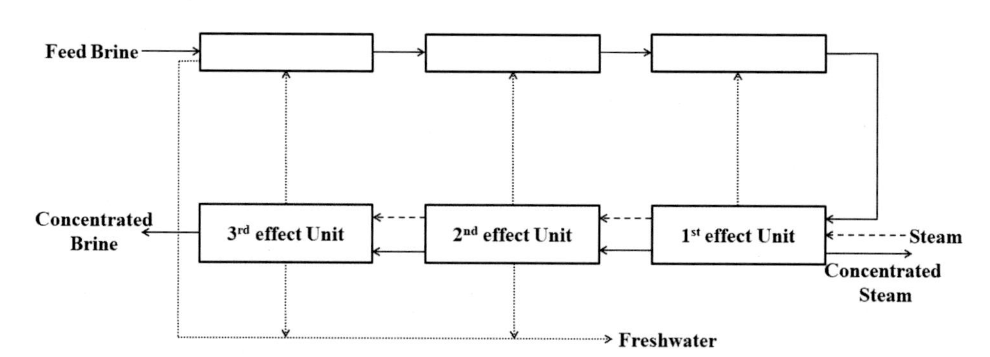

FIGURE 14.13 Working principle of multieffect distillation technology. *From Gautam, A., Dave, T., & Krishnan, S. (2023). Can solar energy help ZLD technologies to reduce their environmental footprint? – A review.* Solar Energy Materials and Solar Cells, 256. *https://doi.org/10.1016/j.solmat.2023.112334.*

The HDH desalination system is found suitable to accomplish the fresh water demand for the cluster of houses or at the community level. Regionally, it is more suitable for remote locations that have electricity shortages and plenty of access to heat through sun, industry, biomass, and others. This is due to the ability of the HDH system to work effectively with low-grade energy at atmospheric/low-pressure conditions. In view of the same, the experts found this technology of fresh water production among the best in terms of sustainability. Accomplishing the thermal energy requirement of the HDH system through clean energy sources that do not influence the environment can play a significant role in achieving sustainability goals. These clean energy sources can be geothermal, solar, wind, biomass, and other hybrid energy generation systems. Usually, solar thermal collectors are found suitable in the case of solar power for this application; however, several studies have also been described on the sustainable feasibility of SPV-based HDH desalination systems.

14.4.4 Renewable energy–based multieffect distillation system

Multieffect distillation (MED) is another thermal-based technology to desalinate brackish and seawater. This technology involves one condenser and numerous effects as shown in Fig. 14.13. The seawater is initially allowed to enter the condenser tubes for preheating and directed toward the stages of the MED process. The seawater is sprinkled over the outer surfaces of the external tubes, where the water is evaporated due to heat absorbed from the low-pressure steam flowing inside the tubes. Hereof, the steam flowing inside the evaporated tubes gets condensed and this condensation occurs only in the first stage as energy produced during the condensation are utilized in the rest of the stages to heat the water (Ghalavand et al., 2015). The production rate of fresh water through the MED method majorly depends on the number of effects and its value of restricted by the minimum temperature difference among the two stages and temperature range (Khawaji et al., 2008).

The lesser value of the operating temperature range (70°C–80°C) in MED methods enables it to integrate with RESs, particularly solar and geothermal energy. Significant development has been observed recently in solar-based MED technology and its market is expected to grow in the coming years (Mezher et al., 2011). Moreover, solar-based MED technology is also found economically feasible compared to conventional fossil fuel–based MED systems (El-Nashar, 2001). There are many studies reported by various researchers to present the advantages of using renewable energy with a MED system and most of the studies concluded that such a system can play a significant role in achieving sustainable goals.

14.5 Different aspects of the energy-water nexus

The aforementioned discussion clarified that there is an inseparable linkage between water and energy and both of them complement each other. Various researchers have also presented the multidimensionality of this nexus, which plays an important role in sustainability. In view of the same, the technical, economic, environmental, political, and social aspects of EWN have been discussed in the following subsections.

14.5.1 Environmental aspect

The environmental aspect secured a major position compared to the other influencing aspects in EWN as the environment is the source of both resources. This aspect provides many links in EWN and substantially influences the approaches for sustainable development. Climate change is caused due to greenhouse gas (GHG) emissions through anthropogenic and other activities. The use of fossil fuels to accomplish energy needs is the primary cause of GHG emissions and hence, climate change. Thereby climate change creates an uncertainty in the water supply that directly hits energy and water security. On the same line, the deforestation associated with the infrastructure of the electrical/energy industry also affects water security in the long term.

The increase in natural calamities like drought and floods are also seen in the last decade and it is also the result of climate change. The insufficiency of water for cooling and boiler in electricity production is a common impact of drought, whereas the excess use of energy to provide relief under the tough conditions of floods also creates an imbalance in the EWN. However, the 100% replacement of fossil fuels with renewable energy can be a milestone in achieving sustainable goals and issues due to climate change can be overcome by the adoption of such sustainable measures in the processes. These sustainable steps also help to protect the aquatic ecosystem and help to maintain the natural balance on the earth.

14.5.2 Technical aspect

The technical aspect of EWN includes the direct or physical link among them that has been already discussed in detail in Sections 14.3 and 14.4. As discussed earlier, a significant amount of water is consumed in the electricity industry for its generation and cooling purposes. Similarly, the water industry is gradually becoming more energy-intensive with technological advancements for wastewater treatment, water extraction, desalination, and water transfer. Moreover, its negative impact is gradually dominating with technical growth and the depletion of traditional sustainable methods in this nexus. This gradual negative impact on the environment starts with natural calamities for regions like drought, and floods and ultimately converts into the global challenge of climate change.

Therefore the choice of technology for all the actions involved under EWN is crucial. The consideration of parameters related to sustainability is essential during the selection of technology to take precautionary steps against climate change and other similar global issues. Moreover, the technologies should be continuously upgraded to magnify the output with minimum input. In view of this, it is necessary to find answers to several questions before we are too late. These questions are as follows: Are energy-efficient technologies always consuming more water and vice versa, is this also true? Are the positive outcomes in terms of sustainability through technological advancement also resulting in the same positive impact over the other? Thus a small deviation in the technological aspect of one block can affect the other block in the EWN and sustainable measures may leave substantial positive impacts in the same for simultaneous growth.

14.5.3 Economic aspect

The economic activity under different sectors is significantly influenced by a few constraints of facilitating the resources and absorbing waste. These constraints are like providing water, clean air, energy, and others that indirectly affect human health as well as monetary calculations. In the case of EWN, the overburden on the environment for the sake of achieving higher output in terms of energy efficiency or water production may substantially affect the economic aspect as well as the sustainability of an organization. Based on the recently released data, the global population is expected to reach 9 billion by 2050 along with a rise in energy, water, and food demand with comparatively more proportionality (Morales-García & Rubio, 2023). Moreover, the existing trends are also alarming about the expected economic risk due to international trade, urbanization, modernization, and technological changes (Gragg et al., 2018).

The economic aspect of EWN from the sustainability point of view may help to reduce a considerable expense along with their demand accomplishment without creating a burden over future generations. The study of EWN can be a useful tool for energy security and the prevention of water scarcity in developing nations that can indirectly help to overcome their economic challenges. Additionally, the local governments may also boost the adaptability of this approach through adequate policies and subsidies. The tariff structure of fresh water and electricity should not be planned linearly, and the price should be increased based on the consumption pattern and volume consumed by the individual. A perception is required to be created about the importance of water and energy through the economic aspect to obtain a substantial impact over the support of people. This point can be better understood by the example of India, where heavily subsidized electricity is resulting in overexploitation of the same. Therefore the suggested steps may play a significant role in the conservation of energy and water.

14.5.4 Political aspect

Similar to the technical, economic, and environmental aspects, the political is also an equal and most undervalued aspect in the role of EWN for sustainable development. The developed policies for industrial reform and environmental management play a crucial role in the economic focus and market creation for the industries related to EWN. All the other influencing aspects of EWN discussed in this section can be controlled through this dimension. More strict regulations can resolve environmental issues like effluent discharge, carbon emissions, and water demand management. On the contrary, overexploitation of energy, irregular use of groundwater, and disposal of waste into water reservoirs without treatment are the negative impacts in case of inappropriate for water and energy.

Therefore the development and implementation of strict policies are necessary steps of the governments for the protection of water and energy. The incorporation of sustainability parameters is the demand of the current situation and policymakers need to consider the same during their development. Such steps not only affect the industries-related electricity and water, it will also provide benefits indirectly to the other industries influenced by EWN like automobiles, building construction, thermal components, and others. Moreover, such appropriate policies can also significantly contribute to the fight against climate change in the world.

14.5.5 Social aspect

As water and electricity (energy) are the two fundamental needs of the global community, their linkage has a strong influence over society and people's behavior toward EWN. Several social sectors are even substantially impacted by the EWN. It can be understood by an example of drought-affected regions as the water allocation for irrigation and hydropower plants is prominently influenced during drought. This insufficiency in the basic need for water and energy becomes the region of disputes among states as well as nations. Moreover, rivers are considered for religious beliefs in many countries, which influences the water drawing consideration from the same. However, a social contribution can play a significant positive role in dealing with a major global issue of the present era, that is, climate change. For example, a significant amount of energy is consumed to accomplish the thermal comfort of a building that requires a significant amount of energy and indirectly water. A small change in social behavior by sensible use of energy may result in significant improvement in the global fight against climate change.

Social awareness about the EWN is today's requirement and it is a much-needed step to change the public perception of water and energy availability. People should be aware of the water and energy value and their support should be collected for achieving sustainability. In view of the same, energy efficiency programs and demand management programs should be regularly conducted at local as well as international levels. Moreover, the social aspect also affects the growth of the energy and water industries. It has been found in many cases that the industrialists or government are dealing with protests against the establishment of desalination plants due to unawareness. On the same grounds, the use of recycled water for drinking purposes and domestic utility is also quite hard for the public to accept. In general, the public is encountered daily with EWN and the related issues can be overcome by public support. Therefore such awareness programs are required at different levels that can play a significant role in achieving sustainability.

14.6 Conclusion

In the present chapter, the conventional hydropower techniques for power generation and its related aspects are discussed in detail, including the various disadvantages of conventional schemes. The chapter also discusses water requirements in power generation using new upcoming technology (hydrokinetic concept) in the hydro sector for sustainable development. It also presents an overview of their production, transportation, and consumption links. Moreover, the discussions on the use of water for power generation in terms of environmental, technical, economic, social, and political concerns are also reported. From the discussion, it has been found that most of the studies on EWN are carried out on narrow aspects such as water or electricity industries. But this topic is multidimensional and includes multiple links, and therefore it is necessary to investigate the same in broader aspects. Moreover, the research methodologies adopted under the reported studies are restricted to unidirectional like econometric or productivity analysis. However, it requires attention to develop a multidirectional framework inclusive of minor characteristics of the relationships.

The present chapter provides a basic insight into EWN and its major aspects. It may be useful for academicians, researchers, industrialists, environmentalists, and other stakeholders working in the field of sustainable development.

References

Ahmed, F. E., Khalil, A., & Hilal, N. (2021). Emerging desalination technologies: Current status, challenges and future trends. *Desalination, 517*, 115183. Available from https://doi.org/10.1016/j.desal.2021.115183.

Alawad, S. M., Mansour, R. B., Al-Sulaiman, F. A., & Rehman, S. (2023). Renewable energy systems for water desalination applications: A comprehensive review. *Energy Conversion and Management, 286*. Available from https://doi.org/10.1016/j.enconman.2023.117035, https://www.journals.elsevier.com/energy-conversion-and-management.

Alhassan, M. O., Opoku, R., Uba, F., Obeng, G. Y., Sekyere, C. K. K., & Nyanor, P. (2023). Techno-economic and environmental estimation assessment of floating solar PV power generation on Akosombo dam reservoir in Ghana. *Energy Reports, 10*, 2740–2755. Available from https://doi.org/10.1016/j.egyr.2023.09.073, http://www.journals.elsevier.com/energy-reports/.

Baniya, R., Talchabhadel, R., Panthi, J., Ghimire, G. R., Sharma, S., Khadka., Shin, S., Pokhrel, Y., Bhattarai, U., Prajapati, R., Thapa, B. R., & Maskey, R. K. (2023). Nepal Himalaya offers considerable potential for pumped storage hydropower. *Sustainable Energy Technologies and Assessments, 60*.

Barelli, L., Liucci, L., Ottaviano, A., & Valigi, D. (2013). Mini-hydro: A design approach in case of torrential rivers. *Energy, 58*, 695–706. Available from https://doi.org/10.1016/j.energy.2013.06.038, http://www.elsevier.com/inca/publications/store/4/8/3/.

Basso, S., & Botter, G. (2012). Streamflow variability and optimal capacity of run-of-river hydropower plants. *Water Resources Research, 48*(10). Available from https://doi.org/10.1029/2012WR012017.

Behrouzi, F., Maimun, A., & Nakisa, M. (2014). Review of various designs and development in hydropower turbines. *International Journal of Mechanical and Mechatronics Engineering, 8*, 293–297.

Belessiotis, V., Kalogirou, S., & Delyannis, E. (2016). *Desalination methods and technologies—Water and energy* (pp. 1–19). Elsevier BV. Available from http://doi.org/10.1016/b978-0-12-809656-7.00001-5.

Bhat, N., Jain, S., Asawa, K., Tak, M., Shinde, K., Singh, A., Gandhi, N., & Gupta, V. V. (2015). Assessment of fluoride concentration of soil and vegetables in vicinity of zinc smelter, Debari, Udaipur, Rajasthan. *Journal of Clinical and Diagnostic Research, 9*(10), ZC63–ZC66. Available from https://doi.org/10.7860/JCDR/2015/13902.6667, http://www.jcdr.net/articles/PDF/6667/13902_CE(Ra1)_F(GH)_PF1(PAK)_PFA(AK)_PF2(PAG).pdf.

Boretti, A., & Rosa, L. (2019). Reassessing the projections of the World Water Development Report. *npj Clean Water, 2*(1). Available from https://doi.org/10.1038/s41545-019-0039-9, https://www.nature.com/npjcleanwater/.

Bragalli, C., Micocci, D., & Naldi, G. (2023). On the influence of net head and efficiency fluctuations over the performance of existing run-of-river hydropower plants. *Renewable Energy, 206*, 1170–1179. Available from https://doi.org/10.1016/j.renene.2023.02.081, http://www.journals.elsevier.com/renewable-and-sustainable-energy-reviews/.

Chauhan, V. K., Shukla, S. K., Tirkey, J. V., & Singh Rathore, P. K. (2021). A comprehensive review of direct solar desalination techniques and its advancements. *Journal of Cleaner Production, 284*. Available from https://doi.org/10.1016/j.jclepro.2020.124719, https://www.journals.elsevier.com/journal-of-cleaner-production.

Couto, T. B. A., & Olden, J. D. (2018). Global proliferation of small hydropower plants – Science and policy. *Frontiers in Ecology and the Environment, 16*(2), 91–100. Available from https://doi.org/10.1002/fee.1746, http://www.esajournals.org/loi/fron.

Dai, J., Wu, S., Han, G., Weinberg, J., Xie, X., Wu, X., Song, X., Jia, B., Xue, W., & Yang, Q. (2018). Water-energy nexus: A review of methods and tools for macro-assessment. *Applied Energy, 210*, 393–408. Available from https://doi.org/10.1016/j.apenergy.2017.08.243, http://www.elsevier.com/inca/publications/store/4/0/5/8/9/1/index.htt.

El-Nashar, A. M. (2001). The economic feasibility of small solar MED seawater desalination plants for remote arid areas. *Desalination, 134*.

Egré, D., & Milewski, J. C. (2002). The diversity of hydropower projects. *Energy Policy, 30*(14), 1225–1230. Available from https://doi.org/10.1016/S0301-4215(02)00083-6.

Eke, J., Yusuf, A., Giwa, A., & Sodiq, A. (2020). The global status of desalination: An assessment of current desalination technologies, plants and capacity. *Desalination, 495*, 114633. Available from https://doi.org/10.1016/j.desal.2020.114633.

Energy Commission of Ghana (2020). 2020 Electricity Supply Plan for the Ghana Power Systems, a Power Supply Outlook with Medium Term Projections. URL: https://www.energycom.gov.gh/files/2020%20Electricity%20Supply%20Plan.pdf.

Esfahani, I. J., Rashidi, J., Ifaei, P., & Yoo, C. K. (2016). Efficient thermal desalination technologies with renewable energy systems: A state-of-the-art review. *Korean Journal of Chemical Engineering, 33*(2), 351–387. Available from https://doi.org/10.1007/s11814-015-0296-3, http://www.springerlink.com/content/120599/.

Esmaeilion, F., Ahmadi, A., Hoseinzadeh, S., Aliehyaei, M., Makkeh, S. A., & Astiaso Garcia, D. (2021). Renewable energy desalination; a sustainable approach for water scarcity in arid lands. *International Journal of Sustainable Engineering, 14*(6), 1916–1942. Available from https://doi.org/10.1080/19397038.2021.1948143, http://www.tandf.co.uk/journals/titles/19397038.asp.

Fiagbe, Y. A. K., & Obeng, D. M. (2007). Optimum operation of hydropower systems in Ghana when Akosomba Dam elevation is below minimum design value. *Journal of Science and Technology (Ghana), 26*(2). Available from https://doi.org/10.4314/just.v26i2.32991.

Fishman, R. (2016). More uneven distributions overturn benefits of higher precipitation for crop yields. *Environmental Research Letters, 11*(2). Available from https://doi.org/10.1088/1748-9326/11/2/024004, http://iopscience.iop.org/article/10.1088/1748-9326/11/2/024004/pdf.

Gautam, A., Dave, T., & Krishnan, S. (2023a). Can solar energy help ZLD technologies to reduce their environmental footprint? – A review. *Solar Energy Materials and Solar Cells, 256*. Available from https://doi.org/10.1016/j.solmat.2023.112334, http://www.sciencedirect.com/science/journal/09270248/100.

Gautam, A., Dave, T., & Krishnan, S. (2023b). Performance investigation of humidification dehumidification desalination with mass extraction and recirculation of rejected water. *Thermal Science and Engineering Progress, 41*. Available from https://doi.org/10.1016/j.tsep.2023.101813, https://www.journals.elsevier.com/thermal-science-and-engineering-progress.

Gautam, A., & Saini, R. P. (2020). Thermal and hydraulic characteristics of packed bed solar energy storage system having spheres as packing element with pores. *Journal of Energy Storage, 30*, 101414. Available from https://doi.org/10.1016/j.est.2020.101414.

Gautam, A., & Saini, R. P. (2021). Development of correlations for Nusselt number and friction factor of packed bed solar thermal energy storage system having spheres with pores as packing elements. *Journal of Energy Storage, 36*, 102362. Available from https://doi.org/10.1016/j.est.2021.102362.

Gautam, A., & Saini, R. P. (2022). Performance analysis and system parameters optimization of a packed bed solar thermal energy storage having spherical packing elements with pores. *Journal of Energy Storage, 48*. Available from https://doi.org/10.1016/j.est.2022.103993, http://www.journals.elsevier.com/journal-of-energy-storage/.

Gernaat, D. E. H. J., Bogaart, P. W., Vuuren, D. P. V., Biemans, H., & Niessink, R. (2017). High-resolution assessment of global technical and economic hydropower potential. *Nature Energy, 2*(10), 821–828. Available from https://doi.org/10.1038/s41560-017-0006-y, http://www.nature.com/nenergy/.

Ghalavand, Y., Hatamipour, M. S., & Rahimi, A. (2015). A review on energy consumption of desalination processes. *Desalination and Water Treatment, 54*(6), 1526–1541. Available from https://doi.org/10.1080/19443994.2014.892837, http://www.deswater.com/openaccess.php.

Gragg, R. S., Anandhi, A., Jiru, M., & Usher, K. M. (2018). A conceptualization of the urban food-energy-water nexus sustainability paradigm: Modeling from theory to practice. *Frontiers in Environmental Science, 6*. Available from https://doi.org/10.3389/fenvs.2018.00133, https://www.frontiersin.org/articles/10.3389/fenvs.2018.00133/full.

Hamiche, A. M., Stambouli, A. B., & Flazi, S. (2016). A review of the water-energy nexus. *Renewable and Sustainable Energy Reviews, 65*, 319–331. Available from https://doi.org/10.1016/j.rser.2016.07.020, https://www.journals.elsevier.com/renewable-and-sustainable-energy-reviews.

Inam Ullah, E., Ahmad, S., Khokhar, M. F., Azmat, M., Khayyam, U., & Qaiser, Fu. R. (2023). Hydrological and ecological impacts of run off river scheme; a case study of Ghazi Barotha hydropower project on Indus River, Pakistan. *Heliyon, 9*(1). Available from https://doi.org/10.1016/j.heliyon.2022.e12659, http://www.journals.elsevier.com/heliyon/.

Khalid, M., Bilal, M., Hassani, D., Zaman, S., & Huang, D. (2017). Characterization of ethno-medicinal plant resources of Karamar valley Swabi, Pakistan. *Journal of Radiation Research and Applied Sciences, 10*(2), 152–163. Available from https://doi.org/10.1016/j.jrras.2017.03.005.

Khan, M. J., Bhuyan, G., Iqbal, M. T., & Quaicoe, J. E. (2009). Hydrokinetic energy conversion systems and assessment of horizontal and vertical axis turbines for river and tidal applications: A technology status review. *Applied Energy, 86*(10), 1823–1835. Available from https://doi.org/10.1016/j.apenergy.2009.02.017, http://www.elsevier.com/inca/publications/store/4/0/5/8/9/1/index.htt.

Khan, M. J., Iqbal, M. T., & Quaicoe, J. E. (2008). River current energy conversion systems: Progress, prospects and challenges. *Renewable and Sustainable Energy Reviews, 12*(8), 2177–2193. Available from https://doi.org/10.1016/j.rser.2007.04.016.

Khawaji, A. D., Kutubkhanah, I. K., & Wie, J. M. (2008). Advances in seawater desalination technologies. *Desalination, 221*(1–3), 47–69. Available from https://doi.org/10.1016/j.desal.2007.01.067.

Lundstrom, J. E. (1979). Water desalting by solar powered electrodialysis. *Desalination, 31*(1–3), 469–488. Available from https://doi.org/10.1016/s0011-9164(00)88551-3.

Magaju, D., Cattapan, A., & Franca, M. (2020). Identification of run-of-river hydropower investments in data scarce regions using global data. *Energy for Sustainable Development, 58*, 30–41. Available from https://doi.org/10.1016/j.esd.2020.07.001, http://www.elsevier.com.

Mezher, T., Fath, H., Abbas, Z., & Khaled, A. (2011). Techno-economic assessment and environmental impacts of desalination technologies. *Desalination, 266*(1–3), 263–273. Available from https://doi.org/10.1016/j.desal.2010.08.035.

Morales-García, M., & Rubio, M. Á. G. (2023). Sustainability of an economy from the water-energy-food nexus perspective. *Environment, Development and Sustainability*. Available from https://doi.org/10.1007/s10668-022-02877-4, https://www.springer.com/journal/10668.

National Research Council. (2004). *Review of the desalination and water purification technology roadmap*. National Academies Press.

Niadas, I. A., & Mentzelopoulos, P. G. (2008). Probabilistic flow duration curves for small hydro plant design and performance evaluation. *Water Resources Management, 22*(4), 509–523. Available from https://doi.org/10.1007/s11269-007-9175-y.

Ortiz, J. M., Expósito, E., Gallud, F., García-García, V., Montiel, V., & Aldaz, V. A. (2008). Desalination of underground brackish waters using an electrodialysis system powered directly by photovoltaic energy. *Solar Energy Materials and Solar Cells, 92*(12), 1677–1688. Available from https://doi.org/10.1016/j.solmat.2008.07.020, http://www.sciencedirect.com/science/journal/09270248/100.

Patel, S., Saxena, P., Choudhary, P., Yadav, A., Rai, V. N., & Mishra, A. (2021). Effect of Li^+ ion substitution on structural and dielectric properties of $Bi_{0.5}Na_{0.5-x}Li_xTiO_3$ nanoceramics. *Journal of Inorganic and Organometallic Polymers and Materials, 31*(2), 851–864. Available from https://doi.org/10.1007/s10904-020-01818-w, http://www.springerlink.com/content/r64737117kr4/http://www.springer.com/east/home/generic/search/results?SGWID = 5-40109-70-35505322-0.

Quaranta, E., Bódis, K., Kasiulis, E., McNabola, A., & Pistocchi, A. (2022). Is there a residual and hidden potential for small and micro hydropower in Europe? A screening-level regional assessment. *Water Resources Management, 36*(6), 1745–1762. Available from https://doi.org/10.1007/s11269-022-03084-6, http://www.wkap.nl/journalhome.htm/0920-4741.

Rahmstorf, S. (2002). Ocean circulation and climate during the past 120,000 years. *Nature, 419*(6903), 207–214. Available from https://doi.org/10.1038/nature01090.

Santolin, A., Cavazzini, G., Pavesi, G., Ardizzon, G., & Rossetti, A. (2011). Techno-economical method for the capacity sizing of a small hydropower plant. *Energy Conversion and Management, 52*(7), 2533–2541. Available from https://doi.org/10.1016/j.enconman.2011.01.001.

Sasthav, C., & Oladosu, G. (2022). Environmental design of low-head run-of-river hydropower in the United States: A review of facility design models. *Renewable and Sustainable Energy Reviews*, *160*. Available from https://doi.org/10.1016/j.rser.2022.112312, https://www.journals.elsevier.com/renewable-and-sustainable-energy-reviews.

Scientific and Cultural Organization. (2020). Water and Climate Change, The United Nations World Water Development Report 2020, the United Nations Educational. URL: https://unesdoc.unesco.org/ark:/48223/pf0000372985.locale = en.

Scientific and Cultural Organization. (2021). Valuing Water, The United Nations World Water Development Report 2021, the United Nations Educational. URL: https://unesdoc.unesco.org/ark:/48223/pf0000375724.

Shi, L., Rossi, R., Son, M., Hall, D. M., Hickner, M. A., Gorski, C. A., & Logan, B. E. (2020). Using reverse osmosis membranes to control ion transport during water electrolysis. *Energy and Environmental Science*, *13*(9), 3138–3148. Available from https://doi.org/10.1039/d0ee02173c, http://pubs.rsc.org/en/journals/journal/ee.

Singal, S. K., & Saini, R. P. (2008a). Analytical approach for development of correlations for cost of canal-based SHP schemes. *Renewable Energy*, *33*(12), 2549–2558. Available from https://doi.org/10.1016/j.renene.2008.02.010.

Singal, S. K., & Saini, R. P. (2008b). Cost analysis of low-head dam-toe small hydropower plants based on number of generating units. *Energy for Sustainable Development*, *12*(3), 55–60. Available from https://doi.org/10.1016/S0973-0826(08)60439-1, http://www.elsevier.com.

Singal, S. K., Saini, R. P., & Raghuvanshi, C. S. (2010). Analysis for cost estimation of low head run-of-river small hydropower schemes. *Energy for Sustainable Development*, *14*(2), 117–126. Available from https://doi.org/10.1016/j.esd.2010.04.001, http://www.elsevier.com.

Sood, M., & Singal, S. K. (2019). Development of hydrokinetic energy technology: A review. *International Journal of Energy Research*, *43*(11), 5552–5571. Available from https://doi.org/10.1002/er.4529, http://onlinelibrary.wiley.com/journal/10.1002/(ISSN)1099-114X.

Sood, M., & Singal, S. K. (2021). A numerical analysis to determine wake recovery distance for the longitudinal arrangement of hydrokinetic turbine in the channel system. *Energy Sources, Part A: Recovery, Utilization and Environmental Effects*. Available from https://doi.org/10.1080/15567036.2021.1979695, http://www.tandf.co.uk/journals/titles/15567036.asp.

Sood, M., & Singal, S. K. (2022). Development of statistical relationship for the potential assessment of hydrokinetic energy. *Ocean Engineering*, *266*. Available from https://doi.org/10.1016/j.oceaneng.2022.112140, http://www.journals.elsevier.com/ocean-engineering/.

Standards of Indian Standards. (1991). *IS 12800-3: Guidelines for Selection of Hydraulic Turbine, Preliminary Dimensioning and Layout of Surface Hydroelectric Power Houses, Part 3.*

Tefera, W. M., & Kasiviswanathan, K. S. (2022). A global-scale hydropower potential assessment and feasibility evaluations. *Water Resources and Economics*, *38*. Available from https://doi.org/10.1016/j.wre.2022.100198, http://www.journals.elsevier.com/water-resources-and-industry/.

United Nations Framework Convention on Climate Change (UNFCCC). (2015). *Adoption of the Paris Agreement Proposal by the President, Paris* (Unpublished content). Available from: <https://unfccc.int/process-and-meetings/the-paris-agreement>.

Varshney, R. S. (1977). *Hydro-power structures*. Nem Chand.

Yildiz, V., & Vrugt, J. A. (2019). A toolbox for the optimal design of run-of-river hydropower plants. *Environmental Modelling and Software*, *111*, 134–152. Available from https://doi.org/10.1016/j.envsoft.2018.08.018, http://www.elsevier.com/inca/publications/store/4/2/2/9/2/1.

Yu, H., Liu, K., Bai, Y., Luo, Y., Wang, T., Zhong, J., Liu, S., & Bai, Z. (2021). The agricultural planting structure adjustment based on water footprint and multi-objective optimisation models in China. *Journal of Cleaner Production*, *297*. Available from https://doi.org/10.1016/j.jclepro.2021.126646, https://www.journals.elsevier.com/journal-of-cleaner-production.

Chapter 15

Water governance, climate change adaptation, and sustainable development: A future perspective

Vahid Karimi[1], **Esmail Karamidehkordi**[1,*] **and Yan Tan**[2]
[1]*Department of Agricultural Extension and Education, Faculty of Agriculture, Tarbiat Modares University (TMU), Tehran, Iran,* [2]*Department of Geography, Environment and Population, School of Social Sciences, The University of Adelaide, Adelaide, South Australia, Australia*
**Corresponding author. e-mail address: e.karamidehkordi@modares.ac.ir.*

15.1 Introduction

Climate change is one of the world's most challenging issues in the 21st century. It affects natural resources, agriculture, biodiversity, society, and many economic activities (Karamidehkordi, 2012; Karimi et al., 2018). The Intergovernmental Panel on Climate Change (IPCC) has comprehensively assessed the anthropogenic causes of climate change and its significant impacts on global development (Pedersen et al., 2022). The 2023 Global Risks Report of the World Economic Forum (WEF, 2023) shows that climate change has been one of the five most damaging or probable global risks every year for the past decade. Catastrophic bushfires in Australia and severe water scarcity in Asia and Africa provide some examples (Karamidehkordi, 2010; Murakami et al., 2020). Over the past two decades, climate-related events have probably had the most considerable socio-economic impacts globally (Karimi et al., 2023). Climate-related impacts on nature and the human systems have been widely reported, one of which is the change in precipitation and snow melting, altering the process of water resources and leading to floods and droughts. Climate change also deteriorates water quality by increasing sedimentation during floods and pollutant concentrations during the dry season. Warming temperature likely raises the water demand, especially in arid and semiarid regions (Berrang-Ford et al., 2015; Blackmore et al., 2016). According to the 2020 Global Climate Risk Index, over 12,000 climate-related events caused economic losses worth USD 3.54 trillion and over 500,000 deaths worldwide between 1999 and 2018 (Eckstein et al., 2019). Most of these climatic events are related to water, such as tsunamis, storms, floods, droughts, and unexpected rainfalls, making the connection between water and climate increasingly critical to address climate change and water governance issues simultaneously (WWAP, 2020).

Climate change imposes an additional burden on many socio-economic issues and challenges that water governance faces across different regions. While some aspects of climate change (e.g., rising rainfall) may bring some local and immediate benefits in some regions, a range of adverse effects, such as declines in water supply in many areas, will be unavoidable. These negative effects may exacerbate current environmental, social, economic, and political crises, especially in water management and governance (Jiménez et al., 2020; Karimi et al., 2022; Akamani, 2023). Effective adaptation can lead to good water governance, managing water scarcity conflicts, and improved livelihood outcomes. Water, as a part of natural resources, is an environmental component and a livelihood resource. Effective water governance and human security planning must consider climate change adaptation (Dinar et al., 2015; Okpara et al., 2018; Pahl-Wostl et al., 2020). Water governance measures and climate change mitigation and adaptation measures must go hand in hand to create social and economic sustainability in all regions (Gustafsson, 2016). The integration of climate change adaptation, water governance, and conflict management in conflict-prone environments is now included in progressive international environmental discourses and development agendas (Subramanian et al., 2014; Pahl-Wostl, 2019).

However, some national and international decision-makers may have neglected many essential concepts. This is especially the case in developing countries, probably due to the lack of ready-for-use decision tools based on climate evidence at all levels of governance (Formiga-Johnsson and Britto, 2020; Romano & Akhmouch, 2019). Understanding

Water Footprints and Sustainable Development. DOI: https://doi.org/10.1016/B978-0-443-23631-0.00015-7

the importance of the challenge of climate adaptation and water governance raises some key questions. Is adaptation necessary? Who is adapted, adapt to what, and how to adapt? Does adaptation vary between countries, regions, sectors, and generally across all levels of governance (Berrang-Ford et al., 2011)? It is generally understood that some high-income countries are more likely to adapt than low- and middle-income countries and that the most vulnerable countries to water scarcity are less likely to adapt (Melo Zurita et al., 2018; Velempini et al., 2018). Studies also argue that limited knowledge of the integration of water governance and climate adaptation impedes the development of adaptation interventions in all dimensions of sustainable development (Garrote, 2017; Joshua et al., 2016).

Moreover, the ability of managers and policymakers to assess the assumptions and monitor the progress of the adaptation and governance is constrained by the lack of measurable outcomes or indicators to judge whether and how adaptation occurs at all levels of water governance (Yang et al., 2014; Sanchez & Roberts, 2014; Oberlack & Eisenack, 2018). Overall, water governance and climate adaptation have created various crises in all dimensions of sustainable development in the present and future. Thus there is a pressing need to examine these concepts from different scientific dimensions. This chapter analyzes the concepts of the adaptation and governance of water resources in the context of climate change. It seeks to identify the influential components in this field's dimensions of sustainable development.

15.2 Climate change

The climate is the average long-time weather conditions (at least 30 years), while the weather is short-time climate conditions. The elements of both weather and climate are the indicators of temperature, pressure, humidity, and precipitation (Werndl, 2016). The primary definition of climate change is the change in atmospheric conditions in a place, which is a combination of average conditions of temperature, precipitation, humidity, and so on. Although the terminologies of climate change and climate variability are not easily distinguishable and cannot be completely separated, both are viewed in the complex evolution of the climate system (Comoé, 2013). These two concepts have a fundamental difference (Hageback et al., 2005). Compared to climate change, climate fluctuations cover a shorter period (Apaydin, 2010). Climatic fluctuations refer to changes in one or more of the climatic variables (e.g., rainfall, temperature, wind) during a certain time, and long-term fluctuations in temperature, precipitation, wind, and other aspects of the earth's climate are related to climate change (Molua, 2002). Climate change denotes changes in meteorological conditions that occur over a long time, sometimes lasting centuries, while climate fluctuations include short-term changes that occur from year to year (Hageback et al., 2005; Islas Vargas, 2020). IPCC defines *climate change* as any type of change in the climate that has occurred over time due to natural changes or human activities (Comoé, 2013). Moreover, long-term variability of temperature, precipitation, wind, and other climatic aspects of the earth are related to the effects of climate change, which have severe effects on the agricultural sector and food security (Molua, 2002).

15.3 Climate change adaptation

Numerous evidence for the consequences of climate change exists (Smith et al., 2009). The Earth continues warming, and no final way has been defined to limit the global temperature increase to 1.5°C above preindustrial (1850–1900) levels (Gao et al., 2017). Temperature rise will likely go above 1.5°C or exceed the 2°C threshold by 2100 (Parry et al., 2009; Warren et al., 2022). If the implementation of the United Nations Framework Convention on Climate Change (UNFCCC) Paris Agreement to alleviate greenhouse gas emissions fails, the baseline scenario of 3.66°C of global warming will likely occur by the end of the century, making adaptation and mitigation inevitable (Berrang-Ford et al., 2011; Warren et al., 2022). Climate change adaptation is considered important in climate policy (Vogel & Henstra, 2015). New literature also shows that climate adaptation research is rapidly increasing in diversity of content and concepts (Ford & Berrang-Ford, 2016). However, defining climate adaptation and what is considered successful adaptation remains challenging (Dilling et al., 2019). Recent literature argues climate change adaptation is a public good (Moser & Boykoff, 2013), a public goal (Persson, 2019), and a public investment (Janetos, 2020).

Adaptation can also be viewed as a process, an adjustment, or an outcome (Schipper, 2020). All these perspectives indicate the necessity to evaluate adaptation measures, particularly considering limited financial resources, existing global policies, and the risk of incompatibility (Singh et al., 2022). Questions related to defining successful adaptation are relevant to all governance levels, where adaptation is planned, designed, and implemented. However, there may be no easy political answers (Moser & Boykoff, 2013), underlying the need for a deep scientific understanding of what constitutes adaptation and its success. A broader discussion emphasizes the necessity to distinguish adaptation from development (Schipper et al., 2020) and identify whether adaptation outcomes should be additional or complementary to the outcomes obtained from development interventions (Martinez & Christiansen, 2018).

Successful adaptation refers to strategies and achieving goals, such as communication and public participation, planning and informed decision-making, integration with other policy objectives, the cost-effectiveness of adaptation and responsibility and accountability, and supporting learning and adaptation management (Moser & Boykoff, 2013). According to the assessment of adaptation measures and the aggregation of relevant information, the UNFCCC has acted as the leading international body for policies and actions at different levels of governance. However, because of the complexity of climate adaptation, it is essential to have a common intuition and understanding of adaptation (Magnan & Ribera, 2016). In other words, this understanding is an essential prerequisite for making a meaningful assessment of successful adaptation. The IPCC provides the most cited definition of climate adaptation, which refers to the process of adaptation to actual or expected climate and explains its effects on human systems. In this definition, adaptation seeks to alleviate or prevent the harm or excessive use of beneficial opportunities. In some natural systems, human interventions may facilitate adaptation to the expected climate and its effects (Guillén Bolaños et al., 2022). "Successful adaptation" denotes any adjustment that reduces climate change-related risks or vulnerability to climate change to a predetermined level without jeopardizing economic, social, and environmental sustainability (Doria et al., 2009).

15.4 Climate change policies

Mitigation and adaptation to climate change represent two general categories of policy responses to climate change. Since the adoption of the UNFCCC in 1992 and the Kyoto Protocol in 1997, international policy on climate change has focused mainly on mitigating climate change to avoid crossing thresholds that could lead to catastrophes (Patterson et al., 2018). The focus of climate change mitigation policies is to reduce greenhouse gas emissions (Füssel, 2007). Such policies are essential in limiting climate-related risks and reducing adaptation costs (McKay et al., 2022). For example, the Paris Agreement seeks to limit global warming below 2°C and possibly below 1.5°C (Iacobuţă et al., 2021). Climate change adaptation policies have also received attention since the early 2000s, especially after the approval of the Paris Agreement in 2015 (Kuyper et al., 2018). However, progress in the implementation of climate change mitigation policies at the global level has been slow (Cole, 2015; Afokpe et al., 2022; Gilmore & Buhaug, 2021). In recent years, climate change adaptation policies have focused on effectiveness and justice, through which societies learn to mitigate the adverse effects of climate change and take advantage of opportunities fairly. Adaptation strategies in specific contexts may be classified as planned or autonomous, predictive, or reactive, or incremental or transformational (Iacobuţă et al., 2021; Eriksen et al., 2011).

15.5 Governance

Governance has become an important and interdisciplinary term in social sciences and politics. It does not have a single definition but contains complex, diverse, and controversial notions. It can be viewed from different perspectives. Governance comprises the rules and structures applicable in the management decisions at the local or the local-systemic level (community, organization, or company). At the government level, the rules and structures are imposed on enterprises or organizations and affect the scope and manner of management decisions (Brooks & Cullinane, 2006; Fukuyama, 2016). Governance may refer to the competencies in management and participatory-consensual decision-making. However, in practice, governance is deeply intertwined with creating incentives and opportunities for political actors (Fukuyama, 2013). Governance refers to the structures and processes by which people in society make decisions, set rules, and share power (Barbazza & Tello, 2014). Others define *governance* as self-organized networks (Bannister & Connolly, 2012).

15.6 Water governance

Since the United Nations Water Conference in 1977, governments, policymakers, and multinational organizations have considered what governance solutions they should adopt to achieve sustainability goals. The Brundtland Report defines *sustainability* as "development that meets the needs of the present without compromising the ability of future generations to meet their own needs" (Brundtland, 1987, p. 43). This definition was elaborated in relevant perspectives, including the definition provided in the Triple Bottom Line approach, which suggests that sustainability consists of three interlinked concepts: economic well-being, environmental quality, and social equity (Elkington, 1998). This framework has been used to measure the impact of each activity on the three dimensions, thus identifying "ideal solutions" that can ensure the long-term benefits of all dimensions. *Sustainability* in the context of the water sector can be defined as "the ability to provide and manage the quantity and quality of water to meet the current needs of humans and environmental ecosystems, while not compromising the needs of future generations to do the same" (Di Vaio et al., 2021). Water sustainability designates that water-related decision-makers or stakeholders consider the impact of their strategies, actions, and practices on current and future generations (Es'haghi & Karamidehkordi, 2023). For a long time, the international political agenda, including the Paris Agreement, the World Water Forum, and the Millennium Development Goals (MDGs), have set institutional frameworks, such as policies, goals, targets and deadlines, and raised concerns about continuous population growth, urbanization, water pollution and scarcity, and climate change. However, the disasters caused by global warming, such as drought, groundwater depletion, and increased flood risk, require elaborating responsible, sustainable, and resilient water use models to respond to these ongoing challenges. Governance plays an important role (OECD, 2015) in managing the complexities of water sustainability. According to the United Nations Educational, Scientific and Cultural Organization (UNESCO), "the water crisis is largely a crisis of governance," which signifies that good governance is essential for the effective long-term management of water resources (Di Vaio et al., 2021).

15.7 Adaptive governance by water resource management organizations

Organizations face continuous change, and organizational change is pervasive as they struggle to adapt their activities and processes in complex, uncertain, and unpredictable environments. For water management organizations, these changes are affected by external factors such as water shortage, natural disasters, laws, political reforms, and technological changes or by internal factors such as changes in leadership and management, politics, and innovation (Saleth & Dinar, 2000). Since climate change is estimated to increase the demand for water and decrease water availability (Jiménez Cisneros et al., 2014), adaptation strategies in the water sector are generally categorized by supply and demand management. Adaptation in the water sector should go beyond structural measures (Stakhiv, 2011) and is expected to include other measures such as organizational change, forecasting and precautionary systems, insurance, and other methods to improve water use productivity and related behavioral changes through economic and financial tools and rules and regulations (Crabbé & Robin, 2006). Therefore water management organizations, for example, organizations and companies that supply water to consumers, need to adapt to prevent adverse effects of climate change and require the use of new opportunities. Organizational change theories differentiate organizational adaptation into three aspects: (1) utility maximization, (2) behavioral change, and (3) institutional change, such as formal and informal legislation and rules (Berkhout, 2012). The utility maximization approach explains that organizations pursue adaptation if the cost of efforts is less than the benefits (Mendelsohn, 2006). Because the costs and benefits of alternatives and the costs of inefficiencies must be identified, utility maximization approaches cannot determine the uncertain nature of climate impacts and the perception, interpretation, and learning processes of organizational adaptation that leads to more reactive or immediate adaptation (Berkhout, 2012). In addition, organizations are shaped by the constraints of external factors, such as laws, regulations, and the social, cultural, political, and economic context in which they are located (Roggero, 2015). Recognizing the necessity to incorporate adaptation in relevant strategies and plans, questions are raised regarding whether the organizations responsible for managing socio-ecological systems have adequate adaptive capacity to new and uncertain conditions (Azhoni et al., 2018).

15.8 Water governance and sustainable development

Following the MDGs, the UN member countries approved UN Sustainable Development Goals (SDGs) to be implemented by nations between 2015 and 2030. SDGs were established as an international agenda to address poverty, global environmental changes, and social transformations toward sustainability (Wiegleb & Bruns, 2018). Its Goal 6 represents a global water program and an important window of opportunity to steer development paths toward a water-secure world by global water governance. The global dimension of water governance is increasingly discussed, as it is argued that local and regional water challenges are influenced by global processes and, in turn, contribute to the global scale (Vörösmarty et al., 2015). Therefore, to develop common norms and understanding of water management, global water governance is a type of collective effort and a global factor for increasing the effectiveness of water policy actions (Gupta & Pahl-Wostl, 2013). Water interlinks human, environmental, and economic aspects and has a substantial role in SDGs. It has multidimensional significance in poverty alleviation, political stability, human and ecosystem health, and socio-economic development (Pahl-Wostl et al., 2013; UN-Water, 2015).

Water, as the basis of life, inherently links humans to the nonhuman world (Bakker, 2012), which is why researchers have often analyzed the relationship between water and society as one of the key aspects of socio-ecological processes (Barnes & Alatout, 2012; Linton & Budds, 2014). Swyngedouw (2004) conceptualizes water and social power as "courses" to understand the relationship between humans and nature in more detail. Water and social power are intrinsically linked and cannot be categorized as either natural or social. Therefore conceptualizing water as a natural-social element becomes a means to examine material, social, and political processes in an integrated manner. The relationship between society and water is also conceptualized in the "social water cycle," which highlights the political nature of water (Linton & Budds, 2014).

The 2030 Agenda for Sustainable Development is expected to govern sustainability and management policies, launch new coalitions and global partnerships, and direct international funding. It is crucial to analyze the relationships of society with nature and understand how complex social-ecological dynamics are formed to foster transformative changes toward sustainability. Ecological-political discourses critically address what perspectives guide the framing of environmental challenges and, ultimately, what defines our understanding of nature (Bakker, 2003; Forsyth, 2004). However, more knowledge is needed about the common understanding of current water challenges among those stakeholders involved in the SDGs negotiation process.

15.9 Reviewing and mapping water governance studies under climate change

We conducted a bibliographic analysis of relevant articles using the Scopus database (https://www.scopus.com/) to understand the scope of studies on water governance under climate change. The clustering analysis reveals that the keywords of these studies are categorized into six distinct clusters:

1. Water policy and governance (red)
2. Transformative climate adaptation and resilience (green)
3. Water resources vulnerability and institutions (dark blue)
4. Adaptive capacity (yellow)
5. Water security and stakeholders (purple)
6. Water scarcity and climate change (light blue).

The prominence of the circles and texts in each cluster signifies the strength of their cooccurrence with the other keywords. The distance of the items and the lines demonstrate the relatedness and linkages of the keywords, respectively (Fig. 15.1). The most adjacency among the clusters is related to the keywords of water governance, climate change, governance, resilience, water management, adaptive capacity, and adaptation to climate change. Moreover, water governance and climate change have the highest centrality among the keywords because of their links and connections with other keywords. Studies have focused on the different dimensions of adaptation, resilience, and resource management, showing the importance and necessity of understanding these concepts (Table 15.1).

A systemic review of 17 selected articles indicates that local opportunities and conditions are essential factors affecting water governance and sustainability transformation (Table 15.2). Water governance needs more coordination,

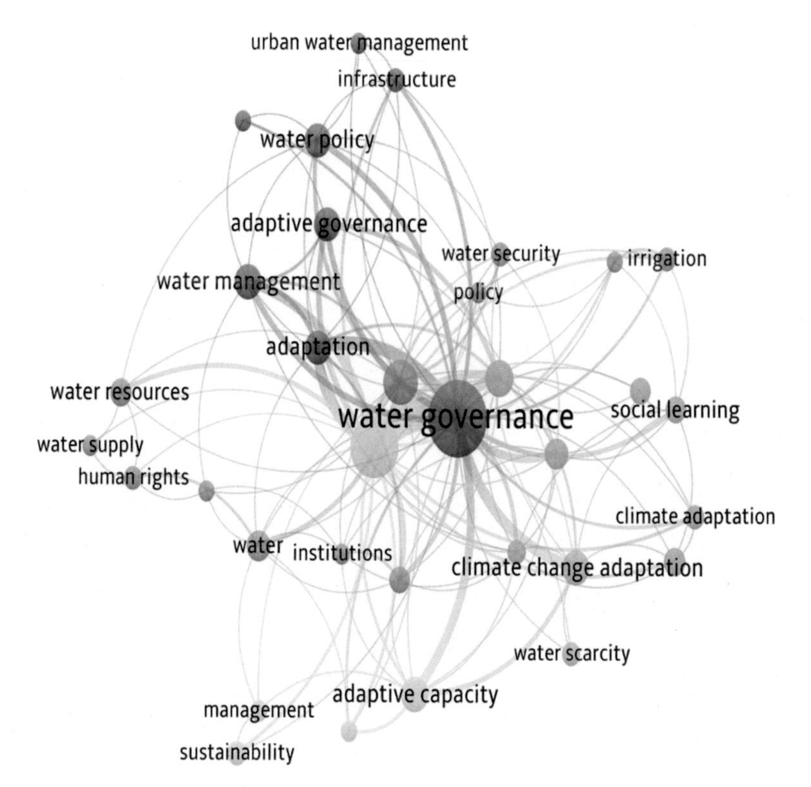

FIGURE 15.1 Six major themes in the literature based on 34 keywords most commonly used by authors (see Table 15.1).

TABLE 15.1 The scope of studies on water governance under climate change: main keywords in 2004–23.

S. no.	Keywords	Cluster	Links	Total link strength	Occurrences	Avg. citations
1	Adaptation	1	10	27	9	9.2
2	Adaptive governance	1	7	16	9	19.1
3	Comparative analysis	1	4	5	3	123.3
4	Infrastructure	1	7	10	4	23.3
5	Urban water management	1	4	4	3	20.3
6	Water governance	1	26	91	62	18.7
7	Water management	1	10	22	10	13.1
8	Water policy	1	10	15	9	16.7
9	Adaptive water governance	2	5	5	5	0.4
10	Climate adaptation	2	5	7	4	15.0
11	Integrated water management	2	6	7	6	15.7
12	Resilience	2	16	28	11	12.3
13	Social learning	2	8	10	5	18.6
14	Sustainable development	2	10	14	7	24.1
15	Transformation	2	12	14	4	18.3
16	Human rights	3	5	6	4	7.5
17	Institutions	3	8	9	3	1.3
18	Resource scarcity	3	5	5	3	14.7
19	Vulnerability	3	9	15	5	27.2
20	Water	3	11	14	7	9.7
21	Water resources	3	6	8	6	5.2
22	Water supply	3	3	3	3	5.3
23	Adaptive capacity	4	10	19	10	29.4
24	Climate change	4	25	77	37	24.1
25	Management	4	4	4	3	5.3
26	Sustainability	4	3	4	4	64.3
27	Governance	5	19	32	17	9.1
28	Irrigation	5	5	8	4	5.5
29	Policy	5	5	6	3	12.7
30	Stakeholders	5	6	8	3	3.3
31	Water security	5	7	8	4	16.8
32	Climate change adaptation	6	9	18	10	37.2
34	Water scarcity	6	3	3	4	18.3

The total link strength refers to the total number of cooccurrences of the node with other nodes (including repeated cooccurrences). Links represent keywords used together in publications with the line thickness proportional to the strength of cooccurrence. Occurrence is the frequency of cooccurrence of the keywords together.
Source: Authors' systematic review of Scopus database by August 10, 2023.

TABLE 15.2 Water governance and sustainable development under climate change in selected literature.

No	References	Journal	Methodology	Country	Action Level		Key results
					Micro	Macro	
1	Breuer and Oswald Spring (2020)	Water	Case study	Mexico	✓		Water governance and sustainability transformation driven based on local opportunities rather than the UN 2030 Agenda.
2	Jiménez et al. (2020)	Water	Systematic review	Sweden		✓	The conceptualization and operationalization of water governance, functions, and attributes and its contribution to stakeholders' values.
3	Li et al. (2021)	Sustainability Science	Quantitative Study	Hong Kong	✓		The involvement of the key actors of water management, actors' ability to perceive water vulnerability, implement policies, and adapt to external changes in a period from 1999 to 2018, and water drinking quality.
4	Islam et al. (2020)	Sustainable Water Resources Management	Mixed Method	Bangladesh	✓		The water governance situation in the nation and gaps in coordination, transparency and citizen involvement.
5	Odei Erdiaw-Kwasie et al. (2020)	World Development Perspectives	Case study	Ghana	✓		The citizens' participation in decision-making processes on water issues and the focus of water companies on technical aspects.
6	Hegga et al. (2020)	Regional Environmental Change	Case study	Namibia	✓		Failure of water governance and resource decentralization when local actors are not supported to participate actively.
7	Agbemor and Smiley (2020)	Journal of Development Studies	Case study	Ghana	✓		Water production situation, the major water source from privately managed boreholes, suggestions for a hybrid governance model to solve the conflicts arising between formal and informal providers.
8	Bayu et al. (2020)	Water Resources Research	Quantitative study	Developing country		✓	The dimensions of water governance and its impact on reducing the inequality of access to water and sanitation services in developing countries.
9	Tosun and Leopold (2019)	Water	Case study	17 nations		✓	The effect of transnational city networks on local water governance and actions.

(Continued)

TABLE 15.2 (Continued)

No	References	Journal	Methodology	Country	Action Level		Key results
					Micro	Macro	
10	Kooy and Walter (2019)	Water	Case study	Indonesia	✓		The impact of the Packaged Drinking Water policy on access inequalities
11	Horne (2020)	International Journal of Water Resources Development	Case study	Australia	✓		The measures adopted in Australian urban areas to meet the goal of water demand reduction.
12	Romano and Akhmouch (2019)	Water	Survey	OECD and non-OECD Countries		✓	How Sustainable Development Goal (SDG) 6 targets can be achieved through the application of OECD's 3Ps framework.
13	Herrera (2019)	World Development	Case study	Latin America and Asia		✓	The barrier and contradictions existing within SDG 6 and between SDG 6 and the UN Agenda.
14	Cisneros (2019)	Ecology and Society	Qualitative study: surveys	Ecuador	✓		The experiences of collaboration and stakeholder participation in water governance.
15	Horne et al. (2018)	International Journal of Water Resources Development	Qualitative Study	Developing country		✓	Evidence on how extreme weather events can impact the achievement of SDG 6 and SDG 11, especially in developing countries.
16	Glass and Newig (2019)	Earth System Governance	Comparative analysis	41 countries		✓	The influence of some governance attributes, such as policy coherence, democracy, and stakeholders' participation, on SDG achievement.
17	Mycoo (2018)	Natural Resources Forum	Case study	Trinidad	✓		Crucial factors that make water governance good to meet SDG 6, considering legislation, technologies, and human and environmental resources.

The Organisation for Economic Co-operation and Development (OECD) consists of 38 member countries: Austria, Australia, Belgium, Canada, Chile, Colombia, Costa Rica, Czech Republic, Denmark, Estonia, Finland, France, Germany, Greece, Hungary, Iceland, Ireland, Israel, Italy, Japan, Korea, Latvia, Lithuania, Luxembourg, Mexico, the Netherlands, New Zealand, Norway, Poland, Portugal, Slovak Republic, Slovenia, Spain, Sweden, Switzerland, Turkey, the United Kingdom and the United States.

integration, transparency, and public participation in decision-making processes. Moreover, water governance and resource decentralization may fail when local actors are not supported to participate actively. Water governance functions and attributes are interlinked with stakeholders' values and interests, and the key actors should be involved in water management. Their ability should be enhanced to perceive water vulnerability, implement policies, adapt to external changes, and manage conflict effectively and successfully. Good water governance can also affect climate change justice and reduce the inequality of access to water and sanitation services in developing countries under climate change. Finally, the successful implementation of Goal 6 of SDGs and its targets should be viewed through linking to other SDGs. It is essential to remove the barriers and contradictions existing within Goal 6 and between Goal 6 and other dimensions of the UN Agenda. Evidence on how extreme weather events can impact the achievement of SDG 6 and SDG 11, especially in developing countries. It means water governance requires further tools and strategies other than the targets of SDG 6, such as legislation, technologies, and human and environmental resources.

15.10 Conclusions and recommendations

The destructive effects of climate change, including droughts and water scarcity, especially in arid and semiarid areas where the livelihoods of local communities are highly dependent on natural resources, have led to the emergence of the concepts of adaptation and water governance as promising approaches to deal with climate change in the social, economic, and environmental dimensions of development. However, managerial and organizational mechanisms to drive transformative change toward adaptation based on good governance have not received sufficient attention. Our review emphasized adaptive governance in meeting the key requirements for the transformation toward water resources management and SDGs under climate change conditions. The concepts of governance and adaptation have been examined from different perspectives. Some challenges highlighted in the emergence of adaptation to climate change and water governance are dependency on traditional institutions and cultural traditions, the role of private and individuals rather than common interests, coordination challenges and the allocation of responsibilities across scales, capacity limitations, and potential inefficiencies in the institutional arrangements of organizations and the potential for outcomes that may be undesirable. Challenges include ignoring power relations and social inequalities due to the inherently political nature of governance in environmental domains, especially water resources management. Hence, the adaptation approach may not be a panacea for all climate change impacts. In some cases, structural options may be necessary alternative or complementary solutions. The application of adaptation-based governance may also be limited in extreme global warming scenarios that lead to irreversible changes in ecosystem structures and functions. Thus we need global climate change policies at all levels of governance that integrate adaptation with mitigation options and combine conservation and broader development. Despite these limitations, climate adaptation policies driven by international organizations along with the participation of all stakeholders at the macro level, which are presented with an adaptive governance approach, can promise to advance efforts toward targeted development against the water shortage crisis in future. However, more work is needed to address the conceptual and implementation deficiencies associated with these challenges and other emerging concepts to overcome compatibility limitations.

To conclude, the results of this text have some practical policy implications which are necessary to move toward adaptation and governance. First, policies should aim to create enabling conditions for the integration of diverse forms of knowledge, develop appropriate indicators to measure the impacts of climate change on water resources and human societies, and examine alternatives for mitigation and adaptation through adaptive governance. Policies should also provide diverse forms of incentives, including economic and noneconomic ones, to motivate collective social responses at different scales. In addition, policies should seek to facilitate capacity building by providing relevant forms of support, from funding and information to technical skills training. Finally, policies aimed at increasing the transition toward adaptation should provide institutional mechanisms for the meaningful participation of different actors at different scales in different stages of decision-making and the implementation of adaptive governance processes under climate change conditions for water resources management. It requires the support of international and national decision-makers to collaborate with stakeholders at the local level and change the current water governance practices toward more adaptive and resilient approaches. We can expect great advances in the understanding of water governance approaches and dynamics of transformation processes over the coming years. We need powerful tools for analysis and methods for evaluating and implementing collaborative approaches to adaptive governance of water resources to reduce the negative effects of climate change (Afokpe et al., 2022; Agbemor & Smiley, 2020; Akamani, 2023; Apaydin, 2010; Azhoni et al., 2018; Bakker, 2003, 2012; Bannister & Connolly, 2012; Barbazza & Tello, 2014; Barnes & Alatout, 2012; Bayu et al., 2020; Berkhout, 2012; Berrang-Ford et al., 2011, 2015; Blackmore et al., 2016; Breuer & Oswald Spring, 2020; Brooks & Cullinane, 2006; Brundtland, 1987; Cisneros, 2019; Cole, 2015; Compagnucci Cunha et al., 2001; Crabbé & Robin, 2006; Di Vaio et al., 2021; Dilling et al., 2019; Dinar et al., 2015; Doria et al., 2009; Elkington, 1998; Eriksen et al., 2011; Es'haghi & Karamidehkordi, 2023; Filho et al., 2022; Ford & Berrang-Ford, 2016; Formiga-Johnsson & Britto, 2020; Forsyth, 2004; Fukuyama, 2013, 2016; Füssel, 2007; Gao et al., 2017; Garrote, 2017; Gilmore & Buhaug, 2021; Glass & Newig, 2019; Guillén Bolaños et al., 2022; Gupta & Pahl-Wostl, 2013; Hageback et al., 2005; Hegga et al., 2020; Herrera, 2019; Horne, 2020; Horne et al., 2018; Iacobuţă et al., 2021; Islam et al., 2020; Islas Vargas, 2020; Janetos, 2020; Jiménez et al., 2020; Joshua et al., 2016; Karamidehkordi, 2010, 2012, Karimi et al., 2018, 2022, 2023; Kooy & Walter, 2019; Kuyper et al., 2018; Li et al., 2021; Linton & Budds, 2014; Magnan & Ribera, 2016; Martinez & Christiansen, 2018; McKay et al., 2022; Mendelsohn, 2006; Molua, 2002;

Moser & Boykoff, 2013; Murakami et al., 2020; Mycoo, 2018; Oberlack & Eisenack, 2018; Odei Erdiaw-Kwasie et al., 2020; Okpara et al., 2018; Pahl-Wostl, 2019; Pahl-Wostl et al., 2013, 2020; Parry et al., 2009; Patterson et al., 2018; Pechlaner et al., 2010; Pedersen et al., 2022; Persson, 2019; Pollitt & Hupe, 2011; Roggero, 2015; Romano & Akhmouch, 2019; Saleth & Dinar, 2000; Sanchez & Roberts, 2014; Schipper, 2020; Schipper et al., 2020; Singh et al., 2022; Smith et al., 2009; Stakhiv, 2011; Subramanian et al., 2014; Swyngedouw, 2004; Tosun & Leopold, 2019; Velempini et al., 2018; Vogel & Henstra, 2015; Vörösmarty et al., 2015; Warren et al., 2022; Werndl, 2016; Wiegleb & Bruns, 2018; Yang et al., 2014; Zurita et al., 2018; Comoé, 2013; OECD, 2015).

References

Afokpe, P. M. K., Phiri, A. T., Lamore, A. A., Toure, H. M. A. C., Traore, R., & Kipkogei, O. (2022). Progress in climate change adaptation and mitigation actions in sub-Saharan Africa farming systems. *Cahiers Agricultures*, *31*. Available from https://doi.org/10.1051/cagri/2021037, http://www.cahiersagricultures.fr/.

Agbemor, B. D., & Smiley, S. L. (2020). Tensions between formal and informal water providers: Receptivity toward mechanised boreholes in the Sunyani West District, Ghana. *Journal of Development Studies*, 1–17. Available from https://doi.org/10.1080/00220388.2020.1786059, http://www.tandf.co.uk/journals/titles/00220388.asp.

Akamani, K. (2023). The roles of adaptive water governance in enhancing the transition towards ecosystem-based adaptation. *Water*, *15*(13), 2341. Available from https://doi.org/10.3390/w15132341.

Apaydin, A. (2010). Response of groundwater to climate variation: Fluctuations of groundwater level and well yields in the Halacli aquifer (Cankiri, Turkey). *Environmental Monitoring and Assessment*, *165*(1–4), 653–663. Available from https://doi.org/10.1007/s10661-009-0976-8.

Azhoni, A., Jude, S., & Holman, I. (2018). Adapting to climate change by water management organisations: Enablers and barriers. *Journal of Hydrology*, *559*, 736–748. Available from https://doi.org/10.1016/j.jhydrol.2018.02.047, http://www.elsevier.com/inca/publications/store/5/0/3/3/4/3.

Bakker, K. (2003). *An uncooperative commodity: Privatizing water in England and Wales*. Oxford University Press.

Bakker, K. (2012). Water: Political, biopolitical, material. *Social Studies of Science*, *42*(4), 616–623. Available from https://doi.org/10.1177/0306312712441396.

Bannister., & Connolly. (2012). Defining e-Governance. *e-Service Journal*, *8*(2), 3. Available from https://doi.org/10.2979/eservicej.8.2.3.

Barbazza, E., & Tello, J. E. (2014). A review of health governance: Definitions, dimensions and tools to govern. *Health Policy (Amsterdam, Netherlands)*, *116*(1), 1–11. Available from https://doi.org/10.1016/j.healthpol.2014.01.007, http://www.elsevier.com/locate/healthpol.

Barnes, J., & Alatout, S. (2012). Water worlds: Introduction to the special issue of Social Studies of Science. *Social Studies of Science*, *42*(4), 483–488. Available from https://doi.org/10.1177/0306312712448524.

Bayu, T., Kim, H., & Oki, T. (2020). Water governance contribution to water and sanitation access equality in developing countries. *Water Resources Research*, *56*(4). Available from https://doi.org/10.1029/2019WR025330, http://agupubs.onlinelibrary.wiley.com/hub/journal/10.1002/(ISSN)1944-7973/.

Berkhout, F. (2012). Adaptation to climate change by organizations. *Wiley Interdisciplinary Reviews: Climate Change*, *3*(1), 91–106. Available from https://doi.org/10.1002/wcc.154, http://onlinelibrary.wiley.com/journal/10.1002/(ISSN)1757-7799.

Berrang-Ford, L., Ford, J. D., & Paterson, J. (2011). Are we adapting to climate change? *Global Environmental Change*, *21*(1), 25–33. Available from https://doi.org/10.1016/j.gloenvcha.2010.09.012.

Berrang-Ford, L., Pearce, T., & Ford, J. D. (2015). Systematic review approaches for climate change adaptation research. *Regional Environmental Change*, *15*(5), 755–769. Available from https://doi.org/10.1007/s10113-014-0708-7, http://springerlink.metapress.com/app/home/journal.asp?wasp = 64cr5a4mwldrxj984xaw&referrer = parent&backto = browsepublicationsresults,451,542;.

Blackmore, C., van Bommel, S., de Bruin, A., de Vries, J., Westberg, L., Powell, N., Foster, N., Collins, K., Roggero, P., & Seddaiu, G. (2016). Learning for transformation of water governance: Reflections on design from the Climate Change Adaptation and Water Governance (CADWAGO) project. *Water*, *8*(11), 510. Available from https://doi.org/10.3390/w8110510.

Breuer, A., & Oswald Spring, U. (2020). The 2030 agenda as agenda setting event for water governance? Evidence from the Cuautla River Basin in Morelos and Mexico. *Water*, *12*(2), 314. Available from https://doi.org/10.3390/w12020314.

Brooks, M. R., & Cullinane, K. (2006). Chapter 18 Governance models defined. *Research in Transportation Economics*, *17*, 405–435. Available from https://doi.org/10.1016/S0739-8859(06)17018-3.

Brundtland, G. H. (1987). Our common future – Call for action. *Environmental Conservation*, *14*(4), 291–294. Available from https://doi.org/10.1017/s0376892900016805.

Cisneros, P. (2019). What makes collaborative water governance partnerships resilient to policy change? A comparative study of two cases in Ecuador. *Resilience Alliance, Ecuador Ecology and Society*, *24*(1). Available from https://doi.org/10.5751/ES-10667-240129, https://www.ecologyandsociety.org/vol24/iss1/art29/ES-2018-10667.pdf.

Cole, D. H. (2015). Advantages of a polycentric approach to climate change policy. *Nature Climate Change*, *5*(2), 114–118. Available from https://doi.org/10.1038/nclimate2490, http://www.nature.com/nclimate/index.html.

Comoé, H. (2013). *Contribution to food security by improving farmers' responses to climate change in northern and central areas of Côte d'Ivoire* [Doctoral dissertation].

Compagnucci Cunha, R., Hanaki, K., Howe, C., Döll, P., Becker, A., & Zhang, J. (2001). *Climate change 2001: Impacts, adaptation, and vulnerability: Contribution of Working Group*. Cambridge University Press.

Crabbé, P., & Robin, M. (2006). Institutional adaptation of water resource infrastructures to climate change in Eastern Ontario. *Climatic Change, 78* (1), 103–133. Available from https://doi.org/10.1007/s10584-006-9087-5.

Di Vaio, A., Trujillo, L., D'Amore, G., & Palladino, R. (2021). Water governance models for meeting sustainable development goals: A structured literature review. *Utilities Policy, 72*, 101255. Available from https://doi.org/10.1016/j.jup.2021.101255.

Dilling, L., Prakash, A., Zommers, Z., Ahmad, F., Singh, N., de Wit, S., Nalau, J., Daly, M., & Bowman, K. (2019). Is adaptation success a flawed concept? *Nature Climate Change, 9*(8), 572–574. Available from https://doi.org/10.1038/s41558-019-0539-0, http://www.nature.com/nclimate/index.html.

Dinar, S., Katz, D., De Stefano, L., & Blankespoor, B. (2015). Climate change, conflict, and cooperation: Global analysis of the effectiveness of international river treaties in addressing water variability. *Political Geography, 45*, 55–66. Available from https://doi.org/10.1016/j.polgeo.2014.08.003, http://www.elsevier.com/inca/publications/store/3/0/4/6/5/index.htt.

Doria, Md. F., Boyd, E., Tompkins, E. L., & Adger, W. N. (2009). Using expert elicitation to define successful adaptation to climate change. *Environmental Science and Policy, 12*(7), 810–819. Available from https://doi.org/10.1016/j.envsci.2009.04.001.

Eckstein, D., Künzel, V., Schäfer, L., & Winges, M. (2019). *Global climate risk index 2020* (pp. 1–50). *Bonn*: *Germanwatch*. Available from https://germanwatch.org/sites/germanwatch.org/files/20-2-01e%20Global%20Climate%20Risk%20Index%202020_14.pdf.

Elkington, J. (1998). Accounting for the triple bottom line. *Measuring Business Excellence, 2*(3), 18–22. Available from https://doi.org/10.1108/eb025539.

Eriksen, S., Aldunce, P., Bahinipati, C. S., Martins, R. D. A., Molefe, J. I., Nhemachena, C., OBrien, K., Olorunfemi, F., Park, J., Sygna, L., & Ulsrud, K. (2011). When not every response to climate change is a good one: Identifying principles for sustainable adaptation. *Climate and Development, 3*(1), 7–20. Available from https://doi.org/10.3763/cdev.2010.0060.

Es'haghi, S. R., & Karamidehkordi, E. (2023). Understanding the structure of stakeholders – Projects network in endangered lakes restoration programs using social network analysis. *Environmental Science and Policy, 140*, 172–188. Available from https://doi.org/10.1016/j.envsci.2022.12.001, https://www.sciencedirect.com/journal/environmental-science-and-policy.

Filho, W. L., Ternova, L., Parasnis, S. A., Kovaleva, M., & Nagy, G. J. (2022). Climate change and zoonoses: A review of concepts, definitions, and bibliometrics. *Journal of Environmental Research and Public Health, 19*(2). Available from https://doi.org/10.3390/ijerph19020893, https://www.mdpi.com/1660-4601/19/2/893/pdf.

Ford, J. D., & Berrang-Ford, L. (2016). The 4Cs of adaptation tracking: Consistency, comparability, comprehensiveness, coherency. *Mitigation and Adaptation Strategies for Global Change, 21*(6), 839–859. Available from https://doi.org/10.1007/s11027-014-9627-7, http://www.wkap.nl/journalhome.htm/1381-2386.

Formiga-Johnsson, R. M., & Britto, A. L. (2020). Water security, metropolitan supply and climate change: Some considerations concerning the Rio de Janeiro case. *Ambiente & Sociedade, 23*, 1–24. Available from https://doi.org/10.1590/1809-4422ASOC20190207R1VU2020L6TD, http://www.scielo.br/scielo.php/script_sci_serial/pid_1414-753X/lng_en/nrm_iso.

Forsyth, T. (2004). *Critical political ecology: The politics of environmental science* (pp. 1–320). Taylor and Francis, United Kingdom. Available from https://doi.org/10.4324/9780203017562, http://www.tandfebooks.com/doi/book/9781134665815.

Fukuyama, F. (2013). What is governance? *Governance, 26*(3), 347–368. Available from https://doi.org/10.1111/gove.12035.

Fukuyama, F. (2016). Governance: What do we know, and how do we know it? *Annual Review of Political Science, 19*, 89–105. Available from https://doi.org/10.1146/annurev-polisci-042214-044240, http://arjournals.annualreviews.org/loi/polisci.

Füssel, H. M. (2007). Adaptation planning for climate change: Concepts, assessment approaches, and key lessons. *Sustainability Science, 2*(2), 265–275. Available from https://doi.org/10.1007/s11625-007-0032-y.

Gao, Y., Gao, X., & Zhang, X. (2017). The 2°C global temperature target and the evolution of the long-term goal of addressing climate change—From the United Nations Framework Convention on Climate Change to the Paris Agreement. *Engineering, 3*(2), 272–278. Available from https://doi.org/10.1016/J.ENG.2017.01.022, http://www.journals.elsevier.com/engineering/.

Garrote, L. (2017). Managing water resources to adapt to climate change: Facing uncertainty and scarcity in a changing context. *Water Resources Management, 31*(10), 2951–2963. Available from https://doi.org/10.1007/s11269-017-1714-6, http://www.wkap.nl/journalhome.htm/0920-4741.

Gilmore, E. A., & Buhaug, H. (2021). Climate mitigation policies and the potential pathways to conflict: Outlining a research agenda. *Wiley Interdisciplinary Reviews: Climate Change, 12*(5). Available from https://doi.org/10.1002/wcc.722, http://onlinelibrary.wiley.com/journal/10.1002/(ISSN)1757-7799.

Glass, L.-M., & Newig, J. (2019). Governance for achieving the Sustainable Development Goals: How important are participation, policy coherence, reflexivity, adaptation and democratic institutions? *Earth System Governance, 2*, 100031. Available from https://doi.org/10.1016/j.esg.2019.100031.

Guillén Bolaños, T., Scheffran, J., & Máñez Costa, M. (2022). Climate adaptation and successful adaptation definitions: Latin American perspectives using the Delphi method. *Sustainability, 14*(9), 5350. Available from https://doi.org/10.3390/su14095350.

Gupta, J., & Pahl-Wostl, C. (2013). Global water governance in the context of global and multilevel governance: Its need, form, and challenges. *Ecology and Society* (4), 18. Available from https://doi.org/10.5751/ES-05952-180453Netherlands, http://www.ecologyandsociety.org/vol18/iss4/art64/ES-2013-6010.pdf.

Gustafsson, M. T. (2016). How do Development Organisations Integrate Climate and Conflict Risks?: Experiences and Lessons Learnt from UK, Germany and the Netherlands. Available from: https://doi.org/10.1016/j.wdp.2020.100205

Hageback, J., Sundberg, J., Ostwald, M., Chen, D., Yun, X., & Knutsson, P. (2005). Climate variability and land-use change in Danangou watershed, China – Examples of small-scale farmers' adaptation. *Climatic Change, 72*(1–2), 189–212. Available from https://doi.org/10.1007/s10584-005-5384-7.

Hegga, S., Kunamwene, I., & Ziervogel, G. (2020). Local participation in decentralized water governance: Insights from north-central Namibia. *Regional Environmental Change, 20*(3). Available from https://doi.org/10.1007/s10113-020-01674-x, http://springerlink.metapress.com/app/home/journal.asp?wasp = 64cr5a4mwldrxj984xaw&referrer = parent&backto = browsepublicationsresults,451,542;.

Herrera, V. (2019). Reconciling global aspirations and local realities: Challenges facing the Sustainable Development Goals for water and sanitation. *World Development, 118*, 106–117. Available from https://doi.org/10.1016/j.worlddev.2019.02.009, http://www.journals.elsevier.com/world-development/.

Horne, J. (2020). Water demand reduction to help meet SDG 6: Learning from major Australian cities. *International Journal of Water Resources Development, 36*(6), 888–908. Available from https://doi.org/10.1080/07900627.2019.1638229, http://www.tandf.co.uk/journals/titles/07900627.asp.

Horne, J., Tortajada, C., & Harrington, L. (2018). Achieving the Sustainable Development Goals: Improving water services in cities affected by extreme weather events. *International Journal of Water Resources Development, 34*(4), 475–489. Available from https://doi.org/10.1080/07900627.2018.1464902, http://www.tandf.co.uk/journals/titles/07900627.asp.

Iacobuţă, G. I., Höhne, N., van Soest, H. L., & Leemans, R. (2021). Transitioning to low-carbon economies under the 2030 agenda: Minimizing trade-offs and enhancing co-benefits of climate-change action for the SDGs. *Sustainability, 13*(19), 10774. Available from https://doi.org/10.3390/su131910774.

Islam, M. R., Jahan, C. S., Rahaman, M. F., & Mazumder, Q. H. (2020). Governance status in water management institutions in Barind Tract, Northwest Bangladesh: An assessment based on stakeholder's perception. *Sustainable Water Resources Management, 6*(2). Available from https://doi.org/10.1007/s40899-020-00371-1, http://springer.com/journal/40899.

Islas Vargas, M. (2020). Adaptación al cambio climático: Definición, sujetos y disputas. *Letras Verdes. Revista Latinoamericana de Estudios Socioambientales* (28), 9–30. Available from https://doi.org/10.17141/letrasverdes.28.2020.4333.

Janetos, A. C. (2020). Why is climate adaptation so important? What are the needs for additional research? *Climatic Change, 161*(1), 171–176. Available from https://doi.org/10.1007/s10584-019-02651-y, http://www.wkap.nl/journalhome.htm/0165-0009.

Jiménez Cisneros, B. E., Oki, T., Arnell, N. W., Benito, G., Cogley, J. G., Doll, P., . . . Mwakalila, S. S. (2014). Freshwater resources.

Jiménez, A., Saikia, P., Giné, R., Avello, P., Leten, J., Liss Lymer, B., Schneider, K., & Ward, R. (2020). Unpacking water governance: A framework for practitioners. *Water, 12*(3), 827. Available from https://doi.org/10.3390/w12030827.

Joshua, M. K., Ngongondo, C., Chipungu, F., Monjerezi, M., Liwenga, E., Majule, A., Stathers, T., & Lamboll, R. (2016). Climate change in semi-arid Malawi: Perceptions, adaptation strategies and water governance. *Jàmbá: Journal of Disaster Risk Studies, 8*(3), 1–10. Available from https://doi.org/10.4102/jamba.v8i3.255, https://hdl.handle.net/10520/EJC191808.

Karamidehkordi, E. (2010). A country report: Challenges facing Iranian agriculture and natural resource management in the twenty-first century. *Human Ecology, 38*(2), 295–303. Available from https://doi.org/10.1007/s10745-010-9309-3.

Karamidehkordi, E. (2012). *Sustainable natural resource management, a global challenge of this century*. InTech. Available from 10.5772/35035.

Karimi, V., Bijani, M., Hallaj, Z., Valizadeh, N., Fallah Haghighi, N., & Karimi, M. (2023). *Adaptation and maladaptation to climate change: Farmers' perceptions* (pp. 113–132). Springer Science and Business Media LLC. Available from http://doi.org/10.1007/978-3-031-32789-6_7.

Karimi, V., Karami, E., & Keshavarz, M. (2018). Climate change and agriculture: Impacts and adaptive responses in Iran. *Journal of Integrative Agriculture, 17*(1), 1–15. Available from https://doi.org/10.1016/S2095-3119(17)61794-5, http://www.elsevier.com/journals/journal-of-integrative-agriculture/2095-3119.

Karimi, V., Valizadeh, N., Rahmani, S., Bijani, M., & Karimi, M. (2022). *Beyond climate change: Impacts, adaptation strategies, and influencing factors climate change: The social and scientific construct* (pp. 49–70). Springer International Publishing. Available from https://doi.org/10.1007/978-3-030-86290-9_4, https://link.springer.com/book/10.1007/978-3-030-86290-9.

Kooy, M., & Walter, C. (2019). Towards a situated urban political ecology analysis of packaged drinking water supply. *Water, 11*(2), 225. Available from https://doi.org/10.3390/w11020225.

Kuyper, J., Schroeder, H., & Linnér, B. O. (2018). The evolution of the UNFCCC. *Annual Review of Environment and Resources, 43*, 343–368. Available from https://doi.org/10.1146/annurev-environ-102017-030119, http://www.annualreviews.org/journal/energy.

Li, W., von Eiff, D., & An, A. K. (2021). Analyzing the effects of institutional capacity on sustainable water governance. *Sustainability Science, 16*(1), 169–181. Available from https://doi.org/10.1007/s11625-020-00842-6.

Linton, J., & Budds, J. (2014). The hydrosocial cycle: Defining and mobilizing a relational-dialectical approach to water. *Geoforum; Journal of Physical, Human, and Regional Geosciences, 57*, 170–180. Available from https://doi.org/10.1016/j.geoforum.2013.10.008, http://www.elsevier.com/inca/publications/store/3/4/4/index.htt.

Magnan, A. K., & Ribera, T. (2016). Global adaptation after Paris. *Science, 352*(6291), 1280–1282. Available from https://doi.org/10.1126/science.aaf5002, http://science.sciencemag.org/content/sci/352/6291/1280.full.pdf.

Martinez, G., & Christiansen, L. (2018). *Adaptation metrics: Perspectives on measuring, aggregating and comparing adaptation results*. UNEP DTU Partnership, Copenhagen.

McKay, A., Staal, D. I., Abrams, A., Winkelmann, J. F., Sakschewski, R., Loriani, B., & Lenton, T. M. (2022). Exceeding 1.5 C global warming could trigger multiple climate tipping points. *Science (New York, N.Y.), 377*(6611). Available from https://doi.org/10.1126/science.abn795.

Melo Zurita, M. D. L., Thomsen, D. C., Holbrook, N. J., Smith, T. F., Lyth, A., Munro, P. G., ... Powell, N. (2018). Global water governance and climate change: Identifying innovative arrangements for adaptive transformation. *Water, 10*(1), 29. Available from https://doi.org/10.3390/w10010029.

Mendelsohn, R. (2006). The role of markets and governments in helping society adapt to a changing climate. *Climatic Change, 78*(1), 203–215. Available from https://doi.org/10.1007/s10584-006-9088-4.

Molua, E. L. (2002). *Global climate change and Cameroon's agriculture: Evaluating the economic impacts*. Cuvillier Verlag.

Moser, S. C., & Boykoff, M. T. (2013). *Successful adaptation to climate change: Linking science and policy in a rapidly changing world* (pp. 1–335). Taylor and Francis, United States. Available from https://doi.org/10.4324/9780203593882, http://www.tandfebooks.com/doi/book/10.4324/9780203593882.

Murakami, H., Delworth, T. L., Cooke, W. F., Zhao, M., Xiang, B., & Hsu, P. C. (2020). Detected climatic change in global distribution of tropical cyclones. *Proceedings of the National Academy of Sciences of the United States of America, 117*(20), 10706–10714. Available from https://doi.org/10.1073/pnas.1922500117, https://www.pnas.org/content/117/20/10706.

Mycoo, M. A. (2018). Achieving SDG 6: Water resources sustainability in Caribbean Small Island Developing States through improved water governance. *Natural Resources Forum, 42*(1), 54–68. Available from https://doi.org/10.1111/1477-8947.12141.

Oberlack, C., & Eisenack, K. (2018). Archetypical barriers to adapting water governance in river basins to climate change. *Journal of Institutional Economics, 14*(3), 527–555. Available from https://doi.org/10.1017/S1744137417000509, http://journals.cambridge.org/action/displayJournal?jid = JOI.

Odei Erdiaw-Kwasie, M., Abunyewah, M., Edusei, J., & Buernor Alimo, E. (2020). Citizen participation dilemmas in water governance: An empirical case of Kumasi, Ghana. *World Development Perspectives, 20*. Available from https://doi.org/10.1016/j.wdp.2020.100242, http://www.journals.elsevier.com/world-development-perspectives.

OECD. (2015). *The OECD principles on water governance*. OECD Publishing.

Okpara, U. T., Stringer, L. C., & Dougill, A. J. (2018). Integrating climate adaptation, water governance and conflict management policies in lake riparian zones: Insights from African drylands. *Environmental Science and Policy, 79*, 36–44. Available from https://doi.org/10.1016/j.envsci.2017.10.002, http://www.elsevier.com/wps/find/journaldescription.cws_home/601264/description#description.

Pahl-Wostl, C. (2019). The role of governance modes and meta-governance in the transformation towards sustainable water governance. *Environmental Science and Policy, 91*, 6–16. Available from https://doi.org/10.1016/j.envsci.2018.10.008, http://www.elsevier.com/wps/find/journaldescription.cws_home/601264/description#description.

Pahl-Wostl, C., Conca, K., Kramer, A., Maestu, J., & Schmidt, F. (2013). Missing links in global water governance: A processes-oriented analysis. *Ecology and Society, 18*(2), 17083087. Available from https://doi.org/10.5751/ES-05554-180233, http://www.ecologyandsociety.org/vol18/iss2/art33/ES-2013-5554.pdf.

Pahl-Wostl, C., Knieper, C., Lukat, E., Meergans, F., Schoderer, M., Schütze, N., Schweigatz, D., Dombrowsky, I., Lenschow, A., Stein, U., Thiel, A., Tröltzsch, J., & Vidaurre, R. (2020). Enhancing the capacity of water governance to deal with complex management challenges: A framework of analysis. *Environmental Science and Policy, 107*, 23–35. Available from https://doi.org/10.1016/j.envsci.2020.02.011, http://www.elsevier.com/wps/find/journaldescription.cws_home/601264/description#description.

Parry, M., Lowe, J., & Hanson, C. (2009). Overshoot, adapt and recover. *Nature, 458*(7242), 1102–1103. Available from https://doi.org/10.1038/4581102a.

Patterson, J. J., Thaler, T., Hoffmann, M., Hughes, S., Oels, A., Chu, E., Mert, A., Huitema, D., Burch, S., & Jordan, A. (2018). Political feasibility of 1.5°C societal transformations: The role of social justice. *Current Opinion in Environmental Sustainability, 31*, 1–9. Available from https://doi.org/10.1016/j.cosust.2017.11.002, http://www.elsevier.com/wps/find/journaldescription.cws_home/718675/description#description.

Pechlaner, H., Ruhanen, L., Scott, N., Ritchie, B., & Tkaczynski, A. (2010). Governance: A review and synthesis of the literature. *Tourism Review, 65*(4), 4–16. Available from https://doi.org/10.1108/16605371011093836.

Pedersen, J. T. S., van Vuuren, D., Gupta, J., Santos, F. D., Edmonds, J., & Swart, R. (2022). IPCC emission scenarios: How did critiques affect their quality and relevance 1990–2022? *Global Environmental Change, 75*. Available from https://doi.org/10.1016/j.gloenvcha.2022.102538, http://www.elsevier.com/inca/publications/store/3/0/4/2/5.

Persson, Å. (2019). Global adaptation governance: An emerging but contested domain. *Wiley Interdisciplinary Reviews: Climate Change, 10*(6). Available from https://doi.org/10.1002/wcc.618, http://onlinelibrary.wiley.com/journal/10.1002/(ISSN)1757-7799.

Pollitt, C., & Hupe, P. (2011). The role of magic concepts. *Public Management Review, 13*(5), 641–658. Available from https://doi.org/10.1080/14719037.2010.532963, http://www.tandf.co.uk/journals/titles/14719037.asp.

Roggero, M. (2015). Adapting institutions: Exploring climate adaptation through institutional economics and set relations. *Ecological Economics, 118*, 114–122. Available from https://doi.org/10.1016/j.ecolecon.2015.07.022, http://www.elsevier.com/inca/publications/store/5/0/3/3/0/5.

Romano, O., & Akhmouch, A. (2019). Water governance in cities: Current trends and future challenges. *Water, 11*(3), 500. Available from https://doi.org/10.3390/w11030500.

Saleth, R. M., & Dinar, A. (2000). Institutional changes in global water sector: Trends, patterns, and implications. *Water Policy, 2*(3), 175–199. Available from https://doi.org/10.1016/S1366-7017(00)00007-6, http://www.iwaponline.com/wp/default.htm.

Sanchez, J. C., & Roberts. (2014). *Transboundary water governance: Adaptation to climate change*. IUCN.

Schipper, E. L. F. (2020). Maladaptation: When adaptation to climate change goes very wrong. *One Earth, 3*(4), 409–414. Available from https://doi.org/10.1016/j.oneear.2020.09.014, http://www.cell.com/one-earth.

Schipper, E. L. F., Tanner, T., Dube, O. P., Adams, K. M., & Huq, S. (2020). The debate: Is global development adapting to climate change? *World Development Perspectives, 18*, 100205. Available from https://doi.org/10.1016/j.wdp.2020.100205.

Singh, C., Iyer, S., New, M. G., Few, R., Kuchimanchi, B., Segnon, A. C., & Morchain, D. (2022). Interrogating 'effectiveness' in climate change adaptation: 11 guiding principles for adaptation research and practice. *Climate and Development, 14*(7), 650–664. Available from https://doi.org/10.1080/17565529.2021.1964937, http://www.tandfonline.com/toc/tcld20/current.

Smith, J. B., Schneider, S. H., Oppenheimer, M., Yohe, G. W., Hare, W., Mastrandrea, M. D., Patwardhan, A., Burton, I., Corfee-Morlot, J., Magadza, C. H. D., Füssel, H.-M., Barrie Pittock, A., Rahman, A., Suarez, A., & van Ypersele, J.-P. (2009). Assessing dangerous climate change through an update of the Intergovernmental Panel on Climate Change (IPCC) "reasons for concern.". *Proceedings of the National Academy of Sciences, 106* (11), 4133–4137. Available from https://doi.org/10.1073/pnas.0812355106.

Stakhiv, E. Z. (2011). Pragmatic approaches for water management under climate change uncertainty. *JAWRA: Journal of the American Water Resources Association, 47*(6), 1183–1196. Available from https://doi.org/10.1111/j.1752-1688.2011.00589.x.

Subramanian, A., Brown, B., & Wolf, A. T. (2014). Understanding and overcoming risks to cooperation along transboundary rivers. *Water Policy, 16* (5), 824–843. Available from https://doi.org/10.2166/wp.2014.010, http://www.iwaponline.com/wp/01605/0824/016050824.pdf.

Swyngedouw, E. (2004). *Social power and the urbanization of water: Flows of power*. OUP.

Tosun, J., & Leopold, L. (2019). Aligning climate governance with urban water management: Insights from transnational city networks. *Water, 11*(4), 701. Available from https://doi.org/10.3390/w11040701.

UN-Water (2015). Indicators and monitoring. UN-Water. http://www.unwater.org/sdgs/indicatorsand-monitoring/en/. Retrieved 28March 2015.

Velempini, K., Smucker, T. A., & Clem, K. R. (2018). Community-based adaptation to climate variability and change: Mapping and assessment of water resource management challenges in the North Pare highlands, Tanzania. *African Geographical Review, 37*(1), 30–48. Available from https://doi.org/10.1080/19376812.2016.1229203, http://www.tandfonline.com/loi/rafg20.

Vogel, B., & Henstra, D. (2015). Studying local climate adaptation: A heuristic research framework for comparative policy analysis. *Global Environmental Change, 31*, 110–120. Available from https://doi.org/10.1016/j.gloenvcha.2015.01.001, http://www.elsevier.com/inca/publications/store/3/0/4/2/5.

Vörösmarty, C. J., Hoekstra, A. Y., Bunn, S. E., Conway, D., & Gupta, J. (2015). Fresh water goes global. *Science (New York, N.Y.), 349*(6247), 478–479. Available from https://doi.org/10.1126/science.aac6009, http://www.sciencemag.org/content/349/6247/478.2.full.pdf.

Warren, R., Andrews, O., Brown, S., Colón-González, F. J., Forstenhäusler, N., Gernaat, D. E. H. J., Goodwin, P., Harris, I., He, Y., Hope, C., Manful, D., Osborn, T. J., Price, J., Van Vuuren, D., & Wright, R. M. (2022). Quantifying risks avoided by limiting global warming to 1.5 or 2 °C above pre-industrial levels. *Climatic Change, 172*(3–4). Available from https://doi.org/10.1007/s10584-021-03277-9, http://www.wkap.nl/journalhome.htm/0165-0009.

WEF (2023). The Global Risks Report 2023 18th Edition. Available from: https://www3.weforum.org/docs/WEF_Global_Risks_Report_2023.pdf

Werndl, C. (2016). On defining climate and climate change. *The British Journal for the Philosophy of Science, 67*(2), 337–364. Available from https://doi.org/10.1093/bjps/axu048.

Wiegleb, V., & Bruns, A. (2018). Hydro-social arrangements and paradigmatic change in water governance: An analysis of the sustainable development goals (SDGs). *Sustainability Science, 13*(4), 1155–1166. Available from https://doi.org/10.1007/s11625-017-0518-1, http://www.springer.com/east/home?SGWID = 5-102-70-144940151-0&changeHeader = true&SHORTCUT = http://www.springer.com/journal/11625.

WWAP (World Water Assessment Programme) (2020). The World Water Development Report 2020: Water and ClimateChange. UNESCO-WWAP, Paris, France. Available from: https://unesdoc.unesco.org/ark:/48223/pf0000372985.locale=en

Yang, Y. C. E., Brown, C., Yu, W., Wescoat, J., & Ringler, C. (2014). Water governance and adaptation to climate change in the Indus river basin. *Journal of Hydrology, 519*, 2527–2537. Available from https://doi.org/10.1016/j.jhydrol.2014.08.055, http://www.elsevier.com/inca/publications/store/5/0/3/3/4/3.

Zurita, Md. L. M., Thomsen, D. C., Holbrook, N. J., Smith, T. F., Lyth, A., Munro, P. G., de Bruin, A., Seddaiu, G., Roggero, P. P., Baird, J., Plummer, R., Bullock, R., Collins, K., & Powell, N. (2018). Global water governance and climate change: Identifying innovative arrangements for adaptive transformation. *Water (Switzerland), 10*(1). Available from https://doi.org/10.3390/w10010029, http://www.mdpi.com/2073-4441/10/1/29/pdf.

Chapter 16

Water conflicts and sustainable development: concepts, impacts, and management approaches

Esmail Karamidehkordi[1,*], Vahid Karimi[1], Gerald Singh[2] and Ladan Naderi[3]

[1]Department of Agricultural Extension and Education, Faculty of Agriculture, Tarbiat Modares University (TMU), Tehran, Iran, [2]Agricultural Extension, Communication and Rural Development Department, Faculty of Agriculture, University of Zanjan, Zanjan, Iran, [3]School of Environmental Studies, University of Victoria, Victoria, British Columbia, Canada

**Corresponding author. e-mail address: e.karamidehkordi@modares.ac.ir.*

16.1 Introduction

Inland water is vital for ecological functions, biodiversity, and human societies (Alexander, 2019). The diversity of freshwater ecosystem services has contributed to its essential functions, such as reducing poverty, maintaining health, economic growth, and the sustainability of natural resources (Karamidehkordi, 2012). On the other hand, water scarcity under the conditions of environmental crises, such as climate change and drought, can be a threatening resource for human beings, which necessitates sustainable management and productivity (Mishra et al., 2021). The occurrence and global impact of climate change on the environment and human societies, especially local communities (Karimi et al., 2018, 2023) have made the water shortage crisis one of the critical global challenges, especially in arid and semiarid regions, and affected sustainable development in various social, economic, political, and environmental dimensions (Karamidehkordi, 2010).

During the last three decades, global water demand has been facing an increasing trend. For this reason, the intensification of water harvesting, destruction of watersheds, and water pollution in recent years have led to an increase in threats to human health, the environment, and its sustainability (Karamidehkordi, 2007). The water crisis is not limited to the shortage of resources, but other factors such as excessive use of resources, land use change, social factors such as population growth, migration, lack of a proper culture of consumption and mismanagement of resources, the inability of responsible institutions to supply water, unstable and disintegrated economic policies, social changes and weak government planning (Gain et al., 2015; Kujinga et al., 2014), the increased overpressure of hydrological changes due to climate change on the sustainable water resources management, particularly areas under water stress (Naderi et al., 2022).

These factors, both quantitatively and qualitatively, have damaged water resources and natural ecosystems, such as wetlands and forests. If unfair access to water resources coincides with other conflict factors such as marginalization or past conflicts. It leads to jeopardizing various aspects of sustainable development at the international and national levels. Therefore, if such increasing competition is accompanied by the reduction of the resilience (i.e., response and recovery ability after disturbances and adaptive capacity) of different relevant actors when they face water scarcity and experience poor water governance, it may lead to tensions and conflicts between different stakeholders. During the last decade, many cases of violent conflicts over water have been reported, especially at the local level (European Union Institute for Security Studies, 2015; Naderi et al., 2022). If this increasing challenge is not managed, it can make national security issues worse in many countries.

Therefore the issue of conflict resolution or governance over water has been the focus of many water planning programmers and managers (Barli et al., 2006). This means that water management and governance fairly and sustainably are vital for peace and the sociopolitical stability of societies (Shan et al., 2020). According to the practical experiences and research on conflict management of natural resources (Ghasemi et al., 2017; Ghasemi & Karamidehkordi, 2017; Petersen-Perlman et al., 2017), some emphasize that conflicts can be resolved, while others believe that conflicts cannot

Water Footprints and Sustainable Development. DOI: https://doi.org/10.1016/B978-0-443-23631-0.00016-9

be resolved, but can only be managed (Hamad, 2005). We use the latter view in this chapter to explain water conflict management, because the experiences of using conflict resolution mechanisms in recent years, especially at the international level, have shown that conflict resolution in the field of water resources is likely idealistic, where these mechanisms have failed to resolve water conflicts at the local level (Gleick & Heberger, 2014; Malla et al., 2022). So, the assessment of the conflicts caused by the water crisis is important from two basic aspects, first from the aspect of proper water resources governance and sustainable water resources use and the other from the aspect of the planning and management for sustainable development and reducing the conflicts caused by water scarcity at local, national, and international levels. This chapter identifies the factors influencing water conflicts and examines innovative approaches, upcoming solutions, and adaptive measures to manage water challenges before they become a source of societal conflict in the present and future.

16.2 Water conflict

The world has been prone to conflicts over scarce natural resources (Novoselov et al., 2016). Some environmental constraints have increased the risk of violent conflicts at different international, regional, and local levels. However, not all challenges related to natural resources have been considered a threat to national or international security (Okpara et al., 2017; Petersen-Perlman et al., 2017; Schillinger et al., 2020). These challenges are viewed as a threat if they affect vulnerable communities (Khilchevskyi & Mezentsev, 2021). Water crisis and conflict are among the most challenging issues of natural resources management (Es'haghi & Karamidehkordi, 2023; Mianabadi et al., 2021; Moorthy & Bibi, 2023; Roudgarmi et al., 2011). The water crisis is a geopolitical challenge that has affected human relations (Gleick & Shimabuku, 2023) and has caused conflicts at the international. national, regional, and local levels over access to water resources. The impact of the global water crisis on world security is assessed to be so high that some people describe this situation as a "Hydro-climatic time bomb" (Bigas, 2012).

Water conflicts have different structures at different levels due to geographical, economic, political, and demographic determinants. Moreover, the duration of conflict between stakeholders has been identified differently depending on the sociocultural context of each region (Fröhlich, 2012). It has been identified that the awareness of the nature of conflicts and increasing cooperation between people in crises can manage conflicts over time (Qin et al., 2022). An example of this conflict management has been seen in intergovernmental water conflicts at international levels as the result of diplomatic relations (Marshall et al., 2017). However, the increase in demand and competition for access to water resources between the service, industry, and agriculture sectors has coincided with the intensification of conflicts at the national, regional, and local levels between different stakeholders, communities, or groups. These conflicts may be varied in different types of climates and regions (Link et al., 2016). Hence, the escalation of conflict, if accompanied by other issues, including poor governance, can increase the risk of internal unrest (Link et al., 2016; Lonergan, 2018; Martinás, 2010; Maru, 2016; Mbonile, 2005). The important issue of whether war (military conflict) is happening over water in different regions of the world or not (Shumilova et al., 2023) poses more acute issues in front of the world, which triggers us to understand what factors cause water scarcity and consequently water and political conflicts. It should be noted that most conflicts caused by water resource scarcity are rarely single-caused and usually a wide range of factors have led to its occurrence (Gleick, 2014).

16.3 Environmental impacts on water conflict

Water conflicts have been created or intensified due to various reasons, including human factors and environmental changes (Es'haghi & Karamidehkordi, 2023; Karimi et al., 2024). In some cases, the origin of water conflicts is caused by environmental factors such as climate change (Nordås & Gleditsch, 2015; Nordqvist & Krampe, 2018; Verhoeven, 2011; Sukanya & Joseph, 2023). The recent droughts along with the reduction of atmospheric precipitation in the world have caused instability in water resources, which is especially noticeable in areas with arid and semiarid climates (Karamidehkordi et al., 2023, 2024; Scanlon et al., 2023; Payus et al., 2020). Some speculations are that climate change can change patterns of human settlements by reducing access to water resources and increasing the risk of violent conflicts between them (Nordqvist & Krampe, 2018; Okpara et al., 2017; Padhy et al., 2015; Phuong et al., 2018). The fluctuation of climate variables (rainfall and temperature) and the rise of drought have directly affected the violent conflicts between the stakeholders or have indirectly affected the relationships between people through the impact on the health and livelihoods of communities highly dependent on water resources and the environment (Abel et al., 2019; Naderi et al., 2022; Padhy et al., 2015).

16.4 Governance and water conflict

Despite the existence of institutions to promote peace, it is observed that there are differences in interactions between individuals and groups in some cases. During the past decades, climate change has had a significant impact on the conflict between people, but it does not mean that climate change is the main driving force of conflict (Hsiang et al., 2013). Various competing theories have been proposed to explain the relationship between climate and human conflict, but none have been convincingly refuted. It is necessary to carry out innovative research to identify this relationship. (Adano et al., 2012), by studying violent conflicts between local institutions, argued that many areas are facing water and land crises, and these factors will cause conflicts and tensions between ethnic groups. The studies conducted on economic, political, and environmental issues, emphasizing poverty and livelihoods, of several regions in Africa highlight how violent behaviors as collective or individual actions can be justified by these socioeconomic factors under special climate conditions. As a result, people's behavioral violence in the face of the water crisis is rooted in not only their environment but also their economy, society, and culture.

16.5 Human impacts on water conflict

Cultural or human factors, directly and indirectly, influence the occurrence of water conflicts between different stakeholders through a set of complex social, political, and economic factors (Bantider et al., 2023). According to empirical studies and global consensus, the challenges of water scarcity from the perspective of human factors can be highlighted by several dimensions: (1) water shortage caused by the advancement of technologies for more exploitation of water resources, such as building dams and digging deep wells.; (2) lifestyle change toward consumerism; and (3) the challenge of a water shortage caused by unfavorable governance and mismanagement (Ansari & Sharma, 2020). The occurrence of each of these challenges has caused concerns and consequently conflicts and tensions between consumers and different sectors related to water resources (Sohrabi et al., 2023). On the other hand, some social factors such as demands, actions, participation, and influence of stakeholders indirectly affect the conflicts between stakeholders (Böhmelt et al., 2014; Lonergan, 2018; Tayia, 2019). They also indirectly create complex challenges, but they are often ignored because they receive less attention among the beneficiaries than the direct factors, which causes the ineffectiveness of some implementation strategies in managing water conflict and disputes (Kapur, 2019; Naderi et al., 2022). Based on a systematic literature review, conflict management in the water supply and demand sector requires simultaneous attention to both human and environmental factors (Table 16.1), so that water disputes and tensions can be reduced by the fair allocation of water resources effectively.

16.6 Impact of water crisis and water conflict

In the past, environmental threats were viewed as critical elements in maintaining human peace, but today these threats are considered a real danger to the security of nation-states (du Plessis, 2019b). The conflicts caused by the water crisis are one of the reactions of human societies in facing resource scarcity. They are complex issues that are often intertwined with other socioeconomic, environmental, and political issues (Brzoska & Fröhlich, 2016). Failure to respond appropriately to water conflicts will have adverse effects on the environment, livelihood assets, and relationships between people (du Plessis, 2019a). One of the consequences of the water crisis is the weakening of international and cross-border relations of countries with a common water border.

The lack of control can quickly lead to international tension (Tatar et al., 2022; Tayia, 2019; Wines, 2014). The US Central Intelligence Agency has warned that some countries will experience instability, government failure, and increased tensions in their region with an increase in the water crisis and the resulting conflicts in the future (du Plessis, 2019b). As an example, the western region of the United States experiences the continuation of the drought and many conflicts have arisen for water supply in different sectors, which has intensified the instability in the region. The conflicts have led to the formation of legal measures, political restrictions, and a new lifestyle change by the United States, as follows: (1) building reservoirs to supply drinking water to the city and preventing water supply to the rice farmers of Houston's southwest coast for three years, (2) blocking the Las Vegas groundwater pipeline from the aquifers on the edge of Nevada and Utah, (3) restricting water transfer through the Rocky Mountains to Denver and other states and preventing local water sale to the suburbs, and (4) stopping groundwater pumping (Wines, 2014).

Water crises and conflicts in the Zayandehrud Basin in Iran have also occurred between the upstream and downstream subbasins and between different sectors of agriculture, industry, and municipality (Naderi et al., 2022). Examples of international water tension are the conflict between Jordan and Syria for the joint exploitation of the

TABLE 16.1 A systematic review of selected articles on human and environmental factors affecting water conflict.

No.	References	Journal	Country	Influencing factor	
				Human	Environmental
1	Abel et al. (2019)	Global Environmental Change	Western Asia		*
2	Al-Muqdadi (2022)	Water	Iraq	*	
3	Zeitoun et al. (2020)	Water International	Syrian	*	*
5	Unfried et al. (2022)	Journal of Environmental Economics and Management	Africa and Central America		*
6	Nadiruzzaman et al. (2022)	Sustainability	International		*
7	Patrick (2020)	Handbook of Climate Change Management	South Africa		*
8	Yates et al. (2017)	Environment and Planning	Canadian	*	
9	Pahl-Wostl (2019)	Environmental science & policy	Germany, Netherlands, Australia, China, and South Africa	*	
10	Liu et al. (2022)	International Journal of Water Resources Development	China	*	
11	Baldwin et al. (2018)	Policy and Governance	Kenya	*	
12	Wolf (2019)	European Journal of International Relations	International		*
13	Bernauer and Siegfried (2012)	Journal of Peace Research	Central Asia		*
14	Bernauer and Böhmelt (2020)	Nature Sustainability	International	*	*

*Asterisk denotes each human or environmental factor affecting water conflicts.

Yarmouk River (Al-Muqdadi, 2022), the conflict between Bolivia and Chile over the Silala River (Muñoz et al., 2023), the conflict between India and Pakistan over the Indus River and signing the Indus Water Treaty in 1960 (Biswas, 1992), disputes and treaty between Iran and Afghanistan over Hirmand (Helmand) River (Mianabadi et al., 2021), and conflicts between Turkey and Iraq and Syria over the Euphrates-Tigris Rivers (El-Fadel et al., 2002).

Another socioeconomic impact of the water crisis and the resulting conflicts is migration. The water crisis has unprecedentedly affected the living conditions of many people in drought-prone areas, flood-prone areas, low-lying coastal plains, and small islands whose livelihoods are dependent on natural resources (Solomon, 2007; Bezu et al., 2020). Despite the adaptive response of many societies to the water crisis, in some areas, the trend of large migrations in search of water and land, or living in better conditions continues (Campbell & Bedford, 2022; Mpandeli et al., 2020). The management of the water crisis and the conflicts arising from it, especially in developing countries, has been very important for the status of water security in the sustainable development goals (SDGs). Nations should use water conflict management approaches to achieve sustainable development.

16.7 Sustainable development

Sustainable development refers to addressing the needs of the present without compromising the ability of future generations to meet their own needs (Chaware, 2018). Sustainable development is associated with institutional developments (Krysovatyy et al., 2018) and is mostly related to three environmental, social, and economic components (Fig. 16.1), which are equally important to successfully protect our planet in the long-term (Purvis et al., 2019).

The advancement of technology, the development of industries, and the transformation of nations from a traditional to modern society have created a wide range of environmental, social, and economic challenges and consequently sustainability (Benson et al., 2020; Bhaduri et al., 2016). Growing global concerns about the challenges of modernization

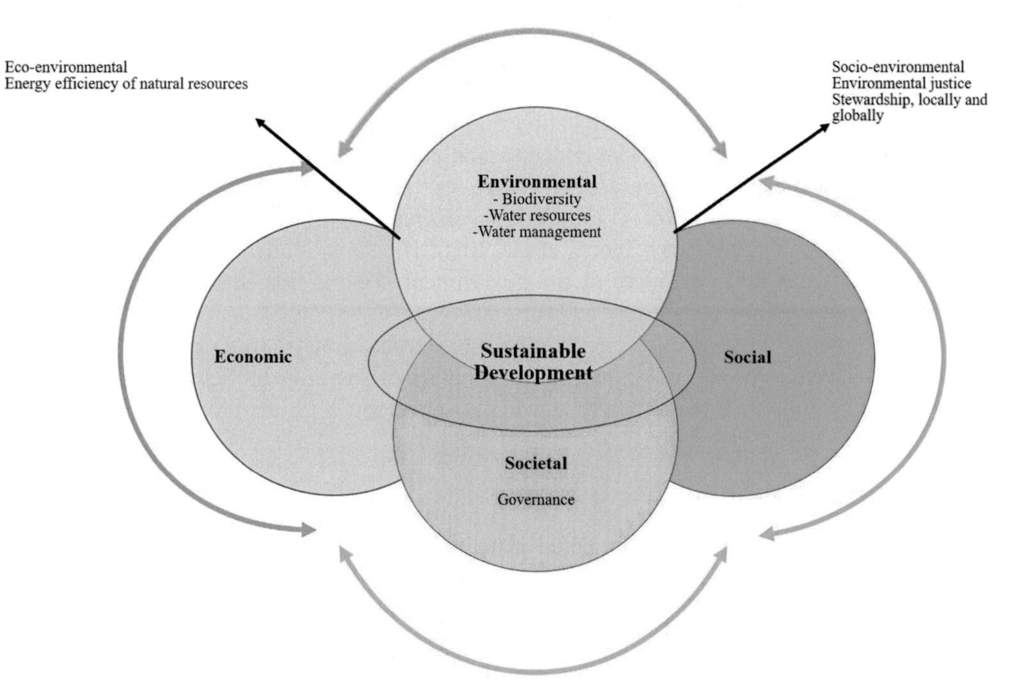

FIGURE 16.1 Sustainable development goals and water governance. *Adapted from (Krysovatyy et al., 2018; Saiefeddine et al., 2018)*

TABLE 16.2 Sustainable development goal (SDG) 6 with emphasis on sustainable water resources conservation and use.

Ensure availability and sustainable management of water and sanitation for all	SDG 6
By 2030, achieve universal and equitable access to safe and affordable drinking water for all	6.1
By 2030, achieve access to adequate and equitable sanitation and hygiene for all, and end open defecation, paying special attention to the needs of women and girls and those in vulnerable situations	6.2
By 2030, improve water quality by reducing pollution, eliminating dumping and minimizing the release of hazardous chemicals and materials, halving the proportion of untreated wastewater, and at least doubling recycling and safe reuse globally	6.3
By 2030, substantially increase water-use efficiency across all sectors, ensure sustainable withdrawals and supply of fresh water to address water scarcity, and substantially reduce the number of people suffering from water scarcity	6.4
By 2030, implement integrated water resources management at all levels, including through transboundary cooperation as appropriate	6.5
By 2020, protect and restore water-related ecosystems, including mountains, forests, wetlands, rivers, aquifers, and lakes	6.6
By 2030, expand international cooperation and capacity-building support to developing countries in water and sanitation-related activities and programs, including water harvesting, desalination, water efficiency, wastewater treatment, recycling, and reuse technologies	6.6a
Support and strengthen the participation of local communities in improving water and sanitation management	6.6b

Source: Adapted from (Ait-Kadi, 2016).

led to the introduction of the SDGs by the UN and their approval by the nations of the world in 2015. These goals cover a wide range of interrelated issues such as economic factors, sustainably increasing production and consumption, managing the water resources crisis, the conservation and restoration of biodiversity, building human capacities, and partnership among stakeholders (Delany-Crowe et al., 2019; Shah, 2016). Among the sustainable development goals, all the targets of SDG 6 emphasize preserving water resources, particularly those related to food, energy, and the environment (Table 16.2). Water is connected to all 17 SDGs and their targets and can no longer be viewed as a separate element from them. This means the conservation and sustainable use of water resources can be achieved if other goals are achieved, and conversely, other SDGs can be achieved if water resource conservation happens (Ait-Kadi, 2016).

When water resources are limited and at the same time water demands are growing beyond water resources, policymakers have the responsibility to determine the allocation of water within the borders of each country in such a way that the dynamic economy is balanced with social concepts and equality. Therefore the government must balance the needs of different stakeholders with the amount of available water resources and resolve water allocation disputes between different actors (Aguilera et al., 2021; Akhmouch & Clavreul, 2016; Almagtome et al., 2020; Armeni, 2015; Berardo & Lubell, 2016; Betts, 2013; Casiano Flores et al., 2016; Hjorth & Madani, 2023). Failure to coordinate and resolve the water conflict is considered a failure of the government. This is related to the role of governance in the dimensions of sustainable development. Therefore the current model of sustainable development needs to be reviewed and modified so that governance and strategies include the multifaceted role of water resources management in economic development, social welfare, and environmental sustainability. Therefore policymakers should consider water resource management as a central and integral part of sustainable development in its proper place.

16.8 Management approaches

Water stress or scarcity in many countries is considered a major challenge that requires appropriate measures and the implementation of new management techniques (Velmurugan et al., 2020). Managing the challenges of water supply and demand in the sectors of agriculture, industry, and municipality requires the use of sustainable water resources management approaches to allocate water resources effectively and fairly along with preserving the environment and reducing socioeconomic and environmental concerns (Kapur, 2019; Sohrabi et al., 2023). The broad nature of water resources management to cope with the water crisis and manage water conflicts between different stakeholders has raised the necessity of an integrated and participatory approach at the national and regional levels (Lukat et al., 2022; Nyika & Dinka, 2022).

An integrated water resources management (IWRM) approach takes into account all socioeconomic and environmental benefits to facilitate achieving the multidimensions of sustainable management (Issaoui et al., 2022). IWRM is also suggested as a practical framework for managing water conflicts in all municipal, industrial, and agricultural sectors and increasing sustainability in the water resources supply (Du et al., 2022). The agricultural sector is known as the

TABLE 16.3 Some studies on water governance approaches.

No.	References	Journal	Country	Result
1	Woodhouse and Muller (2017)	World Development	International	Network governance, emerging practitioner, avoiding normative approaches
2	Basco-Carrera et al. (2017)	Environmental Modelling & Software	Netherlands	Collaborative modeling approach
3	Schillinger et al. (2020)	Wires Water	World	Basic service provision and water governance
4	Furber et al. (2016)	Water	North America	Shared vision planning approach
5	Nyam et al. (2020)	Sustainable Water Resources Management	South Africa	Learning sustainable water management
6	Markowska et al. (2020)	Journal of Environmental Management		Water resources management
7	Yerian et al. (2014)	Environmental Management	Kenya	Governance and women/gender
8	Jayasiri et al. (2023)	Agricultural Systems	Sri Lanka	Overall transparency, participatory decision-making, internal communication
9	Yuan et al. (2023)	Landscape and Urban Planning	China	Local-scale participatory zoning approach
10	Altingoz (2022)	Invisible Borders in a Bordered World	International	Cooperation
11	Jayaram and Sethi (2022)	Science, Policies, and Conflicts of Climate Change	South Asia	Cooperation, treaties, arrangements

major sector of water use, which generally plays a major role in the escalation of water tensions. Therefore the institutions related to this sector should use a fair, efficient, and effective allocation of water and apply techniques for the appropriate use of water. Moreover, all activities for water availability in all seasons, rainwater management, soil and water conservation, and flood management through watershed management techniques, land drainage, and ecosystem protection and restoration should be considered through an integrated and participatory approach (Bertule et al., 2018).

The complexities of water scarcity are expanding and the interdependency of societies on water requires high coordination between social actors to manage the water crisis (Agar, 2016; Araral & Wang, 2013; Es'haghi & Karamidehkordi, 2023; Karimi et al., 2024). Diverse governance is an approach to managing the challenges that stakeholders face. Literature reviews (Table 16.3) show that a set of different governance perspectives are necessary to address complex water management problems and support changes (Akhmouch & Clavreul, 2016; Araral & Wang, 2013). The results indicate that an effective combination of different governance styles can increase cooperation and avoid conflict and it is a determining factor in the success of governance reforms and the evolution of multiple and hybrid governance systems. For example, South Africa is an example of weak government governance, while Germany has a high government capacity for administration, though they believe that such capacity does not automatically lead to successful governance (Biswas & Tortajada, 2010; Pahl-Wostl, 2019). Water governance and management more or less comprise a hierarchical relationship between governments, the self-organization of communities, and competitive markets between different actors. The purposeful interaction between these relationships depends to some extent on how they interact (Grande, 2012). The important components in the institutional sector include laws, policies, and the performance of organizations. According to studies, the performance of organizations and strategies are the core components in water resources governance and management. On the other hand, institutional adaptation around the water crisis is much slower than the water crisis management mechanisms. As a result, ineffective governance can consequently intensify the water crisis (Kumari & Kumar, 2020).

16.9 Conclusion

Water-related disputes have a long history and remain a global and regional problem. Over the past few years, the total number of reports of violent conflict over water has increased. Unlike the situation in the early and mid-20th century, an increasing proportion of reported cases are related to internal disputes, terrorism, and local violence rather than transnational incidents. Many are small-scale, involving local violence over water allocation and use, or violence over local development decisions that affect community-scale environmental and economic conditions. However, more and more reported cases are rooted in water scarcity and competition for access to this limited resource (Gleick & Heberger, 2014).

New challenges in inland and transboundary water management will likely persist and become more complex, particularly with the emergence of an increased diversity of challenges due to climate change and globalization. Organizations should be prepared and participate in conflict management mechanisms and invest in institutional capacity with their neighbors. Integrated and systemic water management can provide an opportunity for more users to participate in programs and meet their needs while increasing water security. To benefit from water resources, the entire water management chain in all regions needs fundamental adjustments. In addition, although many areas of the world once had less water shortage crisis compared to the current decade, they have failed to develop adequate human resources for effective water management. The world, particularly the countries located in arid and semiarid regions, are currently categorized as countries with water stress, and they urgently need to invest in their human resources, to manage the problems of this sector through knowledge and scientific efforts. Therefore governments should go beyond water-related issues in the context of national climate change, environmental sustainability, and national and local water policy to address their water crisis.

Addressing the challenges of water scarcity requires both the development and use of new water resources through an integrated policy reform on supply and demand to encourage more effective and efficient use of existing water resources. New water sources use, such as rainwater harvesting and water storage from flood management during the rainy season through watershed management, can significantly reduce the water shortage in a basin. Therefore the urgent future task of decision-makers should emphasize sustainable water resources management and use to reduce the current water shortage and to help the conservation and restoration of the natural resources and biodiversity of natural ecosystems.

The effective implementation of the water governance system to reduce the conflicts caused by water scarcity requires the integrity, inclusiveness, active participation, coordination, and cooperation of all relevant stakeholders, including local communities, to include pluralism and integration of their interests, values, and considerations systemically. Failure to this systemic perspective causes different stakeholders or organizations to advance and achieve their objectives, regardless of the possible costs and damages to others. Moreover, the weak performance of water governance in many regions of the world can be managed through understanding the vulnerability of existing water management systems and applying integrated and adaptive management approaches, which facilitate the participation and social learning of stakeholders, particularly by emphasizing participatory modeling that involves stakeholders in the planning process, collective decision-making, and managing the conflicts caused by the water crisis.

References

Abel, G. J., Brottrager, M., Crespo Cuaresma, J., & Muttarak, R. (2019). Climate, conflict and forced migration. *Global Environmental Change*, *54*, 239–249. Available from https://doi.org/10.1016/j.gloenvcha.2018.12.003.

Adano, W. R., Dietz, T., Witsenburg, K., & Zaal, F. (2012). Climate change, violent conflict and local institutions in Kenya's drylands. *Journal of peace research*, *49*(1), 65–80. Available from https://doi.org/10.1177/0022343311427344.

Agar, M. (2016). Water governance in the face of global change (Pahl-Wostl, C. 2015). *Water Alternatives*, *9*(1), 165–167. Available from http://www.springer.com/us/book/9783319218540.

Aguilera, R. V., Aragón-Correa, J. A., Marano, V., & Tashman, P. A. (2021). The Corporate Governance of Environmental Sustainability: A Review and Proposal for More Integrated Research. *Journal of Management*, *47*(6), 1468–1497. Available from https://doi.org/10.1177/0149206321991212.

Ait-Kadi, M. (2016). Water for Development and Development for Water: Realizing the Sustainable Development Goals (SDGs) Vision. *Aquatic Procedia*, *6*, 106–110. Available from https://doi.org/10.1016/j.aqpro.2016.06.013.

Akhmouch, A., & Clavreul, D. (2016). Stakeholder Engagement for Inclusive Water Governance: "Practicing What We Preach" with the OECD Water Governance Initiative. *Water*, *8*(5), 204. Available from https://www.mdpi.com/2073-4441/8/5/204.

Alexander, S. (2019). What climate-smart agriculture means to members of the Global Alliance for climate-smart agriculture. *Future of Food: Journal on Food, Agriculture and Society*, *7*(1), 21–30. Available from http://www.thefutureoffoodjournal.com/index.php/FOFJ/article/view/174.

Almagtome, A., Khaghaany, M., & Önce, S. (2020). Corporate governance quality, stakeholders' pressure, and sustainable development: An integrated approach. *International Journal of Mathematical Engineering and Management Sciences*, *5*(6), 1077–1090. Available from https://doi.org/10.33889/ijmems.2020.5.6.082.

Al-Muqdadi, S. W. H. (2022). The spiral of escalating water conflict: The theory of hydro-politics. *Water (Switzerland)*, *14*(21). Available from https://doi.org/10.3390/w14213466, http://www.mdpi.com/journal/water.

Altingoz, M. (2022). *Transboundary water management in separatist regions: Towards a geography of hydro-political tensions. Invisible borders in a bordered world: power, mobility, and belonging* (pp. 62–81). Taylor and Francis, United States. Available from 10.4324/9780429352515-5.

Ansari, M., & Sharma, P. (2020). Water Crisis and Violation of Human Rights: A Special Reference To Women In Chhattisgarh. *Our Heritage*, *68*(1), 1952–1961.

Araral, E., & Wang, Y. (2013). Water governance 2.0: a review and second generation research agenda. *Water Resources Management*, *27*(11), 3945–3957. Available from https://doi.org/10.1007/s11269-013-0389-x.

Armeni, C. (2015). Global experimentalist governance, international law and climate change technologies. *International & Comparative Law Quarterly*, *64*(4), 875–904. Available from https://doi.org/10.1017/S0020589315000408.

Baldwin, E., McCord, P., Dell'Angelo, J., & Evans, T. (2018). Collective action in a polycentric water governance system. *Environmental Policy and Governance*, *28*(4), 212–222. Available from https://doi.org/10.1002/eet.1810, 1756-932X.

Bantider, A., Tadesse, B., Mersha, A. N., Zeleke, G., Alemayehu, T., Nagheeby, M., & Amezaga, J. (2023). Voices in Shaping Water Governance: Exploring Discourses in the Central Rift Valley, Ethiopia. *Water*, *15*(4), 803. Available from https://www.mdpi.com/2073-4441/15/4/803.

Barli, O., Baskent, E. Z., Turker, M. F., & Gedik, T. (2006). Analytical approach for analyzing and providing solutions for the conflicts among forest stakeholders across Turkey. *Forest Policy and Economics*, *9*(3), 219–236. Available from https://doi.org/10.1016/j.forpol.2005.07.009.

Basco-Carrera, L., Warren, A., van Beek, E., Jonoski, A., & Giardino, A. (2017). Collaborative modelling or participatory modelling? A framework for water resources management. *Environmental Modelling & Software*, *91*, 95–110. Available from https://doi.org/10.1016/j.envsoft.2017.01.014, 13648152.

Benson, D., Gain, A. K., & Giupponi, C. (2020). Moving beyond water centricity? Conceptualizing integrated water resources management for implementing sustainable development goals. *Sustainability Science*, *15*(2), 671–681. Available from https://doi.org/10.1007/s11625-019-00733-5.

Berardo, R., & Lubell, M. (2016). Understanding what shapes a polycentric governance system. *Public Administration Review*, *76*(5), 738–751. Available from https://doi.org/10.1111/puar.12532.

Bernauer, T., & Böhmelt, T. (2020). International conflict and cooperation over freshwater resources. *Nature Sustainability*, *3*(5), 350–356. Available from https://doi.org/10.1038/s41893-020-0479-8, 2398-9629.

Bernauer, T., & Siegfried, T. (2012). Climate change and international water conflict in Central Asia. *Journal of Peace Research*, *49*(1), 227–239. Available from https://doi.org/10.1177/0022343311425843, 0022-3433.

Bertule, M., Glennie, P., Koefoed Bjørnsen, P., James Lloyd, G., Kjellen, M., Dalton, J., Rieu-Clarke, A., Romano, O., Tropp, H., Newton, J., & Harlin, J. (2018). Monitoring Water Resources Governance Progress Globally: Experiences from Monitoring SDG Indicator 6.5.1 on Integrated Water Resources Management Implementation. *Water*, *10*(12), 1744. Available from http://doi.org/10.3390/w10121744.

Betts, A. (2013). Survival Migration: Failed Governance and the Crisis of Displacement. Cornell University Press.

Bezu, S., Demissie, T., Abebaw, D., Mungai, C., Samuel, S., Radeny, M. A., Huyer, S., & Solomon, D. (2020). *Climate change, agriculture and international migration nexus: African youth perspective* (CCAFS Working Paper, Issue 324). Available from https://cgspace.cgiar.org/handle/10568/110278.

Bhaduri, A., Bogardi, J., Siddiqi, A., Voigt, H., Vörösmarty, C., Pahl-Wostl, C., Bunn, S. E., Shrivastava, P., Lawford, R., Foster, S., Kremer, H., Renaud, F. G., Bruns, A., & Osuna, V. R. (2016). Achieving Sustainable Development Goals from a Water Perspective [Review]. *Frontiers in Environmental Science*, *4*. Available from https://doi.org/10.3389/fenvs.2016.00064.

Bigas, H. (2012). *The global water crisis: Addressing an urgent security issue*. United Nations University-Institute for Water. *Environment and Health*.

Biswas, A. K. (1992). Indus Water Treaty: the Negotiating Process. *Water International*, *17*(4), 201–209. Available from http://doi.org/10.1080/02508069208686140.

Biswas, A. K., & Tortajada, C. (2010). Future Water Governance: Problems and Perspectives. *International Journal of Water Resources Development*, *26*(2), 129–139. Available from https://doi.org/10.1080/07900627.2010.488853.

Böhmelt, T., Bernauer, T., Buhaug, H., Gleditsch, N. P., Tribaldos, T., & Wischnath, G. (2014). Demand, supply, and restraint: determinants of domestic water conflict and cooperation. *Global Environmental Change*, *29*, 337–348. Available from https://doi.org/10.1016/j.gloenvcha.2013.11.018.

Brzoska, M., & Fröhlich, C. (2016). Climate change, migration and violent conflict: vulnerabilities, pathways and adaptation strategies. *Migration and Development*, *5*(2), 190–210. Available from http://doi.org/10.1080/21632324.2015.1022973.

Campbell, J. R., & Bedford, R. D. (2022). Climate change and migration: Lessons from Oceania. In Routledge Handbook of Immigration and Refugee Studies, (pp. 379–387). Routledge.

Casiano Flores, C., Vikolainen, V., & Bressers, H. (2016). Water governance decentralisation and river basin management reforms in hierarchical systems: do they work for water treatment policy in Mexico's Tlaxcala Atoyac Sub-basin? *Water*, *8*(5), 210. Available from https://doi.org/10.3390/w8050210.

Chaware, B. (2018). Sustainable Development of Water Resources: Concept and Implementation. *Journal of Indian Water Resources Society*, *38*(3), 8–11.

Delany-Crowe, T., Marinova, D., Fisher, M., McGreevy, M., & Baum, F. (2019). Australian policies on water management and climate change: are they supporting the sustainable development goals and improved health and well-being? *Globalization and Health*, *15*(1), 68. Available from https://doi.org/10.1186/s12992-019-0509-3.

du Plessis, A. (2019b). Water as a Source of Conflict and Global Risk. In Water as an Inescapable Risk: Current Global Water Availability, Quality and Risks with a Specific Focus on South Africa, (pp. 115–143). Springer International Publishing. Available from https://doi.org/10.1007/978-3-030-03186-2_6.

du Plessis, A. (2019a). Climate Change and Freshwater Resources: Current Observations, Impacts, Vulnerabilities and Future Risks. In *du Plessis A.*, Water as an Inescapable Risk: Current Global Water Availability, Quality and Risks with a Specific Focus on South Africa, (pp. 55–78). Springer International Publishing. Available from https://doi.org/10.1007/978-3-030-03186-2_4.

Du, E., Tian, Y., Cai, X., Zheng, Y., Han, F., Li, X., Zhao, M., Yang, Y., & Zheng, C. (2022). Evaluating Distributed Policies for Conjunctive Surface Water-Groundwater Management in Large River Basins: Water Uses Versus Hydrological Impacts. *Water Resources Research*, *58*(1), e2021WR031352. Available from https://doi.org/10.1029/2021WR031352.

El-Fadel, M., Sayegh, Y. E., Ibrahim, A. A., Jamali, D., & El-Fadl, K. (2002). The Euphrates–Tigris Basin: A Case Study in Surface Water Conflict Resolution. *Journal of Natural Resources and Life Sciences Education*, *31*(1), 99–110. Available from https://doi.org/10.2134/jnrlse.2002.0099.

Es'haghi, S. R., & Karamidehkordi, E. (2023). Understanding the structure of stakeholders – projects network in endangered lakes restoration programs using social network analysis. *Environmental Science & Policy*, *140*, 172–188. Available from https://doi.org/10.1016/j.envsci.2022.12.001.

European Union Institute for Security Studies. (2015). EUISS Yearbook of European Security (YES). European Union Institute for Security Studies. Available from https://doi.org/10.2815/03961.

Fröhlich, C. J. (2012). Water: Reason for Conflict or Catalyst for Peace ? The Case of the Middle East. *L'Europe en. Formation*, *365*(3), 139–161. Available from https://doi.org/10.3917/eufor.365.0139.

Furber, A., Medema, W., Adamowski, J., Clamen, M., & Vijay, M. (2016). Conflict management in participatory approaches to water management: A case study of Lake Ontario and the St. Lawrence River Regulation. *Water*, *8*(7). Available from https://doi.org/10.3390/w8070280, 2073-4441.

Gain, A. K., Giupponi, C., & Benson, D. (2015). The water–energy–food (WEF) security nexus: the policy perspective of Bangladesh. *Water International*, *40*(5–6), 895–910. Available from https://doi.org/10.1080/02508060.2015.1087616.

Ghasemi, M., & Karamidehkordi, E. (2017). Analyzing Social Actors' Conflict Network on Natural Resources Conservation and Use and its Impact on Rural Households' Livelihoods: A case study in the Dorahan and Cheshme Ali Watersheds. *Iranian Journal of Range and Desert Research*, *24*(1), 39–56. Available from https://doi.org/10.22092/ijrdr.2017.109848.

Ghasemi, M., Karamidehkordi, E., & Ebrahimi, A. (2017). Analyzing Social Actors' Conflict in Natural Resources Management and Its Impact on Rural Communities (Case Study: Borujen County). *Journal of Rural Research*, *8*(4), 635–648. Available from https://doi.org/10.22059/jrur.2017.210178.923.

Gleick, P. H. (2014). Water, drought, climate change, and conflict in Syria. *Weather, climate. and society*, *6*(3), 331–340. Available from https://doi.org/10.1175/WCAS-D-13-00059.1.

Gleick, P. H., & Heberger, M. (2014). Water and Conflict. In P. H. Gleick (Ed.), The World's Water: The Biennial Report on Freshwater Resources (pp. 159–171). Island Press/Center for Resource Economics. Available from https://doi.org/10.5822/978-1-61091-483-3_10.

Gleick, P. H., & Shimabuku, M. (2023). Water-related conflicts: definitions, data, and trends from the water conflict chronology. *Environmental Research Letters*, *18*(3), 034022. Available from https://doi.org/10.1088/1748-9326/acbb8f.

Grande, E. (2012). Governance-Forschung in der Governance-Falle? – Eine kritische Bestandsaufnahme. *Politische Vierteljahresschrift*, *53*(4), 565–592. Available from http://www.jstor.org/stable/24201455.

Hamad, A. A. (2005). The reconceptualisation of conflict management. *Peace, Conflict and Development: An Interdisciplinary Journal*, *7*(July 2005). Available from http://www.peacestudiesjournal.org.uk.

Hjorth, P., & Madani, K. (2023). Adaptive Water Management: On the Need for Using the Post-WWII Science in Water Governance. *Water Resources Management*, *37*(6), 2247–2270. Available from https://doi.org/10.1007/s11269-022-03373-0.

Hsiang, S. M., Burke, M., & Miguel, E. (2013). Quantifying the influence of climate on human conflict. *Science*, *341*(6151), 12353671–123536714. Available from https://doi.org/10.1126/science.1235367.

Issaoui, M., Jellali, S., Zorpas, A. A., & Dutournie, P. (2022). Membrane technology for sustainable water resources management: Challenges and future projections. *Sustainable Chemistry and Pharmacy, 25*, 100590. Available from https://doi.org/10.1016/j.scp.2021.100590.

Jayaram, D., & Sethi, G. (2022). Geopolitics of climate change and water security in South Asia: Conflict and cooperation. In *Science, policies and conflicts of climate change [Springer Climate Series]*, (pp. 77–88). Springer Science and Business Media B.V., India. Available from https://doi.org/10.1007/978-3-031-16254-1_4.

Jayasiri, M. M. J. G. C. N., Dayawansa, N. D. K., & Yadav, S. (2023). Assessing the roles of farmer organizations for effective agricultural water management in Sri Lanka. *Agricultural Systems, 205*, 103587. Available from https://doi.org/10.1016/j.agsy.2022.103587.

Kapur, R. (2019). Management of Water Resources. *Acta Scientific Agriculture, 3*, 100–104.

Karamidehkordi, E. (2007). Knowledge and information systems in watershed management: a study of Bazoft watershed and relevant institutions in Chaharmahal and Bakhtiari Province. Reading, the UK: University of Reading.

Karamidehkordi, E. (2010). A country report: challenges facing Iranian Agriculture and Natural Resource Management in the twenty-first century. *Human ecology, 38*(2), 295–303. Available from https://doi.org/10.1007/s10745-010-9309-3.

Karamidehkordi, E. (2012). Sustainable Natural Resource Management, a Global Challenge of This Century. In K. Abiud (Ed.), *Sustainable Natural Resources Management* (pp. 105–114). IntechOpen. Available from https://doi.org/10.5772/35035.

Karamidehkordi, E., Karimi, V., Hallaj, Z., Karimi, M., & Naderi, L. (2024). Adaptable leadership for arid/semi-arid wetlands conservation under climate change: Using Analytical Hierarchy Process (AHP) approach. *Journal of Environmental Management, 351*, 119860. Available from https://doi.org/10.1016/j.jenvman.2023.119860.

Karamidehkordi, E., Hashemi Sadati, S. A., Tajvar, Y., & Mirmousavi, S. H. (2023). Climate change vulnerability and resilience strategies for citrus farmers. *Environmental and Sustainability Indicators, 20*, 100317. Available from https://doi.org/10.1016/j.indic.2023.100317.

Karimi, V., Bijani, M., Hallaj, Z., Valizadeh, N., Fallah Haghighi, N., & Karimi, M. (2023). Adaptation and Maladaptation to Climate Change: Farmers' Perceptions. In S. A. Bandh (Ed.), Strategizing Agricultural Management for Climate Change Mitigation and Adaptation (pp. 113–132). Springer International Publishing. Available from https://doi.org/10.1007/978-3-031-32789-6_7.

Karimi, V., Karami, E., & Keshavarz, M. (2018). Climate change and agriculture: Impacts and adaptive responses in Iran. *Journal of Integrative Agriculture, 17*(1), 1–15. Available from https://doi.org/10.1016/S2095-3119(17)61794-5.

Karimi, M., Tabiee, M., Karami, S., Karimi, V., & Karamidehkordi, E. (2024). Climate change and water scarcity impacts on sustainability in semi-arid areas: Lessons from the South of Iran. *Groundwater for Sustainable Development, 24*, 101075. Available from https://doi.org/10.1016/j.gsd.2023.101075.

Khilchevskyi, V. K., & Mezentsev, K. V. (2021). Water conflicts and Ukraine. *Donbas region, 2021*(1), 1–5. Available from https://doi.org/10.3997/2214-4609.20215K2004.

Krysovatyy, A. I., Zvarych, I. Y., Zvarych, R. Y., & Zhyvko, M. A. (2018). Preconditions for the tax environment of a alterglobal development. *Comparative Economic Research. Central and Eastern. Europe, 21*(4), 139–154. Available from https://doi.org/10.2478/cer-2018-0031.

Kujinga, K., Vanderpost, C., Mmopelwa, G., & Wolski, P. (2014). An analysis of factors contributing to household water security problems and threats in different settlement categories of Ngamiland, Botswana. *Physics and Chemistry of the Earth, Parts A/B/C, 67*, 187–201. Available from https://doi.org/10.1016/j.pce.2013.09.012.

Kumari, O., & Kumar, M. (2020). Water Governance: A Pragmatic Debate of 21st Century; An Indian Perspective. In M. Kumar, D. D. Snow, & R. Honda (Eds.), Emerging Issues in the Water Environment during Anthropocene: A South East Asian Perspective (pp. 355–365). Singapore: Springer. Available from https://doi.org/10.1007/978-981-32-9771-5_19.

Link, P. M., Scheffran, J., & Ide, T. (2016). Conflict and cooperation in the water-security nexus: a global comparative analysis of river basins under climate change. *Wiley Interdisciplinary Reviews: Water, 3*(4), 495–515. Available from https://doi.org/10.1002/wat2.1151.

Liu, T., Zhang, W., & Wang, R. Y. (2022). How does the Chinese government improve connectivity in water governance? A qualitative systematic review. *International Journal of Water Resources Development, 38*(4), 717–735. Available from https://doi.org/10.1080/07900627.2020.1755955, http://www.tandf.co.uk/journals/titles/07900627.asp.

Lonergan, S. C. (2018). Water and conflict: Rhetoric and reality. In P. Diehl (Ed.), Environmental conflict (pp. 109–124). Routledge.

Lukat, E., Pahl-Wostl, C., & Lenschow, A. (2022). Deficits in implementing integrated water resources management in South Africa: The role of institutional interplay. *Environmental Science & Policy, 136*, 304–313. Available from https://doi.org/10.1016/j.envsci.2022.06.010.

Malla, F. A., Mushtaq, A., Bandh, S. A., Qayoom, I., Hoang, A. T., & Shahid-e-Murtaza. (2022). *Understanding climate change: Scientific opinion and public perspective. Climate change: The social and scientific construct* (pp. 1–20). Springer International Publishing, India. Available from 10.1007/978-3-030-86290-9_1.

Markowska, J., Szalińska, W., Dąbrowska, J., & Brząkała, M. (2020). The concept of a participatory approach to water management on a reservoir in response to wicked problems. *Journal of Environmental Management, 259*. Available from https://doi.org/10.1016/j.jenvman.2019.109626, 03014797.

Marshall, D., Salamé, L., & Wolf, A. T. (2017). A call for capacity development for improved water diplomacy. In S. Islam, & K. Madani (Eds.), Water Diplomacy in Action: Contingent Approaches to Managing Complex Water Problems (pp. 141–154). New York: Anthem Press.

Martinás, K. (2010). Investigating social conflicts linked to water resources through agent-based modelling. In K. Martinás, D. Matika, & A. Srbljinović (Eds.), *Complex Societal Dynamics: Security Challenges and Opportunities*. (75, pp. 142–157). IOS Press. Available from https://doi.org/10.3233/978-1-60750-653-9-142.

Maru, M. (2016). Conflict Early Warning and the Response Nexus: The Case of the African Union-Continental Early Warning System. *Kennesaw State University*. Available from https://digitalcommons.kennesaw.edu/incmdoc_etd/3/.

Mbonile, M. J. (2005). Migration and intensification of water conflicts in the Pangani Basin, Tanzania. *Habitat International*, *29*(1), 41–67. Available from https://doi.org/10.1016/S0197-3975(03)00061-4.

Mianabadi, H., Alioghli, S., & Morid, S. (2021). Quantitative evaluation of 'No-harm' rule in international transboundary water law in the Helmand River basin. *Journal of Hydrology*, *599*, 126368. Available from https://doi.org/10.1016/j.jhydrol.2021.126368.

Mishra, B. K., Kumar, P., Saraswat, C., Chakraborty, S., & Gautam, A. (2021). Water Security in a Changing Environment: Concept, Challenges and Solutions. *Water*, *13*(4), 490. Available from https://www.mdpi.com/2073-4441/13/4/490.

Moorthy, R., & Bibi, S. (2023). Water security and cross-border water management in the Kabul river basin. *Sustainability*, *15*(1), 792. Available from https://doi.org/10.3390/su15010792.

Mpandeli, S., Nhamo, L., Hlahla, S., Naidoo, D., Liphadzi, S., Modi, A. T., & Mabhaudhi, T. (2020). Migration under Climate Change in Southern Africa: A Nexus Planning Perspective. *Sustainability*, *12*(11), 4722. Available from https://www.mdpi.com/2071-1050/12/11/4722.

Muñoz, J. F., Suárez, F., Alcayaga, H., McRostie, V., & Fernández, B. (2023). *Introduction to the Silala River and its hydrology. WIREs Water, n/a(n/a)*, e1644. Available from https://doi.org/10.1002/wat2.1644.

Naderi, L., Karamidehkordi, E., Badsar, M., & Moghadasd, M. (2022). Analyzing the Impact of Human and Environmental Factors on Stakeholders' Conflict in the Zayandehrood Basin. *Journal of Rural Research*, *13*(1), 68–85. Available from https://doi.org/10.22059/jrur.2022.332129.1689.

Naderi, L., Karamidehkordi, E., Moghadas, M., & Badsar, M. (2022). Analyzing the Interaction of Stakeholders' Demands, Power, Participation and Conflicts over the Water Use and Management in the Zayandehrud Basin. *Environmental Researches*, *13*(25), 379–398.

Nadiruzzaman, M., Scheffran, J., Shewly, H. J., & Kley, S. (2022). Conflict-sensitive climate change adaptation: A review. *Sustainability*, *14*(13). Available from https://doi.org/10.3390/su14138060, 2071-1050.

Nordås, R., & Gleditsch, N. P. (2015). Climate Change and Conflict. In S. Hartard, & W. Liebert (Eds.), *Competition and Conflicts on Resource Use. Natural Resource Management and Policy*. In: (46, pp. 21–38). Cham: Springer. Available from https://doi.org/10.1007/978-3-319-10954-1_3.

Nordqvist, P. & Krampe, F. (2018) *Climate change and violent conflict: Sparse evidence from South Asia and South East Asia*. Stockholm International Peace Research Institute. Available from http://www.jstor.org/stable/resrep24462.

Novoselov, A., Potravnii, I., Novoselova, I., & Gassiy, V. (2016). Conflicts management in natural resources use and environment protection on the regional level. *Journal of Environmental Management & Tourism*, *7*(3), 407–415. Available from https://journals.aserspublishing.eu/jemt/article/view/350.

Nyam, Y. S., Kotir, J. H., Jordaan, A. J., Ogundeji, A. A., & Turton, A. R. (2020). Drivers of change in sustainable water management and agricultural development in South Africa: A participatory approach. *Sustainable Water Resources Management*, *6*(4). Available from https://doi.org/10.1007/s40899-020-00420-9, 2363-5037.

Nyika, J., & Dinka, M. O. (2022). Integrated approaches to nature-based solutions in Africa: Insights from a bibliometric analysis. *Nature-Based Solutions*, *2*, 100031. Available from https://doi.org/10.1016/j.nbsj.2022.100031.

Okpara, U. T., Stringer, L. C., & Dougill, A. J. (2017). Using a novel climate–water conflict vulnerability index to capture double exposures in Lake Chad. *Regional Environmental Change*, *17*(2), 351–366. Available from https://doi.org/10.1007/s10113-016-1003-6.

Padhy, S. K., Sarkar, S., Panigrahi, M., & Paul, S. (2015). Mental health effects of climate change. *Indian J Occup Environ Med*, *19*(1), 3–7. Available from https://doi.org/10.4103/0019-5278.156997.

Pahl-Wostl, C. (2019). Governance of the water-energy-food security nexus: A multi-level coordination challenge. *Environmental Science & Policy*, *92*, 356–367. Available from https://doi.org/10.1016/j.envsci.2017.07.017, 14629011.

Patrick, H. O. (2020). Climate Change, Water Security, and Conflict Potentials in South Africa: Assessing Conflict and Coping Strategies in Rural South Africa. In W. Leal Filho, J. Luetz, & D. Ayal (Eds.), Handbook of Climate Change Management: Research, Leadership, Transformation (pp. 1–18). Springer International Publishing. Available from https://doi.org/10.1007/978-3-030-22759-3_84-1.

Payus, C., Ann Huey, L., Adnan, F., Besse Rimba, A., Mohan, G., Kumar Chapagain, S., Roder, G., Gasparatos, A., & Fukushi, K. (2020). Impact of Extreme Drought Climate on Water Security in North Borneo: Case Study of Sabah. *Water*, *12*(4), 1135. Available from https://www.mdpi.com/2073-4441/12/4/1135.

Petersen-Perlman, J. D., Veilleux, J. C., & Wolf, A. T. (2017). International water conflict and cooperation: challenges and opportunities. *Water International*, *42*(2), 105–120. Available from https://doi.org/10.1080/02508060.2017.1276041.

Phuong, L. T. H., Wals, A., Sen, L. T. H., Hoa, N. Q., Van Lu, P., & Biesbroek, R. (2018). Using a social learning configuration to increase Vietnamese smallholder farmers' adaptive capacity to respond to climate change. *Local Environment*, *23*(8), 879–897. Available from http://doi.org/10.1080/13549839.2018.1482859.

Purvis, B., Mao, Y., & Robinson, D. (2019). Three pillars of sustainability: in search of conceptual origins. *Sustainability Science*, *14*(3), 681–695. Available from https://doi.org/10.1007/s11625-018-0627-5.

Qin, J., Duan, W., Chen, Y., Dukhovny, V. A., Sorokin, D., Li, Y., & Wang, X. (2022). Comprehensive evaluation and sustainable development of water–energy–food–ecology systems in Central Asia. *Renewable and Sustainable Energy Reviews*, *157*, 112061. Available from https://doi.org/10.1016/j.rser.2021.112061.

Roudgarmi, P., Anssari, N., & Farahani, E. (2011). Determining effective socio-economic factors on degradation of Natural Resources in Tehran province. *Iranian Journal of Range and Desert Research*, *18*(1), 151–171. Available from http://doi.org/10.22092/ijrdr.2011.102052.

Saiefeddine, A., Bouzoraa, A., & Xiaohu, L. (2018). Zooming survey on the actual situation of the sustainable development in China from 1970 until now. *Environmental Risk Assessment and Remediation*, *2*(2), 51–59. Available from https://doi.org/10.4066/2529-8046.100041.

Scanlon, B. R., Fakhreddine, S., Rateb, A., de Graaf, I., Famiglietti, J., Gleeson, T., Grafton, R. Q., Jobbagy, E., Kebede, S., Kolusu, S. R., Konikow, L. F., Long, D., Mekonnen, M., Schmied, H. M., Mukherjee, A., MacDonald, A., Reedy, R. C., Shamsudduha, M., Simmons, C. T., … Zheng, C.

(2023). Global water resources and the role of groundwater in a resilient water future. *Nature Reviews Earth & Environment*, *4*(2), 87–101. Available from https://doi.org/10.1038/s43017-022-00378-6.

Schillinger, J., Özerol, G., Güven-Griemert, Ş., & Heldeweg, M. (2020). Water in war: Understanding the impacts of armed conflict on water resources and their management. *WIREs Water*, *7*(6). Available from https://doi.org/10.1002/wat2.1480, 2049-1948.

Shah, T. (2016). Increasing water security: the key to implementing the Sustainable Development Goals. *Global Water Partnership (GWP) TEC Background Papers* (22), 1–56.

Shan, V., Singh, S. K., & Haritash, A. K. (2020). Water Crisis in the Asian Countries: Status and Future Trends. In M. Kumar, F. Munoz-Arriola, H. Furumai, & T. Chaminda (Eds.), Resilience, Response, and Risk in Water Systems: Shifting Management and Natural Forcings Paradigms (pp. 173–194). Singapore: Springer. Available from https://doi.org/10.1007/978-981-15-4668-6_10.

Shumilova, O., Tockner, K., Sukhodolov, A., Khilchevskyi, V., De Meester, L., Stepanenko, S., Trokhymenko, G., Hernández-Agüero, J. A., & Gleick, P. (2023). Impact of the Russia–Ukraine armed conflict on water resources and water infrastructure. *Nature Sustainability*, *6*(5), 578–586. Available from https://doi.org/10.1038/s41893-023-01068-x.

Sukanya, S., & Joseph, S. (2023). Climate change impacts on water resources: An overview. In A. Srivastav, A. Dubey, A. Kumar, S. Kumar Narang, & M. Ali Khan (Eds.), *Visualization Techniques for Climate Change with Machine Learning and Artificial Intelligence* (pp. 55–76). Elsevier. Available from https://doi.org/10.1016/B978-0-323-99714-0.00008-X.

Sohrabi, M., Ahani Amineh., Niksokhan, M. H., & Zanjanian, H. (2023). A framework for optimal water allocation considering water value, strategic management and conflict resolution. *Environment, Development and Sustainability*, *25*(2), 1582–1613. Available from https://doi.org/10.1007/s10668-022-02110-2.

Solomon, S. (2007). *Climate change 2007-the physical science basis: Working group I contribution to the fourth assessment report of the* (Vol. 4). IPCC.

Tatar, M., Papzan, A., & Ahmadvand, M. (2022). Understanding factors that contribute to farmers' water conflict behavior. *Water Policy*, *24*(4), 589–607. Available from https://doi.org/10.2166/wp.2022.253.

Tayia, A. (2019). Transboundary Water Conflict Resolution Mechanisms: Substitutes or Complements. *Water*, *11*(7), 1337. Available from https://www.mdpi.com/2073-4441/11/7/1337.

Unfried, K., Kis-Katos, K., & Poser, T. (2022). Water scarcity and social conflict. *Journal of Environmental Economics and Management*, *113*. Available from https://doi.org/10.1016/j.jeem.2022.102633, 00950696.

Velmurugan, A., Swarnam, P., Subramani, T., Meena, B., & Kaledhonkar, M. (2020). Water demand and salinity. In M. H. Davood Abadi Farahani, V. Vatanpour, & A. H. Taheri (Eds.), *Desalination-challenges and opportunities* (pp. 1–11). Intechopen.

Verhoeven, H. (2011). Climate Change, Conflict and Development in Sudan: Global Neo-Malthusian Narratives and Local Power Struggles. *Development and Change*, *42*(3), 679–707. Available from https://doi.org/10.1111/j.1467-7660.2011.01707.x.

Wines, M. (2014). West's drought and growth intensify conflict over water rights. New York Times: New York, NY, USA.

Wolf, R. (2019). Taking interaction seriously: Asymmetrical roles and the behavioral foundations of status. *European Journal of International Relations*, *25*(4), 1186–1211. Available from https://doi.org/10.1177/1354066119837338, 1354-0661.

Woodhouse, P., & Muller, M. (2017). Water governance—An historical perspective on current debates. *World Development*, *92*, 225–241. Available from https://doi.org/10.1016/j.worlddev.2016.11.014, 0305750X.

Yates, J. S., Harris, L. M., & Wilson, N. J. (2017). Multiple ontologies of water: Politics, conflict and implications for governance. *Environment and Planning D: Society and Space*, *35*(5), 797–815. Available from https://doi.org/10.1177/0263775817700395, http://www.sagepub.com/journals/Journal202439.

Yerian, S., Hennink, M., Greene, L. E., Kiptugen, D., Buri, J., & Freeman, M. C. (2014). The role of women in water management and conflict resolution in Marsabit, Kenya. *Environmental Management*, *54*(6), 1320–1330. Available from https://doi.org/10.1007/s00267-014-0356-1, http://link.springer.de/link/service/journals/00267/index.htm.

Yuan, S., Browning, M. H. E. M., McAnirlin, O., Sindelar, K., Shin, S., Drong, G., Hoptman, D., & Heller, W. (2023). A virtual reality investigation of factors influencing landscape preferences: Natural elements, emotions, and media creation. *Landscape and Urban Planning*, *230*, 104616. Available from https://doi.org/10.1016/j.landurbplan.2022.104616.

Zeitoun, M., Mirumachi, N., Warner, J., Kirkegaard, M., & Cascão, A. (2020). Analysis for water conflict transformation. *Water International*, *45*(4), 365–384. Available from https://doi.org/10.1080/02508060.2019.1607479, 0250-8060.

Chapter 17

Governance with principles and standards: water footprint and sustainability in Indonesia

Andi Luhur Prianto[1,*], Tawakkal Baharuddin[2] and Nina Yuslaini[3]

[1]Department of Government Studies, Universitas Muhammadiyah, Makassar, Indonesia, [2]Department of Government Studies, Universitas Muhammadiyah Makassar, Indonesia, [3]Department of Government Science, Universitas Islam Riau, Indonesia

**Corresponding author. e-mail address: luhur@unismuh.ac.id.*

17.1 Introduction

Water is a critical resource, and if it is not managed properly, its negative impacts can be sustained (Herslund & Mguni, 2019). Unsustainable exploitation of water resources can cause serious environmental impacts, such as habitat destruction, water quality degradation, and climate change (Fatahi Nafchi et al., 2021; Liu & Mao, 2020; Prianto et al., 2021; Prianto et al., 2023; Kurniasih et al., 2023; Malik et al., 2023). In addition to serious environmental impacts, poor water management also has direct implications for human well-being. Uncertainty in clean water supply can threaten safe drinking water supply, water availability for agriculture, and industrial needs (Anderson et al., 2019; Boltz et al., 2019). Prolonged drought can also threaten the food security and economy of a region, while increasingly severe flooding due to climate change can cause significant material and public health losses (Ijaodola et al., 2019; Kurniasih et al., 2023; Rasul, 2021). Thus water management becomes crucial to protect water, human life, and the environment for future generations. This indicates the need for sustainable water management efforts.

Water footprint is one of the key concepts in sustainable water management efforts (Berger et al., 2021; Novoa et al., 2019; Wang et al., 2020). In the context of the negative impacts that may result from a lack of good water management, water footprints provide a framework for understanding and quantifying how human activities, such as agriculture, industry, and consumption, may contribute to water use (Chen et al., 2021; Zhao et al., 2021). In a situation where drought and water scarcity are ongoing threats, monitoring and reducing the water footprint are crucial (Dai et al., 2019; Wu et al., 2019). Measuring the water footprint of different activities and products can identify patterns that require change (Dai et al., 2019; Nouri et al., 2019). In general, water footprint is a concept that measures how much water is used throughout the life cycle of a particular product, process, or activity. The water footprint includes both directly used water (such as water used for cooking or bathing) and indirectly used water (such as water used to produce raw materials in certain products). This concept is important because it links human consumption to water use and its impact on the environment (Hoekstra, 2017; Sriyakul et al., 2019; Vanham & Bidoglio, 2013).

Indonesia is an archipelago rich in water resources, yet challenges related to sustainable water management are increasingly pressing (Prianto et al., 2023; Umami et al., 2022). In recent years, awareness of the water footprint, that is the amount of water used in various human activities, has increased worldwide (Ali et al., 2018; Jermsittiparsert et al., 2021; 2023; Moros-Ochoa et al., 2022). Indonesia, as a country with a large population and diverse economic activities, is not exempt from this challenge (Congge & Gohwong, 2023; Handayani et al., 2019; Nathaniel, 2021). As such, the adoption of national principles and standards in water footprint governance and sustainability is required to effectively deal with this deepening problem. One of the main reasons for having national principles and standards is to ensure that water use in Indonesia can be properly monitored, evaluated, and managed. With clear guidelines in place, governments, companies, and communities can work together to measure, report, and mitigate the impacts of their water footprint on the environment and society (BSN, 2019). This is important in the context of sustainability, where water resources can be maintained for future generations.

Water Footprints and Sustainable Development. DOI: https://doi.org/10.1016/B978-0-443-23631-0.00017-0

While research on water footprint has become an increasingly important topic in the context of global water sustainability, there is a lack of research that specifically and simultaneously combines aspects of water footprint with the importance of national principles and standards in water governance, especially in the Indonesian context. To achieve sustainable water management in Indonesia, combining research on water footprints with national principles and standards can provide a holistic view of the challenges and opportunities in managing increasingly limited water resources. Research focusing on the integration between water footprints and national regulatory frameworks would be valuable in helping to formulate more sustainable water management policies and practices in the country. However, there are still some results of previous studies that are still considered relevant. Firstly, water footprints help understand the extent to which human activities affect water availability and quality in a region (Daniels et al., 2011; Ma et al., 2020). Second, by looking at the water footprint, it is possible to identify areas that are vulnerable to water scarcity or water pollution (Wöhler et al., 2020; Xu et al., 2019). Third, strict regulations related to the water footprint are crucial in achieving sustainable water management (Liu et al., 2020; Zhang et al., 2019).

This research aims to bridge the gap of previous research by combining the concept of water footprint with the importance of national principles and standards in water governance in Indonesia in a holistic manner. The research questions mapped out include (1) How can national principles and standards implemented by the government provide support to issues related to water footprint and sustainability in Indonesia? (2) What are the barriers and challenges in supporting these issues? Through an in-depth analysis of these two questions, this research can provide a more comprehensive view of how the government can play a key role in promoting sustainable water management and protecting water resources for Indonesia's future. In addition, the findings in this study can be taken into consideration by the government and other parties to minimize obstacles that may occur in the future.

17.2 Research methods

The research method applied in this study is a qualitative method with a thematic analysis approach. This approach was chosen as it allows for an in-depth understanding of national principles and standards relevant to water footprint and sustainability issues in Indonesia. The first step in this research is data collection. The data used in this study are official government documents published by the Environmental and Forestry Instrument Standardization Agency, which is part of the Ministry of Environment and Forestry. These documents include environmental management, water footprint, principles, requirements, and guidelines.

Next, thematic analysis was used to analyze the content of these documents. The process of thematic analysis began with a careful reading of the content of the documents to identify emerging patterns, themes, and concepts. Relevant data related to national principles and standards that support water footprint and sustainability issues were identified and extracted. In data analysis, Nvivo 12 Plus software was used to help organize and understand the data systematically. This tool allows researchers to categorize data based on certain themes and make connections between relevant concepts.

The results of the thematic analysis will be used to answer the two research questions that have been posed, namely regarding the national principles and standards applied by the government in supporting water footprint and sustainability issues in Indonesia, as well as identifying barriers and challenges that may arise in supporting these issues. Through this approach, this research aims to provide a comprehensive insight into the role of national principles and standards in water governance in Indonesia, as well as provide an in-depth understanding of the potential barriers and challenges that need to be overcome to achieve better water management sustainability.

Data validity in this study was assured by official government documents published by the Environmental and Forestry Instrument Standardization Agency, which is the official authority on regulations and guidelines related to the environment and water governance in Indonesia. These documents include regulations, policies, guidelines, and other relevant documents relevant to water footprint and sustainability. In addition, a thorough content analysis of the documents was conducted to ensure that the information extracted was in line with the research objectives. Additional validation was done through triangulation with other relevant sources and through reflection on potential biases. Through this approach, the data used in this research can be considered valid and reliable in supporting the research findings and conclusions.

17.3 Results and discussion

17.3.1 National principles and standards: water footprint management and sustainability

Sustainable water management is an urgent need in Indonesia, a country with high levels of water diversity but also facing serious challenges in the maintenance and protection of its water resources. To address issues related to water

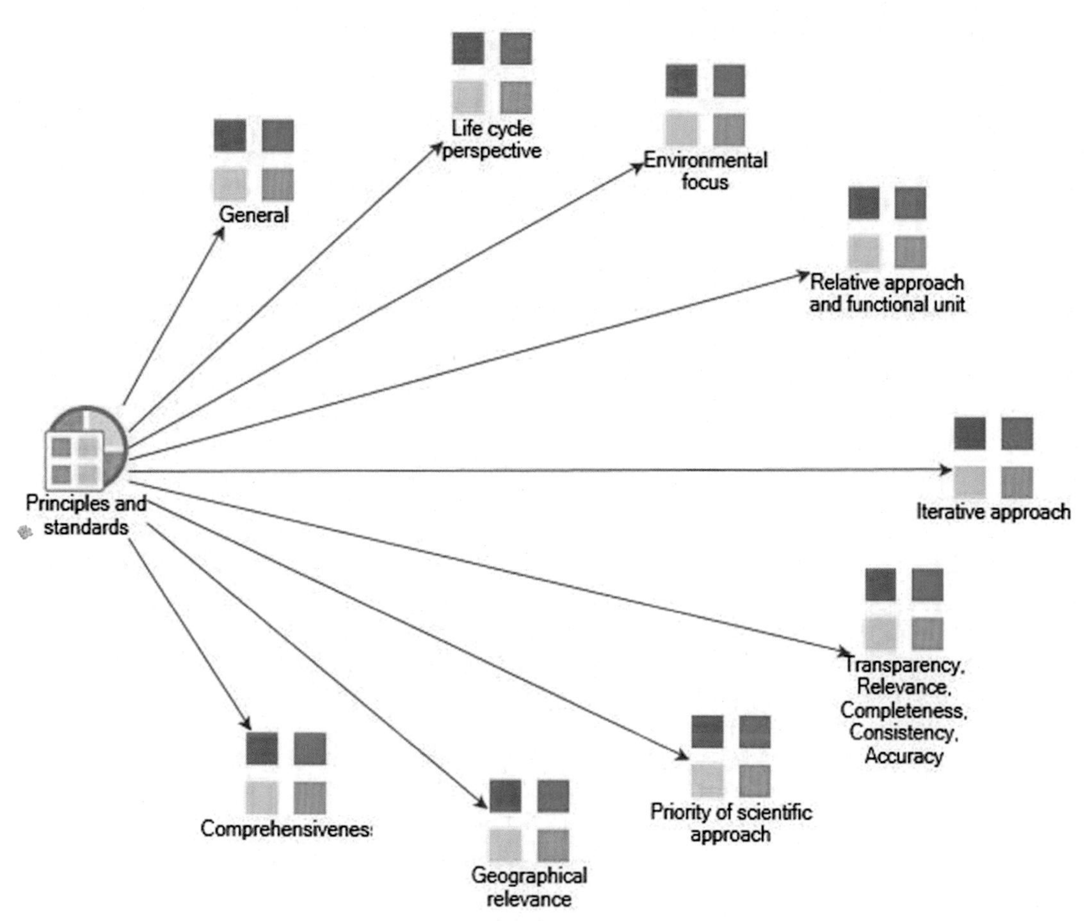

FIGURE 17.1 Principles considered by the Indonesian government in water footprint management. *Processed by researchers using Nvivo 12 Plus, 2023.*

footprint and sustainability in Indonesia, the government has taken strategic steps by implementing relevant national principles and standards. These measures not only play a role in ensuring adequate water availability for communities and industries but also contribute to the preservation of freshwater and marine ecosystems that are critical to Indonesia's biodiversity and national economy.

From Fig. 17.1 it can be seen that the principles and standards applied by the government include generality, life cycle perspective, environmental focus, relative approach and functional units, iterative approach, transparency, relevance, completeness, consistency, accuracy, priority scientific approach, geographical relevance, and comprehensiveness (BSN, 2019). These principles and standards applied by governments in water management are essential to ensure that water footprint assessments and sustainability efforts are of high quality and effective. One of the emerging principles is the life cycle perspective, which recognizes the importance of considering the full life cycle of a product or process in evaluating the water footprint. This means not only thinking about water impacts at the use stage but also during production, transport, and raw material selection. This approach helps to identify hidden impacts that may be missed in more narrow evaluations.

The environmental focus principle emphasizes the importance of focusing on environmental impacts in water footprint assessment. This means that the evaluation should specifically relate to impacts on aquatic ecosystems and the environment in general. The principle of relative approach and functional unit suggests that relative comparisons between products or processes should be taken into account, and water footprint evaluation should be conducted based on relevant functional units, such as the quantity of products produced. Other principles such as transparency, relevance, completeness, consistency, accuracy, and priority of scientific approach support the quality and credibility of the water footprint evaluation. The principle of geographical relevance reminds us that evaluations should consider geographical differences in water characteristics and policies. Finally, the principle of comprehensiveness emphasizes that water footprint evaluations should be comprehensive and consider all relevant aspects, from environmental aspects to human health and resources, to provide a complete and holistic understanding of water management impacts. By following these principles, the government can ensure that water management is more sustainable and environmentally orientated.

In general, it states that government-owned documents are fundamental and should be used to guide decision-making regarding the planning, implementation, and reporting of water footprint evaluations. Water footprint evaluations, in accordance with this standard, may be conducted and reported as standalone assessments that evaluate only potential environmental impacts related to water or as part of a broader life cycle assessment that considers all relevant environmental impacts, not just those related to water. This water footprint evaluation should be comprehensive and consider all relevant attributes or aspects relating to the natural environment, human health, and resources. By considering all relevant attributes and aspects in one study with a cross-media perspective, this evaluation enables the identification and assessment of potential trade-offs that may occur in water management, which is important in sustainable decision-making (BSN, 2019).

The principles and standards considered by the Indonesian government in water footprint management have significant relevance to the results of previous research in several key aspects. First, the life cycle perspective principle that emphasizes the importance of considering the entire life cycle of a product or process in water footprint evaluation directly supports efforts to comprehensively understand water impacts (Gu et al., 2015; Pacetti et al., 2015). This reflects the understanding that water impacts occur not only at the use stage, but also during production, transport, and decision-making related to resource use. Second, the environmental focus principle explicitly emphasizes the importance of prioritizing environmental impacts in water footprint evaluation (Bello et al., 2018; Ercin & Hoekstra, 2014). This is in line with previous research findings that have identified negative impacts on freshwater and marine ecosystems and climate change due to unsustainable water exploitation (Scherer & Pfister, 2016). The principle of prioritizing environmental aspects allows the government to protect vulnerable water ecosystems and ensure good water quality for environmental sustainability.

Considering the principles, the Indonesian government has an important opportunity to strengthen and optimize its approach to water footprint management. Integrating all elements of the principles, water footprint evaluation can become a more powerful tool to measure overall water-related environmental and resource impacts. This will not only result in more accurate and comprehensive data but will also provide a firmer foundation for formulating more sustainable policies and actions in the management of Indonesia's precious water resources. Drawing on lessons learnt from previous studies that emphasize the need to protect water resources for ecosystems and human well-being, the implementation of these principles can contribute positively to maintaining water sustainability for the country's future. In this case, the collaboration of the government and other parties is also needed.

From Fig. 17.2, it is clear that collaboration between the Indonesian government and other parties is a key factor in efforts to manage the water footprint and ensure its sustainability. The government has an important role in designing and implementing policies and regulations that support the principles and standards outlined above. They also can supervise and control water use at national and regional levels. However, others, such as the private sector, civil society organizations, and academic institutions, have valuable knowledge and resources that can be used to assist in water

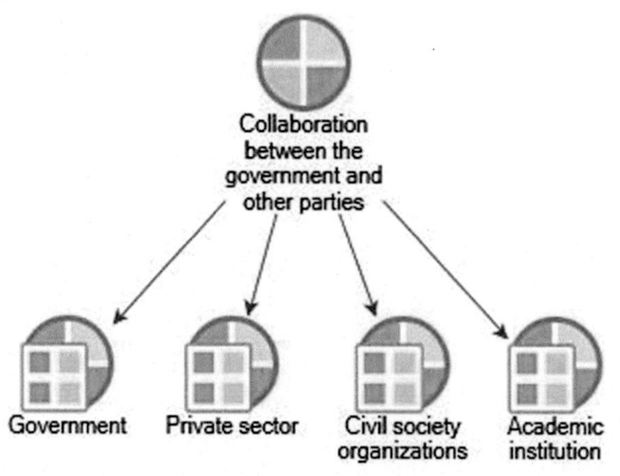

FIGURE 17.2 Collaboration between government and other parties is needed in water trail management. *Processed by researchers using Nvivo 12 Plus, 2023.*

footprint management. This collaboration can yield several benefits. Firstly, the private sector can assist in identifying and implementing sustainable practices in their operations involving water use. They can also play a role in developing technologies and innovations that are more efficient in water use.

Civil society organizations can play an important role in monitoring and advocacy, ensuring that government and private parties adhere to established principles and standards in water footprint management. Academic institutions can provide in-depth research and analyses to support evidence-based decision-making. However, this collaboration also faces a number of challenges. One is effective coordination between the various parties, given the differences in their goals and interests. In addition, there is a need for transparency and accountability in water footprint management for this collaboration to be successful. Therefore the government needs to facilitate a platform for dialog and cooperation between these sectors and ensure that the role of each party is recognized and valued in the joint effort to maintain water sustainability in Indonesia. With strong collaboration between the government and other parties, efforts to manage the water footprint and its sustainability can be more effective and have a positive impact in the long run.

Close collaboration between the Indonesian government, private sector, civil society organizations, and academic institutions has significant implications for water footprint management in Indonesia. In the context of countries with serious water resource management challenges, this collaboration can strengthen the capacity and resources available to formulate and implement sustainable policies and practices. The private sector can play a major role in encouraging environmentally friendly and water-efficient business practices. Civil society organizations can provide independent control and oversight and ensure that government policies and industry practices comply with sustainable water footprint principles. Academic institutions can contribute with research and an in-depth understanding of the challenges and solutions in water footprint management. As such, this strong collaboration will not only increase the effectiveness of water footprint management but will also help create a stronger culture of environmental awareness and support water sustainability across Indonesia (Prianto & Abdillah, 2023; Prianto et al., 2021).

17.3.2 Barriers and challenges in supporting sustainable water footprint management

Sustainable water footprint management is a complex challenge faced by many countries, including Indonesia. Despite efforts to understand and address water footprint–related issues, there are several barriers and challenges faced. One of the main obstacles is the lack of accurate data and information on water footprints (Mohammad Sabli et al. (2017)). In many cases, data on water use, water quality, and environmental impacts associated with water are not always available or are incomplete (Fang et al., 2014; Jeswani & Azapagic, 2011). This makes it difficult to make evidence-based decisions and formulate effective policies in water management. In addition, problems with consistent monitoring and immature reporting systems can also hinder efforts to understand and measure the full water footprint (Galli et al., 2012).

In addition, climate change is also a serious challenge in water footprint management. Climate change can result in variability in rainfall, flood patterns, and drought patterns, all of which affect water footprint (Hoekstra & Chapagain, 2007; Shrestha et al., 2017). This necessitates the adoption of more adaptive and flexible strategies in water management to cope with the increasingly unpredictable impacts of climate change (Yuslaini et al., 2023). Another challenge is public awareness and understanding of the importance of sustainable water footprint management. Motivating people to participate in sustainable practices in water use and considering their impact on the water footprint can be challenging. Education and awareness campaigns are important elements in addressing this challenge. Finally, coordination between various parties, including governments, the private sector, civil society organizations, and academic institutions, often requires considerable effort. Ensuring that all stakeholders work together to achieve the goal of sustainable water footprint management can be a challenge in itself.

In addition to the obstacles and challenges previously outlined, there are still several other important aspects that need to be considered in supporting sustainable water footprint management in Indonesia. One of them is the issue of consistent policies and regulations. In some cases, regulations related to water management are not aligned or not assertive enough in addressing water footprint issues. Efforts are needed to formulate consistent and robust policies that support sustainable practices in water use, including water waste reduction and water use efficiency. Funding can also be a barrier, especially in implementing the technologies and infrastructure needed to reduce the water footprint. Significant investments are required to adopt more water-efficient technologies, and these initiatives require adequate financial support.

In addition, it is necessary to pay attention to the issue of fair and equitable access to water. Some communities, especially marginalized ones, may face difficulties in accessing clean water, while certain sectors have greater access to

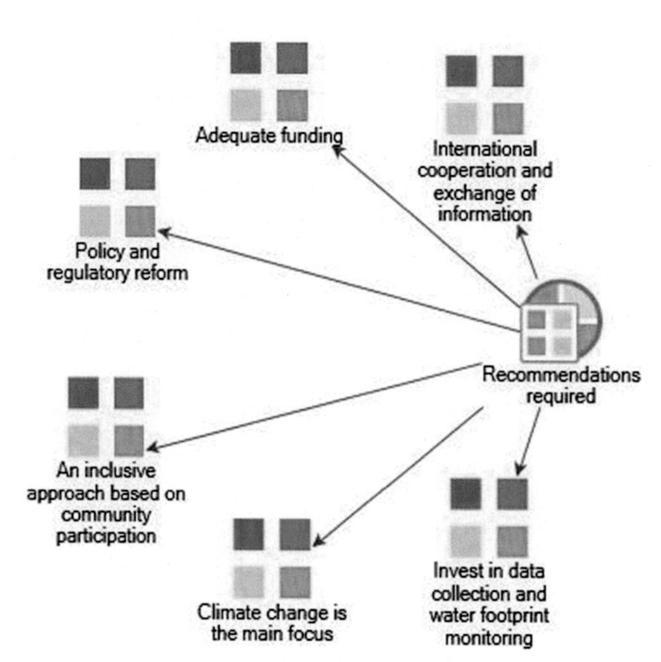

FIGURE 17.3 Recommendations needed to reduce barriers to water management and water footprint. *Processed by researchers using Nvivo 12 Plus, 2023.*

water resources. Sustainable water footprint management must take into account aspects of social equity in the distribution of and access to water. Furthermore, it is important to recognize that sustainable water footprint management is also closely linked to global issues. Indonesia's efforts to reduce its water footprint can also impact neighboring countries and the world at large. Therefore international cooperation and information exchange need to be enhanced in a joint effort to maintain water sustainability globally. By understanding and addressing these barriers and challenges, Indonesia can move toward more sustainable water footprint management and ensure adequate water availability for future generations.

In the face of barriers and challenges to sustainable water footprint management, a set of recommendations and ways forward are needed that can help minimize negative impacts and strengthen water management efforts in Indonesia.

From Fig. 17.3, it can be seen that the Indonesian government needs to increase investment in data collection and water footprint monitoring. This cannot be separated from involving the development of more robust systems to measure water use, water quality, and associated environmental impacts. Accurate data is the foundation for formulating effective policies and measuring the impact of implemented water footprint management efforts. On the other hand, climate change should be a major focus in future water footprint management planning. Governments need to adopt more adaptive and flexible strategies in the face of increasingly unpredictable variations in rainfall, flood patterns, and droughts. This could include investments in climate-resilient infrastructure and the development of planning models that take climate change into account. In addition, an inclusive approach based on community participation is key. The government needs to collaborate with civil society organizations, the private sector, and academic institutions in water footprint management. Education and awareness campaigns should also be enhanced to motivate people in sustainable water use.

Water-related policy and regulatory reforms also need to be carried out to ensure consistency and decisiveness in addressing water footprint issues. This includes improved monitoring and reporting to achieve sustainable practices in water use. In addition, adequate funding needs to be addressed. The government needs to find adequate sources of funding to support the implementation of sustainable technologies and practices in water management. Finally, international cooperation and information exchange need to be enhanced. Indonesia can learn from other countries' experiences in sustainable water footprint management and can also contribute to global efforts in maintaining water sustainability. By implementing these recommendations, Indonesia can minimize barriers and challenges to sustainable water footprint management and ensure adequate water availability for future generations. This is an important step in maintaining the sustainability of water and natural ecosystems in Indonesia.

17.4 Conclusion

Water is a very important and strategic resource for the country of Indonesia, affecting many sectors including drinking water supply, agriculture, and industry. However, unsustainable water use can lead to ongoing negative impacts, such as drought, pollution, and water resource conflicts. It is important to adopt national principles and standards that guide effective and sustainable water footprint management. The use of national principles and standards in water footprint governance is key to maintaining water sustainability in Indonesia. These principles include a life cycle perspective, environmental focus, relative approach, transparency, and more. Applying these principles helps ensure that water footprint evaluations cover all relevant aspects, from environmental impacts to human health. However, collaboration between the government and other stakeholders, including the private sector and civil society, is essential in designing, implementing, and monitoring sustainable practices in water management.

However, various barriers and challenges also exist in supporting sustainable water footprint management. From data limitations to unpredictable climate change, these obstacles require serious attention. In addition, collaboration between different parties can also face obstacles such as differences in goals and effective coordination. Recommendations include increased investment in data collection, adjustment to climate change, increased public awareness, policy reform, and adequate funding. Sustainable water footprint management is an important challenge that needs to be addressed in Indonesia. Following national principles and standards, and through collaboration between the government and others, Indonesia can overcome the barriers and challenges and move toward more sustainable water management with positive long-term impacts.

References

Ali, Y., Pretaroli, R., Socci, C., & Severini, F. (2018). Carbon and water footprint accounts of Italy: A multi-region input-output approach. *Renewable and Sustainable Energy Reviews*, *81*, 1813–1824. Available from https://doi.org/10.1016/j.rser.2017.05.277. Available from: https://www.journals.elsevier.com/renewable-and-sustainable-energy-reviews.

Anderson, E. P., Jackson, S., Tharme, R. E., Douglas, M., Flotemersch, J. E., Zwarteveen, M., Lokgariwar, C., Montoya, M., Wali, A., Tipa, G. T., Jardine, T. D., Olden, J. D., Cheng, L., Conallin, J., Cosens, B., Dickens, C., Garrick, D., Groenfeldt, D., Kabogo, J., ... Arthington, A. H. (2019). Understanding rivers and their social relations: A critical step to advance environmental water management. *Interdisciplinary Reviews: Water*, *6*(6). Available from https://doi.org/10.1002/WAT2.1381. Available from: http://wires.wiley.com/WileyCDA/WiresJournal/wisId-WAT2.html.

Bello, M. O., Solarin, S. A., & Yen, Y. Y. (2018). The impact of electricity consumption on CO_2 emission, carbon footprint, water footprint and ecological footprint: The role of hydropower in an emerging economy. *Journal of Environmental Management*, *219*, 218–230. Available from https://doi.org/10.1016/j.jenvman.2018.04.101. Available from: https://www.sciencedirect.com/journal/journal-of-environmental-management.

Berger, M., Campos, J., Carolli, M., Dantas, I., Forin, S., Kosatica, E., Kramer, A., Mikosch, N., Nouri, H., Schlattmann, A., Schmidt, F., Schomberg, A., & Semmling, E. (2021). Advancing the water footprint into an instrument to support achieving the SDGs – Recommendations from the "Water as a Global Resources" Research Initiative (GRoW). *Water Resources Management*, *35*(4), 1291–1298. Available from https://doi.org/10.1007/s11269-021-02784-9. Available from: http://www.wkap.nl/journalhome.htm/0920-4741.

Boltz, F., LeRoy Poff, N., Folke, C., Kete, N., Brown, C. M., St. George Freeman, S., Matthews, J. H., Martinez, A., & Rockström, J. (2019). Water is a master variable: Solving for resilience in the modern era. *Water Security*, *8*. Available from https://doi.org/10.1016/j.wasec.2019.100048. Available from: http://www.journals.elsevier.com/water-security/.

Chen, J., Gao, Y., Qian, H., Jia, H., & Zhang, Q. (2021). Insights into water sustainability from a grey water footprint perspective in an irrigated region of the Yellow River Basin. *Journal of Cleaner Production*, *316*128329. Available from https://doi.org/10.1016/j.jclepro.2021.128329.

Congge, U., & Gohwong, S. (2023). Local development and environmental governance: Experienced from extractive industry in East Luwu, Indonesia. *Otoritas: Jurnal Ilmu Pemerintahan*, *13*(1), 156–168. Available from https://doi.org/10.26618/ojip.v13i1.10681.

Dai, D., Sun, M., Xu, X., & Lei, K. (2019). Assessment of the water resource carrying capacity based on the ecological footprint: A case study in Zhangjiakou City, North China. *Environmental Science and Pollution Research*, *26*(11), 11000–11011. Available from https://doi.org/10.1007/s11356-019-04414-9. Available from: http://www.springerlink.com/content/0944-1344.

Daniels, P. L., Lenzen, M., & Kenway, S. J. (2011). The ins and outs of water use – A review of multi-region input-output analysis and water footprints for regional sustainability analysis and policy. *Economic Systems Research*, *23*(4), 353–370. Available from https://doi.org/10.1080/09535314.2011.633500.

Ercin, A. E., & Hoekstra, A. Y. (2014). Water footprint scenarios for 2050: A global analysis. *Environment International*, *64*, 71–82. Available from https://doi.org/10.1016/j.envint.2013.11.019. Available from: http://www.elsevier.com/locate/envint.

Fang, K., Heijungs, R., & De Snoo, G. R. (2014). Theoretical exploration for the combination of the ecological, energy, carbon, and water footprints: Overview of a footprint family. *Ecological Indicators*, *36*, 508–518. Available from https://doi.org/10.1016/j.ecolind.2013.08.017. Available from: http://www.elsevier.com/locate/ecolind.

Fatahi Nafchi, R., Yaghoobi, P., Reaisi Vanani, H., Ostad-Ali-Askari, K., Nouri, J., & Maghsoudlou, B. (2021). Eco-hydrologic stability zonation of dams and power plants using the combined models of SMCE and CEQUALW2. *Applied Water Science*, *11*(7). Available from https://doi.org/10.1007/s13201-021-01427-z. Available from: http://www.springer.com/earth + sciences + and + geography/hydrogeology/journal/13201.

Galli, A., Wiedmann, T., Ercin, E., Knoblauch, D., Ewing, B., & Giljum, S. (2012). Integrating ecological, carbon and water footprint into a "footprint Family" of indicators: Definition and role in tracking human pressure on the planet. *Ecological Indicators*, *16*, 100–112. Available from https://doi.org/10.1016/j.ecolind.2011.06.017.

Gu, Y., Xu, J., Keller, A. A., Yuan, D., Li, Y., Zhang, B., Weng, Q., Zhang, X., Deng, P., Wang, H., & Li, F. (2015). Calculation of water footprint of the iron and steel industry: A case study in Eastern China. *Journal of Cleaner Production*, *92*, 274–281. Available from https://doi.org/10.1016/j.jclepro.2014.12.094. Available from: https://www.journals.elsevier.com/journal-of-cleaner-production.

Handayani, W., Kristijanto, A. I., & Hunga, A. I. R. (2019). A water footprint case study in Jarum village, Klaten, Indonesia: The production of natural-colored batik. *Environment, Development and Sustainability*, *21*(4), 1919–1932. Available from https://doi.org/10.1007/s10668-018-0111-5. Available from: http://www.wkap.nl/journalhome.htm/1387-585X.

Herslund, L., & Mguni, P. (2019). Examining urban water management practices – Challenges and possibilities for transitions to sustainable urban water management in Sub-Saharan cities. *Sustainable Cities and Society*, *48*101573. Available from https://doi.org/10.1016/j.scs.2019.101573.

Hoekstra, A. Y. (2017). Water footprint assessment: Evolvement of a new research field. *Water Resources Management*, *31*(10), 3061–3081. Available from https://doi.org/10.1007/s11269-017-1618-5.

Hoekstra, A. Y., & Chapagain, A. K. (2007). Water footprints of nations: Water use by people as a function of their consumption pattern. *Water Resources Management*, *21*, 35–48. Available from https://doi.org/10.1007/s11269-006-9039-x.

Ijaodola, O. S., El- Hassan, Z., Ogungbemi, E., Khatib, F. N., Wilberforce, T., Thompson, J., & Olabi, A. G. (2019). Energy efficiency improvements by investigating the water flooding management on proton exchange membrane fuel cell (PEMFC). *Energy*, *179*, 246–267. Available from https://doi.org/10.1016/j.energy.2019.04.074. Available from: http://www.elsevier.com/inca/publications/store/4/8/3/.

Jermsittiparsert, K., Chankoson, T., Prianto, A. L., & Thaicharoen, W. (2021). Business ethical values as a mechanism linking CSR and internal outcomes on job performance. *Journal of Legal, Ethical and Regulatory Issues*, *24*(1), 1–11. Available from: http://www.alliedacademies.org/public/Journals/JournalDetails.aspx?jid = 14.

Jeswani, H. K., & Azapagic, A. (2011). Water footprint: Methodologies and a case study for assessing the impacts of water use. *Journal of Cleaner Production*, *19*(12), 1288–1299. Available from https://doi.org/10.1016/j.jclepro.2011.04.003.

Liu, J., Zhao, D., Mao, G., Cui, W., Chen, H., & Yang, H. (2020). Environmental sustainability of water footprint in Mainland China. *Geography and Sustainability*, *1*(1), 8–17. Available from https://doi.org/10.1016/j.geosus.2020.02.002. Available from: http://www.journals.elsevier.com/geography-and-sustainability.

Kurniasih, D., Prianto, A. L., Abdillah, A., Congge, U., & Akib, E. (2023). Strengthening Climate Change Governance & Smart City Through Smart Education in Bandung Indonesia. In U. Chatterjee, A. Antipova, S. Ghosh, S. Majumdar, & M. D. Setiawati (Eds.), Urban Environment and Smart Cities in Asian Countries. Human Dynamics in Smart Cities. Cham: Springer. Available from https://doi.org/10.1007/978-3-031-25914-2_16

Liu, Y., & Mao, D. (2020). Integrated assessment of water quality characteristics and ecological compensation in the Xiangjiang River, south-central China. *Ecological Indicators*, *110*105922. Available from https://doi.org/10.1016/j.ecolind.2019.105922.

Ma, W., Opp, C., & Yang, D. (2020). Past, present, and future of virtual water and water footprint. *Water (Switzerland)*, *12*(11), 1–20. Available from https://doi.org/10.3390/w12113068. Available from: http://www.mdpi.com/journal/water.

Malik, I., Prianto, A. L., Roni, N. I., Yama, A., & Baharuddin, T. (2023). *Multi-level governance and digitalization in climate change: A bibliometric analysis. Digital technologies and applications. ICDTA 2023. [Lecture notes in networks and systems (LNNS), Volume 669]* (pp. 95–104). Springer, https://www.springer.com/series/15179. doi: 10.1007/978-3-031-29860-8_10.

Mohammad Sabli, N. S., Zainon Noor, Z., Kanniah, K. A., Kamaruddin, S. N., & Mohamed Rusli, N. (2017). Developing a methodology for water footprint of palm oil based on a methodological review. *Journal of Cleaner Production*, *146*, 173–180. Available from https://doi.org/10.1016/j.jclepro.2016.06.149.

Moros-Ochoa, M. A., Castro-Nieto, G. Y., Quintero-Español, A., & Llorente-Portillo, C. (2022). Forecasting biocapacity and ecological footprint at a worldwide level to 2030 using neural networks. *Sustainability (Switzerland)*, *14*(17). Available from https://doi.org/10.3390/su141710691. Available from: http://www.mdpi.com/journal/sustainability/.

Nathaniel, S. P. (2021). Ecological footprint, energy use, trade, and urbanization linkage in Indonesia. *GeoJournal*, *86*(5), 2057–2070. Available from https://doi.org/10.1007/s10708-020-10175-7. Available from: http://www.kluweronline.com/issn/0343-2521/.

Nouri, H., Chavoshi Borujeni, S., & Hoekstra, A. Y. (2019). The blue water footprint of urban green spaces: An example for Adelaide, Australia. *Landscape and Urban Planning*, *190*103613. Available from https://doi.org/10.1016/j.landurbplan.2019.103613.

Novoa, V., Ahumada-Rudolph, R., Rojas, O., Sáez, K., de la Barrera, F., & Arumí, J. L. (2019). Understanding agricultural water footprint variability to improve water management in Chile. *Science of the Total Environment*, *670*, 188–199. Available from https://doi.org/10.1016/j.scitotenv.2019.03.127. Available from: http://www.elsevier.com/locate/scitotenv.

Pacetti, T., Lombardi, L., & Federici, G. (2015). Water-energy nexus: A case of biogas production from energy crops evaluated by Water Footprint and Life Cycle Assessment (LCA) methods. *Journal of Cleaner Production*, *101*, 278–291. Available from https://doi.org/10.1016/j.jclepro.2015.03.084. Available from: https://www.journals.elsevier.com/journal-of-cleaner-production.

Prianto, A. L., & Abdillah, A. (2023). *Vulnerable countries, resilient communities: Climate change governance in the coastal communities in Indonesia. Climate change, community response and resilience: Insight for socio-ecological sustainability* (pp. 135–152). Elsevier. Available from: https://www.sciencedirect.com/book/9780443187070. doi: 10.1016/B978-0-443-18707-0.00007-2.

Prianto, A. L., Nurmandi, A., Qodir, Z., & Jubba, H. (2021). Climate change and religion: From ethics to sustainability action. *E3S Web of Conferences*, *277*. Available from https://doi.org/10.1051/e3sconf/202127706011, 22671242, http://www.e3s-conferences.org/.

Prianto, A. L., Usman, S., Amri, A. R., Nurmandi, A., Qodir, Z., Jubba, H., & Ilik, G. (2023). Faith-Based Organizations' Humanitarian Work from the Disaster Risk Governance Perspective: Lessons from Covid-19 Pandemic in Indonesia. *Mazahib*, *22*(1), 129–174. Available from https://doi.org/10.21093/mj.v22i1.6317.

Rasul, G. (2021). Twin challenges of COVID-19 pandemic and climate change for agriculture and food security in South Asia. *Environmental Challenges*, 2100027. Available from https://doi.org/10.1016/j.envc.2021.100027.

Scherer, L., & Pfister, S. (2016). Global water footprint assessment of hydropower. *Renewable Energy*, *99*, 711–720. Available from https://doi.org/10.1016/j.renene.2016.07.021. Available from: http://www.journals.elsevier.com/renewable-and-sustainable-energy-reviews/.

Shrestha, S., Chapagain, R., & Babel, M. S. (2017). Quantifying the impact of climate change on crop yield and water footprint of rice in the Nam Oon Irrigation Project, Thailand. *Science of the Total Environment*, *599–600*, 689–699. Available from https://doi.org/10.1016/j.scitotenv.2017.05.028. Available from: http://www.elsevier.com/locate/scitotenv.

Sriyakul, T., Jermsittiparsert, K., Phanwichit, S., & Prianto, A. L. (2019). Improving the perceived partnership synergy and sustainability through the social and political context in Indonesia: Business law compliance as a mediator. *International Journal of Innovation, Creativity and Change*, *8* (8), 142–159. Available from: https://www.ijicc.net/images/vol8iss8/8810_Sriyakaul_2019_E_R.pdf.

Umami, A., Sukmana, H., Wikurendra, E. A., & Paulik, E. (2022). A review on water management issues: Potential and challenges in Indonesia. *Sustainable Water Resources Management*, *8*(3). Available from https://doi.org/10.1007/s40899-022-00648-7. Available from: http://springer.com/journal/40899.

Vanham, D., & Bidoglio, G. (2013). A review on the indicator water footprint for the EU28. *Ecological Indicators*, *26*, 61–75. Available from https://doi.org/10.1016/j.ecolind.2012.10.021.

Wang, H., Huang, J., Zhou, H., Deng, C., & Fang, C. (2020). Analysis of sustainable utilization of water resources based on the improved water resources ecological footprint model: A case study of Hubei Province, China. *Journal of Environmental Management*, *262*110331. Available from https://doi.org/10.1016/j.jenvman.2020.110331.

Wöhler, L., Niebaum, G., Krol, M., & Hoekstra, A. Y. (2020). The grey water footprint of human and veterinary pharmaceuticals. *Water Research X*, *7*100044. Available from https://doi.org/10.1016/j.wroa.2020.100044.

Wu, X., Xia, J., Guan, B., Liu, P., Ning, L., X, Yi, Yang, L., & Hu, S. (2019). Water scarcity assessment based on estimated ultimate energy recovery and water footprint framework during shale gas production in the Changning play. *Journal of Cleaner Production*, *241*118312. Available from https://doi.org/10.1016/j.jclepro.2019.118312.

Xu, Z., Chen, X., Wu, S. R., Gong, M., Du, Y., Wang, J., Li, Y., & Liu, J. (2019). Spatial-temporal assessment of water footprint, water scarcity and crop water productivity in a major crop production region. *Journal of Cleaner Production*, *224*, 375–383. Available from https://doi.org/10.1016/j.jclepro.2019.03.108. Available from: https://www.journals.elsevier.com/journal-of-cleaner-production.

Yuslaini, N., Suwaryo, U., Deliarnoor, N. A., & Sri Kartini, D. (2023). Palm oil industry and investment development in Dumai City, Indonesia: A focus on local economy development and sustainability. *Cogent Social Sciences*, *9*(1). Available from https://doi.org/10.1080/23311886.2023.2235780. Available from: http://www.cogentoa.com/journal/social-sciences.

Zhang, L., Dong, H., Geng, Y., & Francisco, M. J. (2019). China's provincial grey water footprint characteristic and driving forces. *Science of the Total Environment*, *677*, 427–435. Available from https://doi.org/10.1016/j.scitotenv.2019.04.318. Available from: http://www.elsevier.com/locate/scitotenv.

Zhao, D., Liu, J., Yang, H., Sun, L., & Varis, O. (2021). Socioeconomic drivers of provincial-level changes in the blue and green water footprints in China. *Resources, Conservation and Recycling*, *175*105834. Available from https://doi.org/10.1016/j.resconrec.2021.105834.

Chapter 18

Agro-industrial water conservation by water footprint and Sustainable Development Goals

Tiziana Crovella*, **Giovanni Lagioia and Annarita Paiano**
Department of Economics, Management and Business Law, University of Bari Aldo Moro, Largo Abbazia Santa Scolastica, Bari, Italy
**Corresponding author. e-mail address: tiziana.crovella@uniba.it.*

18.1 Introduction

Globally, water demand is rising sharply and 11.11% of people, almost 1 billion do not have access to uncontaminated water (Sela, 2023) and agriculture used 70% of freshwater (Alberti et al. 2022; Crovella et al., 2022). Moreover, global agricultural water consumption equivalent of 2,700 billion cubic meters can surpass 3,200 billion cubic meters in a few years due to the population growth (He et al., 2021) that which will reach 9.7 billion people in 2050.

At European level, freshwater resources are depleted: Europeans use billions of cubic meters of water every year for agriculture, manufacturing, drinking, heating, refrigeration, electricity production, tourism, and other service activities. Likewise, constant population growth, urbanization, pollution, and the effects of climate change, such as the persistent droughts that affecting Italy, are putting a strain on Europe's freshwater resource and its quality (EEA, 2023).

According to the United Nations (UN) World Water Development Report, it has been tested a water stress condition considering the annual water resources less than 1700 m^3 per inhabitant. Among the EU Member States in Poland, Czech Republic, Cyprus, and Malta persist in this circumstance (Eurostat, 2023).

Particularly, Italy is hit by an extreme drought, causing negative impacts on the productivity and competitiveness of the entire agri-food chain. For this reason, it is crucial the adoption of extraordinary measures to reduce water stress. In 2022, the long drought has drastically reduced the water resources available in Southern Europe, particularly in Italy. The rainfall reduction, also due to climate change, is jeopardizing the water supply for domestic, agricultural, and energy production uses. This dangerous circumstance of drought and, consequently, the related water emergency are putting 300,000 Italian agricultural firms in serious difficulty (European Parliament, 2023). Currently, the water crisis that is affecting Italy mainly distresses agriculture, because the current withdrawal satisfies mainly urban and domestic demands (Benedini & Rossi, 2021).

In addition, climate change and pollution increased the pressure on water resources and infrastructures, already stressed by urban activities and economic development (ISTAT, 2022). For this reason, it is essential to strengthen the resilience of the water system, bringing processes towards greater efficiency, mainly in vulnerable areas by water and affected by drought.

In this regard, the UN published the Sustainable Development Goals (SDGs), the global guidance for addressing actions towards an urgent global challenge, particularly to reduce natural resources consumption, respect ethnic groups, eliminating unequal distributionof food and to provide improving solutions for all (He et al., 2022). Therefore, due to the growing need of natural resources supply, and in particular water resources, more restrictive standards for water quality access and use have been envisaged. Particularly, these standards allow the pursuit of SDG 6 *"Ensure availability and sustainable management of water and sanitation for all"* (UN, 2015).

Among the impact calculation indicators, water footprint (WF) theorized by the scholar Arjen Hoekstra in 2002 represents a widely used tool to analyze the consumption of water and the consequent local impacts caused during agricultural and industrial production (Hoekstra & Hung, 2002). Particularly, the WF represents an effective guidance tool for achieving, in particular, the SDG 6 (Berger et al., 2021).

Water Footprints and Sustainable Development. DOI: https://doi.org/10.1016/B978-0-443-23631-0.00018-2

Methodologically, this current replicable study aims to raise awareness of the use of the WF indicator to analyze the potential of water and orient the decision-making process towards improving the management of the resources themselves in the light of SDG 6 operating according to an innovative process of evaluation. Therefore, in order to address this purpose, the authors adopted a dual methodological approach, initially conducting an analysis of the scientific literature, and simultaneously an overview of the associated statistical data.

After the introductory section and the methodological pathway description, this chapter reviews what current research revealed about the state of the art of water resource in Section 18.3. After the debate about the WF indicator in Section 18.4, the scholars focused on the interaction between water conservation and reuse for a circular and resilience context in Section 18.5 and on the level of achievement of the SDG 6 related to water consumption in Section 18.6. This chapter concludes underlying which actions (natural, economic/finance, political/programming, industrial, approach, or research) can be undertaken by stakeholders and practitioners to minimize water consumption, reduce WF and water stress, and for increasing water reuse strategies. Finally, the authors developed a roadmap that synthetizes the necessary actions towards sustainability in water management.

18.2 Methodology

The aim of this chapter is a structured analysis of the water use in agriculture, focusing on agri-food industry. Particularly, the general objective of this study is the development of a combined analysis of literature and statistical data to verify the level of achievement of the sustainability at European and national level.

The authors proposed a dual methodological framework that combines literature review analysis and multicriteria evaluation to address the problem of water resource reduction. This application is also planned for supporting the decision-making process of the Italian National Recovery and Resilience Plan funding by European Union (EU) and is based on Lami and Todella's methodology (Lami and Todella, 2023). Theoretically, the scholars put in relation the WF concept and the water stress as consumption in agricultural context. Moreover, statistically this chapter compared the water efficiency and the state of the art of water stress at European and national level (Fig. 18.1).

Particularly, the authors applied a dual methodological framework to a tangible case (water stress in the agro-industry context) interacting with real data collected statistically by international organizations. Moreover, it was

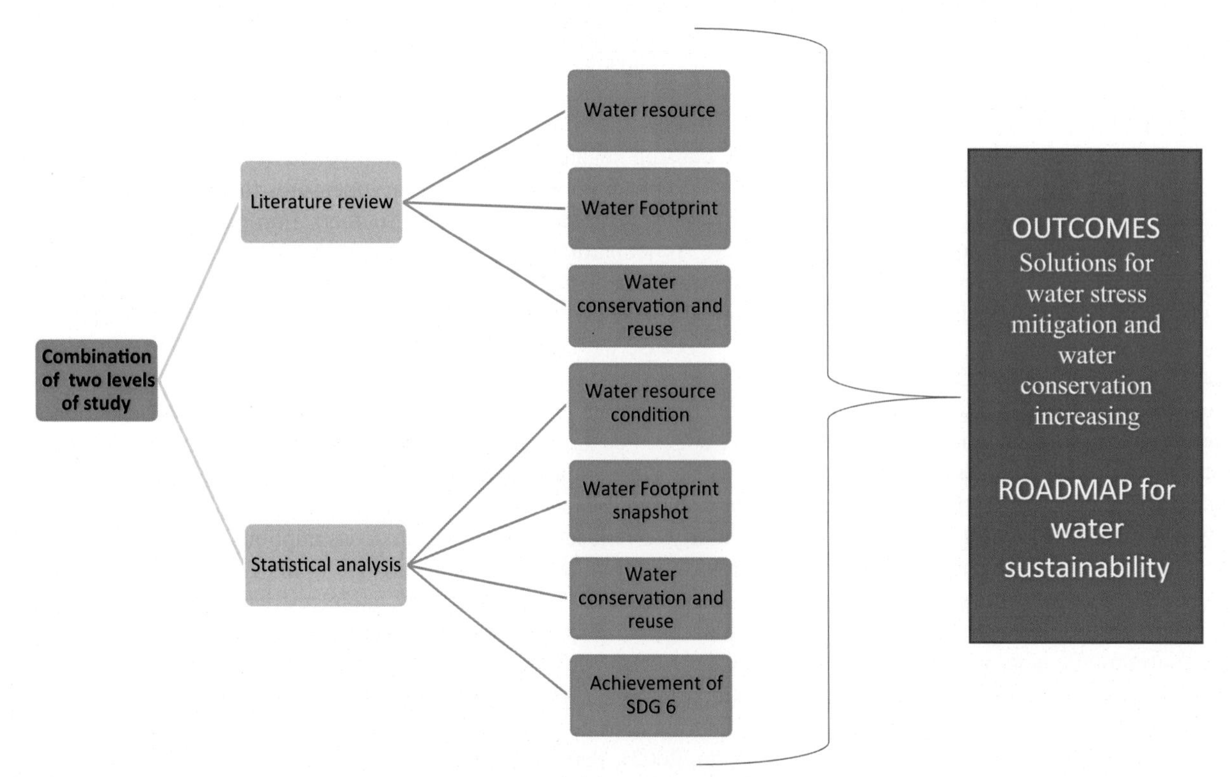

FIGURE 18.1 Structure of the study conducted and presented in this chapter. *SDG*, Sustainable Development Goal.

relevant to analyze statistical data validated by international organizations to offer a better reproduction of the water stress phenomenon.

Through this dual benchmarking methodology of water stress, the scholars carried out a literature review on the topics of water resource, considering WF and water use related to agricultural context and agri-food industry. Particularly, they observed the phenomena at European and national level, focusing on Italy, the country involved in the Italian National Recovery and Resilience Plan granted by the EU. Subsequently, the abovementioned topic was included in a statistical analysis, with the aim to collect the main data on water resource condition, focusing on the water consumption at the European and national level and to provide a snapshot of the WF indicator for main crops and sectors. After these two steps, the authors analyzed the context of water reuse, focusing the need to implement some circularity practices to best achieve sustainability objectives. In the end, the scholars presented the level of sustainability achievement of some indicators included in SDG 6.

Finally, this structured analysis proposed an overview of the main solutions for reducing water stress and WF and for increasing water conservation. Moreover, this study can stimulate a dual methodological approach, consulting scientific literature and statistical data. Hence, the authors developed a roadmap for water resource sustainability useful for all stakeholders and practitioners involved in the water resource management.

In order to carry out an in-depth review, the authors checked the main platforms and databases, such as Science Direct, Web of Science and Scopus. Simultaneously, for the same topics scrutinized in the literature review, the scholars analyzed the correlated statistics and database checking the data in International Organization platform (e.g., UN), in international statistical data-warehouse (e.g., OECD, Eurostat, ISTAT), or collected by interactive tools (e.g., WF Assessment Tool).

18.3 State of the art of water resource

Globally, agro-industry is a sector highly dependent on the availability of water. In particular, the sometimes-irregular rainfall and drought must be replaced with the irrigation necessary to increase vitality, crops yield, and quality of the harvest (European Courts of Auditor, 2021). According to the World Bank (2023), over the last 55 years, there has been a 17% decrease in renewable water resources per capita at community level due to the population growth, the pressure exerted by economic activity, and climate change, which is worsening water scarcity seasonally and annually in some area. Particularly, irrigation consumes almost 95% of total withdrawals (Lovelace et al., 2020) and is the main responsible of water-level changes. Besides, the current withdrawal water resource cannot be defined sustainable and, in the future, this condition will exacerbate the climate change (Ou et al., 2018; Whittemore et al., 2023).

Furthermore, the increasing intensity and frequency of droughts due to climate change are significantly influencing the absorption of carbon by ecosystems, also leading to increase the global temperature (Anderegg et al., 2020; Cartwright et al., 2020; Piao et al., 2019; Wang et al., 2021; Xi & Yuan, 2022).

As underlined by several scholars, there are many consequences caused by drought that lead to plant mortality, difficult in agricultural production, risk of fires, epidemics of parasites and pathogens that affect vegetation (Xi & Yuan, 2022; (Wang et al., 2022). At the European level, the Commission stressed that in the last three years, the extreme drought affecting Central and Western Europe has caused extensive damage (European Commission, 2021).

In terms of general availability, for the EU Member States, the average water resources are around 4,000–5,000 m^3 per inhabitant. In particular, in countries rich in water, the share of an inhabitant can reach up to around 30,000 m^3 (Croatia) or even more than 60,000 m^3 as in Norway (Eurostat, 2023). Furthermore, with global warming at 3°C, the absolute annual losses due to drought could amount to 40 billion €/year (European Commission, 2021).

Statistically, agro-industry used 70% of freshwater globally as underlined by He et al. (2021). Conversely, at European level, agriculture is responsible for 24% of water withdrawal in the EU, although a fair reduction has been documented over the last 30 years, thanks to the reduced pressure on resources and sustainable practices adopted by some EU members (EEA, 2020). Furthermore, the demand of water for crops cultivation, food production, and processing at urban and local scales still remains a crucial problem to be addressed nowadays (Alberti et al., 2022).

As shown in Table 18.1, the annual freshwater abstraction for agricultural use from 2016 to 2020 increased in several countries. Among these countries included in the Table 18.1, the withdrawal increased by 77.47% in Belgium, 5.10% in Cyprus, and 0.44% in Denmark. Conversely, in Bulgaria and Czechia, it decreased by 17% and 30.24%, respectively. Other countries as Croatia, Greece, and Malta, presented constant values of annual freshwater abstraction for agricultural use.

For the assessment, the scholars, authors of the current chapter, considered Italy as a country involved in the "PNRR GRINS - Growing, Resilient, Inclusive and Sustainable Programme". From a first consideration, the main

TABLE 18.1 Annual freshwater abstraction by agriculture, forestry, and fishing (in million cubic meters).

Country	2016	2017	2018	2019	2020
Austria	–	–	–	–	–
Belgium	4.66	5.53	8.17	7.09	8.27
Bulgaria	898.45	823.59	713.28	787.02	746.00
Croatia	30.00	30.00	30.00	30.00	30.00
Cyprus	37.30	33.20	22.30	31.90	39.20
Czechia	33.40	31.90	32.40	27.30	23.30
Denmark	198.86	213.64	230.15	211.64	200.74
Estonia	0.16	0.30	0.33	0.06	0.05
Finland	–	–	–	–	–
France	1,935.77	1,826.18	1,744.04	1,888.05	1,946.41
Germany	71.19	88.39	105.59	122.79	-
Greece	2,801.64	2,801.64	2,801.64	2,801.64	2,801.64
Hungary	–	–	–	–	–
Ireland	–	–	32.81	32.96	35.33
Italy	–	–	–	–	–
Latvia	56.10	52.72	51.85	51.70	53.85
Lithuania	55.78	55.19	53.47	54.10	56.13
Luxembourg	0.01	0.01	0.00	0.04	0.02
Malta	1.25	1.25	1.25	1.25	1.25
Netherlands	20.06	22.73	76.83	54.60	93.00
Poland	1,039.94	1,015,54	953.47	847.41	801.62
Portugal	–	–	–	–	–
Romania	1,203.00	1,444.00	1,292.00	1,538.00	2,095.00
Slovakia	9.40	17.60	12.80	14.50	14.40
Slovenia	2.70	3.00	2.59	2.90	2.40
Spain	15.722,00	15.300,00	14.878,00	14.780,00	14.710,00
Sweden	45.40	49.80	54.20	58.60	63.00
United Kingdom of Great Britain and North Ireland	–	–	–	–	–

(–), data not available.
Source: Authors' elaboration on data from Eurostat. (2023). Water statistics—annual freshwater abstraction by source and sector. https://ec.europa.eu/eurostat/databrowser/view/ENV_WAT_ABS/default/table?lang = en.

problem that affects water resources is the lack of rainfall and the resulting drought. In 2020, the Italian regions and metropolitan capitals collected a total annual rainfall equal of approximately 661 mm, which was one of the lowest values of the last decade (as in 2011). This is a decrease of 132 mm of rainfall compared to the average for the period 2006–2015 (ISTAT, 2022).

Considering the consumption at national level, from 2016 to 2020, a constant withdrawal of water resources equal to 88% of the existing renewable resource was documented (AQUASTAT, 2023).

As shown in Table 18.2, all the water withdrawn was used in agriculture (a constant value of 10E + 09 m^3/year) and in some years, it was necessary to consume more than the available resources, including rainwater and purified water.

TABLE 18.2 Overview of water state in Italy (2016–2020).

Year	Agricultural water withdrawal as percentage of total water withdrawal	Percentage of area equipped for irrigation by direct use of treated municipal wastewater
2016	49.850	0.097249067
2017	499.764	0.0855249091
2018	501.031	0.0790670093
2019	50.132	0.0834724541
2020	50.162	0.1045478306

Source: Authors' elaboration on AQUASTAT. (2023). FAO AQUASTAT dissemination system. https://data.apps.fao.org/aquastat/?lang = en&share = f-5f0a35cd-c72f-4650-945e-384af030dd64.

Despite this, Italy was not yet ready for wastewater reuse considering that less than 0.1% of agricultural areas are equipped for irrigation by direct use of treated municipal wastewater.

18.4 Water footprint

Considering that freshwater is needful for the functioning of human activities, especially for agriculture and food transformation, its quantification through statistical tools and assessment approaches is necessary (Siyal et al., 2023). Among the available tools, WF is able to evaluate water consumption in production chains by accounting for all the freshwater necessary to produce a good or service (Mekonnen & Hoekstra, 2010).

Particularly, as originally theorized, the WF concept was developed to create an indicator of water use relative to people's consumption. Actually, the first definition of WF focused on the volume of water necessary for a country to address the production needs of goods and services consumed by its inhabitants (Chapagain & Hoekstra, 2004).

Briefly, the WF indicator of Hoekstra et al. (2011) is expressed according to Eq. (18.1):

$$\text{Water Footprint} = \text{Blue Water} + \text{Green Water} + \text{Gray Water}, \quad (18.1)$$

and considering that the Water Footprint (of a process or a of growing crops/trees) is represented by the sum of the green, blue and gray components (WF = WFblue + WFgreen + WF gray) (Hoekstra et al., 2009), each component is represented by:

$$WF_{\text{blue}} = CWU_{\text{blue}}/Y, \quad (18.2)$$

$$WF_{\text{green}} = CWU_{\text{green}}/Y, \quad (18.3)$$

$$WF_{\text{gray}} = L/(C_{\text{max}} - C_{\text{nat}}). \quad (18.4)$$

Particularly, *CWU* is the crop water utilization, *Y* represents the crop yeld, *L* is the pollutant load, C_{max} is the maximum acceptable concentration (mass/volume), C_{nat} is the natural concentration of the receiving body (mass/volume), and CWU_{green} and CWU_{blue} (m^3/ha) were calculated from the accumulated corresponding actual crop evapotranspiration (Crovella et al., 2022; Hoekstra et al., 2011).

Particularly, blue water flow generally includes surface runoff, interflow, and subsurface runoff of the water. Conversely, green water, important for rain-fed agriculture, refers to ground water or water in the atmosphere (Falkenmark, 2006; Ke et al., 2023; Liu et al., 2016). Furthermore, gray water quantifies polluted freshwater (Wen et al., 2023).

In order to provide a snapshot of the WF related to the most representative crops at global level (Table 18.3), the scholars accessed the platform WF Assessment Tool.

Considering Table 18.3, for the agricultural sector at global level, fodder crops, maize, rice, and wheat present the largest green WF and rice, wheat, coconuts and cotton the highest values of blue WF. Conversely, the largest gray WF is related to coconuts, maize, and wheat production (Fig. 18.2).

As Fig. 18.3 showed the water stress at European level, it emerged a level of WF among significant and severe levels, on a medium range of risk, probably due to agricultural and industrial sector that characterize the continent.

TABLE 18.3 WF for the most representative crops and main sectors at global level (2022).

Crop or sector	Green WF (m^3/year)	Percentage of green WF	Blue WF (m^3/year)	Percentage of blue WF	Gray WF (m^3/year)	Percentage of gray WF	Total WF (m^3/year)	Percentage of total WF
Barley	170E + 09	3	11E + 09	1	18E + 09	1	200E + 09	3
Coconuts	140E + 09	2	84E + 06	<1	820E + 06	<1	140E + 09	2
Cotton	130 + E09	2	75E + E09	8	25E + 09	2	230E + 09	3
Domestic	0 + E00	<1	42E + 09	4	280E + 09	21	320 + E09	4
Fodder Crops	550E + 09	10	72E + 09	7	51E + 109	4	680 + E09	8
Industrial	0 + E00	<	38E + 09	4	370E + 09	27	410 + E09	5
Maize	590E + 09	10	51E + 09	5	120E + 09	9	760 + E09	9
Millets	120E + 09	2	1.6E + 09	<1	3.1E + 09	<1	130E + 09	2
Oil Palm Fruit	140E + 09	2	30E + 06	<1	5.2E + 09	<1	140E + 09	2
Rice	670E + 09	12	200E + 09	21	110E + 09	8	990 + E09	12
Sorghum	170E + 09	3	6.2E + 09	<1	5.2E + 09	<1	180E + 09	2
Soya Beans	350 + E09	6	12E + 09	1	6.4E + 09	<1	370 + E09	5
Sugar Cane	180 + E09	3	74E + 09	8	17E + 09	1	270E + 09	3
Wheat	730E + 09	13	200E + 09	21	120E + 09	9	1100E + 09	13

WF, water footprint.
Source: Authors' elaboration on Water Footprint Assessment Tool. (2023). Accounting global WF. https://www.waterfootprintassessmenttool.org/world/scope.

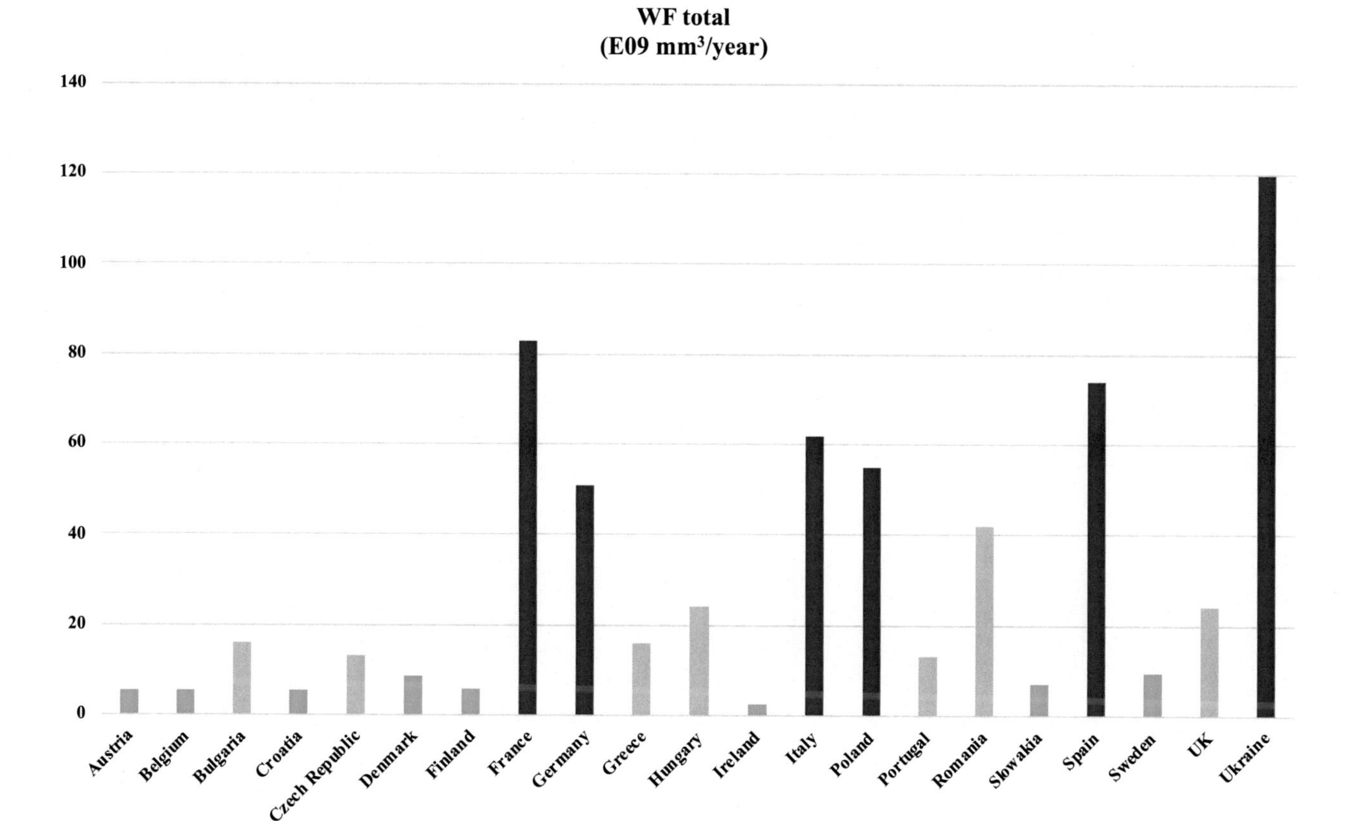

FIGURE 18.2 Infographic of total water footprint at European level. Green (moderate); orange (significant); red (very severe). *WF*, water footprint. *Source: Authors' elaboration on Water Footprint Assessment Tool. (2023). Accounting global WF (Water Footprint Assessment Tool, 2023). https://www.waterfootprintassessmenttool.org/world/scope.*

Therefore, it can be stressed that the WF of crop production has been widely accepted as a comprehensive indicator of agricultural water consumption. However, this indicator has been widely debated and extended, as underlined by Feng et al. (2021) in their review. Particularly, starting from the evolution of the WF concept from the last 20 years until today, launched by Hoekstra and Hung in 2002 (Hoekstra and Hung, 2002), WF assessments are able to support different stakeholders in the agricultural and agro-industrial production chain towards the achievment of the SDGs, and in particular the SDG 6 and their targets included. Furthermore, this indicator (the WF) supports all practitioners in achieving determining environmental sustainability objectives set by the planning, production, and consumption financing policies of the agricultural and agro-industrial sector (Berger et al., 2021).

18.5 Water conservation and reuse in a circularity context

Among the first studies on the linkage between water conservation and water reuse, Matsumura and Mierzwa (2008) focused on these concepts in food industry demonstrating that through rational water use strategies can achieve a reduction of almost 31% in water consumption. This percentage was used to estimate a hypothetical decrease at European level (Table 18.4), considering for a baseline scenario, the highest value from 2016 to 2020 for each country included in Table 18.1. Particularly, the baseline scenario was compared with the best scenario (the worst reduced by 31% as proposed by Matsumura & Mierzwa, 2008).

Specifically, Table 18.4 contains six columns: the first two columns are derived from Table 18.1, indicating data of freshwater abstraction in 2016 and 2020. The third column indicates the difference (positive or negative) verified from these years in the real context. Later, the scholars, authors of this chapter, proposed an estimation of the reduction. Particularly, they used as baseline scenario the highest value over the years 2016–2020 for each country. Subsequently, from this value, they reduced for the percentage (31%) calculated by Matsumura and Mierzwa (2008) as rate of reduction of water withdrawal through the water conservation and reuse practices. Thus, the last column indicated the value

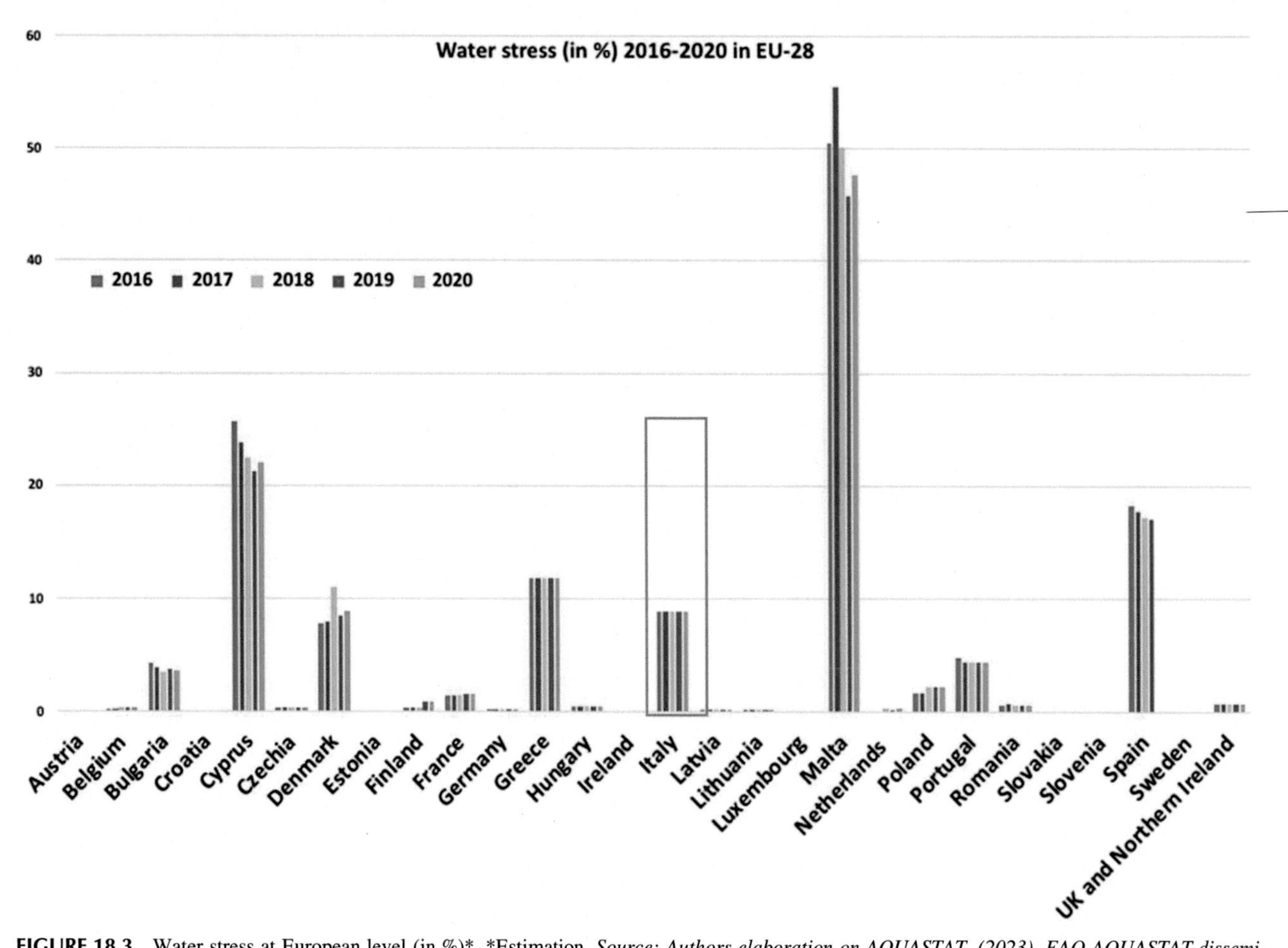

FIGURE 18.3 Water stress at European level (in %)*. *Estimation. *Source: Authors elaboration on AQUASTAT. (2023). FAO AQUASTAT dissemination system. https://data.apps.fao.org/aquastat/?lang = en&share = f-5f0a35cd-c72f-4650-945e-384af030dd64.*

achieved to be compared with the value included in the third column. It emerged that the most countries would achieve excellent water withdrawal reduction performance if they implemented sustainable water conservation and reuse practices. Amomgst these practices, for example, countries can implement wastewater reuse as recently proposed by Crovella et al. (Crovella et al., 2024; Crovella & Paiano, 2023) towards a reduction of water resource withdrawals and a more circular, resilient and sustainable agro-industry chain.

Conversely, Estonia presented a decrease of 0.11 million cubic meters using the water conservation and reuse practices.

Therefore, in the current context of extreme consumption of water resources, water conservation symbolizes an essential option to ensure sustainability of the terrestrial ecosystem, mitigating surface runoff and replenishing groundwater (Zhao et al., 2023).

Consequently, for recovering more water resources for civil activities, nonconventional resources can be used, such as desalinated sea water, brackish water, and urban and industrial wastewater treated with the different available and sustainable systems (Benedini & Rossi, 2021). Moreover, in order to ensure greater sustainability it is also possible to use Life Cycle Assessment (LCA) approaches in order to evaluate the impacts and hypothesize the reuse of wastewater (Crovella et al., 2024).

Nevertheless, for addressing the ongoing need for water in the agricultural and processing sectors, also to address the needs of the growing world population, it is important to treat wastewater for reuse in other applications (Mazumder et al., 2023). Particularly, for water treatment, various membrane-based separation processes are available for the treatment, recycling, reuse of industrial effluents, and the use of photolysis and photodegradation processes (Crovella & Paiano, 2023).

From a political point of view, at European level, *"The Green Deal"* proposed an ambitious agenda for transforming the continent into a prosperous civilization by operating a resource-efficient and competitive economy, with zero net

TABLE 18.4 Estimation of water consumption and water conservation and reuse in the baseline and best scenarios at European level.

Country	Real data derived from Table 18.1			Estimation reduction		
	2016	2020	Difference	Baseline	Best (−31%)	Difference
	In million cubic meters					
Austria	–	–	–	–	–	–
Belgium	4.66	8.27	+3.61	8.27	5.70	−2.57
Bulgaria	898.45	746.00	−152.45	898.45	619.93	−278.52
Croatia	30.00	30.00	0.00	30.00	20.7	−9.3
Cyprus	37.30	39.20	+1.90	39.20	27.05	−11.15
Czechia	33.40	23.30	−10.10	33.40	23.05	−10.35
Denmark	198.86	200.74	+1.88	230.15	158.80	−71.35
Estonia	0.16	0.05	−0.11	0.33	0.23	−0.10
Finland	–	–	–	–	–	–
France	1935.77	1946.41	−10.64	1946.41	1343.02	−603.39
Germany	71.19	–	–	122.79	84.73	−38.06
Greece	2801.64	2801.64	0.00	2801.64	1933.13	−868.51
Hungary	–	–	–	–	–	–
Ireland	–	35.33	–	35.33	24.38	−10.95
Italy	–	–	–	–	–	–
Latvia	56.10	53.85	−2.25	56.10	38.71	−17.40
Lithuania	55.78	56.13	1.93	56.13	38.73	−17.40
Luxembourg	0.01	0.02	+0.01	0.04	0.007	−0.003
Malta	1.25	1.25	0.00	1.25	0.86	−0.39
Netherlands	20.06	93.00	+72.94	93.00	64.17	−28.83
Poland	1039.94	801.62	−238.32	1039.94	717.56	−322.28
Portugal	–	–	–	–	–	–
Romania	1203.00	2095.00	+892	2095.00	1445.55	−639.45
Slovakia	9.40	14.40	+5	17.60	12.14	−5.46
Slovenia	2.70	2.40	−0.30	3.00	2.07	−0.93
Spain	15.722.00	14.710.00	−1012.00	15,722.00	10,848.18	−4,873.82
Sweden	45.40	63.00	+17.60	63.00	43.47	−19.53
United Kingdom of Great Britain and North Ireland	–	–	–	–	–	–

(–), data not available.
Source: Authors' elaboration.

greenhouse gas emissions (GHG) (European Commission, 2019). According to this strategy, the VIII Environment Action Program (European Commission, 2022) aims to *"accelerate"* the transition towards a climate-neutral, resource-efficient, clean, and circular economy in a fair and inclusive way.

Lastly, water reuse has become a fundamental process for the management of water resources, especially in countries that suffer from greater stress and where low costs favor reuse (EEA, 2022). Furthermore, for the regions where water resources are abundant or, in less stressed conditions, water reuse is also favored by conservation of underground water resources, cost reduction, and the precautionary principle, that is, the prevention of possible damage (EEA, 2019). Moreover, the increased scarcity of water resources for agro-industrial use can stimulate stakeholders and practitioners interest in water reuse practices for reducing impacts in the context of a sustainable, resilent and circular economy (Crovella et al., 2024).

Generally, Europe is seeing a higher percentage of the population connected to wastewater treatment. This share has generally increased in recent decades and was above 80% in half of the EU Member States. Currently, the share of the population connected to at least one secondary wastewater treatment plant has risen to 95% or more in six Member States (Denmark, Germany, Greece, the Netherlands, Austria, and Sweden), as well as in Switzerland and the United Kingdom. On the contrary, however, less than one in two households was connected to at least secondary urban wastewater treatment plants (Malta and Croatia; Eurostat, 2023).

18.6 The achievement of Sustainable Development Goal 6

As underlined in the earlier sections, an affective water management is fundamental to achieve sustainable development (economic, social, and environmental), poverty reduction, and supportable use of natural resources (Ansorge et al., 2020). As matter of fact, the UN thought the adoption of the 17 SDGs in 2015 and imposed the actions for ending poverty, protecting the planet and improving the lives and prospects of everyone. Furthermore, the UN invited countries to plan an access to clean and safe water resources (UN, 2023). In particular, water and sanitation represent at the core of sustainable development and their improvement is fundamental to reducing poverty, increasing economic growth, and stimulating environmental sustainability globally (Çankaya, 2023).

Although all SDGs are interconnected with the sustainable use of water (Weerasooriya et al., 2021), SDG 6 is directly linked to access to clean and safe water.

Particularly, SDG 6, titled "*Ensure availability and sustainable management of water and sanitation for all,*" focuses on the uneven distribution of water resources that causes unnecessary disease and death. This SDG includes eight targets: 6.1 (safe and affordable drinking water), 6.2 (end open defecation and provide access to sanitation and hygiene), 6.3 (improve water quality, wastewater treatment, and safe reuse), 6.4 (increase water-use efficiency and ensure freshwater supplies), 6.5 (implement integrated water resources management), 6.6 (protect and restore water-related ecosystems), 6.7 (expand water and sanitation support to developing countries), and 6.8 (support local engagement in water and sanitation management; Global Goals, 2023).

In this assessment, particularly the authors of this chapter considered the target 6.4 and, in general, as underlined by UN (2023) in the definition "*By 2030, substantially increase water-use efficiency across all sectors and ensure sustainable withdrawals and supply of freshwater to address water scarcity and substantially reduce the number of people suffering from water scarcity",* they considered the general purpose of the mentioned goal.

The main source of information considered by the authors for evaluating the SDG achievement related to water resource is the AQUASTAT (2023) dataset. Particularly, analyzing the results achieved for target 6.4, the water use efficiency has increased by 9% from 17.4 $/m^3 in 2015 to 18.9 $/m^3 globally in recent years. Furthermore, in 2020, approximately 57% of countries had a water use efficiency of $20/m^3 or less, reducing values by one percentage point compared to 2015. Furthermore, at global level, considering the water stress indicator, constantly average values of 18.2% were noted in 2020, albeit with very different regional variations and an increase of 1.2% from 2015 to 2020 (SDGS, 2023).

Among several indicators included in target 6.4, in the current application the scholars considered:

a. 6.4.1: Change in water-use efficiency over time, focusing on: *industrial water use efficiency*, *irrigated agriculture water use efficiency*, and *services water use efficiency;*
b. 6.4.2: Level of water stress: freshwater withdrawal as a proportion of available freshwater resources, converging on *Water Stress*.

Specifically, the indicator SDG 6.4.1 describes the value added (in US dollars) per volume of water used (in cubic meters), by a given economic activity over time. In particular, SDG 6.4.1 considers the use of water by all economic activities, with particular focus on agriculture, industry and the service sector (Tables 18.5–18.7).

Observing the phenomena of water use for the industrial sector at European level (considering EU 28 with the United Kingdom) for the period 2016–2020, it emerged that Belgium presented a constant industrial water efficiency

TABLE 18.5 Industrial water use efficiency at European level (in US$/m³).[a]

Country	2016	2017	2018	2019	2020
Austria	33.09	33.82	34.84	35.32	32.74
Belgium	23.8	23.32	23.66	24.32	23.08
Bulgaria	2.65	2.67	2.70	2.82	2.93
Croatia	52.76	48.37	12.61	12.96	13.22
Cyprus	458.00	125.24	145.73	162.02	153.5
Czechia	58.14	60.99	66.61	74.42	80.01
Denmark	1510.65	1731.2	1309.37	1271.94	1320.4
Estonia	2.87	2.97	3.44	5.84	6.53
Finland	33.6	36.16	26.96	27.00	26.48
France	20.66	20.51	20.72	22.34	19.13
Germany	48.54	49.28	49.43	48.60	45.66
Greece	60.19	56.87	60.95	61.85	61.49
Hungary	8.77	9.09	9.14	10.34	9.08
Ireland	248.02	240.13	246.32	248.2	278.15
Italy	45.56	46.55	47.22	47.21	42.99
Latvia	123.07	129.72	125.6	140.53	138.18
Lithuania	84.37	147.71	176.79	182.91	179.46
Luxembourg	5248.84	3856.00	3386.37	1616.83	1483.59 (I)
Malta	1182.65	1320.81	1447.03	1556.73	1538.52
Netherlands	20.35	21.24	22.88	21.58	21.77
Poland	16.88	18.09	20.85	22.02	22.68
Portugal	17.88	18.58	19.16	19.36	18.19
Romania	12.26	12.14	13.34	13.23	12.34
Slovakia	102.86	103.25	107.29	108.98	96.71
Slovenia	15.38	15.71	15.83	17.06	15.21
Spain	34.55	38.23	37.53	39.64	36.24
Sweden	74.25	78.87	81.21	84.03	78.78
United Kingdom of Great Britain and Northern Ireland	455.16	471.8	485.16	491.86	421.82

[a]*Estimation.*
Source: Author's elaboration from AQUASTAT. (2023). FAO AQUASTAT dissemination system. https://data.apps.fao.org/aquastat/?lang = en&share = f-5f0a35cd-c72f-4650-945e-384af030dd64.

use equal to 23$/m³, similar to Romania and Slovenia. Conversely, from 2016 to 2020, Italy collected a decrease of 5.60% and Luxembourg of 71.73%. Czechia and Estonia have increased, respectively, by 7 and 11%nd 127% (Table 18.5).

Instead, in the irrigated agricultural sector, Italy recorded a decrease of 3.5% from 2016 to 2020, the United Kingdom and Northern Ireland, after a pick in 2019, presented a constant value around 0.70$/m³ and Denmark, an increase of 36%. The lowest value is documented by Finland with 0.02$/m³ of water use for irrigation (Table 18.6).

Analyzing the efficiency related to the use of water resource at European scale services, it emerged that the values (Table 18.7) were higher than those for industrial water use efficiency (Table 18.5) and irrigated agriculture water use (Table 18.6). Particularly, over the years 2016–2020, Austria presented constant values around 352$/m³, similar to

TABLE 18.6 Irrigated agriculture water use efficiency at European level (in US$/m^3).[a]

Country	2016	2017	2018	2019	2020
Austria	2.64	2.95	2.83	2.75	2.60
Belgium	1.35	1.44	1.12	1.26	1.14
Bulgaria	0.10	0.12	0.11	0.1	0.11
Croatia	0.39	0.48	0.61	0.68	0.68
Cyprus	0.79	0.71	0.73	0.85	0.80
Czechia	1.47	1.51	1.33	1.48	1.53
Denmark	0.97	1.31	0.80	1.65	1.32
Estonia	0.55	0.65	0.60	0.69	0.57
Finland	0.09	0.05	0.05	0.02	0.02
France	1.36	1.72	1.98	1.87	1.93
Germany	3.91	4.94	4.17	4.73	4.65
Greece	0.39	0.43	0.4	0.42	0.42
Hungary	0.45	0.40	0.37	0.40	0.43
Ireland	–	–	–	–	–
Italy	0.86	0.98	1.05	0.97	0.83
Latvia	0.01	0.02	0.02	0.02	0.03
Lithuania	0.04	0.04	0.04	0.05	0.05
Luxembourg	–	–	–	–	–
Malta	2.02	1.41	1.77	1.37	1.27
Netherlands	21.55	26.7	18.42	19.73	19.88
Poland	0.25	0.23	0.10	0.10	0.11
Portugal	0.45	0.52	0.52	0.54	0.50
Romania	0.18	0.27	0.32	0.32	0.29
Slovakia	2.83	1.93	2.28	1.47	1.70
Slovenia	4.66	4.09	6.95	6.52	8.64
Spain	0.49	0.51	0.54	0.52	0.57
Sweden	3.20	3.50	3.54	3.81	3.62
United Kingdom of Great Britain and Northern Ireland	0.71	0.76	0.75	0.81	0.70

(–), data not available.
[a]*Estimation.*
Source: Author's elaboration from AQUASTAT. (2023). FAO AQUASTAT dissemination system. https://data.apps.fao.org/aquastat/?lang = en&share = f-5f0a35cd-c72f-4650-945e-384af030dd64.

Belgium, Hungary, Latvia, and the United Kingdom. Moreover, many countries noted a decrease from 2016 to 2020, particularly, Denmark: 0.47%, Finland: 18.24%, France: 3.67%, and Greece: 5.45%.

Instead, SDG 6.4.2, mentioned earlier, quantifies the amount of freshwater withdrawn by all economic activities, compared to the total renewable freshwater resources available, also considering environmental flow requirements. Fig. 18.3 shows that Malta presented the highest percentage of water stress between 45% and 55%, and Italy recorded constant values around 9%. Conversely, Austria, Estonia, Ireland, Luxembourg, Slovakia, Slovenia, and Sweden recorded the lowest values around 0.1%.

TABLE 18.7 Services water use efficiency at European level (in US$/m^3).[a]

Country	2016	2017	2018	2019	2020
Austria	352.54	359.95	369.65	375.9	352.89
Belgium	453.92	460.36	469.18	478.91	453.99
Bulgaria	36.49	37.11	41.16	42.85	42.37
Croatia	70.27	74.51	76.28	78.12	72.58
Cyprus	143.25	154.24	159.40	172.10	173.41
Czechia	185.47	193.07	192.67	201.41	198.81
Denmark	543.09	560.31	547.85	572.18	540.49
Estonia	246.11	268.59	269.29	223.43	274.75
Finland	379.22	387.74	392.50	320.21	310.03
France	346.39	338.54	349.12	356.14	333.66
Germany	212.21	217.92	221.2	224.35	219.07
Greece	84.42	85.81	86.61	88.26	79.82
Hungary	125.22	124.3	129.75	134.69	125.27
Ireland	265.04	226.13	206.39	220.61	223.21
Italy	136.48	139.57	142.10	143.12	131.63
Latvia	219.47	228.49	219.74	229.13	219.09
Lithuania	205.05	214.22	214.68	223.73	225.05
Luxembourg	1195.06	1190.02	1056.42	1214.92	1124.19
Malta	241.95	266.54	277.27	283.88	266.91
Netherlands	446.30	341.82	301.32	306.00	281.59
Poland	144.96	153.64	154.07	163.55	174.98
Portugal	157.41	157.31	167.20	172.34	158.09
Romania	104.00	115.24	117.47	125.29	123.56
Slovakia	195.23	195.87	200.95	208.76	206.94
Slovenia	163.93	163.59	168.65	175.39	169.91
Spain	179.23	191.79	205.27	208.44	184.49
Sweden	388.09	419.41	455.25	493.16	524.50
United Kingdom of Great Britain and Northern Ireland	356.64	363.30	368.77	375.54	347.40

[a]*Estimation.*
Source: Author's elaboration from AQUASTAT. (2023). FAO AQUASTAT dissemination system. https://data.apps.fao.org/aquastat/?lang = en&share = f-5f0a35cd-c72f-4650-945e-384af030dd64.

After this assessment, it has been stressed that the monitoring for the achievement of SDG 6 supports each country in the water management actions for a more sustainable consumption (UNwater, 2023). It has also been shown that there are many positive synergies between the SDGs and water (Garcia et al., 2023; Ho & Goethals, 2019). However, there are also some negative correlations that represent difficulties in achieving certain sustainable development objectives through improving water quality, as in the water–energy nexus (Alcamo, 2018; Garcia et al., 2023; Karnib, 2017). Finally, as underlined by Garcia et al. (2023) the use and the consultation of SDG 6 indicators allows companies, stakeholders, practitioners, scholars and water system managers for evaluating the achievement of sustainability, incorporating new challenges to improve sustainability, and designing new solutions to reduce water stress. Therefore, water management data are necessary for stronger accountability, increased commitment and investments, and more effective decision-making.

18.7 Main recent solutions for water stress mitigation and water conservation

Through this assessment, the scholars confirmed what is supposed by the literature. Particularly, the considerable reduction in stored water reserves due to irrigation practices and the decreased rainfall in many major global aquifers are putting future food production at risk. For this reason, new water conservation measures must be used to reduce withdrawal and extend the lifespan of aquifers (Whittemore et al., 2023). These good practices can stimulate the achievement of the targets included in SDG number 6 and, in general, the goals included in Agenda 2030.

Hence, the implementation of the water conservation approach requires an extensive understanding of trends in climate change and rainfall on temporal (annual, monthly, and daily), spatial, and ecosystem scales, also using joint approaches such as WF, GIS, LCA and others, as indicated in the first chapter of this volume.Therefore, to date, it is very complex to proceed with quantitative evaluations due to the unavailability of data at country level that is not updated and systematized, sometimes not collected in digital form, as also underlined by Crovella et al. (2021). Nevertheless, several Member States introduced some pricing mechanisms that incentivize efficient water use on agriculture and for all water users (European Courts of Auditor, 2021). Some Italian regions have planned a system of variable water prices according to the efficiency of the irrigation system (e.g. Emilia-Romagna), others European countries applied a water resource tax based on the measured volume of use (e.g. Germany [Berlin-Brandenburg], Hungary, and Portugal). Moreover, other countries as Belgium (Flanders) used progressive pricing for certain types of groundwater (the greater the volume abstracted, the higher the price).

TABLE 18.8 Classification of the main solutions for water scarcity mitigation and water conservation.

Intervention area	Solutions	Outcomes	References
Natural solutions	Nature-based solutions: constructed wetland, adequate maintenance of wetlands, vegetation management, and cleaning plans to remove duckweed	Reliable effluent concentrations and low concentrations of COD, TSS, and other microbial indicators	Gonzalez-Flo et al. (2023)
Economic/ finance mechanism	Pricing mechanisms	System of variable water prices according to the efficiency of the irrigation system; reduction of water consumption per hectare increase in volumetric prices	European Courts of Auditor (2021); Pronti and Berbel (2022)
Economic/ Finance mechanism	Withdrawal of water incorporated into the trading of raw materials	Reuse of virtual water flows linked to the circulation of goods	Wei et al. (2023)
Political/ Programming initiatives	State governments can direct funds towards water-stressed areas, suggesting selective use of public funds towards regions with the highest returns	In the presence of water stress, politicians could include sustainability strategies in political programs to increase funds towards water scarcity	Mahadevan and Shenoy (2023)
Industrial technologies	Membranes based on green polymers for various water treatment and reuse processes	Versatility of synthesis methods and treatment processes are advantageous in the separation of different chemical species (dyes, heavy metals, or pharmaceutical and personal care products)	Voicu and Thakur (2023)
	Phodegradation and photocatalysis of wastewater	Water treated reserve	Crovella and Paiano (2023)
Approaches	Introduction water footprint and virtual water methods to study sustainable water resources management	Reduction of indirect impact of local water stress on the global economic output that exceeds the direct impact	Chen et al. (2022); Crovella et al. (2022)
Research	Encourage research about alternative sources of water	Water treatment in loco, production systems with closed irrigation systems	Alberti et al. (2022)

Source: Authors' elaboration based upon on literature review.

Conversely, other European Member States have introduced a price differentiation rule to discourage/encourage the use of water from various sources. Among these practices it can mention those used in the area where prices are higher and water is scarcer (Belgium, France, Hungary, and Portugal), in the area where groundwater is more expensive than surface water (Bulgaria, Germany, and France), or where freshwater is more expensive than recycled water (Cyprus).

Other solutions implementable are included in natural intervention, building wetland, or proposing a selective use of public funds towards regions with the highest returns.

Generally, there are several solutions/practices/programs for reducing water stress, mitigating water scarcity, and increasing water conservation (Table 18.8). All solutions can be applied by countries or by agro-industries, and more in general by industrial chain, for operating in a resilience context of water consumption.

18.8 Conclusion and implications in resilience context

This chapter underlined that an operational framework, also based on the digitalisation of agriculture and on the data systematization, can push public administrations to build financing programs and sustainable strategies to implement the best circular economy (CE) model. In fact, only through a complete digital transformation of agriculture, circular strategies will be implemented for agricultural companies and manufacturing sector (Crovella et al., 2021).

In order to assess water resources, the availability of data, collected over a long sequence of years, and by several stakeholders involved in this chain as the farms, agro-industries, by the responsible structures belonging to the central government and regional administrations, is needed. Unfortunately, this data is often not systematized and updated, and sometimes not digitized and not comparable due to the use of different indicators. Especially in Italy, this discrepancy negatively influences the possibility of implementing sustainable, reuse, reduction, and conversion strategies of water resources in the short term (Benedini & Rossi, 2021).

Especially, the recycling of treated wastewater can be a promising solution to reduce the demand for freshwater (Crovella & Paiano, 2023).

In conclusion, countries and industries must consider the general purpose that includes an optimal agro-industrial water conservation and use several tools for managing and supervising the implementation of strategies and solutions for address this goal. Among the assessment approaches available, life cycle assessment, WF, GIS, material flow analysis can be considered (Fig. 18.4).

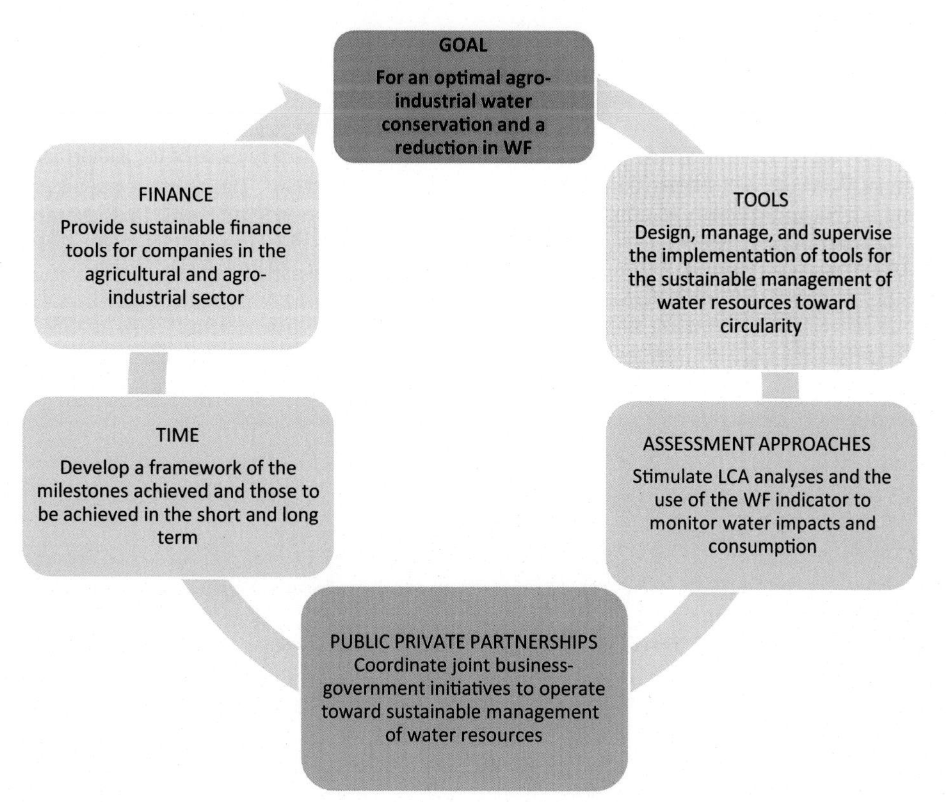

FIGURE 18.4 Roadmap for water resource sustainability. *LCA*, life cycle assessment; *WF*, water footprint. *Source: Authors' elaboration.*

Furthermore, this study confirmed what was highlighted by Ansorge et al. (2020) and (Crovella et al., 2024) that the optimal solution for greater sustainability and circularity in the use of water resources lies in the launch and use of technologies that allow the use of derivatives of the treatment of water resources as fertilizers in agriculture and other industries, or in the implementation of principles of the circular economy.

Finally, it was recognized that the analysis model suggested by the authors of this chapter can support the decision-making process of politicians and professionals involved in water resources management. It is applicable to water context managed by a public–private partnership and in other specific contexts, which are, for example, the use of energy resources, the use of the territory, and urban transformations. This replicable methodology can be used even in the case of joint methodologies (WF and GIS), because it limits the generalizability of the results, use updated data, is very punctual, and directs the results towards a more sustainable practices.

Author contributions

Tiziana Crovella: Conceptualization, methodology, software; validation, formal analysis, investigation, resources, data curation, writing—original draft preparation, visualization, supervision.

Giovanni Lagioia: Writing—review and editing, supervision.

Annarita Paiano: Methodology, validation, supervision, project administration, and funding acquisition.

All authors have read and agreed to the published version of the manuscript.

Funding information

"This study was funded by the European Union - NextGenerationEU, in the framework of the GRINS -Growing Resilient, INclusive and Sustainable project (GRINS PE00000018 – CUP H93C22000650001). The views and opinions expressed are solely those of the authors and do not necessarily reflect those of the European Union, nor can the European Union be held responsible for them".

Data availability statement

Not applicable.

Acknowledgments

"This study was funded by the European Union - NextGenerationEU, in the framework of the GRINS -Growing Resilient, INclusive and Sustainable project (GRINS PE00000018 – CUP H93C22000650001). The views and opinions expressed are solely those of the authors and do not necessarily reflect those of the European Union, nor can the European Union be held responsible for them." This research was undertaken thanks to a PNRR M4C2 - Investiment 1.3 Enlarged Partnership GRINS - Growing Resilient, INclusive and Sustainable by University of Bari Aldo Moro, Department of Economics, Management and Business Law and to address the objectives of Spoke 1. CUP: H93C22000650001, Code PNRR_PE_71, Identified Code PE0000018, Thematic: Building of datasets for the circular economy of the main Italian production systems.

Conflicts of interest

The authors declare no conflict of interest.

References

Alberti, M. A., Blanco, I., Vox, G., Scarascia-Mugnozza, G., Schettini, E., & da Svilva, L. P. (2022). The challenge of urban food production and sustainable water use: current situation and future perspectives of the urban agriculture in Brazil and Italy. *Sustainable Cities and Society*, *83*, 103961. Available from https://doi.org/10.1016/j.scs.2022.103961.

Alcamo, J. (2018). Water quality and its interlinkages with the Sustainable Development Goals. Available from https://doi.org/10.1016/j.cosust.2018.11.005.

Anderegg, W. R. L., Trugman, A. T., Badgley, G., Konings, A. G., & Shaw, J. (2020). Divergent forest sensitivity to repeated extreme droughts. *Nature Climate Change*, *10*(12), 1091. Available from https://doi.org/10.1038/s41558-020-00919-1, –U1019.

Ansorge, L., Stejskalová, L., & Dlabal, J. (2020). Effect of WWTP size on grey water footprint—Czech Republic case study. *Environmental Research Letters*, *15*(10), 104020. Available from https://doi.org/10.1088/1748-9326/aba6ae.

AQUASTAT. (2023). FAO AQUASTAT dissemination system. https://data.apps.fao.org/aquastat/?lang = en&share = f-5f0a35cd-c72f-4650-945e-384af030dd64.

Benedini, M., & Rossi, G. (2021). Water resources of Italy, chapter 1, In *Water law, policy and economics in Italy* (pp. 1–3), Springer.

Berger, M., Campos, J., Carolli, M., Dantas, I., Forin, S., Kosatica, E., Kramer, A., Mikosch, N., Nouri, H., Schlattmann, A., Schmidt, F., Schomberg, A., & Semmling, E. (2021). Advancing the water footprint into an instrument to support achieving the SDGs—recommendations from the "Water as a Global Resources" Research Initiative (GRoW). *Water Resources Management*, *35*, 1291–1298. Available from https://doi.org/10.1007/s11269-021-02784-9.

Çankaya, S. (2023). Evaluation of the impact of water reclamation on blue and grey water footprint in a municipal wastewater treatment plant. *Science of the Total Environment*, *903*, 166196. Available from https://doi.org/10.1016/j.scitotenv.2023.166196.

Cartwright, J. M., Littlefield, C. E., Michalak, J. L., Lawler, J. J., & Dobrowski, S. Z. (2020). Topographic, soil, and climate drivers of drought sensitivity in forests and shrublands of the Pacific Northwest, USA. *Scientific Reports*, *10*(1). Available from https://doi.org/10.1038/s41598-020-75273-5.

Chapagain, A.K., & Hoekstra, A.Y. (2004). Water footprints of nations—value of water, research report series no. 16. Volume 1: main report, edited by UNESCO-IHE. Available from https://www.waterfootprint.org/resources/Report16Vol1.pdf.

Chen, X. I., Zhao, B. U., Shuai, C., Qu, S., & Xu, M. (2022). Global spread of water scarcity risk through trade. *Resources, Conservation and Recycling*, *187*, 106643. Available from https://doi.org/10.1016/j.resconrec.2022.106643.

Crovella, T., & Paiano, A. (2023). Assessing the sustainability of photodegradation and photocatalysis for wastewater reuse in an agricultural resilience context. *Water*, *5*(15), 2758. Available from https://doi.org/10.3390/w15152758.

Crovella, T., Paiano, A., Falciglia, P. P., Lagioia, G., & Ingrao, C. (2024). Wastewater recovery for sustainable agricultural systems in the circular economy – A systematic literature review of Life Cycle Assessments. *Science of The Total Environment*, *912*, 169310. Available from https://doi.org/10.1016/j.scitotenv.2023.169310.

Crovella, T., Paiano, A., & Lagioia, G. (2022). A meso-level water use assessment in the Mediterranean agriculture. Multiple applications of water footprint for some traditional crops. *Journal of Cleaner Production*, *330*, 129886. Available from https://doi.org/10.1016/j.jclepro.2021.129886.

Crovella, T., Paiano, A., Lagioia, G., Cilardi, A. M., & Trotta, L. (2021). Modelling digital circular economy framework in the agricultural sector. an application in Southern Italy. *Eng. Proc.*, *9*(1), 15. Available from https://doi.org/10.3390/engproc2021009015.

EEA. (2019). European Environment Agency—urban waste water treatment for 21st century challenges. EEA Briefing, European Environment Agency. https://www.eea.europa.eu/publications/urban-waste-water-treatment-for.

EEA. (2020). European Environment Agency, Water and Agriculture: towards sustainable solutions, EEA Report No 17/2020, https://www.eea.europa.eu/publications/water-and-agriculture-towards-sustainable-solutions.

EEA (2022). European Environment Agency—EEA report "beyond water quality - sewage treatment in a circular economy," 5, Luxembourg: Publications Office of the European Union, ISBN 978-92-9480-478-5, ISSN 1977–8449. Available from https://doi.org/10.2800/897113, https://www.eea.europa.eu/publications/beyond-water-quality-sewage-treatment.

EEA. (2023). European Environment Agency, Water. Available from https://www.eea.europa.eu/en/topics/in-depth/water.

European Commission. (2019). Communication from the Commission to the European Parliament, the European Council, the Council, the European Economic and Social Committee and the Committee of the Regions. The European Green Deal (COM(2019) 640 final of 11 December 2019). Available from https://ec.europa.eu/info/sites/default/files/european-green-deal-communication_en.pdf.

European Commission. (2021). Communication from the Commission to the European Parliament, the Council, the European Economic and Social Committee and the Committee of the Regions: forging a climate-resilient Europe—the new EU Strategy on Adaptation to Climate Change (COM (2021) 82 final), Available from https://eur-lex.europa.eu/legal-content/EN/TXT/?uri = COM%3A2021%3A82%3AFIN.

European Commission. (2022). Decision (EU) 2022/591 of the European Parliament and of the Council of 6 April 2022 on a General Union Environment Action Programme to 2030. Available from https://eur-lex.europa.eu/legal-content/EN/TXT/?uri = CELEX:32022D0591.

European Courts of Auditor. (2021). Special report. Sustainable water use in agriculture: CAP funds more likely to promote greater rather than more efficient water use. Available from https://www.eca.europa.eu/Lists/ECADocuments/SR21_20/SR_CAP-and-water_EN.pdf.

European Parliament. (2023). Extreme drought in Italy, jeopardising the autonomy and competitiveness of the entire agri-food sector. The Commission should take extraordinary measures—question for written answer E-001264/2023 to the Commission. Available from https://www.europarl.europa.eu/doceo/document/E-9–2023-001264_EN.html.

Eurostat. (2023). Water statistics—annual freshwater abstraction by source and sector. Available from https://ec.europa.eu/eurostat/databrowser/view/ENV_WAT_ABS/default/table?lang = en.

Falkenmark, M. (2006). The new blue and green water paradigm: breaking new ground for water resources planning and management. *Journal of Water Resources Planning and Management*, *132*(3), 129–132. Available from https://doi.org/10.1061/(ASCE)0733–9496(2006)132:3(129).

Feng, B., Zhuo, L., Xie, D., Mao, Y., Gao, J., Xie, P., & Wu, P. (2021). A quantitative review of water footprint accounting and simulation for crop production based on publications during 2002–2018. *Ecological Indicators*, *120*, 106962. Available from https://doi.org/10.1016/j.ecolind.2020.106962.

Garcia, C., López-Jiminenéz, P. A., Sánchez-Romero, F. J., & Pérez Sánchez, M. (2023). Assessing water urban systems to the compliance of SDGs through sustainability indicators. Implementation in the valencian community. *Sustainable Cities and Society*, *96*, 104704. Available from https://doi.org/10.1016/j.scs.2023.104704.

Global Goals. (2023). Clean water and sanitation. Available from https://www.globalgoals.org/goals/6-clean-water-and-sanitation/.

Gonzalez-Flo, E., Xavier, R., & Garcìa, J. (2023). Nature based-solutions for water reuse: 20 years of performance evaluation of a full-scale constructed wetland system. *Ecological Engineering, 188*, 106876. Available from https://doi.org/10.1016/j.ecoleng.2022.106876.

He, G., Geng, C., Zhai, J., Zhao, Y., Wang, Q., Jiang, S., et al. (2021). Impact of food consumption patterns change on agricultural water requirements: an urban-rural comparison in China. *Agricultural Water Management, 243*, 106504. Available from https://doi.org/10.1016/j.agwat.2020.106504.

He, J., Yang, Y., Liao, Z., Xu, A., & Fang, K. (2022). Linking SDG 7 to assess the renewable energy footprint of nations by 2030. *Applied Energy, 317*, 119167. Available from https://doi.org/10.1016/j.apenergy.2022.119167.

Ho, L.T., Goethals, P.L.M. (2019). Opportunities and challenges for the sustainability of lakes and reservoirs in relation to the sustainable development goals (SDGs). Available from https://doi.org/10.3390/w11071462.

Hoekstra, A.Y. and Hung, P.Q. (2002). Virtual water trade: a quantification of virtual water flows between nations in relation to international crop trade. Value of Water Research Report Series No. 11, UNESCO-IHE Institute for Water Education, Delft, the Netherlands. Available from http://www.waterfootprint.org/Reports/Report11.pdf.

Hoekstra, A.Y., Chapagain, A.K., Aldaya, M.M., & Mekonnen, M.M. (2011). The water footprint assesstment manual. Setting the global standard. Earthscan. Available from https://waterfootprint.org/media/downloads/TheWaterFootprintAssessmentManual_2.pdf.

ISTAT. (2022). ISTAT water statistics, years 2019–2021. Available from https://www.istat.it/it/files//2022/04/Report_ISTAT-WATER-STATISTICS.pdf.

Hoekstra, A. Y., Chapagain, A. K., Aldaya, M. M., & Mekonnen, M. M. (2009). *Water Footprint Manual - State of the Art 2009*. Enschede, The Netherlands: Water Footprint Network.

Hoekstra, A. Y., & Hung, P. Q. (2002). Virtual water trade: A quantification of virtual water flows between nations in relation to international crop trade. *Value of Water Research Report, 11*.

Karnib, A. (2017). Mapping the direct and indirect interlinkages across the sustainable development goals: a qualitative nexus approach water as cross-cutting factor in the SDGs view project quantitative water–energy–food nexus: the Q-nexus model view project mapping the direct and indirect interlinkages across the sustainable development goals: a qualitative nexus approach. *International Journal of Development and Sustainability, 6* (9), 1150–1158.

Ke, Z., Xiaoqi, L., Chuanfu, Z., Yiwen, L., Xintomg, Q., & Miaolin, D. (2023). Transformation characteristics and mechanism of blue and green water flows at watershed and typical ecosystem scale in China. *Ecohydrology & Hydrobiology*. Available from https://doi.org/10.1016/j.ecohyd.2023.09.002.

Lami, I. M., & Todella, E. (2023). A multi-methodological combination of the strategic choice approach and the analytic network process: from facts to values and vice versa. *European Journal of Operational Research, 307*(2), 802–812. Available from https://doi.org/10.1016/j.ejor.2022.10.029.

Liu, C.M., Li, Y.Z., Liu, X.M. et al. (2016). Impact of vegetation change on water transformation in the Middle Yellow River, *38*(10), 7–12. Available from https://doi.org/10.3969/j.issn.1000-1379.2016.10.002.

Lovelace, J. K., Nielsen, M. G., Read, A. L., Murphy, C. J., & Maupin, M. A. (2020). Estimated groundwater withdrawals from principal aquifers in the United States, 2015 (ver. 1.2, October 2020). *U.S. Geological Survey Circular Archives, 1464*, 70. Available from http://doi.org/10.3133/cir1464.

Mahadevan, M., & Shenoy, A. (2023). The political consequences of resource scarcity: targeted spending in a water-stressed democracy. *Journal of Public Economics, 220*, 104842. Available from https://doi.org/10.1016/j.jpubeco.2023.104842.

Matsumura, E. M., & Mierzwa, J. C. (2008). Water conservation and reuse in poultry processing plant—a case study. *Resources, Conservation and Recycling, 52*(6), 835–842. Available from https://doi.org/10.1016/j.resconrec.2007.10.002.

Mazumder, A., Sarkar, S., Sen, D., & Bhattacharjee, C. (2023). Membranes for industrial wastewater recovery and reuse, chapter 1. *Resource Recovery in Industrial Waste Waters*, 1–21. Available from https://doi.org/10.1016/B978-0-323-95327-6.00029-4.

Mekonnen, M.M., & Hoekstra, A.Y. (2010). The green, blue and grey water footprint of crops and derived crop products. Value of water research report series no. 47, UNESCO-IHE, Delft, the Netherlands. Available from http://www.waterfootprint.org/Reports/Report47-WaterFootprintCrops-Vol1.pdf.

Ou, G., Munoz-Arriola, F., Uden, D. R., Martin, D., Allen, C. R., & Shank, N. (2018). Climate change implications for irrigation and groundwater in the Republican River Basin, USA. *Climatic Change, 151*, 303–316. Available from https://doi.org/10.1007/s10584-018-2278-z.

Piao, S., Zhang, X., Chen, A., Liu, Q., Lian, X., Wang, X., et al. (2019). The impacts of climate extremes on the terrestrial carbon cycle: a review. *Science China Earth Sciences, 62*(10), 1551–1563. Available from https://doi.org/10.1007/s11430-018-9363-5.

Pronti, A., & Berbel, J. (2022). The impact of volumetric water tariffs in irrigated agriculture in Northern Italy. *Environmental Impact Assessment Review, 98*, 106922. Available from https://doi.org/10.1016/j.eiar.2022.106922.

SDGS. (2023). Progress and info of SDG 6. Available from https://sdgs.un.org/goals/goal6.

Sela, R. (2023). Global water usage and resources, Cropaia.

Siyal, A. W., Geerbens-Leenes, P. W., & Vaca-Jiménez, S. D. (2023). Freshwater competition among agricultural, industrial, and municipal sectors in a water-scarce country. Lessons of Pakistan's fifty-year development of freshwater consumption for other water-scarce countries. *Water Resources and Industry, 29*, 100206. Available from https://doi.org/10.1016/j.wri.2023.100206.

UN. (2015). United Nations—Goal 6 ensure availability and sustainable management of water and sanitation for all. Available from https://sdgs.un.org/goals/goal6.

UN. (2023). Sustainable development goals. Available from https://www.un.org/sustainabledevelopment/development-agenda.

UNwater. (2023). Integrated monitoring initiative for SDG 6. Available from https://www.unwater.org/our-work/integrated-monitoring-initiative-sdg-6.

Voicu, S. I., & Thakur, V. K. (2023). Green polymers-based membranes for water reuse in a circular economy context. *Current Opinion in Green and Sustainable Chemistry, 43*, 100852. Available from https://doi.org/10.1016/j.cogsc.2023.100852.

Wang, Y., Fu, B., Liu, Y., Li, Y., Feng, X., & Wang, S. (2021). Response of vegetation to drought in the Tibetan Plateau: elevation differentiation and the dominant factors. *Agricultural and Forest Meteorology, 306*. Available from https://doi.org/10.1016/j.agrformet.2021.108468.

Water Footprint Assessment Tool. (2023). Accounting global WF. https://www.waterfootprintassessmenttool.org/world/scope.

Wang, C., Wang, X., Jin, Z., et al. (2022). Occurrence of crop pests and diseases has largely increased in China since 1970. *Nature Food, 3*, 57–65. Available from https://doi.org/10.1038/s43016-021-00428-0.

Weerasooriya, R. R., Liyanage, L. P. K., Rathnappriya, R. H. K., Bandara, W. B. M. A. C., Perera, T. A. N. T., Gunarathna, M. H. J. P., & Jayasinghe, G. Y. (2021). Industrial water conservation by water footprint and sustainable development goals: a review. *Environment, Development and Sustainability*, 1–49. Available from https://doi.org/10.1007/s10668-020-01184-0.

Wei, J., Lei, Y., Liu, L., & Yao, H. (2023). Water scarcity risk through trade of the Yellow River Basin in China. *Ecological Indicators, 154*, 110893. Available from https://doi.org/10.1016/j.ecolind.2023.110893.

Wen, L., Lei, M., Zhang, B., Kong, X., Liao, Y., & Chen, W. (2023). Significant increase in gray water footprint enhanced the degradation risk of cropland system in China since 1990. *Journal of Cleaner Production, 423*, 138715. Available from https://doi.org/10.1016/j.jclepro.2023.138715.

Whittemore, D. O., Butler, J., Jr, Bohling, G. C., & Wilson, B. B. (2023). Are we saving water? Simple methods for assessing the effectiveness of groundwater conservation measures. *Agricultural Water Management, 287*, 108408. Available from https://doi.org/10.1016/j.agwat.2023.108408.

World Bank. (2023). Renewable internal freshwater resources per capita. https://databank.worldbank.org/metadataglossary/millennium-development-goals/series/ER.H2O.INTR.PC.

Xi, X., & Yuan, X. (2022). Significant water stress on gross primary productivity during flash droughts with hot conditions. *Agricultural and Forest Meteorology, 324*, 109100. Available from https://doi.org/10.1016/j.agrformet.2022.109100.

Zhao, G., Thian, S., Jing, Y., Cao, Y., Lliang, S., Han, B., Cheng, X., & Liu, B. (2023). Establishing a quantitative assessment methodology framework of water conservation based on the water balance method under spatiotemporal and different discontinuous ecosystem scales. *Journal of Environmental Management, 346*, 119006. Available from https://doi.org/10.1016/j.jenvman.2023.119006.

Chapter 19

Empowering sustainable water management: the confluence of artificial intelligence and Internet of Things

Achintya Das[1,*] and Ananya Roy Chowdhury[2]

[1]*Department of Physics, Mahadevananda Mahavidyalaya, Barrackpore, West Bengal, India,* [2]*Department of Botany, Chakdaha College, Nadia, West Bengal, India*

**Corresponding author. e-mail address: achintya.bappa@gmail.com.*

19.1 Navigating the complex landscape of water management: an introduction

Nowadays, sustainability requires a circular economy (Das & Chowdhury, 2023), effective water management (Nova, 2023), and more. Artificial intelligence (AI) and the Internet of Things (IoT) contributions to sustainable water management will be discussed in this chapter. This important section explores the intricate web of water management in depth, revealing the challenges and complexities that characterise this vital domain. In light of the current juncture between technological advancement and environmental sustainability, it is critical to develop an in-depth understanding of the intricate facets of water resources. This expedition shall encompass the worldwide water crisis, the paradigm-shifting capabilities of digital technologies, and the auspicious prospects presented by the IoT and AI. Commence an exhaustive exploration of the complexities inherent in water management, as we explore the pivotal obstacles, innovative solutions, and the strategic trajectory necessary for ensuring a water future that is both secure and sustainable.

19.1.1 Global water challenges in a nutshell

19.1.1.1 Introduction to the global water crisis

Life-sustaining water is currently facing an imminent threat. The present condition of water resources on a global scale is ominous. As the demand for water rises in response to factors such as industrialization, urbanization, and population expansion, the disparity in water availability becomes more pronounced (Mauter et al., 2018).

19.1.1.2 Impact of climate change

Climate change is a tangible phenomenon that is having an impact on our water resources. Water scarcity and quality issues are being made more severe in numerous regions of the globe due to shifting weather patterns, glacier melting, and rising sea levels (Troy et al., 2011).

19.1.1.3 Socioeconomic implications

Water scarcity has repercussions that extend beyond the environment. Water-scarce regions encounter agricultural obstacles that ultimately result in food insecurity. Consequently, this may give rise to migration, conflicts, and economic recessions (Moe & Rheingans, 2006). Assessing the economic ramifications of water scarcity in a world, the article "Evaluating the economic impact of water scarcity in a changing world" examines these issues in light of climate change, water resources, and adaptive capacities. The methodology employs a holistic strategy by integrating a worldwide hydrologic model, a global human-Earth system model, and an economic surplus loss metric (Dolan et al., 2021). Constraints of global trade and regional reactions to scarcity affect the economic effects on main hydrologic basins,

Water Footprints and Sustainable Development. DOI: https://doi.org/10.1016/B978-0-443-23631-0.00019-4

according to the findings. Economic uncertainty can occasionally be heightened to a greater extent than hydrologic uncertainty due to market adaptations. This study highlights the criticality of considering various determinants of uncertainty when designing complex adaptive systems. An example that illustrates this point is the Lower Colorado River Basin, which exhibits significant economic impact uncertainty, particularly when considering factors such as limited groundwater resources, inadequate agricultural output, and specific socioeconomic paths (Dolan et al., 2021).

19.1.2 The digital transformation in environmental solutions

19.1.2.1 *The Fourth Industrial Revolution*

Currently, the world is immersed in the Fourth Industrial Revolution, during which digital technologies are assuming a leading role in tackling worldwide issues. In the realm of water management, technology assumes a critical role in the pursuit of resolutions (Schwab, 2017). Traditional water management approaches, while once efficacious, are currently confronted with constraints when compared to contemporary challenges. Innovative digital solutions are more urgently required than ever before. Already, contemporary technology has demonstrated its capacity to aid in environmental preservation. For example, the application of nanotechnology in tackling worldwide water challenges has demonstrated encouraging outcomes (Mauter et al., 2018).

19.1.2.2 *Global water challenges and digital technologies: case studies*

The incorporation of digital technologies has surfaced as a crucial resolution in tackling worldwide water predicaments, augmenting the management of water resources in diverse regions. Digital tools and solutions for water resource management as an element of a green economy in rural areas are the subject of the article "Digital Tools for Water Resource Management as a Part of a Green Economy in Rural Areas" (Józefowicz & Michniewicz-Ankiersztajn, 2023). The paper's objectives are to optimize water consumption and enhance communication with local communities. The research additionally employs k-means clustering to ascertain the extent of technological integration and digitalization in water resource administration within the Bydgoszcz District. Three clusters are identified, each exhibiting a distinct degree of technological integration and digitalization (Józefowicz & Michniewicz-Ankiersztajn, 2023).

19.1.3 The promise of artificial intelligence and Internet of Things in addressing water issues

19.1.3.1 *Interconnected solutions*

IoT and AI are not merely technical jargon. Regarding water management, they possess tremendous potential. These technologies can deliver concrete resolutions to practical challenges. The integration of AI and IoT provides comprehensive solutions. By facilitating data collection and analysis, enabling real-time interventions, and enabling long-term planning, these technologies have the potential to significantly transform water management (Savio et al., 2021).

19.1.3.2 *A glimpse of success*

The incorporation of AI and IoT into water management transcends mere speculation. Exemplary instances, such as the implementation of IoT in municipal water management systems, serve to illustrate the profound capacity of these technologies to effect change (Khan & See, 2016).

19.1.4 Insights and reflections

We confront urgent global water challenges. Nonetheless, optimism exists with the integration of AI and IoT. These technologies have the potential to significantly impact water management. To achieve their utmost capabilities, however, innovation, cooperation, and a concerted endeavor are necessary to establish a sustainable water future.

19.2 Technological deep dive

This section explores the complexities of AI and the IoT, two technological advancements that are bringing about significant transformations in various sectors worldwide. The incorporation of these technologies into water management represents not merely a passing trend but a paradigm shift that improves our capacity to monitor, assess, and maximize water resources.

19.2.1 Artificial intelligence: beyond the hype—what it really means and its relevance to water management

Fundamentally, AI pertains to the replication of human cognitive processes through the utilization of computer systems. These processes consist of self-correction, reasoning, and learning. AI empowers machines to reason and make decisions comparable to those of humans, frequently with greater speed and precision. The AI toolbox comprises an assortment of methodologies and tools, including but not limited to natural language processing, machine learning (ML), neural networks, and deep learning, each possessing distinct merits and applications (Russell & Norvig, 2016).

The potential of AI in water management is extensive and can bring about significant changes. By utilizing algorithms to detect contaminants in real time, predictive analytics, which is a component of AI, can forecast water demand and utilization patterns (Guo et al., 2019). This ensures a safe and clean water supply. The incorporation of AI into water management is transforming our strategy for preserving and optimizing water resources, resulting in more sustainable and effective processes.

19.2.2 Internet of Things: understanding its components and potential

The term "Internet of Things" encompasses a network of tangible objects, including vehicles, appliances, and other items, that are equipped with connectivity, sensors, and software. This integration allows these objects to mutually communicate, gather, and exchange data. Sensors for data collection, connectivity for communication, data processing for analysis, and a user interface for interaction and control are the fundamental elements of an IoT system (Das & Roy Chowdhury, 2022).

The utilization of IoT devices, specifically sensors, is critical in water management. They furnish up-to-date information on a multitude of parameters, including utilization patterns, flow rates, and water quality. When integrated with data analysis, whether performed manually or by artificial intelligence, this data offers priceless insights that motivate effective and environmentally responsible water management strategies (Adeleke et al., 2023; Ray, 2016).

19.2.3 The power duo: how artificial intelligence and Internet of Things complement each other

The combination of AI and IoT generates a synergistic relationship in which AI contributes its computational capabilities to data analysis and decision-making, while IoT utilizes its network of interconnected devices to provide sensory input. The outcome of this integration is a resilient system that possesses the ability to observe, forecast, and react to a multitude of obstacles in real time.

Already, the integration of AI and IoT is advancing water management. As an illustration, alterations in water quality can be detected by IoT sensors, and the analysis of this data by AI algorithms can ascertain the origin of contamination, thereby enabling prompt and efficient remedial measures. Through the incorporation of AI and IoT, water management, along with numerous other industries, stands to undergo a paradigm shift that will result in systems that are more robust, effective, and environmentally friendly.

19.2.4 Case studies: artificial intelligence and Internet of Things in water management

In the realm of water management, the integration of AI and the IoT has been a game changer.

19.2.4.1 Improving agricultural water utilization

Water availability plays a pivotal role in supporting agriculture, which is an indispensable sector of the economy. In a time when water scarcity and the unpredictability of precipitation make efficient irrigation systems more vital than ever. To tackle this concern, municipal wastewater treatment plants have integrated an innovative AI-based system with IoT technologies. Using an ESP 8266 Wi-Fi processor and an Atmega 328p Microcontroller, the system tracks and uploads data regarding a variety of water parameters to a cloud server. The data is subsequently analyzed by AI algorithms to ascertain the optimal utilization of water, be it for irrigation in gardens or reuse in agriculture. This process not only streamlines water usage but also promotes sustainable water management (Lakshmi Narayanan et al., 2023).

19.2.4.2 Enhancing water quality monitoring

Even though access to pure water is considered a fundamental human right, conventional approaches to monitoring water quality remain antiquated and labor-intensive. To address this issue, a system prototype for water storage stations has been created utilizing machine learning and the Internet of Things. The aforementioned system employs artificial neural network (ANN) and support vector machine (SVM) algorithms to effectively forecast water contaminant levels while measuring a diverse range of water parameters. In addition, automated corrective actions for water treatment can be triggered by the system in response to detected contamination levels, thereby guaranteeing the sanitation and safety of water supplies (Maria et al., 2023).

Although these developments provide substantial advantages, they are not devoid of obstacles and constraints. Concerns on the dependability of systems and data security must be resolved to safeguard the integrity of water management systems. In addition, validation and ongoing development are necessary to ensure the accuracy and efficacy of the predictive models in use. In summary, it is crucial to maintain a continuous focus on the challenges posed by AI and IoT technologies to completely exploit their potential and promote water management practices that are both sustainable and efficient.

19.3 Artificial intelligence–driven forecasting in water management

The application of AI in the field of water management has presented novel prospects for forecasting and resolving challenges associated with water. In light of the global challenges posed by water scarcity, pollution, and the repercussions of climate change, there has never been a greater demand for sophisticated forecasting tools. The proactive nature of AI-driven forecasting permits organizations and authorities to anticipate water-related issues and act accordingly.

19.3.1 The imperative for predictive insights

Water management is a fundamental component of rural and urban sustainability. Global communities are confronted with a variety of complex issues, including fluctuating demand brought about by urbanization and population expansion and unforeseeable supply fluctuations resulting from climate variations. Conventional approaches to water management frequently hinge on reactive strategies that respond to emerging challenges. Nevertheless, due to the growing complexity of these obstacles, it is critical to proactively identify potential issues to prevent their escalation (Bartram & Cairncross, 2010). Predictive insights powered by AI are instrumental in this regard, providing a proactive methodology for water management.

19.3.2 The science and mechanics of artificial intelligence in forecasting

AI-powered forecasting generates informed predictions by utilizing enormous quantities of data. Anomalous data, historical water utilization patterns, and real-time inputs from IoT sensors are utilized in the training of AI models to identify patterns and anomalies. As new data becomes available, these models iteratively improve their predictions to maintain their accuracy and relevance. This is the result of sophisticated algorithms and machine learning models capable of processing and analyzing data at rates incomparable to those of a human (Wood et al., 2011).

19.3.3 Enhancing water management strategies with artificial intelligence

AI possesses enormous transformative potential in water management. By forecasting demand using a variety of variables (population growth, industrial activities, and climatic conditions), AI can optimize water allocation. In addition, real-time detection of contaminants in water by algorithms powered by AI ensures the safety and sanctity of water resources. AI further guarantees uninterrupted water supply and mitigates expenses by forecasting potential maintenance needs of infrastructure, such as treatment facilities or pipelines.

19.3.4 Real-world applications and success stories

One noteworthy instance is to utilize AI to predict water levels. By incorporating IoT sensors dispersed throughout its water distribution network with AI models, the municipality could anticipate surges in demand, identify pipeline leaks, and guarantee perfect water quality. AI-driven interventions of this nature resulted in substantial water conservation, decreased waste, and increased public contentment (Donbosco & Chakraborty, 2021; Narendar Singh et al., 2022; Singh & Ahmed, 2021).

19.3.5 Advantages, limitations, and the path forward

The integration of AI-driven forecasting in water management has ushered in a new era of proactive and efficient solutions. Here, we delve into the multifaceted implications of this integration.

19.3.5.1 Advantages of artificial intelligence–driven forecasting in water management

19.3.5.1.1 Efficiency and proactivity

Traditional water management frequently employs a reactive strategy, reacting to problems as they become apparent. In contrast, AI-powered forecasting takes a proactive approach by predicting impending challenges before their escalation. This transition not only guarantees optimal water usage but also mitigates the likelihood of potential emergencies.

19.3.5.1.2 Minimized wastage

AI contributes to the optimization of water allocation, reduction of wastage, and prudent utilization of water resources through the prediction of demand and supply patterns.

19.3.5.2 Limitations of artificial intelligence–driven forecasting

19.3.5.2.1 Data dependency

The effectiveness of AI predictions is significantly impacted by both the quality and quantity of data supplied into the system. Erroneous predictions may result from inaccurate or insufficient data, which may exacerbate water management challenges.

19.3.5.2.2 Technological and training investments

The implementation of AI-driven solutions requires substantial technological investments. Furthermore, the training of personnel to operate and maintain these sophisticated systems can be a demanding task in terms of resources.

19.3.5.2.3 The path forward

The ongoing development of technology holds the potential for future water management integrations to be even more sophisticated. As society progresses, it is imperative to prioritize the resolution of existing constraints and guarantee universal access to AI-powered solutions across regions, regardless of their economic standing. International cooperation, collaborative endeavors, and ongoing research will be crucial to fully exploit the capabilities of AI and establish a sustainable water future.

19.4 Internet of Things: the eyes and ears of modern water systems

The integration of the IoT into water resource monitoring and management has been transformed. Sensors and IoT devices function as the "eyes and ears" of contemporary water systems, delivering vital real-time data and insights for the implementation of sustainable and effective water management practises. These technologies facilitate a proactive methodology by enabling instantaneous reactions to fluctuations in water quality, discharge rates, and other pivotal parameters.

19.4.1 The essence and role of sensors in real-time monitoring

Water management is heavily reliant on sensors, which serve as the foundation of the IoT ecosystem and monitor a variety of parameters associated with water distribution and quality. Continuously, they gather data pertaining to reservoir levels, flow rates, water quality, and flow rates. These insights are crucial in facilitating well-informed decision-making. The ability to monitor water systems in real-time guarantees the timely detection and resolution of any anomalies or issues that may arise, thereby averting the escalation of potential problems. Research has underscored the significance of sensor integration in water management, citing it as a critical technology for assuring the sustainability and efficacy of water systems (Oliveira et al., 2023).

19.4.2 Feedback loops, automated response systems, and the power of proactivity

In water management, the true potential of the IoT transcends simple data collection. Frequently in conjunction with AI algorithms, integrated feedback loops and automated response systems interpret sensor data to instigate proactive interventions. For example, if a sensor identifies impurities in the water supply, the system can autonomously sever the impacted supply line, thereby averting possible risks to human health. By adopting this proactive approach, water safety is effectively maintained and water resources are managed more efficiently. Recently, flood management systems with IoT sensors have been the subject of some research. A recent publication illustrates the potential application of IoT, CNN, and related technologies in the management of water storage systems and flood prevention (Smys et al., 2020).

19.4.3 Spotlight on transformation: from urban centers to rural settings

The potential advantages of incorporating IoT into water management extend beyond urban environments to encompass rural sectors as well. These technologies offer substantial benefits. IoT has the potential to enhance water distribution efficiency for densely populated urban areas and guarantee a reliable and secure water supply for rural regions. One prominent illustration of this paradigm shift can be observed in rural areas where IoT technologies have been implemented to streamline water distribution, minimize waste, and promptly resolve problems. This guarantees that inhabitants of remote regions are provided with access to dependable and secure water sources.

As a result of the incorporation of IoT into water management, the monitoring and administration of water resources have been substantially improved. In contrast to the proactive interventions made possible by automated response systems and feedback loops, sensors furnish real-time data and insights. These technologies are revolutionising water management in both urban and rural environments by assuring safety, sustainability, and efficiency.

19.5 Ensuring water purity: artificial intelligence and Internet of Things to the rescue

19.5.1 Rapid detection and proactive monitoring

The implementation of IoT sensors within water management systems has precipitated a paradigm shift in the way water quality concerns are monitored and resolved. These sophisticated sensors, outfitted with real-time monitoring capabilities, serve as vigilant guardians of our water resources, guaranteeing their integrity and safety.

19.5.1.1 Monitoring in real-time using Internet of Things sensors

The use of IoT sensors permits the continuous monitoring of bodies of water, ensuring that any changes in water quality are promptly identified. An investigation carried out on fish ponds underscored the significant importance of water quality monitoring systems that employ underwater sensors and the IoT to maintain ideal conditions for aquatic life and deliver real-time data (Manoj et al., 2022).

19.5.1.2 Immediate detection of contaminants

The fundamental efficacy of these sensors resides in their capacity to promptly detect impurities present in the water. If irregularities in water quality are identified, these sensors can initiate notifications, thereby guaranteeing prompt measures to rectify the situation. The significance of intelligent water quality monitoring systems, as demonstrated by research utilizing IoT and electronic sensors, has been emphasized, especially in light of the escalating levels of pollution and industrial waste (Koditala et al., 2018; Yadav, 2022).

19.5.1.3 Proactive monitoring for prompt responses

The perpetual vigilance facilitated by IoT sensors guarantees the preservation of water quality in a safe and usable state. When anomalies are identified, the information gathered by these sensors can be applied swiftly to corrective actions, thus reducing the likelihood of potential damage. As an illustration, a cloud-based intelligent water quality monitoring system utilized a combination of IoT sensors and machine learning algorithms to forecast water quality and verify adherence to mandatory standards (Troy et al., 2011).

In conclusion, the implementation of IoT sensors has significantly enhanced our capacity to continuously monitor water quality. The integration of AI-driven analysis with their continuous monitoring capabilities is of the utmost importance in protecting our water resources, guaranteeing their continued purity and suitability for human consumption.

19.5.2 Artificial intelligence–driven analysis: from immediate response to predictive modeling

Beyond immediate detection and response, AI possesses the capacity to bring about significant changes in water management. The analytical capabilities of AI provide profound insights into present water conditions and predict potential future hazards. This segment explores how AI analyzes sensor data to generate a comprehensive assessment of water quality and forecast possible areas of high pollution.

19.5.2.1 Immediate detection and analysis

The criticality of AI's immediate response capability in detection and analysis cannot be overstated. Embedded AI algorithms enable sensors to detect anomalies in water quality in real time, thereby generating alerts that demand immediate attention. The implementation of this rapid response mechanism guarantees the prompt detection of contaminants, thereby protecting both aquatic ecosystems and human health (Lu et al., 2023; Vilupuru et al., 2022).

19.5.2.2 Modeling predictions and future threats

Predictive modeling represents the true prowess of AI, surpassing immediate detection. Through the analysis of extensive datasets, AI models can predict potential areas of high pollution, thereby facilitating preventative measures (Li et al., 2023; Lv et al., 2023). An example of this is a study that examined the progress made in water quality prediction using AI, utilizing models such as ANN, FUZZY, and SVM (Kang et al., 2017). An additional study showcased the effectiveness of a stacked model constructed using H_2O AutoML and Explainable AI methodologies, which successfully predicted water quality with a remarkable 97% accuracy rate (Aldhyani et al., 2020).

19.5.2.3 Identification of pollution hotspots

The detection of pollution hotspots is a task that AI models excel at accomplishing with exceptional proficiency. AI can ascertain the effects of various parameters, including temperature, pH, ions, nutrients, and microbes, on the health of water (Egbueri & Agbasi, 2022). This comprehension facilitates the surveillance of water behavior, thereby guaranteeing prompt interventions in areas prone to pollution.

19.5.2.4 Preemptive measures

Predicting prospective threats enables the implementation of proactive measures. Through proactive identification of potential pollution sources or areas of concern, governing bodies can execute preventative measures against contamination. This practice not only guarantees the integrity of the water source but also diminishes the expenditures linked to remediation and treatment.

As a whole, the incorporation of AI into water management provides a comprehensive strategy, encompassing both real-time identification and forecasting. We can pave the way for a secure future by ensuring the sustainability and safety of our water resources through the application of AI.

19.5.3 Integrated systems: the synergy of artificial intelligence and Internet of Things

The convergence of AI and the IoT has enabled significant progress in numerous industries, including water management. Through the strategic utilization of these technologies in concert, integrated systems have materialised, providing all-encompassing and proactive resolutions to urgent challenges.

19.5.3.1 Proactive contaminant detection

The real-time detection of contaminants in water supply systems is among the most revolutionary applications of AI and IoT integration. By leveraging sophisticated sensors and AI-powered analysis, these systems are capable of rapidly identifying irregularities in water quality. After prospective pollutants are detected, the system can independently deactivate the impacted water supply segment, thereby preventing the distribution of contaminated water to consumers.

19.5.3.2 Proposed optimal treatment guidelines

In addition to simple detection, AI-powered systems can analyze the characteristics of the contaminants and recommend the most effective remediation approaches. By utilizing extensive datasets and ML algorithms, these systems can recommend remediation processes that are optimized for the particular contaminants that are present (Lakshmi Narayanan et al., 2023).

19.5.3.3 *Intelligent irrigation*

The convergence of AI and the IoT has produced intelligent irrigation systems. By employing AI algorithms to optimize water consumption and sensors to monitor soil moisture levels, these systems guarantee that crops receive the appropriate quantity of water at the appropriate moment. Research has demonstrated that the implementation of such systems can substantially enhance the well-being of crops while reducing water wastage (Qazi et al., 2022).

19.5.3.4 *Wastewater treatment*

In addition to comprehensive water quality monitoring, the integration of AI and IoT has proven advantageous for wastewater treatment facilities. Integrated systems have been developed that offer real-time water quality monitoring, ensuring that the treated water satisfies the required standards before being released into the environment or reused (Martínez et al., 2020).

As a result of the integration of AI and IoT, water management has gained access to an infinite number of opportunities. Technology plays a leading role in guaranteeing potable water resources for all, encompassing both preventative measures and comprehensive solutions.

19.5.4 Spotlight on transformation: real-world success stories

A paradigm shift has occurred in water management with the advent of AI and the IoT, which have revolutionised effectiveness, sustainability, and creativity. By utilizing concrete illustrations from the real world, this segment emphasizes the concrete advantages of this integration.

19.5.4.1 *Reviving a contaminated water source*

The implementation of AI and IoT in the administration of municipal wastewater treatment facilities is one of the most compelling success stories. Utilizing IoT (Lakshmi Narayanan et al., 2023), a research study implemented an innovative AI-based system to effectively manage water consumption in municipal wastewater treatment plants. By implementing this methodology, not only was the effective treatment of wastewater guaranteed, but it also significantly contributed to the conversion of a polluted water source into a sustainable and secure resource.

19.5.4.2 *Intelligent water distribution*

In water distribution systems, the significance of the IoT cannot be exaggerated. The IoT architecture for intelligent water networks, which is intended to facilitate the monitoring and control of water distribution networks in real time, was unveiled in an exhaustive study (Narendar Singh et al., 2022). These systems optimize water resource allocation, thereby reducing wastage and guaranteeing sufficient supplies for all regions.

19.5.4.3 *Expert water management*

An additional noteworthy implementation is the Integrated Expert Water Management system, which optimizes the utilization of water resources across multiple applications by integrating IoT and AI algorithms (Ni Nyoman et al., 2023). These kinds of systems offer significant advantages to areas experiencing water scarcity, as they guarantee the prudent and sustainable utilization of resource availability.

19.5.4.4 *Wastewater management*

In addition, AI and IoT innovations have played a crucial role in wastewater management. An investigation underscored the application of technology in the examination of water quality and wastewater remediation via IoT (Iswanto et al., 2023). These systems effectively treat effluent, thereby mitigating its environmental impact and safeguarding uncontaminated water sources.

In a nutshell, substantial progress has been made in the field of water management as a result of the integration of AI and IoT. The aforementioned instances of practical achievement serve to emphasize the profound capacity that technology possesses to guarantee sustainable and effective water management.

19.5.5 The path forward: embracing innovation for a cleaner tomorrow

Technological advancements, particularly the incorporation of AI and the IoT, will impact the future of water management. As one contemplates the forthcoming era, it becomes evident that these technological advancements will play a crucial role in guaranteeing universal access to potable water, rather than regarding it as a privilege.

Constant advancements in AI and IoT hold the potential to revolutionize water management procedures. For example, in regions such as southern Africa, the implementation of intelligent water management technologies—such as instruments for monitoring soil moisture and nutrients—has already yielded substantial enhancements in water resource management (Qazi et al., 2022).

The strength of AI resides in its capability to analyze and forecast potential threats from immense quantities of data. The identification of pollution sites in advance enables the implementation of proactive measures aimed at protecting our water resources. Recent studies highlight the potential of AI in forecasting water quality and stress the need for continuous innovation in this domain (Alghadeer, 2023).

A collaborative approach is imperative to effectively tackle the complex challenges associated with water management and ensure a sustainable future. The promotion of collaborations between policymakers, researchers, and industry experts can expedite the implementation of groundbreaking solutions, thereby guaranteeing universal access to potable water (Gade & Aithal, 2022).

The overarching goal is to ensure that all individuals, irrespective of geographic location or socioeconomic standing, are provided with uncontaminated and potable water. By capitalising on the potential of artificial intelligence and the IoT, this aspiration can be materialized, thereby establishing a foundation for a more environmentally friendly and sustainable future.

In conclusion, it is apparent how to proceed. By advocating for innovation, fostering research, and promoting collaboration, it is possible to utilize technological progress to guarantee universal access to clean water in the future.

19.6 Behavioral insights: understanding water consumption with machine learning

To utilize ML for water consumption, customized algorithms must be selected and trained on historical data sets to identify consumption patterns (Boudhaouia & Wira, 2021). In light of emerging data, these models undergo ongoing refinement to guarantee precision and pertinence. With skill, they discern patterns in consumption and classify users into distinct categories, such as high consumers and efficient users (Choi & Kim, 2018). ML models can also forecast future consumption trends, which facilitates effective resource allocation, by incorporating variables including significant events and meteorological conditions.

19.6.1 The imperative of data-driven insights

The investigation of water usage patterns is of the utmost importance when it comes to sustainable water administration. Effective conservation strategies can be developed by policymakers and water management authorities through a comprehension of these behaviors. In this regard, ML presents a resilient resolution, with the ability to analyze extensive and diverse consumption data and deliver practical insights (Boudhaouia & Wira, 2021). These insights not only facilitate prompt decision-making but also influence the development of enduring water conservation policies.

19.6.2 Delving into consumption patterns with machine learning

ML possesses sophisticated analytic capabilities that enable it to effectively decipher intricate datasets. Models that are trained on water consumption data can detect complex patterns, trends, and anomalies (Choi & Kim, 2018). These models can classify consumers into distinct categories according to their utilization patterns: high consumers, efficient users, or wasteful entities. Precise comprehension of consumption patterns is crucial for the implementation of focused conservation initiatives.

19.6.3 Tailored recommendations for sustainable consumption

The adaptability of ML is one of its defining characteristics. ML can generate customized water conservation strategies for particular communities or industries through the analysis of consumption data (Ahansal et al., 2022). In addition, future consumption patterns can be predicted utilizing the insights gleaned from present data. This capability of prediction enables authorities to proactively implement preventive measures, thereby guaranteeing long-term water sustainability.

19.6.4 Spotlight: promoting water conservation through artificial intelligence insights

Practical applications of machine learning in the domain of water consumption analysis have generated concrete advantages. Communities, water councils, and municipalities have utilized these insights to advocate for water conservation. For example, specific geographic areas have effectively implemented intelligent water management technologies, including tools for monitoring soil moisture and nutrients, to optimize water consumption (Rahim et al., 2020). The aforementioned case studies highlight the capacity of ML to promote sustainable water consumption practices.

19.6.5 Challenges, ethical considerations, and the path forward

Despite the tremendous potential of ML in water consumption analysis, it is critical to be aware of the obstacles. Exclusively depending on ML analysis could result in the neglect of specific subtleties. Moreover, there are ethical concerns raised by the use of data, particularly concerning data security and privacy (Abbas et al., 2022). In light of the ongoing development of ML algorithms, it is imperative to maintain a delicate equilibrium between harnessing their potential and guaranteeing their responsible and ethical application.

19.6.6 Case studies: machine learning insights into Seville's water consumption

Seville, Spain, undertook an innovative endeavor by adopting ML to construct water demand models at the census tract level. Random Forest (RF) and Classification and Regression Trees (CART) were utilized for this purpose. With an error of 22 L/day/inhabitant, the CART method estimated water demand and identified key variables that influence consumption. In contrast, the predictions generated by RF were more accurate, exhibiting an error range of 18.89–26.91 L/day/inhabitant. The case study not only showcased the effectiveness of ML in interpreting intricate water consumption patterns but also emphasized its capacity to improve water management strategies by providing a more nuanced comprehension and a pragmatic substitute for conventional modeling methodologies (Villarin & Rodriguez-Galiano, 2019).

19.7 Building resilient water infrastructure

Water management is confronted with unprecedented challenges in the modern era, which require not only robust but also intelligent infrastructure. The convergence of AI and the IoT is causing a paradigm shift in water infrastructure design, monitoring, and maintenance.

19.7.1 Proactive maintenance with Internet of Things and artificial intelligence

Historically, water infrastructure maintenance consisted primarily of reactive measures; restorations were only undertaken after the occurrence of problems, resulting in reduced efficiency and escalated expenses (Venkatasubramanian et al., 2020). On the contrary, the implementation of IoT has enabled sensors integrated into the infrastructure to consistently monitor the health of the system, promptly identifying irregularities such as pressure decreases or leaks (Kumar et al., 2022). In addition to augmenting this capability, AI algorithms analyze the data gathered from these sensors to forecast possible maintenance requirements. By adopting a proactive stance, concerns can be effectively resolved before their escalation, leading to substantial savings in both resources and costs (Abbas et al., 2022).

19.7.2 Smart systems for efficient water flow and reduced wastage

Throughout history, water infrastructure systems have frequently been beset by inefficiencies, which have led to significant water wastage. Contemporary infrastructures, made possible by AI and IoT, are transforming this domain. By optimizing water flow in accordance with demand, they can guarantee efficient distribution while minimizing wastage. Furthermore, in accordance with real-time data, these intelligent systems are capable of autonomously regulating flow rates, halting sections in the event of leakage, and redirecting water to ensure maximum efficiency and security (Mazhar et al., 2022).

19.7.3 Spotlight: a success story of minimizing water loss

One notable illustration of the profound impact that technology can have on water management is evident in specific metropolitan areas that have experienced substantial water wastage as a consequence of antiquated infrastructure (Palsodkar et al., 2022). The incorporation of AI and IoT was instrumental in the transformation of this infrastructure. Including leak detection and water distribution optimization, the technological intervention was exhaustive. The results were as follows: a significant decrease in water wastage, considerable financial savings, and an evident enhancement in public contentment with the water infrastructure (Ayoub et al., 2023).

19.7.4 Challenges and solutions in building resilient water infrastructure

Compatibility issues must be resolved before integrating AI and IoT into water infrastructure, especially when combining legacy systems with contemporary technologies. It is crucial to establish interoperable systems to facilitate smooth communication and optimize the advantages offered by these technologies (Luzolo et al., 2023). In addition to integration, safeguarding data security is of the utmost importance, particularly as IoT devices that process sensitive information proliferate. AgriSecure, an innovative framework that integrates fog computing and blockchain, serves as a benchmark in the realm of IoT and smart computing security, guaranteeing resilient protection against potential intrusions (Padhy et al., 2023).

In addition to a proficient labor force, effectively tackling these technical obstacles underscores the importance of significant financial commitments toward training initiatives that furnish local governments and communities with the requisite knowledge. Moreover, the integration of IoT and AI into emergency response systems has the potential to improve their effectiveness and knowledge of how to handle crises. This is exemplified by systems such as the Flood Emergency Response System with Face Analytics (Mardaid et al., 2023).

19.8 Engaging communities and shaping policies

The advent of the digital age, which is marked by the widespread adoption of AI and the IoT, is significantly altering how water management communities participate and policies are formulated. Technological advancements are facilitating the democratization of decision-making processes by ensuring that they are transparent, inclusive, and grounded in data-driven insights.

19.8.1 Harnessing data for public sentiment analysis

Effective water management is predicated on the comprehension and resolution of the concerns expressed by the community, which is its direct recipient. Throughout history, assessing public sentiment has proven to be a formidable task, frequently dependent on inadequate feedback mechanisms. The emergence of AI, particularly Natural Language Processing (NLP), has revolutionised this domain. Massive quantities of unstructured data from various sources, including social media, community forums, and feedback forms, can be combed through by NLP algorithms. This functionality enables a comprehensive comprehension of public opinion concerning water matters, encompassing the apprehensions, recommendations, and sentiments of the community (Roose & Panez, 2020). By utilizing these perspectives, policymakers and water management authorities can develop approaches that align with the desires and requirements of the community.

19.8.2 Informed and transparent policymaking with artificial intelligence and Internet of Things insights

Conventional approaches to water management policy formulation frequently encountered censure for their perceived detachment from practical circumstances, predominantly attributable to their dependence on restricted or obsolete data. At times, policies resulting from this disparity failed to correspond with the genuine requirements or apprehensions of the populace. The implementation of AI and IoT technologies enables an uninterrupted flow of up-to-the-minute information pertaining to a multitude of aspects of water management, including infrastructure health, consumption patterns, and water source quality. These current, granular insights guarantee that policies remain pertinent and promptly. Moreover, the dissemination of these data-driven insights cultivates an atmosphere of confidence and openness among members of the community. Individuals are inclined to endorse and comply with policies that are supported by data and comprehend the reasoning behind them.

19.8.3 Spotlight: a participatory approach to water management using technology

The active pursuit of community involvement in water management decisions has been observed in specific urban regions. These geographical areas have utilized an array of technological resources to enable this interaction. Mobile applications collect feedback from residents, chatbots powered by AI respond to inquiries in real time, and monitoring systems facilitated by the IoT furnish the public with up-to-date information regarding the condition of infrastructure and water sources (von Korff et al., 2012). This technologically advanced, multifaceted strategy has produced remarkable results. A sense of ownership and accountability is fostered within communities about their water resources. In addition, water management practices have significantly improved, public confidence has increased, and the relationship between authorities and the community has been strengthened.

19.8.4 Challenges in community engagement

Although community participation in water management initiatives is vital, it also poses a number of obstacles. A dearth of knowledge may result in passivity; therefore, it is imperative to execute educational initiatives that emphasize the significance of water management. Another obstacle that frequently arises is resistance to change, which is frequently motivated by concerns about disruption or expense. Approaches to surmount this challenge encompass presenting instances of accomplishment and furnishing explicit, quantifiable advantages of engagement.

To ensure that all communities can participate in water management initiatives, it is critical to reconcile the digital divide that exists, especially in developing nations where access to technology is uneven. Moreover, water management can introduce a wide range of viewpoints and priorities, underscoring the criticality of collaborative strategies and inclusive participation. These strategies not only mitigate the difficulties but also foster the implementation of water management practices that are both fair and efficient.

19.9 Navigating the challenges

A new era of opportunities begins with the incorporation of AI and the IoT into water management. Nevertheless, similar to any revolutionary technology, it presents intrinsic difficulties. Addressing the challenges posed by technological intricacies and ethical dilemmas is crucial to effectively leverage the potential of AI and IoT in the field of water management.

19.9.1 Addressing technological hurdles

When it comes to integrating modern technologies into water management, interoperability is one of the greatest obstacles. The integration of legacy infrastructure with modern IoT devices or AI platforms frequently gives rise to compatibility challenges, necessitating complex resolutions to guarantee uninterrupted functioning.

Furthermore, scalability is a concern. Regions confronted with varied water management requirements necessitate flexible solutions. The development of systems capable of accommodating the distinct needs of urban metropolises and desolate rural areas is of the utmost importance to guarantee efficiency in diverse topographies.

Furthermore, the incorporation of IoT devices gives rise to apprehensions regarding data breaches and security. Due to the processing and transmission of immense quantities of sensitive data, the systems are susceptible to cyberattacks. This highlights the criticality of establishing strong cybersecurity protocols to protect information and preserve the overall system's integrity (Sharma et al., 2020).

19.9.2 Ethical dilemmas in data collection and artificial intelligence decisions

Particularly as it relates to the water consumption of individuals or households, the accumulation of vast quantities of data raises substantial privacy concerns. It is imperative to ensure the anonymity of data and acquire informed consent from individuals before data collection to uphold ethical standards.

An additional obstacle pertains to the possible biases present in AI algorithms. The utilization of skewed or incomplete datasets during the training process of these algorithms may result in unintended biases or prejudices toward particular groups or regions. It is of utmost importance to avoid biases and uphold impartiality to preserve the confidence and effectiveness of AI-powered solutions (Franke, 2022).

Finally, due to the opaque character of certain AI models, an emphasis on accountability and transparency is required. The decision-making processes of AI must be transparent to stakeholders, and accountability mechanisms should be established to hold these systems responsible for any erroneous or detrimental choices (Mitchell et al., 2019).

19.9.3 Overcoming financial and infrastructural barriers

Initial investments in the integration of cutting-edge technologies are frequently substantial. Obtaining and deploying AI and IoT solutions can have significant financial ramifications, particularly for regions that have limited financial resources.

Infrastructure deficiencies exacerbate the difficulty. Phosphorus-depleted or obsolete water management systems may render a phased integration strategy more viable in certain regions. in addition, public–private collaborations may be of great significance in bridging these infrastructure gaps.

Fundamental to the effective implementation of these technologies is the development of necessary capabilities. By organizing seminars and training programmes, communities, and local governments can be effectively-prepared to utilize the advantages of AI and IoT in the realm of water management.

19.10 Gazing into the crystal ball: what's next?

The nascent stages of AI and IoT integration are being observed in water management. Given the rapid pace of technological advancement, it is crucial to proactively anticipate imminent developments and obstacles. This proactive approach guarantees that we maintain our position at the forefront of sustainable water management for all.

19.10.1 Upcoming innovations in artificial intelligence and Internet of Things for water management

Present-day AI and IoT technologies, although revolutionary, are not without their constraints (Qazi et al., 2022). Continuous research is actively striving to overcome these limitations to expand the realm of possibility. An instance of such attention is devoted to the advancement of predictive models. The proposed models aim to predict water trends over extended periods by incorporating various factors such as industrial expansion, urbanization rates, and global climate change (Dunkel, 2023).

Notably, the IoT environment is also positioned to undergo substantial progress. It is anticipated that forthcoming sensors will possess prolonged operational times, improved precision, and the ability to track a wider range of parameters (Lee et al., 2022). These advancements will serve to enhance the level of detail and accuracy in the gathering of data, thereby facilitating more knowledgeable approaches to decision-making.

19.10.2 Preparing for future water challenges with adaptive technologies

Water issues are ever-changing, transforming in reaction to variables such as the increase in sea levels and the worsening of water contamination caused by industrial processes (Cheng et al., 2022). AI and IoT systems must be adaptable, and capable of adjusting to these shifting circumstances, to maintain their efficacy. The ability to adapt guarantees the continued relevance of technological solutions, enabling them to tackle both present and forthcoming challenges. At the core of this adaptability lies the tenet of perpetual improvement and learning. Foremost in forthcoming water management strategies will be systems capable of assimilating novel data and improving their forecasts and reactions (Wang & Wei, 2019).

19.10.3 The global collaboration imperative

Water management is an issue of international concern that surpasses geographical and political limitations. To confront this challenge, international cooperation and collaboration are required. The exchange of information, innovations, and optimal methodologies can expedite advancements, guaranteeing that regions gain from the combined expertise. By extending policymaking to technology development, international collaborations can enhance the effectiveness of individual initiatives and guarantee that the benefits of artificial intelligence and the IoT are felt universally (Dunkel, 2023).

19.10.4 Sustainability

The convergence of AI and the IoT is critical for the achievement of sustainable water management objectives, as it guarantees the optimal utilization and availability of resources for posterity. By utilizing predictive analytics and real-time monitoring, these technologies improve decision-making processes, leading to more efficient water distribution and decreased wastage. In addition, they encourage responsible water usage and foster confidence by promoting accountability and transparency. Nevertheless, to completely harness their capabilities, obstacles such as interoperability, data security, and infrastructural development must be confronted. It is critical to prioritize cross-sector collaboration and innovation to effectively utilize AI and IoT for sustainable water management and to make a meaningful contribution to global sustainable development objectives.

19.11 Wrapping up: the road ahead

As one contemplates the profound impact that AI and IoT could have on the water management sector, it becomes apparent that while the future appears auspicious, it necessitates a combined endeavor, comprehension, and promotion (Delgado et al., 2019).

19.11.1 Embracing the artificial intelligence and Internet of Things revolution for water sustainability

The incorporation of artificial intelligence and the IoT into water management represents an impending transition towards guaranteeing uninterrupted water resources for posterity. These technologies hold the potential to provide numerous advantages, including improved community engagement, proactive contamination detection, and effective water allocation and waste reduction (Popkova et al., 2023). Communities, industries, and governments must embrace this technological revolution, nevertheless, to fully exploit its capabilities. Sustainable futures require that water management strategies incorporate AI and IoT prioritization and investment in research and development (Maraveas et al., 2023).

19.11.2 Encouraging readers to be advocates and informed consumers

The cumulative impact of individual actions and decisions, although seemingly insignificant, can have a substantial effect on sustainable water management (De Ramos & Galang, 2020). It is of the utmost importance that individuals remain well-informed regarding how AI and IoT are influencing the water management domain and how they affect routine activities (Rolle et al., 2021). Individuals can positively influence their communities and promote sustainable practises by increasing their knowledge and understanding. Furthermore, in support of collective perspectives, community-driven initiatives can sway policy decisions and effect change, highlighting the significance of each individual's contribution to this endeavor toward sustainable water management.

References

Abbas, S., Ehsan, A., Ahmed, S., Khan, S. A., Jadoon, T. M., & Alizai, M. H. (2022). ASHRAY: enhancing water-usage comfort in developing regions using data-driven IoT retrofits. *ACM Transactions on Cyber-Physical Systems*, *6*(2), 1–28. Available from https://doi.org/10.1145/3491242.

Adeleke, I. A., Nwulu, N. I., & Ogbolumani, O. A. (2023). A hybrid machine learning and embedded IoT-based water quality monitoring system. *Internet of Things*, *22*, 100774. Available from https://doi.org/10.1016/j.iot.2023.100774.

Ahansal, Y., Bouziani, M., Yaagoubi, R., Sebari, I., Sebari, K., & Kenny, L. (2022). Toward smart irrigation: a literature review on the use of geospatial technologies and machine learning in the management of water resources in arboriculture. *Agronomy*, *12*(2), 297. Available from https://doi.org/10.3390/agronomy12020297.

Aldhyani, T. H. H., Al-Yaari, Mohammed, Alkahtani, Hasan, Maashi, Mashael, & Abd Algalil, Fahd (2020). Water quality prediction using artificial intelligence algorithms. *Applied Bionics and Biomechanics*, *2020*, 1–12. Available from https://doi.org/10.1155/2020/6659314.

Alghadeer, S. G. (2023). The use of artificial intelligence in water management projects in the Kingdom of Saudi Arabia. *American Journal of Technology*, *2*(1), 54–65. Available from https://doi.org/10.58425/ajt.v2i1.208.

Ayoub, N. A., Aziz, A. A., & Mustafa, W. A. (2023). FloodIntel: advancing flood disaster forecasting through comprehensive intelligent system approach. *Journal of Autonomous Intelligence*, *7*(1). Available from https://doi.org/10.32629/jai.v7i1.870.

Bartram, J., & Cairncross, S. (2010). Hygiene, sanitation, and water: forgotten foundations of health. *PLoS Medicine*, *7*(11), e1000367. Available from https://doi.org/10.1371/journal.pmed.1000367.

Boudhaouia, A., & Wira, P. (2021). A real-time data analysis platform for short-term water consumption forecasting with machine learning. *Forecasting*, *3*(4), 682–694. Available from https://doi.org/10.3390/forecast3040042, http://www.mdpi.com/journal/forecasting.

Cheng, Z., Yan, S., Song, T., Cheng, L., & Wang, H. (2022). Adaptive water governance research in social sciences journals: a bibliometric analysis. *Water Policy*, *24*(12), 1951–1970. Available from https://doi.org/10.2166/wp.2022.196.

Choi, J., & Kim, J. (2018). Analysis of water consumption data from smart water meter using machine learning and deep learning algorithms. *Journal of the Institute of Electronics and Information Engineers*, *55*(7), 31–39. Available from https://doi.org/10.5573/ieie.2018.55.7.31.

Das, A., & Roy Chowdhury, A. (2022). *Algal cultivation in the pursuit of emerging technology for sustainable development Valorization of Microalgal Biomass and Wastewater Treatment* (pp. 357–366). India: Elsevier. Available from https://www.sciencedirect.com/book/9780323918695, https://doi.org/10.1016/B978-0-323-91869-5.00014-4.

Das, A., & Chowdhury A. R. (2023). *Energy decarbonization via material-based circular economy* (pp. 263–295). Springer Science and Business Media LLC. Available from https://doi.org/10.1007/978-3-031-42220-1_15.

Delgado, J. A., Short, N. M., Roberts, D. P., & Vandenberg, B. (2019). Big data analysis for sustainable agriculture on a geospatial cloud framework. *Frontiers in Sustainable Food Systems*, *3*. Available from https://doi.org/10.3389/fsufs.2019.00054.

Dolan, F., Lamontagne, J., Link, R., Hejazi, M., Reed, P., & Edmonds, J. (2021). Evaluating the economic impact of water scarcity in a changing world. *Nature Communications*, *12*(1). Available from https://doi.org/10.1038/s41467-021-22194-0.

Donbosco, I. S., & Chakraborty, U. K. (2021). An IoT-based water management system for smart cities. *Lecture Notes in Civil Engineering*, *115*, 247–259. Available from https://doi.org/10.1007/978-981-15-9805-0_21, http://www.springer.com/series/15087.

Dunkel, M. (2023). Unconventional plays: water management's evolution and forecast. Society of Petroleum Engineers (SPE), United States JPT. *Journal of Petroleum Technology*, *75*(1), 38–43. Available from https://doi.org/10.2118/0123-0038-JPT, https://doi.org/10.2118/0123-0038-JPT.

Egbueri, J. C., & Agbasi, J. C. (2022). Combining data-intelligent algorithms for the assessment and predictive modeling of groundwater resources quality in parts of southeastern Nigeria. *Environmental Science and Pollution Research*, *29*(38), 57147–57171. Available from https://doi.org/10.1007/s11356-022-19818-3, https://link.springer.com/journal/11356.

Franke, U. (2022). First- and second-level bias in automated decision-making. *Philosophy and Technology*, *35*(2). Available from https://doi.org/10.1007/s13347-022-00500-y, http://www.springer.com/philosophy/epistemology + and + philosophy + of + science/journal/13347.

Gade, D. S., & Aithal, P. S. (2022). ICT and digital technology based solutions for smart city challenges and opportunities. *International Journal of Applied Engineering and Management Letters*, 1–21. Available from https://doi.org/10.47992/ijaeml.2581.7000.0116.

Guo, X., Shen, Z., Zhang, Y., & Wu, T. (2019). Review on the application of artificial intelligence in smart homes. *Smart Cities*, *2*(3), 402–420. Available from https://doi.org/10.3390/smartcities2030025, https://www.mdpi.com/2624-6511/2/3/25.

Józefowicz, I., & Michniewicz-Ankiersztajn, H. (2023). Digital tools for water resource management as a part of a green economy in rural areas. *Sustainability*, *15*(6), 5231. Available from https://doi.org/10.3390/su15065231.

Kang, G., Gao, J.Z., & Xie, G. (2017). Data-driven water quality analysis and prediction: a survey. In *Proceedings—3rd IEEE International Conference on Big Data Computing Service and Applications, BigDataService 2017*, Institute of Electrical and Electronics Engineers Inc., United States, pp. 224–232. 9781509063185. Available from https://doi.org/10.1109/BigDataService.2017.40.

Lee, J., Abid, A., Le Gall, F., & Song, J.S. (2022). Recent trends on artificial intelligence-enabled internet of things platform and standard technologies. In *IEEE 8th World Forum on Internet of Things, WF-IoT 2022*. Institute of Electrical and Electronics Engineers Inc., South Korea. 9781665491532. Available from https://doi.org/10.1109/WF-IoT54382.2022.10152226, http://ieeexplore.ieee.org/xpl/mostRecentIssue.jsp?punumber = 10151825.

Nova, K. (2023). AI-enabled water management systems: an analysis of system components and interdependencies for water conservation. *Eigenpub Review of Science and Technology*, *7*, 105–124.

Khan, Y., & See, C.S. (2016). Predicting and analyzing water quality using Machine Learning: a comprehensive model. In *IEEE Long Island Systems, Applications and Technology Conference, LISAT* 2016. Institute of Electrical and Electronics Engineers Inc., Malaysia. 9781467384902. Available from https://doi.org/10.1109/LISAT.2016.7494106.

Koditala, N. & Pandey, P. (2018). Water quality monitoring system using IoT and machine learning. 1-5. Available from https://doi.org/10.1109/RICE.2018.8509050.

von Korff, Y., Daniell, K. A., Moellenkamp, S., Bots, P., & Bijlsma, R. M. (2012). Implementing participatory water management: recent advances in theory, practice, and evaluation. *Ecology and Society*, *17*(1). Available from https://doi.org/10.5751/ES-04733-170130Netherlands, http://www.ecologyandsociety.org/vol17/iss1/art30/ES-2012-4733.pdf.

Kumar, B.S., Soumiya, S., Ramalingam, S., Yogeswari, S., & Balamurugan, S. (2022). Water management and control systems for smart city using IoT and Artificial Intelligence. In *Proceedings of the International Conference on Edge Computing and Applications, ICECAA 2022*. Institute of Electrical and Electronics Engineers Inc. India, pp. 653–657. 9781665482325. http://ieeexplore.ieee.org/xpl/mostRecentIssue.jsp?punumber = 9935819. Available from https://doi.org/10.1109/ICECAA55415.2022.9936166.

Lakshmi Narayanan, K., Karthik Ganesh, R., Bharathi, S.T., Srinivasan, A., Santhana Krishnan, R., & Sundararajan, S. (2023). AI enabled IoT based intelligent waste water management system for municipal waste water treatment plant. In *Proceedings of the 6th International Conference on Inventive Computation Technologies, ICICT 2023*. Institute of Electrical and Electronics Engineers Inc. India, pp. 361–365. 9798350398496. http://ieeexplore.ieee.org/xpl/mostRecentIssue.jsp?punumber = 10133935. doi:10.1109/ICICT57646.2023.10134075.

Li, Y., Li, X., Xu, C., & Tang, X. (2023). Dissolved oxygen prediction model for the Yangtze River Estuary Basin using IPSO-LSSVM. *Water*, *15*(12), 2206. Available from https://doi.org/10.3390/w15122206.

Luzolo, P.H., Galland, S., Elrawashdeh, Z., Outay, F., & Tchappi, I. (2023). Combining multiagent systems with IoT for smarter buildings and cities: a literature review. Research Square, France. https://www.researchsquare.com/browse. Available from https://doi.org/10.21203/rs.3.rs-3081374.

Lu, H., Ding, A., Zheng, Y., Jiang, J., Zhang, J., Zhang, Z., Xu, P., Zhao, X., Quan, F., Gao, C., Jiang, S., Xiong, R., Men, Y., & Shi, L. (2023). Securing drinking water supply in smart cities: an early warning system based on online sensor network and machine learning. *AQUA—Water Infrastructure, Ecosystems and Society*, *72*(5), 721–738. Available from https://doi.org/10.2166/aqua.2023.007.

Lv, M., Niu, X., Zhang, D., Ding, H., Lin, Z., Zhou, S., & Zhu, Y. (2023). A data-driven framework for spatiotemporal analysis and prediction of river water quality: a case study in Pearl River, China. *Water*, *15*(2), 257. Available from https://doi.org/10.3390/w15020257.

Manoj, M., Dhilip Kumar, V., Arif, M., Bulai, E.-R., Bulai, P., et al. (2022). Oana Geman: state of the art techniques for water quality monitoring systems for fish ponds using IoT and underwater sensors: a review. *Sensors*, *22*(6).

Maraveas, C., Karavas, C.-S., Loukatos, D., Bartzanas, T., Arvanitis, K. G., & Symeonaki, E. (2023). Agricultural greenhouses: resource management technologies and perspectives for zero greenhouse gas emissions. *Agriculture*, *13*(7), 1464. Available from https://doi.org/10.3390/agriculture13071464.

Mardaid, E., Abidin, Z. Z., Asmai, S. A., & Abas, Z. A. (2023). Implementation of flood emergency response system with face analytics. *Malaysia International Journal of Advanced Computer Science and Applications*, *14*(1), 400–406. Available from https://doi.org/10.14569/IJACSA.2023.0140143, http://thesai.org/Publications/Archives?code = IJACSA, Science and Information Organization.

Martínez, R., Vela, N., el Aatik, A., Murray, E., Roche, P., & Navarro, J. M. (2020). On the use of an IoT integrated system for water quality monitoring and management in wastewater treatment plants. *Water*, *12*(4), 1096. Available from https://doi.org/10.3390/w12041096.

Mauter, M. S., Zucker, I., Perreault, F., Werber, J. R., Kim, J. H., & Elimelech, M. (2018). The role of nanotechnology in tackling global water challenges. *Nature Sustainability*, *1*(4), 166–175. Available from https://doi.org/10.1038/s41893-018-0046-8, http://www.nature.com/natsustain/, Nature Publishing Group, United States.

Mazhar, T., Malik, M. A., Haq, I., Rozeela, I., Ullah, I., Khan, M. A., Adhikari, D., Ben Othman, M. T., & Hamam, H. (2022). The role of ML, AI and 5G technology in smart energy and smart building management. *Electronics*, *11*(23), 3960. Available from https://doi.org/10.3390/electronics11233960.

Mitchell, M., Wu, S., Zaldivar, A., Barnes, P., Vasserman, L., Hutchinson, B., Spitzer, E., Raji, I.D., Gebru, T. (2019). Model cards for model reporting FAT* 2019. In *Proceedings of the 2019 Conference on Fairness, Accountability, and Transparency*, pp. 220–229. Association for Computing Machinery, Inc. 9781450361255. http://dl.acm.org/citation.cfm?id = 3287560. Available from https://doi.org/10.1145/3287560.3287596.

Moe, C. L., & Rheingans, R. D. (2006). Global challenges in water, sanitation and health. *Journal of Water and Health*, *4*(1), 41–58. Available from https://doi.org/10.2166/wh.2006.0043, http://www.iwaponline.com/jwh/default.htm.

Narendar Singh, D., Murugamani, C., Kshirsagar, P. R., Tirth, V., Islam, S., Qaiyum, S., Suneela, B., Al Duhayyim, M., Waji, Y. A., & Pallikonda Rajasekaran, M. (2022). IOT based smart wastewater treatment model for Industry 4.0 using artificial intelligence. *Scientific Programming*, *2022*, 1–11. Available from https://doi.org/10.1155/2022/5134013.

Ni Nyoman, P., Nyoman, I., Parwata, S., Antara, I. M. O. G., Kazumi, K., & Rivai, A. (2023). Development of IoT-based real-time monitoring system and LFA to improve the efficiency and performance of wastewater treatment plant in Udayana University Hospital. *Journal of the Civil Engineering Forum*, 109–116. Available from https://doi.org/10.22146/jcef.5122.

Padhy, S., Alowaidi, M., Dash, S., Alshehri, M., Malla, P. P., Routray, S., & Alhumyani, H. (2023). AgriSecure: a fog computing-based security framework for Agriculture 4.0 via Blockchain. *Processes*, *11*(3). Available from https://doi.org/10.3390/pr11030757, http://www.mdpi.com/journal/processes.

Palsodkar, P., Shrivastav, R., Ayangar, A., Atkare, S., Yadav, S., & Palsodkar, P. (2022). Sensor cloudlet interconnecting system for water reservoirs security IEEE Region 10 Humanitarian Technology Conference, R10-HTC, pp. 67–70. Institute of Electrical and Electronics Engineers Inc. India. 9781665401562. https://ieeexplore.ieee.org/xpl/mostRecentIssue.jsp?punumber = 8625998 Available from https://doi.org/10.1109/R10-HTC54060.2022.9929471.

Popkova, E. G., Bogoviz, A. V., & Sergi, B. S. (2023). Editorial: smart grids and EnergyTech as a way for sustainable and environmental development of energy economy. *Frontiers in Energy Research*, *11*. Available from https://doi.org/10.3389/fenrg.2023.1145234.

Qazi, S., Khawaja, B. A., & Farooq, Q. U. (2022). IoT-equipped and AI-enabled next generation smart agriculture: a critical review, current challenges and future trends. *IEEE Access*, *10*, 21219–21235. Available from https://doi.org/10.1109/ACCESS.2022.3152544, http://ieeexplore.ieee.org/xpl/RecentIssue.jsp?punumber = 6287639.

Rahim, M. S., Nguyen, K. A., Stewart, R. A., Giurco, D., & Blumenstein, M. (2020). Machine learning and data analytic techniques in digitalwater metering: a review. *Water (Switzerland)*, *12*(1). Available from https://doi.org/10.3390/w12010294, https://res.mdpi.com/d_attachment/water/water-12-00294/article_deploy/water-12-00294-v2.pdf.

De Ramos, Z., & Galang, G. (2020). Sustainable development practices implemented by the community partners of San Beda University (SBU). *Bedan Research Journal*, *5*(1), 160–190. Available from https://doi.org/10.58870/berj.v5i1.16.

Ray, P. P. (2016). A survey on Internet of Things architectures. *EAI Endorsed Transactions on Internet of Things*, *2*(5), 151714. Available from https://doi.org/10.4108/eai.1-12-2016.151714.

Rolle, M. L., Garba, D. L., & Ekedede, M. (2021). Re-branding global neurosurgery in paradise. *British Journal of Neurosurgery*, *35*(4), 375–376. Available from https://doi.org/10.1080/02688697.2021.1879013, http://www.tandfonline.com/loi/ibjn20.

Roose, I., & Panez, A. (2020). Social innovations as a response to dispossession: community water management in view of socio-metabolic rift in Chile. *Water*, *12*(2), 566. Available from https://doi.org/10.3390/w12020566.

Russell, S. J., & Norvig, P. (2016). *Artificial intelligence: a modern approach.* Pearson.

Schwab, K. (2017). *The fourth industrial revolution*. Portfolio Penguin.

Sharma, S. K., Bhushan, B., & Debnath, N. C. (2020). *Security and privacy issues in IoT devices and sensor networks* (pp. 1–318). India: Elsevier. Available from 10.1016/C2019-0-03189-5, https://www.sciencedirect.com/book/9780128212554.

Singh, M., & Ahmed, S. (2021). IoT based smart water management systems: a systematic review. *Materials Today: Proceedings, 46*, 5211–5218. Available from https://doi.org/10.1016/j.matpr.2020.08.588.

Smys., Basar., & Wang, Dr (2020). CNN based flood management system with IoT sensors and cloud data. *Journal of Artificial Intelligence and Capsule Networks, 2*. Available from https://doi.org/10.36548/jaicn.2020.4.001.

Venkatasubramanian, N., Davis, C.A., & Eguchi, R.T. (2020). Designing community-based intelligent systems for water infrastructure resilience. In *Proceedings of the 3rd ACM SIGSPATIAL International Workshop on Advances in Resilient and Intelligent Cities, ARIC 2020*, pp. 62–65. Association for Computing Machinery, Inc United States. 9781450381659. Available from https://doi.org/10.1145/3423455.3430318, http://dl.acm.org/citation.cfm?id = 3423455.

Villarin, M. C., & Rodriguez-Galiano, V. F. (2019). Machine learning for modeling water demand. *Journal of Water Resources Planning and Management, 145*(5). Available from https://doi.org/10.1061/(asce)wr.1943-5452.0001067.

Vilupuru, J.R., Amuluru, D.C., Ghousiya Begum, K. (2022). Water quality analysis using artificial intelligence algorithms. In *Proceedings of the 4th International Conference on Inventive Research in Computing Applications, ICIRCA 2022*. pp. 1193–1199. Institute of Electrical and Electronics Engineers Inc. India. 9781665497077 Available from http://ieeexplore.ieee.org/xpl/mostRecentIssue.jsp?punumber = 9985450, https://doi.org/10.1109/RICE.2018.8509050.

Wang, S., & Wei, Y. (2019). Water resource system risk and adaptive management of the Chinese Heihe River Basin in Asian arid areas. *Mitigation and Adaptation Strategies for Global Change, 24*(7), 1271–1292. Available from https://doi.org/10.1007/s11027-019-9839-y, http://www.wkap.nl/journalhome.htm/1381-2386.

Wood, E. F., Roundy, J. K., Troy, T. J., van Beek, L. P. H., Bierkens, M. F. P., Blyth, E., de Roo, A., Döll, P., Ek, M., Famiglietti, J., Gochis, D., van de Giesen, N., Houser, P., Jaffé, P. R., Kollet, S., Lehner, B., Lettenmaier, D. P., Peters-Lidard, C., ... Whitehead, P. (2011). Hyperresolution global land surface modeling: meeting a grand challenge for monitoring Earth's terrestrial water. *Water Resources Research, 47*(5). Available from https://doi.org/10.1029/2010wr010090.

Yadav, H. (2022). Smart water quality monitoring system using IOT and electronic sensors. *International Journal for Research in Applied Science and Engineering Technology, 10*(6), 2873–2875. Available from https://doi.org/10.22214/ijraset.2022.44271.

Index

Note: Page numbers followed by "*f*" and "*t*" refer to figures and tables, respectively

F

G

H

I